Synchronization and Linearity

WITHDRAWN
CARNEGIE MELLON

Synchronization and Linearity

An Algebra for Discrete Event Systems

Francois Louis Baccelli
Institut National de Recherche en Informatique et Automatique (INRIA), Sophia Antipolis, France

Guy Cohen
Ecole des Mines de Paris, Fontainebleau, France

Geert Jan Olsder
Delft University of Technology, The Netherlands

Jean-Pierre Quadrat
Institut National de Recherche en Informatique et Automatique (INRIA), Rocquencourt, France

JOHN WILEY & SONS
Chichester · New York · Brisbane · Toronto · Singapore

Other Wiley Editorial Offices

John Wiley & Sons, Inc., 605 Third Avenue,
New York, NY 10158-0012, USA

Jacaranda Wiley Ltd, G.P.O. Box 859, Brisbane,
Queensland 4001, Australia

John Wiley & Sons (Canada) Ltd, 22 Worcester Road,
Rexdale, Ontario M9W 1L1, Canada

John Wiley & Sons (SEA) Pte Ltd, 37 Jalan Pemimpin #05-04,
Block B, Union Industrial Building, Singapore 2057

Library of Congress Cataloging-in-Publication Data

Synchronization and linearity : an algebra for discrete event systems
/ Francois Louis Baccelli . . . [et al.].
p. cm.
Includes bibliographical references and index.
ISBN 0 471 93609 X
1. System analysis. 2. Discrete-time systems. 3. Algebra.
I. Baccelli, F. (François), 1954
T57.6.S953 1992
003—dc20 92-542
CIP

British Library Cataloguing in Publication Data

A catalogue record for this book is available from the British Library

ISBN 0 471 93609 X

Produced from camera-ready copy supplied by the authors
Printed and bound in Great Britain by Biddles Ltd, Guildford, Surrey

To

our

families

Contents

III Deterministic System Theory 217

Preface

The mathematical theory developed in this book finds its initial motivation in the modeling and the analysis of the time behavior of a class of dynamic systems now often referred to as 'discrete event (dynamic) systems' (DEDS). This class essentially contains man-made systems that consist of a finite number of resources (processors or memories, communication channels, machines) shared by several users (jobs, packets, manufactured objects) which all contribute to the achievement of some common goal (a parallel computation, the end-to-end transmission of a set of packets, the assembly of a product in an automated manufacturing line). The coordination of the user access to these resources requires complex control mechanisms which usually make it impossible to describe the dynamic behavior of such systems in terms of differential equations, as in physical phenomena. The dynamics of such systems can in fact be described using the two (Petri net like) paradigms of 'synchronization' and 'concurrency'. Synchronization requires the availability of several resources or users at the same time, whereas concurrency appears for instance when, at a certain time, some user must choose among several resources. The following example only contains the synchronization aspect which is the main topic of this book.

Consider a railway station. A departing train must wait for certain incoming trains so as to allow passengers to change, which reflects the synchronization feature. Consider a network of such stations where the traveling times between stations are known. The variables of interest are the arrival and departure times, assuming that trains leave as soon as possible. The departure time of a train is related to the maximum of the arrival times of the trains conditioning this departure. Hence the max operation is the basic operator through which variables interact. The arrival time at a station is the sum of the departure time from the previous station and the traveling time. There is no concurrency since it has tacitly been assumed that each train has been assigned a fixed route.

The thesis developed here is that there exists an algebra in which DEDS that do not involve concurrency can naturally be modeled as linear systems. A linear model is a set of equations in which variables can be added together and in which variables can also be multiplied by coefficients which are a part of the data of the model. The train example showed that the max is the essential operation that captures the synchronization phenomenon by operating on arrival times to compute departure times. Therefore the basic idea is to treat the max as the 'addition' of the algebra (hence this max will be written $\oplus$ to suggest 'addition'). The same example indicates that we also need conventional addition to transform variables from one end of an arc of the network to the other end (the addition of the traveling time, the data, to the departure time). This is why $+$ will be treated as multiplication in this algebra (and it will be denoted $\otimes$). The operations $\oplus$ and $\otimes$ will play their own roles and in other examples they are not necessarily confined to operate as max and $+$, respectively.

The basic mathematical feature of $\oplus$ is that it is idempotent: $x \oplus x = x$. In practice, it may be the max or the min of numbers, depending on the nature of the variables which are handled (either the times at which events occur, or the numbers of events during given intervals). But the main feature is again idempotency of addition. The role of $\otimes$ is generally played by conventional addition, but the important thing is that it behaves well with respect to addition (e.g. that it distributes with respect to $\oplus$). The algebraic structure outlined is known under the name of 'dioid', among other names. It has connections with standard linear algebra with which it shares many combinatorial properties (associativity and commutativity of addition, etc.), but also with lattice-ordered semigroup theory for, speaking of an idempotent addition is equivalent to speaking of the 'least upper bound' in lattices.

Conventional system theory studies networks of integrators or 'adders' connected in series, parallel and feedback. Similarly, queuing theory or Petri net theory build up complex systems from elementary objects (namely, queues, or transitions and places). The theory proposed here studies complex systems which are made up of elementary systems interacting through a basic operation, called synchronization, located at the nodes of a network.

The mathematical contributions of the book can be viewed as the first steps toward the development of a theory of *linear* systems on dioids. Both deterministic and stochastic systems are considered. Classical concepts of system theory such as 'state space' recursive equations, input-output (transfer) functions, feedback loops, etc. are introduced. Overall, this theory offers a unifying framework for systems in which the basic 'engine' of dynamics is synchronization, when these systems are considered from the point of view of performance evaluation. In other words, dioid algebra appears to be the right tool to handle synchronization in a *linear* manner, whereas this phenomenon seems to be very nonlinear, or even nonsmooth, 'through the glasses' of conventional algebraic tools. Moreover, this theory may be a good starting point to encompass other basic features of discrete event systems such as concurrency, but at the price of considering systems which are *nonlinear* even in this new framework. Some perspectives are opened in this respect in the last chapter.

Although the initial motivation was essentially found in the study of discrete event systems, it turns out that this theory may be appropriate for other purposes too. This happens frequently with mathematical theories which often go beyond their initial scope, as long as other objects can be found with the same basic features. In this particular case the common feature may be expressed by saying that the input-output relation has the form of an inf- (or a sup-) convolution. In the same way, the scope of conventional system theory is the study of input-output relations which are convolutions. In Chapter 1 it is suggested that this theory is also relevant for some systems which either are continuous or do not involve synchronization. Systems which mix fluids in certain proportions and which involve flow constraints fall in the former category. Recursive 'optimization processes', of which dynamic programming is the most immediate example, fall in the latter category. All these systems involve max (or min) and $+$ as the basic operations. Another situation where dioid algebra naturally shows up is the asymptotic behavior of exponential functions. In mathematical terms, the conventional operations $+$ and $\times$ over positive numbers, say, are transformed into max and $+$, respectively, by the mapping: $x \mapsto \lim_{s\to+\infty} \exp(sx)$. This is relevant, for example, in the theory of large deviations, and, coming back to conventional system theory, when outlining Bode diagrams by their asymptotes.

There are numerous concurrent approaches for constructing a mathematical framework for discrete event systems. An important dichotomy arises depending on whether the framework is intended to assess the logical behavior of the system or its temporal behavior. Among the first class, we would quote theoretical computer science languages like CSP or CCS and recent system-theoretic extensions of automata theory [114]. The algebraic approach that is proposed here is clearly of the latter type, which makes it comparable with such formalisms as timed (or stochastic) Petri nets [1], generalized semi-Markov processes [63] and in a sense queuing network theory. Another approach, that emphasizes computational aspects, is known as Perturbation Analysis [70].

A natural question of interest concerns the scope of the methodology that we develop here. Most DEDS involve concurrency at an early stage of their design. However, it is often necessary to handle this concurrency by choosing certain priority rules (by specifying routing and/or scheduling, etc.), in order to completely specify their behavior. The theory developed in this book may then be used to evaluate the consequences of these choices in terms of performance. If the delimitation of the class of queuing systems that admit a max-plus representation is not an easy task within the framework of queuing theory, the problem becomes almost transparent within the setting of Petri networks developed in Chapter 2: stochastic event graphs coincide with the class of discrete event systems that have a representation as a max-plus linear system in a random medium (i.e. the matrices of the linear system are random); any topological violation of the event graph structure, be it a competition like in multiserver queues, or a superimposition like in certain Jackson networks, results in a min-type nonlinearity (see Chapter 9). Although it is beyond the scope of the book to review the list of queuing systems that are stochastic event graphs, several examples of such systems are provided ranging from manufacturing models (e.g. assembly/disassembly queues, also called fork-join queues, jobshop and flowshop models, production lines, etc.) to communication and computer science models (communication blocking, wave front arrays, etc.)

Another important issue is that of the design gains offered by this approach. The most important structural results are probably those pertaining to the existence of periodic and stationary regimes. Within the deterministic setting, we would quote the interpretation of the pair (cycle time, periodic regime) in terms of eigenpairs together with the polynomial algorithms that can be used to compute them. Moreover, because bottlenecks of the systems are explicitly revealed (through the notion of critical circuits), this approach provides an efficient way not only to evaluate the performance but also to assess certain design choices made at earlier stages. In the stochastic case, this approach first yields new characterizations of throughput or cycle times as Lyapunov exponents associated with the matrices of the underlying linear system, whereas the steady-state regime receives a natural characterization in terms of 'stochastic eigenvalues' in max-plus algebra, very much in the flavor of Oseledeç's multiplicative ergodic theorems. Thanks to this, queuing theory and timed Petri nets find some sort of (linear) garden where several known results concerning small dimensional systems can be derived from a few classical theorems (or more precisely from the max-plus counterpart of classical theorems).

The theory of DEDS came into existence only at the beginning of the 1980s, though it is fair to say that max-plus algebra is older, see [49], [130], [67]. The field of DEDS is in full development and this book presents in a coherent fashion the results obtained so far by this algebraic approach. The book can be used as a textbook, but it also presents the current state of the theory. Short historical notes and other remarks are given in the note sections at the end of most chapters. The book should be of interest to (applied) mathematicians, operations researchers, electrical engineers, computer scientists, probabilists, statisticians, management scientists and in general to those with a professional interest in parallel and distributed processing, manufacturing, etc. An undergraduate degree in mathematics should be sufficient to follow the flow of thought (though some parts go beyond this level). Introductory courses in algebra, probability theory and linear system theory form an ideal background. For algebra, [61] for instance provides suitable background material; for probability theory this role is for instance played by [20], and for linear system theory it is [72] or the more recent [122].

The heart of the book consists of four main parts, each of which consists of two chapters. Part I (Chapters 1 and 2) provides a natural motivation for DEDS, it is devoted to a general introduction and relationships with graph theory and Petri nets. Part II (Chapters 3 and 4) is devoted to the underlying algebras. Once the reader has gone through this part, he will also appreciate the more abstract approach presented in Parts III and IV. Part III (Chapters 5 and 6) deals with deterministic system theory, where the systems are mostly DEDS, but continuous max-plus linear systems also are discussed in Chapter 6. Part IV (Chapters 7 and 8) deals with stochastic DEDS. Many interplays of comparable results between the deterministic and the stochastic framework are shown. There is a fifth part, consisting of one chapter (Chapter 9), which deals with related areas and some open problems. The notation introduced in Parts I and II is used throughout the other parts.

The idea of writing this book took form during the summer of 1989, during which the third author (GJO) spent a mini-sabbatical at the second author's (GC's) institute. The other two authors (FB and JPQ) joined in the fall of 1989. During the process of writing, correcting, cutting, pasting, etc., the authors met frequently, mostly in Fontainebleau, the latter being situated close to the center of gravity of the authors' own home towns. We acknowledge the working conditions and support of our home institutions that made this project possible. The Systems and Control Theory Network in the Netherlands is acknowledged for providing some financial support for the necessary travels. Mr. J. Schonewille of Delft University of Technology is acknowledged for preparing many of the figures using Adobe Illustrator. Mr. G. Ouanounou of INRIA-Rocquencourt deserves also many thanks for his help in producing the final manuscript using the high-resolution equipment of this Institute. The contents of the book have been improved by remarks of P. Bougerol of the University of Paris VI, and of A. Jean-Marie and Z. Liu of INRIA-Sophia Antipolis who were all involved in the proofreading of some parts of the manuscript. The authors are grateful to them. The second (GC) and fourth (JPQ) authors wish to acknowledge the permanent interaction with the other past or present members of the so-called Max Plus working group at INRIA-Rocquencourt. Among them, M. Viot and S. Gaubert deserve special mention. Moreover, S. Gaubert helped us to check some examples included in this book, thanks to his handy computer software MAX manipulating the $\mathcal{M}_{\mathrm{in}}^{\mathrm{ax}}[\![\gamma,\delta]\!]$ algebra. Finally, the publisher, in the person of Ms. Helen Ramsey, is also to be thanked, specifically because of her tolerant view with respect to deadlines.

We would like to stress that the material covered in this book has been and is still in fast evolution. Owing to our different backgrounds, it became clear to us that many different cultures within mathematics exist with regard to style, notation, etc. We did our best to come up with one, uniform style throughout the book. Chances are, however, that, when the reader notices a higher density of Theorems, Definitions, etc., GC and/or JPQ were the primary authors of the corresponding parts[1]. As a last remark, the third author can always be consulted on the problem of coping with three French co-authors.

François Baccelli, Sophia Antipolis
Guy Cohen, Fontainebleau
Geert Jan Olsder, Delft
Jean-Pierre Quadrat, Rocquencourt

June 1992

[1]GC: I do not agree. FB is more prone to that than any of us!

PART I

Discrete Event Systems and Petri Nets

CHAPTER 1

Introduction and Motivation

1.1 PRELIMINARY REMARKS AND SOME NOTATION

Probably the most well-known equation in the theory of difference equations is

$$x(t+1) = Ax(t) \ , \quad t = 0, 1, 2, \ldots . \tag{1.1}$$

The vector $x \in \mathbb{R}^n$ represents the 'state' of an underlying model and this state evolves in time according to this equation; $x(t)$ denotes the state at time t. The symbol A represents a given $n \times n$ matrix. If an initial condition

$$x(0) = x_0 \tag{1.2}$$

is given, then the whole future evolution of (1.1) is determined.

Implicit in the text above is that (1.1) is a vector equation. Written out in scalar equations it becomes

$$x_i(t+1) = \sum_{j=1}^{n} A_{ij} x_j(t) \ , \quad i = 1, \ldots, n \ ; \quad t = 0, 1, \ldots . \tag{1.3}$$

The symbol x_i denotes the i-th component of the vector x; the elements A_{ij} are the entries of the square matrix A. If $A_{ij}, i, j = 1, \ldots, n$, and $x_j(t), j = 1, \ldots, n$, are given, then $x_j(t+1), j = 1, \ldots, n$, can be calculated according to (1.3).

The only operations used in (1.3) are multiplication ($A_{ij} \times x_j(t)$) and addition (the $\sum$ symbol). Most of this book can be considered as a study of formulæ of the form (1.1), in which the operations are changed. Suppose that the two operations in (1.3) are changed in the following way: addition becomes maximization and multiplication becomes addition. Then (1.3) becomes

$$\begin{aligned} x_i(k+1) &= \max(A_{i1} + x_1(k), A_{i2} + x_2(k), \ldots, A_{in} + x_n(k)) \\ &= \max_j (A_{ij} + x_j(k)) \ , \quad i = 1, \ldots, n \ . \end{aligned} \tag{1.4}$$

If the initial condition (1.2) also holds for (1.4), then the time evolution of (1.4) is completely determined again. Of course the time evolutions of (1.3) and (1.4) will be different in general. Equation (1.4), as it stands, is a nonlinear difference equation. As an example take $n = 2$, such that A is a 2×2 matrix. Suppose

$$A = \begin{pmatrix} 3 & 7 \\ 2 & 4 \end{pmatrix} \tag{1.5}$$

and that the initial condition is

$$x_0 = \begin{pmatrix} 1 \\ 0 \end{pmatrix}. \tag{1.6}$$

The time evolution of (1.1) becomes

$$x(0) = \begin{pmatrix} 1 \\ 0 \end{pmatrix}, \; x(1) = \begin{pmatrix} 3 \\ 2 \end{pmatrix}, \; x(2) = \begin{pmatrix} 23 \\ 14 \end{pmatrix}, \; x(3) = \begin{pmatrix} 167 \\ 102 \end{pmatrix}, \ldots$$

and the time evolution of (1.4) becomes

$$x(0) = \begin{pmatrix} 1 \\ 0 \end{pmatrix}, \; x(1) = \begin{pmatrix} 7 \\ 4 \end{pmatrix}, \; x(2) = \begin{pmatrix} 11 \\ 9 \end{pmatrix}, \; x(3) = \begin{pmatrix} 16 \\ 13 \end{pmatrix}, \ldots. \tag{1.7}$$

We are used to thinking of the argument t in $x(t)$ as time; at time t the state is $x(t)$. With respect to (1.4) we will introduce a different meaning for this argument. In order to emphasize this different meaning, the argument t has been replaced by k. For this new meaning we need to think of a network, which consists of a number of nodes and some arcs connecting these nodes. The network corresponding to (1.4) has n nodes; one for each component x_i. Entry A_{ij} corresponds to the arc from node j to node i. In terms of graph theory such a network is called a directed graph ('directed' because the individual arcs between the nodes are one-way arrows). Therefore the arcs corresponding to A_{ij} and A_{ji}, if both exist, are considered to be different.

The nodes in the network can perform certain activities; each node has its own kind of activity. Such activities take a finite time, called the activity time, to be performed. These activity times may be different for different nodes. It is assumed that an activity at a certain node can only start when all preceding ('directly upstream') nodes have finished their activities and sent the results of these activities along the arcs to the current node. Thus, the arc corresponding to A_{ij} can be interpreted as an output channel for node j and simultaneously as an input channel for node i. Suppose that this node i starts its activity as soon as all preceding nodes have sent their results (the rather neutral word 'results' is used, it could equally have been messages, ingredients or products, etc.) to node i, then (1.4) describes when the activities take place. The interpretation of the quantities used is:

- $x_i(k)$ is the earliest epoch at which node i becomes active for the k-th time;
- A_{ij} is the sum of the activity time of node j and the traveling time from node j to node i (the rather neutral expression 'traveling time' is used instead of, for instance, 'transportation time' or 'communication time').

The fact that we write A_{ij} rather than A_{ji} for a quantity connected to the arc from node j to node i has to do with matrix equations which will be written in the classical way with column vectors, as will be seen later on. For the example given above, the network has two nodes and four arcs, as given in Figure 1.1. The interpretation of the number 3 in this figure is that if node 1 has started an activity, the next

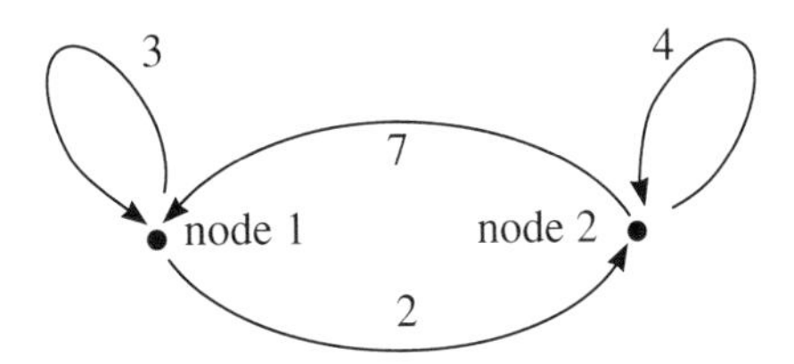

Figure 1.1: Network corresponding to Equation (1.5)

activity cannot start within the next 3 time units. Similarly, the time between two subsequent activities of node 2 is at least 4 time units. Node 1 sends its results to node 2 and once an activity starts in node 1, it takes 2 time units before the result of this activity reaches node 2. Similarly it takes 7 time units after the initiation of an activity of node 2 for the result of that activity to reach node 1. Suppose that an activity refers to some production. The production time of node 1 could for instance be 1 time unit; after that, node 1 needs 2 time units for recovery (lubrication say) and the traveling time of the result (the final product) from node 1 to node 2 is 1 time unit. Thus the number $A_{11} = 3$ is made up of a production time 1 and a recovery time 2 and the number $A_{21} = 2$ is made up of the same production time 1 and a traveling time 1. Similarly, if the production time at node 2 is 4, then this node does not need any time for recovery (because $A_{22} = 4$), and the traveling time from node 2 to node 1 is 3 (because $A_{12} = 7 = 4 + 3$).

If we now look at the sequence (1.7) again, the interpretation of the vectors $x(k)$ is different from the initial one. The argument k is no longer a time instant but a counter which states how many times the various nodes have been active. At time 14, node 1 has been active twice (more precisely, node 1 has started two activities, respectively at times 7 and 11). At the same time 14, node 2 has been active three times (it started activities at times 4, 9 and 13). The counting of the activities is such that it coincides with the argument of the x vector. The initial condition is henceforth considered to be the zeroth activity.

In Figure 1.1 there was an arc from any node to any other node. In many networks referring to more practical situations, this will not be the case. If there is no arc from node j to node i, then node i does not need any result from node j. Therefore node j does not have a direct influence on the behavior of node i. In such a situation it is useful to consider the entry A_{ij} to be equal to $-\infty$. In (1.4) the term $-\infty + x_j(k)$ does not influence $x_i(k+1)$ as long as $x_j(k)$ is finite. The number $-\infty$ will occur frequently in what follows and will be indicated by ε.

For reasons which will become clear later on, Equation (1.4) will be written as

$$x_i(k+1) = \bigoplus_j A_{ij} \otimes x_j(k) \ , \quad i = 1, \ldots, n \ ,$$

or in vector notation,

$$x(k+1) = A \otimes x(k) \ . \tag{1.8}$$

The symbol $\bigoplus_j c(j)$ refers to the maximum of the elements $c(j)$ with respect to all appropriate j, and $\otimes$ (pronounced 'o-times') refers to addition. Later on the symbol $\oplus$ (pronounced 'o-plus') will also be used; $a \oplus b$ refers to the maximum of the scalars a and b. If the initial condition for (1.8) is $x(0) = x_0$, then

$$\begin{aligned} x(1) &= A \otimes x_0 \ , \\ x(2) &= A \otimes x(1) = A \otimes (A \otimes x_0) = (A \otimes A) \otimes x_0 = A^2 \otimes x_0 \ . \end{aligned}$$

It will be shown in Chapter 3 that indeed $A \otimes (A \otimes x_0) = (A \otimes A) \otimes x_0$. For the example given above it is easy to check this by hand. Instead of $A \otimes A$ we simply write A^2. We obtain

$$x(3) = A \otimes x(2) = A \otimes (A^2 \otimes x_0) = (A \otimes A^2) \otimes x_0 = A^3 \otimes x_0 \ ,$$

and in general

$$x(k) = (\underbrace{A \otimes A \otimes \cdots \otimes A}_{k \text{ times}}) \otimes x_0 = A^k \otimes x_0 \ .$$

The matrices A^2, A^3,..., can be calculated directly. Let us consider the A-matrix of (1.5) again, then

$$A^2 = \begin{pmatrix} \max(3+3, 7+2) & \max(3+7, 7+4) \\ \max(2+3, 4+2) & \max(2+7, 4+4) \end{pmatrix} = \begin{pmatrix} 9 & 11 \\ 6 & 9 \end{pmatrix} .$$

In general

$$(A^2)_{ij} = \bigoplus_l A_{il} \otimes A_{lj} = \max_l (A_{il} + A_{lj}) \ . \tag{1.9}$$

An extension of (1.8) is

$$\left.\begin{aligned} x(k+1) &= (A \otimes x(k)) \oplus (B \otimes u(k)) \ , \\ y(k) &= C \otimes x(k) \ . \end{aligned}\right\} \tag{1.10}$$

The symbol $\oplus$ in this formula refers to componentwise maximization. The m-vector u is called the input to the system; the p-vector y is the output of the system. The components of u refer to nodes which have no predecessors. Similarly, the components of y refer to nodes with no successors. The components of x now refer to internal nodes, i.e. to nodes which have both successors and predecessors. The matrices $B = \{B_{ij}\}$ and $C = \{C_{ij}\}$ have sizes $n \times m$ and $p \times n$, respectively. The traditional way of writing (1.10) would be

$$\begin{aligned} x_i(k+1) &= \max(A_{i1} + x_1(k), \ldots, A_{in} + x_n(k), \\ &\qquad B_{i1} + u_1(k), \ldots, B_{im} + u_m(k)) \ , \quad i = 1, \ldots, n \ ; \\ y_i(k) &= \max(C_{i1} + x_1(k), \ldots, C_{in} + x_n(k)) \ , \quad i = 1, \ldots, p \ . \end{aligned}$$

Sometimes (1.10) is written as

$$\left.\begin{aligned} x(k+1) &= A \otimes x(k) \oplus B \otimes u(k) \ , \\ y(k) &= C \otimes x(k) \ , \end{aligned}\right\} \tag{1.11}$$

where it is understood that multiplication has priority over addition. Usually, however, (1.10) is written as

$$\left.\begin{array}{rcl} x(k+1) & = & Ax(k) \oplus Bu(k) \ , \\ y(k) & = & Cx(k) \ . \end{array}\right\} \quad (1.12)$$

> If it is clear where the '$\otimes$'-symbols are used, they are sometimes omitted, as shown in (1.12). This practice is exactly the same one as with respect to the more common multiplication ' $\times$ ' or ' . ' symbol in conventional algebra. In the same vein, in conventional algebra $1 \times x$ is the same as $1x$, which is usually written as x. Within the context of the $\otimes$ and $\oplus$ symbols, $0 \otimes x$ is exactly the same as x. The symbol ε is the neutral element with respect to maximization; its numerical value equals $-\infty$. Similarly, the symbol e denotes the neutral element with respect to addition; it assumes the numerical value 0. Also note that $1 \otimes x$ is different from x.

If one wants to think in terms of a network again, then $u(k)$ is a vector indicating when certain resources become available for the k-th time. Subsequently it takes B_{ij} time units before the j-th resource reaches node i of the network. The vector $y(k)$ refers to the epoch at which the final products of the network are delivered to the outside world.

Take for example

$$\left.\begin{array}{rcl} x(k+1) & = & \begin{pmatrix} 3 & 7 \\ 2 & 4 \end{pmatrix} x(k) \oplus \begin{pmatrix} \varepsilon \\ 1 \end{pmatrix} u(k) \ , \\ y(k) & = & \begin{pmatrix} 3 & \varepsilon \end{pmatrix} x(k) \ . \end{array}\right\} \quad (1.13)$$

The corresponding network is shown in Figure 1.2. Because $B_{11} = \varepsilon \ (= -\infty)$, the

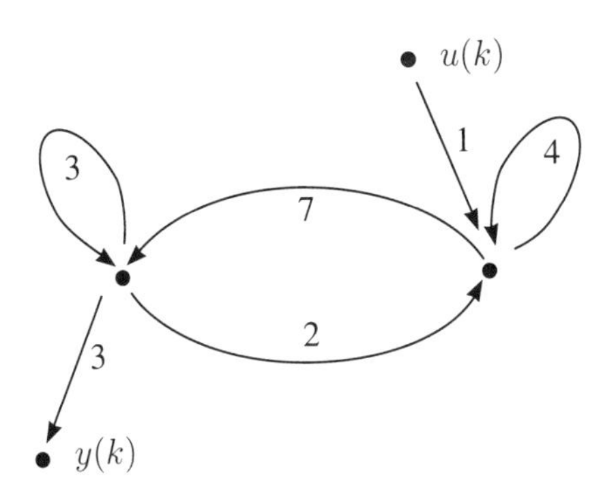

Figure 1.2: Network with input and output

input $u(k)$ only goes to node 2. If one were to replace B by $\begin{pmatrix} 2 & 1 \end{pmatrix}'$ for instance, where the prime denotes transposition, then each input would 'spread' itself over the two nodes. In this example from epoch $u(k)$ on, it takes 2 time units for the input to reach node 1 and 1 time unit to reach node 2. In many practical situations an input will enter the network through one node. That is why, as in this example, only one B-entry per column is different from ε. Similar remarks can be made with respect to the output. Suppose that we have (1.6) as an initial condition and that

$$u(0) = 1 \ , \quad u(1) = 7 \ , \quad u(2) = 13 \ , \quad u(3) = 19 \ ,\ldots,$$

then it easily follows that

$$x(0) = \begin{pmatrix} 1 \\ 0 \end{pmatrix}, \ x(1) = \begin{pmatrix} 7 \\ 4 \end{pmatrix}, \ x(2) = \begin{pmatrix} 11 \\ 9 \end{pmatrix}, \ x(3) = \begin{pmatrix} 16 \\ 14 \end{pmatrix}, \ldots,$$

$$y(0) = 4 \ , \quad y(1) = 10 \ , \quad y(2) = 14 \ , \quad y(3) = 19 \ ,\ldots.$$

We started this section with the difference equation (1.1), which is a first-order linear vector difference equation. It is well known that a higher order linear scalar difference equation

$$z(k+1) = a_1 z(k) + a_2 z(k-1) + \cdots + a_n z(k-n+1) \tag{1.14}$$

can be written in the form of Equation (1.1). If we introduce the vector $(z(k), z(k-1), \ldots, z(k-n+1))'$, then (1.14) can be written as

$$\begin{pmatrix} z(k+1) \\ z(k) \\ \vdots \\ \vdots \\ z(k-n+2) \end{pmatrix} = \begin{pmatrix} a_1 & a_2 & \ldots & \ldots & a_n \\ 1 & 0 & \ldots & \ldots & 0 \\ 0 & & & & \\ \vdots & & & & \\ 0 & \ldots & 0 & 1 & 0 \end{pmatrix} \begin{pmatrix} z(k) \\ z(k-1) \\ \vdots \\ \vdots \\ z(k-n+1) \end{pmatrix} . \tag{1.15}$$

This equation has exactly the form of (1.1). If we change the operations in (1.14) in the standard way, addition becomes maximization and multiplication becomes addition; then the numerical evaluation of (1.14) becomes

$$z(k+1) = \max(a_1 + z(k), a_2 + z(k-1), \ldots, a_n + z(k-n+1)) \ . \tag{1.16}$$

This equation can also be written as a first-order linear vector difference equation. In fact this equation is almost Equation (1.15), which must now be evaluated with the operations maximization and addition. The only difference is that the 1's and 0's in (1.15) must be replaced by e's and ε's, respectively.

1.2 MISCELLANEOUS EXAMPLES

In this section, seven examples from different application areas are presented, with a special emphasis on the modeling process. The examples can be read independently. It is shown that all problems formulated lead to equations of the kind (1.8), (1.10), or related ones. Solutions to the problems which are formulated are not given in this section. To solve these problems, the theory must first be developed and that will be done in the next chapters. Although some of the examples deal with equations with the look of (1.8), the operations used will again be different. The mathematical expressions are the same for many applications. The underlying algebra, however, differs. The emphasis of this book is on these algebras and their relationships.

1.2.1 Planning

Planning is one of the traditional fields in which the max-operation plays a crucial role. In fact, many problems in planning areas are more naturally formulated with the min-operation than with the max-operation. However, one can easily switch from minimization to maximization and vice versa. Two applications will be considered in this subsection; the first one is the shortest path problem, the second one is a scheduling problem. Solutions to such problems have been known for some time, but here the emphasis is on the notation introduced in §1.1 and on some analysis pertaining to this notation.

1.2.1.1 Shortest Path

Consider a network of n cities; these cities are the nodes in a network. Between some cities there are road connections; the distance between city j and city i is indicated by A_{ij}. A road corresponds to an arc in the network. If there is no road from j to i, then we set $A_{ij} = \varepsilon$. In this example $\varepsilon = +\infty$; nonexisting roads get assigned a value $+\infty$ rather than $-\infty$. The reason is that we will deal with minimization rather than maximization. Owing to the possibility of one-way traffic, it is allowed that $A_{ij} \neq A_{ji}$. Matrix A is defined as $A = (A_{ij})$.

The entry A_{ij} denotes the distance between j and i if only one link is allowed. Sometimes it may be more advantageous to go from j to i via k. This will be the case if $A_{ik} + A_{kj} < A_{ij}$. The shortest distance from j to i using exactly two links is

$$\min_{k=1,\dots,n} (A_{ik} + A_{kj}) \ . \tag{1.17}$$

When we use the shorthand symbol $\oplus$ for the minimum operation, then (1.17) becomes

$$\bigoplus_k A_{ik} \otimes A_{kj} \ .$$

Note that $\bigoplus$ has been used for both the maximum and the minimum operation. It should be clear from the context which is meant. The symbol $\oplus$ will be used similarly. The reason for not distinguishing between these two operations is that $(\mathbb{R} \cup \{-\infty\}, \max, +)$ and $(\mathbb{R} \cup \{+\infty\}, \min, +)$ are isomorphic algebraic structures. Chapters 3 and 4 will deal with such structures. It is only when the operations max and min appear in the same formula that this convention would lead to ambiguity. This situation will occur in Chapter 9 and different symbols for the two operations will be used there. Expression (1.17) is the ij-th entry of A^2:

$$(A^2)_{ij} = \bigoplus_k A_{ik} \otimes A_{kj} \ .$$

Note that the expression A^2 can have different meanings also. In (1.9) the max-operation was used whereas the min-operation is used here.

If one is interested in the shortest path from j to i using one *or* two links, then the length of the shortest path becomes

$$(A \oplus A^2)_{ij} \ .$$

If we continue, and if one, two or three links are allowed, then the length of the shortest path from j to i becomes

$$(A \oplus A^2 \oplus A^3)_{ij} \ ,$$

where $A^3 = A^2 \otimes A$, and so on for more than three links. We want to find the shortest path whereby any number of links is allowed. It is easily seen that a road connection consisting of more than $n - 1$ links can never be optimal. If it were optimal, the traveler would visit one city at least twice. The road from this city to itself forms a part of the total road connection and is called a circuit. Since it is (tacitly) assumed that all distances are nonnegative, this circuit adds to the total distance and can hence be disregarded. The conclusion is that the length of the shortest path from j to i is given by

$$(A \oplus A^2 \oplus \cdots \oplus A^{n-1})_{ij} \ .$$

Equivalently one can use the following infinite series for the shortest path (the terms $A^k,\ k \geq n$ do not contribute to the sum):

$$A^+ \stackrel{\text{def}}{=} A \oplus A^2 \oplus \cdots \oplus A^n \oplus A^{n+1} \oplus \cdots \ . \tag{1.18}$$

The matrix A^+, sometimes referred to as the *shortest path matrix*, also shows up in the scheduling problem that we define below shortly.

Note that $(A^+)_{ii}$ refers to a path which first leaves node i and then comes back to it. If one wants to include the possibility of staying at a node, then the shortest path matrix should be defined as $e \oplus A^+$, where e denotes the identity matrix of the same size as A. An identity matrix in this set-up has zeros on the diagonal and the other entries have the value $+\infty$. In general, e is a unit matrix of appropriate size.

The shortest path problem can also be formulated according to a difference equation of the form (1.8). To that end, consider an $n \times n$ matrix X: the ij-th entry of X refers to a connection from city j to city i, $X_{ij}(k)$ is the minimum length with respect to all roads from j to i with k links. Then it is not difficult to see that this vector satisfies the equation

$$X_{ij}(k) = \min_{l=1,\ldots,n} (X_{il}(k-1) + A_{lj}) \ , \quad i,j = 1,\ldots,n \ . \tag{1.19}$$

Formally this equation can be written as

$$X(k) = X(k-1)A = X(k-1) \otimes A \ ,$$

but it cannot be seen from this equation that the operations to be used are minimization and addition. The principle of dynamic programming can be recognized in (1.19). The following formula gives exactly the same results as (1.19):

$$X_{ij}(k) = \min_{l=1,\dots,n} (A_{il} + X_{lj}(k-1)) \ , \quad i,j = 1,\dots,n \ .$$

The difference between this formula and (1.19) is that one uses the principle of forward dynamic programming and the other one uses backward dynamic programming.

1.2.1.2 *Scheduling*

Consider a project which consists of various tasks. Some of these tasks cannot be started before some others have been finished. The dependence of these tasks can be given in a directed graph in which each node coincides with a task (or, equivalently, with an activity). As an example, consider the graph of Figure 1.3. There are six nodes, numbered $1,\dots,6$. Node 1 represents the initial activity and

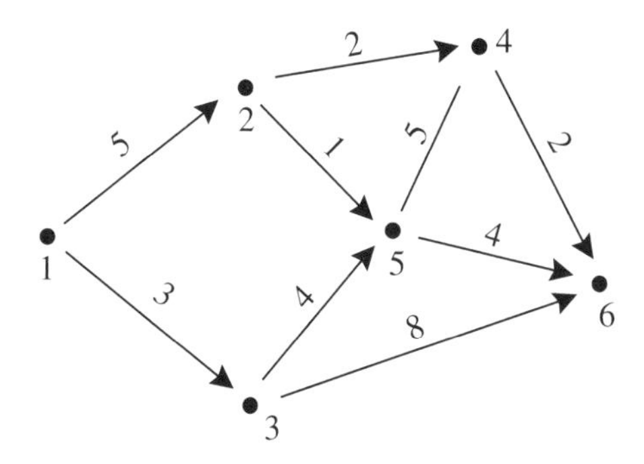

Figure 1.3: Ordering of activities in a project

node 6 represents the final activity. It is assumed that the activities, except the final one, take a certain time to be performed. In addition, there may be traveling times. The fact that the final activity has a zero cost is not a restriction. If it were to have a nonzero cost, a fictitious node 7 could be added to node 6. Node 7 would represent the final activity. The arcs between the nodes in Figure 1.3 denote the precedence constraints. For instance, node 4 cannot start before nodes 2 and 5 have finished their activities. The number A_{ij} associated with the arc from node j to node i denotes the minimum time that should elapse between the beginning of an activity at node j and the beginning of an activity at node i.

By means of the principle of dynamic programming it is not difficult to calculate the *critical path* in the graph. Critical here refers to 'slowest'. The total duration of the overall project cannot be smaller than the summation of all numbers A_{ij} along the critical path.

Another way of finding the time at which the activity at node i can start at the earliest, which will be denoted x_i, is the following. Suppose activity 1 can start at epoch u at the earliest. This quantity u is an input variable which must be given from the outside. Hence $x_1 = u$. For the other x_i's we can write

$$x_i = \max_{j=1,\ldots,6} (A_{ij} + x_j) \ .$$

If there is no arc from node i to j, then A_{ij} gets assigned the value ε $(= -\infty)$. If $x = (x_1, \ldots, x_6)'$ and $A = (A_{ij})$, then we can compactly write

$$x = Ax \oplus Bu \ , \tag{1.20}$$

where

$$A = \begin{pmatrix} \varepsilon & \varepsilon & \varepsilon & \varepsilon & \varepsilon & \varepsilon \\ 5 & \varepsilon & \varepsilon & \varepsilon & \varepsilon & \varepsilon \\ 3 & \varepsilon & \varepsilon & \varepsilon & \varepsilon & \varepsilon \\ \varepsilon & 2 & \varepsilon & \varepsilon & 5 & \varepsilon \\ \varepsilon & 1 & 4 & \varepsilon & \varepsilon & \varepsilon \\ \varepsilon & \varepsilon & 8 & 2 & 4 & \varepsilon \end{pmatrix} , \quad B = \begin{pmatrix} e \\ \varepsilon \\ \varepsilon \\ \varepsilon \\ \varepsilon \\ \varepsilon \end{pmatrix} .$$

Note that e in B equals 0 in this context. Here we recognize the form of (1.11), although in (1.20) time does not play a role; x in the left-hand side equals the x in the right-hand side. Hence (1.20) is an implicit equation for the vector x. Let us see what we obtain by repeated substitution of the complete right-hand side of (1.20) into x of this same right-hand side. After one substitution:

$$\begin{aligned} x &= A^2x \oplus ABu \oplus Bu \\ &= A^2x \oplus (A \oplus e)Bu \ , \end{aligned}$$

and after n substitutions:

$$x = A^n x \oplus (A^{n-1} \oplus A^{n-2} \oplus \cdots \oplus A \oplus e)Bu \ .$$

In the formulæ above, e refers to the identity matrix; zeros on the diagonal and ε's elsewhere. The symbol e will be used as the identity element for all spaces that will be encountered in this book. Similarly, ε will be used to denote the zero element of any space to be encountered.

Since the entries of A^n denote the weights of paths with length n in the corresponding graph and A does not have paths of length greater than 4, we get $A^n = -\infty$ for $n \geq 5$. Therefore the solution x in the current example becomes

$$x = (A^4 \oplus A^3 \oplus A^2 \oplus A \oplus e)Bu \ , \tag{1.21}$$

for which we can write

$$x = (e \oplus A^+)Bu \ ,$$

where A^+ was defined in (1.18).

In (1.21), we made use of the series

$$A^* \stackrel{\text{def}}{=} e \oplus A \oplus \cdots \oplus A^n \oplus A^{n+1} \oplus \cdots \ , \tag{1.22}$$

although it was concluded that $A^k, k > n$, does not contribute to the sum. With the conventional matrix calculus in mind one might be tempted to write for (1.22):

$$(e \oplus A \oplus A^2 \oplus \cdots) = (e \ominus A)^{-1} . \tag{1.23}$$

Of course, we have not defined the inverse of a matrix within the current setting and so (1.23) is an empty statement. It is also strange to have a 'minus' sign $\ominus$ in (1.23) and it is not known how to interpret this sign in the context of the max-operation at the left-hand side of the equation. It should be the reverse operation of $\oplus$. If we dare to continue along these shaky lines, one could write the solution of (1.20) as

$$(e \ominus A)x = Bu \Rightarrow x = (e \ominus A)^{-1}Bu .$$

Quite often one can guide one's intuition by considering formal expressions of the kind (1.23). One tries to find formal analogies in the notation using conventional analysis. In Chapter 3 it will be shown that an inverse as in (1.23) does not exist in general and therefore we get 'stuck' with the series expansion.

There is a dual way to analyze the critical path of Figure 1.3. Instead of starting at the initial node 1, one could start at the final node 6 and then work backward in time. This latter approach is useful when a target time for the completion of the project has been set. The question then is: what is the latest moment at which each node has to start its activity in such a way that the target time can still be fulfilled? If we call the starting times x_i again, then it is not difficult to see that

$$x_i = \min\left[\min_j\left(\left(\widehat{A}\right)_{ij} + x_j\right), \left(\widehat{B}\right)_i + u\right], \quad i = 1, \ldots, 6 ,$$

where

$$\widehat{A} = \begin{pmatrix} \varepsilon & 5 & 3 & \varepsilon & \varepsilon & \varepsilon \\ \varepsilon & \varepsilon & \varepsilon & 2 & 1 & \varepsilon \\ \varepsilon & \varepsilon & \varepsilon & \varepsilon & 4 & 8 \\ \varepsilon & \varepsilon & \varepsilon & \varepsilon & \varepsilon & 2 \\ \varepsilon & \varepsilon & \varepsilon & 5 & \varepsilon & 4 \\ \varepsilon & \varepsilon & \varepsilon & \varepsilon & \varepsilon & \varepsilon \end{pmatrix}, \quad \widehat{B} = \begin{pmatrix} \varepsilon \\ \varepsilon \\ \varepsilon \\ \varepsilon \\ \varepsilon \\ e \end{pmatrix} .$$

It is easily seen that $\widehat{A}$ is equal to the transpose of A in (1.20); x_6 has been chosen as the completion time of the project. In matrix form, we can write

$$x = \left(\widehat{A} \odot x\right) \wedge \left(\widehat{B} \odot u\right) ,$$

where $\odot$ is a new matrix multiplication using min as addition of scalars and $+$ as multiplication, whereas $\wedge$ is the min of vectors, componentwise. This topic of target times will be addressed in §5.6.

1.2.2 Communication

This subsection focuses on the Viterbi algorithm. It can conveniently be described by a formula of the form (1.1). The operations to be used this time are maximization and multiplication.

The stochastic process of interest in this section, $\nu(k)$, $k \geq 0$, is a time homogeneous Markov chain with state space $\{1, 2, \ldots, n\}$, defined on some probability space $(\Omega, \mathbb{F}, \mathbb{P})$. The Markov property means that

$$\mathbb{P}\left[\nu(k+1) = i_{k+1} \mid \nu(0) = i_0, \ldots, \nu(k) = i_k\right] = \mathbb{P}\left[\nu(k+1) = i_{k+1} \mid \nu(k) = i_k\right] ,$$

where $\mathbb{P}\left[\mathcal{A} \mid \mathcal{B}\right]$ denotes the conditional probability of the event $\mathcal{A}$ given the event $\mathcal{B}$ and $\mathcal{A}$ and $\mathcal{B}$ are in $\mathbb{F}$. Let M_{ij} denote the transition probability[1] from state j to i. The initial distribution of the Markov chain will be denoted p.

The process $\nu = (\nu(0), \ldots, \nu(K))$ is assumed to be observed with some noise. This means that there exists a sequence of $\{1, 2, \ldots, n\}$-valued random variables $z(k), k = 0, \ldots, K$, called the *observation*, and such that $N_{i_k j_k} \stackrel{\text{def}}{=} \mathbb{P}[z(k) = i_k \mid \nu(k) = j_k]$ does not depend on k and such that the joint law of (ν, z), where $z = (z(0), \ldots, z(K))$, is given by the relation

$$\mathbb{P}[\nu = j, z = i] = \left(\prod_{k=1}^{K} N_{i_k j_k} M_{j_k j_{k-1}} \right) N_{i_0 j_0} p_{j_0} \ , \tag{1.24}$$

where $i = (i_0, \ldots, i_K)$ and $j = (j_0, \ldots, j_K)$.

Given such a sequence z of observations, the question to be answered is to find the sequence j for which the probability $\mathbb{P}\left[\nu = j \mid z\right]$ is maximal. This problem is a highly simplified version of a text recognition problem. A machine reads handwritten text, symbol after symbol, but makes mistakes (the observation errors). The underlying model of the text is such that after having read a symbol, the probability of the occurrence of the next one is known. More precisely, the sequence of symbols is assumed to be produced by a Markov chain.

We want to compute the quantity

$$x_{j_K}(K) = \max_{j_0, \ldots, j_{K-1}} \mathbb{P}\left[\nu = j, z = i\right] \ . \tag{1.25}$$

This quantity is also a function of i, but this dependence will not be made explicit. The argument that achieves the maximum in the right-hand side of (1.25) is the most likely text up to the $(K-1)$-st symbol for the observation i; similarly, the argument j_K which maximizes $x_{j_K}(K)$ is the most likely K-th symbol given that the first K observations are i. From (1.24), we obtain

$$\begin{aligned} x_{j_K}(K) &= \max_{j_0, \ldots, j_{K-1}} \left(\prod_{k=1}^{K} \left(N_{i_k j_k} M_{j_k j_{k-1}} \right) N_{i_0 j_0} p_{j_0} \right) \\ &= \max_{j_{K-1}} \left(\left(N_{i_K j_K} M_{j_K j_{K-1}} \right) x_{j_{K-1}}(K-1) \right) \ , \end{aligned}$$

with initial condition $x_{j_0}(0) = N_{i_0 j_0} p_{j_0}$. The reader will recognize the above algorithm as a simple version of (forward) dynamic programming. If $N_{i_k j_k} M_{j_k j_{k-1}}$ is denoted $A_{j_k j_{k-1}}$, then the general formula is

$$x_m(k) = \max_{\ell = 1, \ldots, n} \left(A_{m\ell} x_\ell(k-1) \right), \quad m = 1, \ldots, n \ . \tag{1.26}$$

[1]Classically, in Markov chains, M_{ij} would rather be denoted M_{ji}.

Table 1.1: Processing times

	P_1	P_2	P_3
M_1		1	5
M_2	3	2	3
M_3	4	3	

This formula is similar to (1.1) if addition is replaced by maximization and multiplication remains multiplication.

The Viterbi algorithm maximizes $\mathbb{P}[\nu, z]$ as given in (1.24). If we take the logarithm of (1.24), and multiply the result by -1, (1.26) becomes

$$-\ln(x_m(k)) = \min_{\ell=1,\ldots,n} \left[-\ln(A_{m\ell}) - \ln(x_\ell(k-1))\right], \quad m = 1,\ldots,n \; .$$

The form of this equation exactly matches (1.19). Thus the Viterbi algorithm is identical to an algorithm which determines the shortest path in a network. Actually, it is this latter algorithm—minimizing $-\ln(\mathbb{P}[\nu \mid z])$—which is quite often referred to as the Viterbi algorithm, rather than the one expressed by (1.26).

1.2.3 Production

Consider a manufacturing system consisting of three machines. It is supposed to produce three kinds of parts according to a certain product mix. The routes to be followed by each part and each machine are depicted in Figure 1.4 in which $M_i, i = 1, 2, 3$, are the machines and $P_i, i = 1, 2, 3$, are the parts. Processing

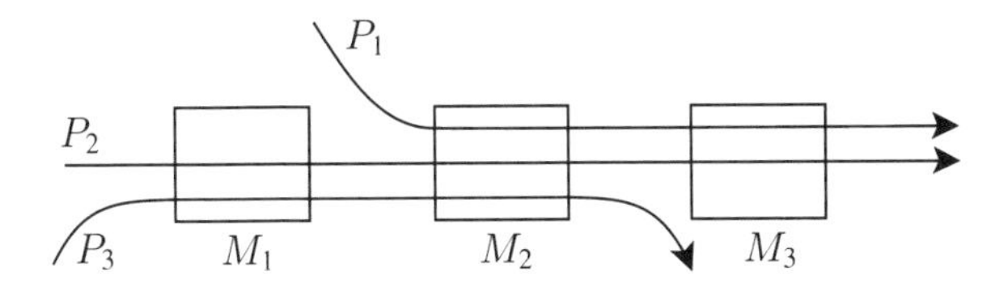

Figure 1.4: Routing of parts along machines

times are given in Table 1.1. Note that this manufacturing system has a flow-shop structure, i.e. all parts follow the same sequence on the machines (although they may skip some) and every machine is visited at most once by each part. We assume that there are no set-up times on machines when they switch from one part type to another. Parts are carried on a limited number of pallets (or, equivalently, product carriers). For reasons of simplicity it is assumed that

1. only one pallet is available for each part type;

2. the final product mix is balanced in the sense that it can be obtained by means of a periodic input of parts, here chosen to be P_1, P_2, P_3;
3. there are no set-up times or traveling times;
4. the sequencing of part types on the machines is known and it is (P_2, P_3) on M_1, (P_1, P_2, P_3) on M_2 and (P_1, P_2) on M_3.

The last point mentioned is not for reasons of simplicity. If any machine were to start working on the part which arrived first instead of waiting for the appropriate part, the modeling would be different. Manufacturing systems in which machines start working on the first arriving part (if it has finished its current activity) will be dealt with in Chapter 9. We can draw a graph in which each node corresponds to a combination of a machine and a part. Since M_1 works on 2 parts, M_2 on 3 and M_3 on 2, this graph has seven nodes. The arcs between the nodes express the precedence constraints between operations due to the sequencing of operations on the machines. To each node i in Figure 1.5 corresponds a number x_i which denotes the earliest epoch at which the node can start its activity. In order to be able to calculate these quantities, the epochs at which the machines and parts (together called the resources) are available must be given. This is done by means of a six-dimensional input vector u (six since there are six resources: three machines and three parts). There is an output vector also; the components of the six-dimensional vector y denote the epochs at which the parts are ready and the machines have finished their jobs (for one cycle). The model becomes

$$x = Ax \oplus Bu \ ; \tag{1.27}$$

$$y = Cx \ , \tag{1.28}$$

in which the matrices are

$$A = \begin{pmatrix} \varepsilon & \varepsilon & \varepsilon & \varepsilon & \varepsilon & \varepsilon & \varepsilon \\ 1 & \varepsilon & \varepsilon & \varepsilon & \varepsilon & \varepsilon & \varepsilon \\ \varepsilon & \varepsilon & \varepsilon & \varepsilon & \varepsilon & \varepsilon & \varepsilon \\ 1 & \varepsilon & 3 & \varepsilon & \varepsilon & \varepsilon & \varepsilon \\ \varepsilon & 5 & \varepsilon & 2 & \varepsilon & \varepsilon & \varepsilon \\ \varepsilon & \varepsilon & 3 & \varepsilon & \varepsilon & \varepsilon & \varepsilon \\ \varepsilon & \varepsilon & \varepsilon & 2 & \varepsilon & 4 & \varepsilon \end{pmatrix} ; \quad B = \begin{pmatrix} e & \varepsilon & \varepsilon & \varepsilon & e & \varepsilon \\ \varepsilon & \varepsilon & \varepsilon & \varepsilon & \varepsilon & e \\ \varepsilon & e & \varepsilon & e & \varepsilon & \varepsilon \\ \varepsilon & \varepsilon & \varepsilon & \varepsilon & \varepsilon & \varepsilon \\ \varepsilon & \varepsilon & \varepsilon & \varepsilon & \varepsilon & \varepsilon \\ \varepsilon & \varepsilon & e & \varepsilon & \varepsilon & \varepsilon \\ \varepsilon & \varepsilon & \varepsilon & \varepsilon & \varepsilon & \varepsilon \end{pmatrix} ;$$

$$C = \begin{pmatrix} \varepsilon & 5 & \varepsilon & \varepsilon & \varepsilon & \varepsilon & \varepsilon \\ \varepsilon & \varepsilon & \varepsilon & \varepsilon & 3 & \varepsilon & \varepsilon \\ \varepsilon & \varepsilon & \varepsilon & \varepsilon & \varepsilon & \varepsilon & 3 \\ \varepsilon & \varepsilon & \varepsilon & \varepsilon & \varepsilon & 4 & \varepsilon \\ \varepsilon & \varepsilon & \varepsilon & \varepsilon & \varepsilon & \varepsilon & 3 \\ \varepsilon & \varepsilon & \varepsilon & \varepsilon & 3 & \varepsilon & \varepsilon \end{pmatrix} .$$

Equation (1.27) is an implicit equation in x which can be solved as we did in the subsection on Planning;

$$x = A^* Bu \ .$$

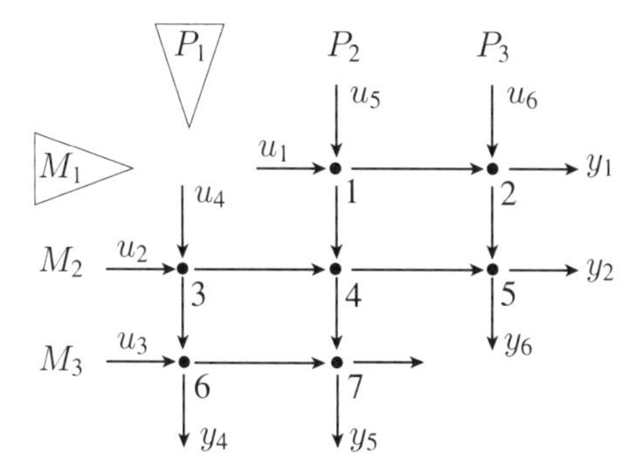

Figure 1.5: The ordering of activities in the flexible manufacturing system

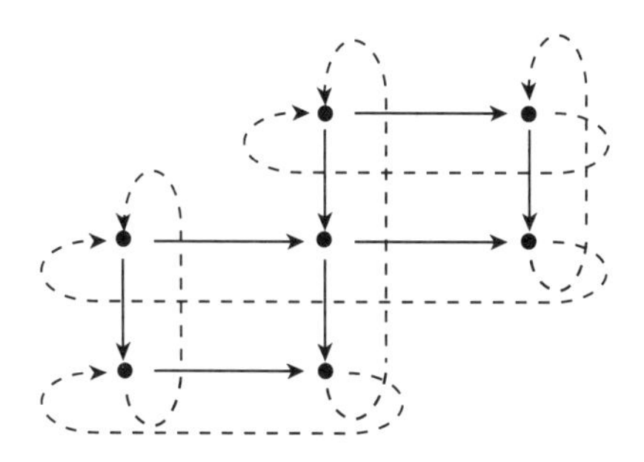

Figure 1.6: Production system with feedback arcs

Now we add feedback arcs to Figure 1.5 as illustrated in Figure 1.6. In this graph the feedback arcs are indicated by dotted lines. The meaning of these feedback arcs is the following. After a machine has finished a sequence of products, it starts with the next sequence. If the pallet on which product P_i was mounted is at the end, the finished product is removed and the empty pallet immediately goes back to the starting point to pick up a new part P_i. If it is assumed that the feedback arcs have zero cost, then $u(k) = y(k-1)$, where $u(k)$ is the k-th input cycle and $y(k)$ the k-th output. Thus we can write

$$\begin{aligned} y(k) &= Cx(k) = CA^*Bu(k) \\ &= CA^*By(k-1) . \end{aligned} \tag{1.29}$$

The transition matrix from $y(k-1)$ to $y(k)$ can be calculated (it can be done by hand, but a simple computer program does the job also):

$$M \stackrel{\text{def}}{=} CA^*B = \begin{pmatrix} 6 & \varepsilon & \varepsilon & \varepsilon & 6 & 5 \\ 9 & 8 & \varepsilon & 8 & 9 & 8 \\ 6 & 10 & 7 & 10 & 6 & \varepsilon \\ \varepsilon & 7 & 4 & 7 & \varepsilon & \varepsilon \\ 6 & 10 & 7 & 10 & 6 & \varepsilon \\ 9 & 8 & \varepsilon & 8 & 9 & 8 \end{pmatrix} .$$

This matrix M determines the speed with which the manufacturing system can work. We will return to this issue in §1.3

1.2.4 Queuing System with Finite Capacity

Let us consider four servers, $S_i, i = 1, \ldots, 4$, in series (see Figure 1.7). Each

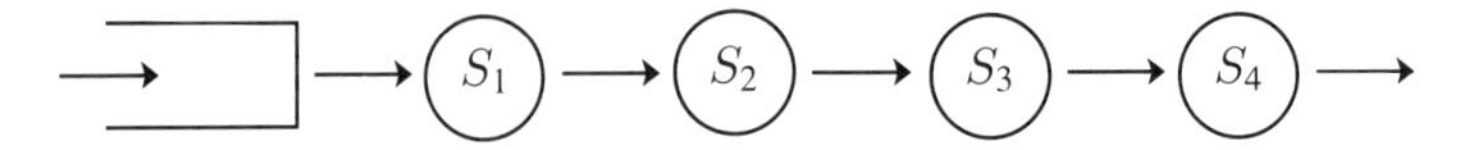

Figure 1.7: Queuing system with four servers

customer is to be served by S_1, S_2, S_3 and S_4, and specifically in this order. It takes $\tau_i(k)$ time units for S_i to serve customer k ($k = 1, 2, \ldots$). Customer k arrives at epoch $u(k)$ into the buffer associated with S_1. If this buffer is empty and S_1 is idle, then this customer is served directly by S_1. Between the servers there are no buffers. The consequence is that if $S_i, i = 1, 2, 3$, has finished serving customer k, but S_{i+1} is still busy serving customer $k - 1$, then S_i cannot start serving the new customer $k + 1$. He must wait. To complete the description of the queuing system, it is assumed that the traveling times between the servers are zero. Let $x_i(k)$ denote the beginning of the service of customer k by server S_i. Before S_i can start serving customer $k + 1$, the following three conditions must be fulfilled:

- S_i must have finished serving customer k;
- S_{i+1} must be idle (for $i = 4$ this condition is an empty one);
- S_{i-1} must have finished serving customer $k + 1$ (for $i = 1$ this condition is an empty one and must be related to the arrival of customer $k + 1$ in the queuing system).

It is not difficult to see that the vector x, consisting of the four x-components, satisfies

$$
\begin{aligned}
x(k+1) &= \begin{pmatrix} \varepsilon & \varepsilon & \varepsilon & \varepsilon \\ \tau_1(k+1) & \varepsilon & \varepsilon & \varepsilon \\ \varepsilon & \tau_2(k+1) & \varepsilon & \varepsilon \\ \varepsilon & \varepsilon & \tau_3(k+1) & \varepsilon \end{pmatrix} x(k+1) \\
&\oplus \begin{pmatrix} \tau_1(k) & e & \varepsilon & \varepsilon \\ \varepsilon & \tau_2(k) & e & \varepsilon \\ \varepsilon & \varepsilon & \tau_3(k) & e \\ \varepsilon & \varepsilon & \varepsilon & \tau_4(k) \end{pmatrix} x(k) \oplus \begin{pmatrix} e \\ \varepsilon \\ \varepsilon \\ \varepsilon \end{pmatrix} u(k+1) \ .
\end{aligned}
\tag{1.30}
$$

We will not discuss issues related to initial conditions here. For those questions, the reader is referred to Chapters 2 and 7. Equation (1.30), which we write formally as

$$x(k+1) = A_2(k+1, k+1)x(k+1) \oplus A_1(k+1, k)x(k) \oplus Bu(k+1) \ ,$$

is an implicit equation in $x(k+1)$ which can be solved again, as done before. The result is

$$x(k+1) = \left(A_2(k+1,k+1)\right)^* \left(A_1(k+1,k)x(k) \oplus Bu(k+1)\right) ,$$

where $\left(A_2(k+1,k+1)\right)^*$ equals

$$\begin{pmatrix} e & \varepsilon & \varepsilon & \varepsilon \\ \tau_1(k+1) & e & \varepsilon & \varepsilon \\ \tau_1(k+1)\tau_2(k+1) & \tau_2(k+1) & e & \varepsilon \\ \tau_1(k+1)\tau_2(k+1)\tau_3(k+1) & \tau_2(k+1)\tau_3(k+1) & \tau_3(k+1) & e \end{pmatrix} .$$

The customers who arrive in the queuing system and cannot directly be served by S_1, wait in the buffer associated with S_1. If one is interested in the buffer contents, i.e. the number of waiting customers, at a certain moment, one should use a counter (of customers) at the entry of the buffer and one at the exit of the buffer. The difference of the two counters yields the buffer contents, but this operation is nonlinear in the max-plus algebra framework. In §1.2.6 we will return to the 'counter'-description of discrete event systems. The counters just mentioned are nondecreasing with time, whereas the buffer contents itself is fluctuating as a function of time.

The design of buffer sizes is a basic problem in manufacturing systems. If the buffer contents tends to go to ∞, one speaks of an unstable system. Of course, an unstable system is an example of a badly designed system. In the current example, buffering between the servers was not allowed. Finite buffers can also be modeled within the max-plus algebra context as shown in the next subsection and more generally in §2.6.2. Another useful parameter is the utilization factor of a server. It is defined by the 'busy time' divided by the total time elapsed.

Note that we did not make any assumptions on the service time $\tau_i(k)$. If one is faced with unpredictable breakdowns (and subsequently a repair time) of the servers, then the service times might be modeled stochastically. For a deterministic and invariant ('customer invariant') system, the serving times do not, by definition, depend on the particular customer.

1.2.5 Parallel Computation

The application of this subsection belongs to the field of VLSI array processors (VLSI stands for 'Very Large Scale Integration'). The theory of discrete events provides a method for analyzing the performances of so-called systolic and wavefront array processors. In both processors, the individual processing nodes are connected together in a nearest neighbor fashion to form a regular lattice. In the application to be described, all individual processing nodes perform the same basic operation.

The difference between systolic and wavefront array processors is the following. In a systolic array processor, the individual processors, i.e. the nodes of the network, operate synchronously and the only clock required is a simple global clock. The wavefront array processor does not operate synchronously, although the required processing function and network configuration are exactly the same as for the systolic processor. The operation of each individual processor in the wavefront case is controlled locally. It depends on the necessary input data available and on the output of the previous cycle having been delivered to the appropriate (i.e. directly downstream) nodes. For this reason a wavefront array processor is also called a data-driven net.

We will consider a network in which the execution times of the nodes (the individual processors) depend on the input data. In the case of a simple multiplication, the difference in execution time is a consequence of whether at least one of the operands is a zero or a one. We assume that if one of the operands is a zero or a one, the multiplication becomes trivial and, more importantly, faster. Data driven networks are at least as fast as systolic networks since in the latter case the period of the synchronization clock must be large enough to include the slowest local cycle or largest execution time.

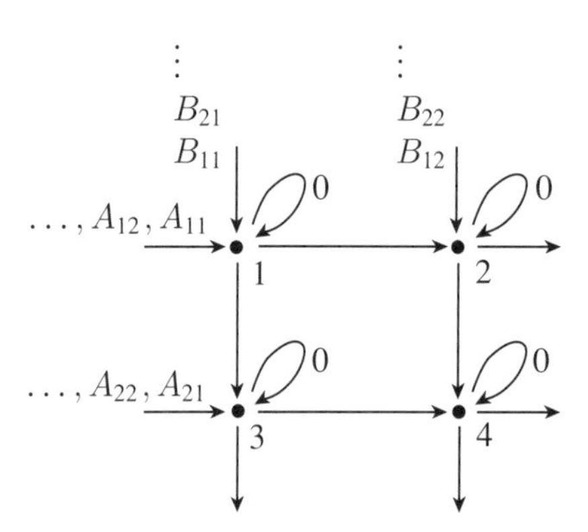

Figure 1.8: The network which multiplies two matrices

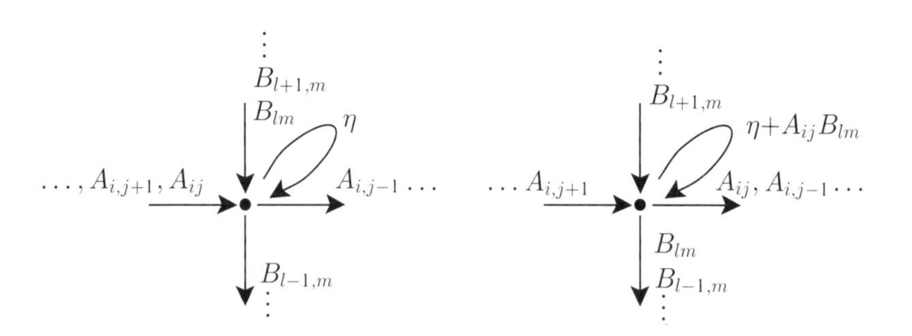

Figure 1.9: Input and output of individual node, before and after one activity

Consider the network shown in Figure 1.8. In this network four nodes are connected. Each of these nodes has an input/output behavior as given in Figure 1.9. The purpose of this network is to multiply two matrices A and B; A has size $2 \times n$

and B has size $n \times 2$. The number n is large ($\gg 2$), but otherwise arbitrary. The entries of the rows of A are fed into the network as indicated in Figure 1.8; they enter the two nodes on the left. Similarly, the entries of B enter the two top nodes. Each node will start its activity as soon as each of the input channels contains a data item, provided it has finished its previous activity and sent the output of that previous activity downstream. Note that in the initial situation—see Figure 1.8—not all channels contain data. Only node 1 can start its activity immediately. Also note that initially all loops contain the number zero.

The activities will stop when all entries of A and B have been processed. Then each of the four nodes contains an entry of the product AB. Thus for instance, node 2 contains $(AB)_{12}$. It is assumed that each node has a local memory at each of its input channels which functions as a buffer and in which at most one data item can be stored if necessary. Such a necessity arises when the upstream node has finished an activity and wants to send the result downstream while the downstream node is still busy with an activity. If the buffer is empty, the output of the upstream node can temporarily be stored in this buffer and the upstream node can start a new activity (provided the input channels contain data). If the upstream node cannot send out its data because one or more of the downstream buffers are full, then it cannot start a new activity (it is 'blocked').

Since it is assumed that each node starts its activities as soon as possible, the network of nodes can be referred to as a wavefront array processor. The execution time of a node is either τ_1 or τ_2 units of time. It is τ_1 if at least one of the input items, from the left and from above (A_{ij} and B_{ij}), is a zero or a one. Then the product to be performed becomes a trivial one. The execution time is τ_2 if neither input contains a zero or a one.

It is assumed that the entry A_{ij} of A equals zero or one with probability p, $0 \le p \le 1$, and that A_{ij} is neither zero nor one with probability $1-p$. The entries of B are assumed to be neither zero nor one (or, if such a number would occur, it will not be detected and exploited).

If $x_i(k)$ is the epoch at which node i becomes active for the k-th time, then it follows from the description above that

$$\left.\begin{aligned} x_1(k+1) &= \alpha_1(k)x_1(k) \oplus x_2(k-1) \oplus x_3(k-1) \ , \\ x_2(k+1) &= \alpha_1(k)x_2(k) \oplus \alpha_1(k+1)x_1(k+1) \oplus x_4(k-1) \ , \\ x_3(k+1) &= \alpha_2(k)x_3(k) \oplus \alpha_1(k+1)x_1(k+1) \oplus x_4(k-1) \ , \\ x_4(k+1) &= \alpha_2(k)x_4(k) \oplus \alpha_1(k+1)x_2(k+1) \\ &\quad \oplus \alpha_2(k+1)x_3(k+1) \ . \end{aligned}\right\} \tag{1.31}$$

In these equations, the coefficients $\alpha_i(k)$ are either τ_1 (if the entry is either a zero or a one) or τ_2 (otherwise); $\alpha_i(k+1)$ have the same meaning with respect to the next entry. Systems of this type will be considered in Chapter 7.

There is a correlation among the coefficients of (1.31) during different time steps. By replacing $x_2(k)$ and $x_3(k)$ by $x_2(k+1)$ and $x_3(k+1)$, respectively, $x_4(k)$ by

$x_4(k+2)$ and $\alpha_2(k)$ by $\alpha_2(k+1)$, we obtain

$$\left.\begin{aligned} x_1(k+1) &= \alpha_1(k)x_1(k) \oplus x_2(k) \oplus x_3(k) \ , \\ x_2(k+1) &= \alpha_1(k-1)x_2(k) \oplus \alpha_1(k)x_1(k) \oplus x_4(k) \ , \\ x_3(k+1) &= \alpha_2(k)x_3(k) \oplus \alpha_1(k)x_1(k) \oplus x_4(k) \ , \\ x_4(k+1) &= \alpha_2(k-1)x_4(k) \oplus \alpha_1(k-1)x_2(k) \oplus \alpha_2(k)x_3(k) \ . \end{aligned}\right\} \tag{1.32}$$

The correlation between some of the α_i-coefficients still exists. The standard procedure to avoid problems connected to this correlation is to augment the state vector x. Two new state variables are introduced: $x_5(k+1) = \alpha_1(k)$ and $x_6(k+1) = \alpha_2(k)$. Equation (1.32) can now be written as

$$\left.\begin{aligned} x_1(k+1) &= \alpha_1(k)x_1(k) \oplus x_2(k) \oplus x_3(k) \ , \\ x_2(k+1) &= x_5(k)x_2(k) \oplus \alpha_1(k)x_1(k) \oplus x_4(k) \ , \\ x_3(k+1) &= \alpha_2(k)x_3(k) \oplus \alpha_1(k)x_1(k) \oplus x_4(k) \ , \\ x_4(k+1) &= x_6(k)x_4(k) \oplus x_5(k)x_2(k) \oplus \alpha_2(k)x_3(k) \ , \\ x_5(k+1) &= \alpha_1(k) \ , \\ x_6(k+1) &= \alpha_2(k) \ . \end{aligned}\right\} \tag{1.33}$$

The correlation in time of the coefficients α_i has disappeared at the expense of a larger state vector. Also note that Equation (1.33) has terms $x_j(k) \otimes x_l(k)$, which cause the equation to become nonlinear (actually, bilinear). For our purposes of calculating the performance of the array processor, this does not constitute basic difficulties, as will be seen in Chapter 8. Equation (1.32) is non-Markovian and linear, whereas (1.33) is Markovian and nonlinear.

1.2.6 Traffic

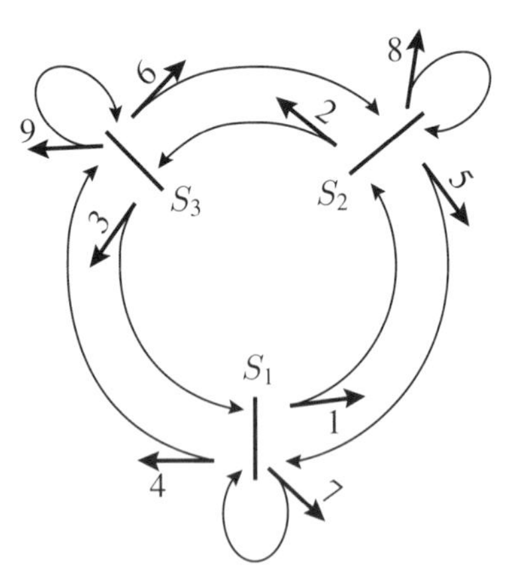

Figure 1.10: The railway system, routing no. 1

In a metropolitan area there are three railway stations, S_1, S_2, and S_3, which are connected by a railway system as indicated in Figure 1.10. The railway system consists of two inner circles, along which the trains run in opposite direction, and of three outer circles. The trains on these outer circles deliver and pick up passengers at local stations. These local stations have not been indicated in Figure 1.10 since they do not play any role in the problem to be formulated.

There are nine railway tracks. Track S_iS_j denotes the direct railway connection from station S_i to station S_j; track S_iS_i denotes the outer circle connected to station S_i. Initially, a train runs along each of these nine tracks. At each station the trains must wait for the trains arriving from the other directions (except for the trains coming from the direction the current train is heading for) in order to allow for transfers. Another assumption to be satisfied is that trains on the same circle cannot bypass one another. If $x_i(k)$ denotes the k-th departure time of the train in direction i (see Figure 1.10) then these departure times are described by $x(k+1) = A_1 \otimes x(k)$, where A_1 is the 9×9 matrix

$$A_1 = \begin{pmatrix} e & \varepsilon & s_{31} & \varepsilon & \varepsilon & \varepsilon & s_{11} & \varepsilon & \varepsilon \\ s_{12} & e & \varepsilon & \varepsilon & \varepsilon & \varepsilon & \varepsilon & s_{22} & \varepsilon \\ \varepsilon & s_{23} & e & \varepsilon & \varepsilon & \varepsilon & \varepsilon & \varepsilon & s_{33} \\ \varepsilon & \varepsilon & \varepsilon & e & s_{21} & \varepsilon & s_{11} & \varepsilon & \varepsilon \\ \varepsilon & \varepsilon & \varepsilon & \varepsilon & e & s_{32} & \varepsilon & s_{22} & \varepsilon \\ \varepsilon & \varepsilon & \varepsilon & s_{13} & \varepsilon & e & \varepsilon & \varepsilon & s_{33} \\ \varepsilon & \varepsilon & s_{31} & \varepsilon & s_{21} & \varepsilon & s_{11} & \varepsilon & \varepsilon \\ s_{12} & \varepsilon & \varepsilon & \varepsilon & \varepsilon & s_{32} & \varepsilon & s_{22} & \varepsilon \\ \varepsilon & s_{23} & \varepsilon & s_{13} & \varepsilon & \varepsilon & \varepsilon & \varepsilon & s_{33} \end{pmatrix} .$$

An entry s_{ij} refers to the traveling time on track S_iS_j. These quantities include transfer times at the stations. The diagonal entries e prevent trains from bypassing one another on the same track at a station. The routing of the trains was according to Figure 1.10; trains on the two inner circles stay on these inner circles and keep the same direction; trains on the outer circles remain there. Other routings of the trains are possible; two such different routings are given in Figures 1.11 and 1.12. If $x_i(k)$ denotes the k-th departure time from the same station as given in Figure 1.10, then the departure times are described again by a model of the form $x(k+1) = A \otimes x(k)$. The A-matrix corresponding to Figure 1.11 is indicated by A_2 and the A-matrix corresponding to Figure 1.12 by A_3. If we define matrices F_i of size 3×3 in the following way:

$$F_1 = \begin{pmatrix} \varepsilon & \varepsilon & \varepsilon \\ \varepsilon & \varepsilon & \varepsilon \\ \varepsilon & \varepsilon & \varepsilon \end{pmatrix}, \quad F_2 = \begin{pmatrix} \varepsilon & \varepsilon & s_{31} \\ s_{12} & \varepsilon & \varepsilon \\ \varepsilon & s_{23} & \varepsilon \end{pmatrix},$$

$$F_3 = \begin{pmatrix} s_{11} & \varepsilon & \varepsilon \\ \varepsilon & s_{22} & \varepsilon \\ \varepsilon & \varepsilon & s_{33} \end{pmatrix}, \quad F_4 = \begin{pmatrix} \varepsilon & s_{21} & \varepsilon \\ \varepsilon & \varepsilon & s_{32} \\ s_{13} & \varepsilon & \varepsilon \end{pmatrix},$$

then the matrices A_1, A_2 and A_3 can be compactly written as

$$A_1 = \begin{pmatrix} e \oplus F_2 & F_1 & F_3 \\ F_1 & e \oplus F_4 & F_3 \\ F_2 & F_4 & F_3 \end{pmatrix}, \quad A_2 = \begin{pmatrix} e \oplus F_2 & F_4 & F_3 \\ F_2 & e \oplus F_4 & F_3 \\ F_2 & F_4 & F_3 \end{pmatrix},$$

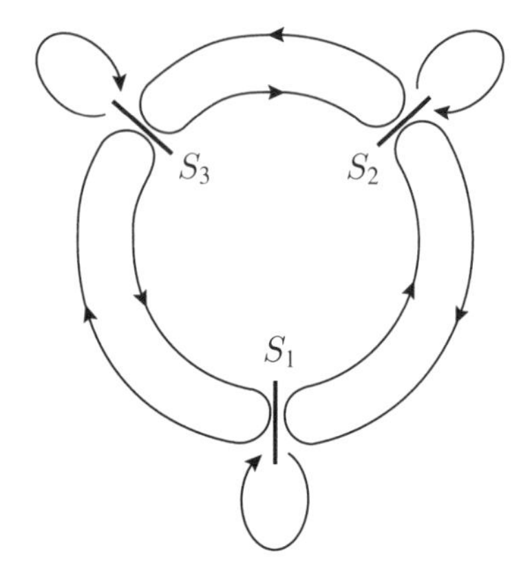

Figure 1.11: Routing no. 2

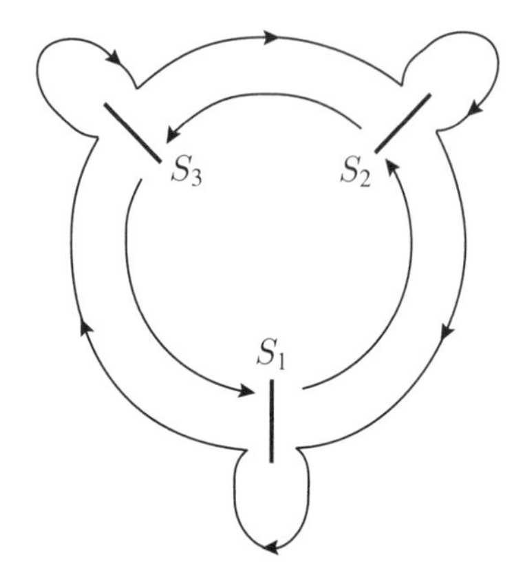

Figure 1.12: Routing no. 3

$$A_3 = \begin{pmatrix} e \oplus F_2 & F_1 & F_3 \\ F_1 & e & F_3 \\ F_2 & F_4 & e \end{pmatrix} .$$

It is seen that for different routing schemes, different models are obtained. In general, depending on the numerical values of s_{ij}, the time evolutions of these three models will be different. One needs a criterion for deciding which of the routing schemes is preferable.

Depending on the application, it may be more realistic to introduce stochastic s_{ij}-quantities. Suppose for instance that there is a swing bridge in the railway track from S_3 to S_1 (only in this track; there is no bridge in the track from S_1 to S_3). Each time a train runs from S_3 to S_1 there is a probability p, $0 \leq p \leq 1$, that the train will be delayed, i.e. s_{31} becomes larger. Thus the matrices A_i become k-dependent; $A_i(k)$. The system has become stochastic. In this situation one may also have a preference for one of the three routings, or for another one.

The last part of the discussion of this example will be devoted to the deterministic model of routing no. 1 again. However, to avoid some mathematical subtleties which are not essential at this stage, we assume that there is at least a difference of τ time units between two subsequent departures of trains in the directions 1 to 6 ($\tau > e$). Consequently, the equations are now $x(k+1) = \widehat{A}_1 \otimes x(k)$, where $\widehat{A}_1$ equals the previous A_1 except for the first six diagonal entries that are replaced by τ (instead of e earlier), and the last three diagonal entries s_{ii} that are replaced by $s_{ii} \oplus \tau$ respectively.

We then introduce a quantity $\chi_i(t)$ which is related to $x_i(k)$. The argument t of $\chi_i(t)$ refers to the actual clock time and $\chi_i(t)$ itself refers to the number of trains departures in the direction i which have occurred up to (and including) time t. The quantity χ_i can henceforth only assume the values $0, 1, 2, \ldots$. At an arbitrary time t, the number of trains which left in the direction 1 can exceed at most by one:

- the same number of trains τ units of time earlier;

- the number of trains that left in the direction 3 s_{31} time units earlier (recall that initially a train was already traveling on each track);
- the number of trains that left in the direction 7 s_{11} time units earlier.

Therefore, we have

$$\chi_1(t) = \min\left(\chi_1(t-\tau)+1, \chi_3(t-s_{31})+1, \chi_7(t-s_{11})+1\right) .$$

For χ_2 one similarly obtains

$$\chi_2(t) = \min\left(\chi_1(t-s_{12})+1, \chi_2(t-\tau)+1, \chi_8(t-s_2)+1\right) ,$$

etc. If all quantities s_{ij} and τ are equal to 1, then one can compactly write

$$\chi(t) = \overline{A} \otimes \chi(t-1) ,$$

where $\chi(t) = (\chi_1(t), \ldots, \chi_9(t))'$, and where the matrix $\overline{A}$ is derived from A_1 by replacing all entries equal to ε by $+\infty$ and the other entries by 1. This equation must be read in the min-plus algebra setting. In case we want something more general than s_{ij} and τ equal to 1, we will consider the situation when all these quantities are integer multiples of a positive constant g_c; then g_c can be interpreted as the time unit along which the evolution will be expressed. One then obtains

$$\chi(t) = \overline{A}_1 \otimes \chi(t-1) \oplus \overline{A}_2 \otimes \chi(t-2) \oplus \cdots \oplus \overline{A}_l \otimes \chi(t-l) \tag{1.34}$$

for some finite l. The latter equation (in the min-plus algebra) and $x(k+1) = \widehat{A}_1 \otimes x(k)$ (in the max-plus algebra) describe the same system. Equation (1.34) is referred to as the *counter description* and the other one as the *dater description*. The word 'dater' must be understood as 'timer', but since the word 'time' and its declinations are already used in various ways in this book, we will stick to the word 'dater'. The awareness of these two different descriptions for the same problem has far-reaching consequences as will be shown in Chapter 5.

The reader should contemplate that the stochastic problem (in which some of the s_{ij} are random) is more difficult to handle in the counter description, since the delays in (1.34) become stochastic (see §8.2.4).

1.2.7 Continuous System Subject to Flow Bounds and Mixing

So far, the examples have been related to the realm of discrete event systems, and dynamic equations have been obtained under the form of recursive equations involving max (or min) and +-operations. We close this section by showing that a continuous-time system may naturally lead to essentially the same type of equations whenever some flow limitation and mixing phenomena are involved. Had we adopted the point of view of conventional system theory, the model of such a system in terms of differential equations would have exhibited very complex nonlinearities. With the following approach, we will see that this system is 'linear' in a precise mathematical sense (see Chapter 6). An analogous discrete event model will be discussed in Example 2.48.

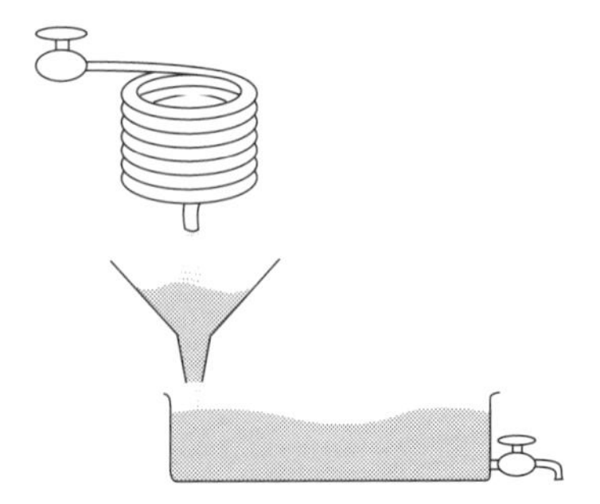

Figure 1.13: A continuous system

In Figure 1.13, a fluid is poured through a long pipe into a first reservoir (empty at time $t = 0$). The input $u(t)$ denotes the *cumulated* flow at the inlet of the pipe up to time t (hence $u(t)$ is a nondecreasing time function and $u(t) = 0$ for $t \leq 0$). It is assumed that it takes a delay of 2 units of time (say, 2 seconds) for the fluid to travel through the pipe. From the first reservoir, the fluid drops into a second reservoir through an aperture which limits the instantaneous flow to a maximum value of, say, 0.5 liter per second. The volume of fluid at time t in this second reservoir is denoted $y(t)$, and it is assumed that $y(0) = 3$ liters.

Let us establish dynamic equations for such a system relating the output y to the input u. Because of the flow limitation into the second reservoir, we have:

$$\forall t \ , \quad \forall s \geq 0 \ , \quad y(t+s) \leq y(t) + 0.5s \ . \tag{1.35}$$

On the other hand, since there is a delay of 2 seconds caused by the pipe, $y(t)$ should be compared with $u(t-2)$, and because there is a stock of 3 liters in the second reservoir at $t = 0$, we have:

$$\forall t \ , \quad y(t) \leq u(t-2) + 3 \ . \tag{1.36}$$

It follows that

$$\forall t \ , \quad \forall s \geq 0 \ , \quad \begin{aligned} y(t) &\leq y(t-s) + 0.5s \\ &\leq u(t-2-s) + 3 + 0.5s \ , \end{aligned}$$

hence,

$$\forall t \ , \quad \begin{aligned} y(t) &\leq \inf_{s \geq 0} [u(t-2-s) + 3 + 0.5s] \\ &= \inf_{\tau \geq 2} [u(t-\tau) + 3 + 0.5(\tau - 2)] \ . \end{aligned} \tag{1.37}$$

Let

$$h(t) \stackrel{\text{def}}{=} \begin{cases} 3 & \text{if } t \leq 2; \\ 3 + 0.5(t-2) & \text{otherwise.} \end{cases} \tag{1.38}$$

and consider

$$\forall t \ , \quad \overline{y}(t) \stackrel{\text{def}}{=} \inf_{\tau \in \mathbb{R}} [u(t-\tau) + h(\tau)] \ . \tag{1.39}$$

Indeed, in (1.39), the range of τ may be limited to $\tau \geq 2$ since, for $\tau < 2$, $h(\tau)$ remains equal to 3 whereas $u(t-\tau) \geq u(t-2)$ (remember that $u(\cdot)$ is nondecreasing). Therefore, comparing (1.39) with (1.37), it is clear that $y(t) \leq \overline{y}(t), \forall t$.

Moreover, choosing $\tau = 2$ at the right-hand side of (1.39), we see that $\overline{y}$ satisfies (1.36). In addition, since for all s and all $\vartheta \geq 0, h(s+\vartheta) \leq h(s) + 0.5\vartheta$, then

$$\begin{aligned}
\forall t \ , \quad \forall \theta \geq 0 \ , \quad \overline{y}(t+\vartheta) &= \inf_{\tau \in \mathbb{R}} [u(t+\vartheta-\tau) + h(\tau)] \\
&= \inf_{s \in \mathbb{R}} [u(t-s) + h(s+\vartheta)] \\
&\leq \inf_{s \in \mathbb{R}} [u(t-s) + h(s)] + 0.5\vartheta \\
&= \overline{y}(t) + 0.5\vartheta \ .
\end{aligned}$$

Thus, $\overline{y}$ satisfies (1.35).

Finally, we have proved that $\overline{y}$ is the maximum solution of (1.35)–(1.36). It can also be checked that (1.39) yields $\overline{y}(t) = 3, \ \forall t \leq 2$. Therefore, $\overline{y}$ is the solution which will be physically realized if we assume that, subject to (1.37)–(1.39), the fluid flows as fast as possible. This output trajectory is related to the input history u by an 'inf-convolution' (see Equation (1.39)). In order to make this inf-convolution more visible, the inf-operator should be viewed as an 'integration' (which is nothing else than the $\oplus$-operator ranging over the real numbers). If moreover $+$ in (1.39) is replaced by $\otimes$ one obtains the appearance of the conventional convolution. The same kind of input-output relationship (indeed, a 'sup-convolution' in that context) can be obtained from the recursive equations (1.12) by developing the recursion from any initial condition.

As a final remark, observe that if we have two systems similar to the one shown in Figure 1.13, one producing a red fluid and the other producing a white fluid, and if we want to produce a pink fluid by mixing them in equal proportions, then the new output is related to the two inputs by essentially the same type of equations. More specifically, let $y_{\mathrm{r}}(t)$ and $y_{\mathrm{w}}(t)$ be the quantities of red and white fluids that have been produced in the two downstream reservoirs up to time t (including the initial reserves). Suppose that the two taps at their outlets are opened so that the same (very large) outflow of red and white liquids can be obtained unless one of the two reservoirs is empty, in which case the two taps are closed immediately. Then, $\min(y_{\mathrm{r}}(t), y_{\mathrm{w}}(t))$ is directly related to the quantity $y_{\mathrm{p}}(t)$ of pink fluid produced up to time t. Therefore, this mixing operation does not introduce new mathematical operators.

1.3 ISSUES AND PROBLEMS IN PERFORMANCE EVALUATION

In the previous sections we dealt with equations of the form $x(k+1) = Ax(k)$, or more generally $x(k+1) = A(k)x(k) \oplus B(k)u(k)$. In the applications three different interpretations in terms of the operations were given: maximization and addition, minimization and addition and lastly, maximization and multiplication. In this section only the first interpretation will be considered (we will say that the system under consideration is in the max-plus algebra framework). Before that a brief introduction to the solution of the conventional equation (1.1) is needed.

Assume that the initial vector (1.2) equals an eigenvector of A; the corresponding eigenvalue is denoted by λ. The solution of (1.1) can be written as

$$x(t) = \lambda^t x_0 \ , \quad t = 0, 1, \ldots . \tag{1.40}$$

More generally, if the initial vector can be written as a linear combination of the set of linearly independent eigenvectors,

$$x_0 = \sum_j c_j v_j \ , \tag{1.41}$$

where v_j is the j-th eigenvector with corresponding eigenvalue λ_j, the c_j are coefficients, then

$$x(t) = \sum_j c_j \lambda_j^t v_j \ .$$

If the matrix A is diagonalizable, then the set of linearly independent eigenvectors spans $\mathbb{R}^n$, and any initial condition x_0 can be expressed as in (1.41). If A is not diagonalizable, then one must work with generalized eigenvectors and the formula which expresses $x(t)$ in terms of eigenvalues and x_0 is slightly more complicated. This complication does not occur in the max-plus algebra context and therefore will not be dealt with explicitly.

In Chapter 3 it will be shown that under quite general conditions an eigenvalue (λ) and corresponding eigenvector (v) also exist in the max-plus algebra context for a square matrix (A). The definition is

$$A \otimes v = \lambda \otimes v \ .$$

To exclude degenerate cases, it is assumed that not all components of v are identical to ε. For example,

$$\begin{pmatrix} 3 & 7 \\ 2 & 4 \end{pmatrix} \begin{pmatrix} 2.5 \\ e \end{pmatrix} = 4.5 \begin{pmatrix} 2.5 \\ e \end{pmatrix} .$$

Thus it is seen that the matrix A of (1.5) has an eigenvalue 4.5. Equation (1.40) is also valid in the current setting. If x_0 is an eigenvector of A, with corresponding eigenvalue λ, then the solution of the difference equation (1.8) can be written as

$$x(k) = \lambda^k x_0 \quad (= \lambda^k \otimes x_0) \ , \quad k = 0, 1, \ldots . \tag{1.42}$$

The numerical evaluation of λ^k in this formula equals $k\lambda$ in conventional analysis. The eigenvalue λ can be interpreted as the cycle time (defined as the inverse of the throughput) of the underlying system; each node of the corresponding network becomes active every λ units of time, since it follows straightforwardly from (1.42).

Also, the relative order in which the nodes become active for the k-th time, as expressed by the components $x_i(k)$, is exactly the same as the relative order in which the nodes become active for the $(k+1)$-st time. More precisely, Equation (1.42) yields

$$x_l(k+1) - x_j(k+1) = x_l(k) - x_j(k) \ , \quad j,l = 1,\ldots,n \ .$$

Thus the solution (1.42) exhibits a kind of periodicity. Procedures exist for the calculation of eigenvalues and eigenvectors; an efficient one is the procedure known as Karp's algorithm for which the reader is referred to Chapter 2. More discussion about related issues can be found in Chapter 3. Under suitable conditions the eigenvalue turns out to be unique (which differs from the situation in conventional analysis). It can be shown for instance that A of (1.5) has only one eigenvalue. Similarly, the matrix M of §1.2.3 also has a unique eigenvalue:

$$\begin{pmatrix} 6 & \varepsilon & \varepsilon & \varepsilon & 6 & 5 \\ 9 & 8 & \varepsilon & 8 & 9 & 8 \\ 6 & 10 & 7 & 10 & 6 & \varepsilon \\ \varepsilon & 7 & 4 & 7 & \varepsilon & \varepsilon \\ 6 & 10 & 7 & 10 & 6 & \varepsilon \\ 9 & 8 & \varepsilon & 8 & 9 & 8 \end{pmatrix} \begin{pmatrix} e \\ 3 \\ 3.5 \\ 0.5 \\ 3.5 \\ 3 \end{pmatrix} = 9.5 \begin{pmatrix} e \\ 3 \\ 3.5 \\ 0.5 \\ 3.5 \\ 3 \end{pmatrix} .$$

It follows that the eigenvalue equals 9.5, which means in more practical terms that the manufacturing system 'delivers' an item (a product or a machine) at all of its output channels every 9.5 units of time. The eigenvector of this example is also unique, apart from adding the same constant to all components. If v is an eigenvector, then cv, where c is a scalar, also is an eigenvector, since it follows directly from the definition of eigenvalue. It is possible that several eigenvectors can be associated with the only eigenvalue of a matrix, i.e. eigenvectors may not be identical up to an additional constant.

Suppose that we deal with the system characterized by the matrix of (1.5); then it is known from earlier that the 'cycle time' is $9/2$ units of time. The throughput is defined as the inverse of the cycle time and equals $2/9$. If we had the choice of reducing one arbitrary entry of A by 2, which entry should we choose such that the cycle time becomes as small as possible? To put it differently, if a piece of equipment were available which reduces the traveling time at any connection by 2, where should this piece of equipment be placed? By trial and error it is found that either A_{12} or A_{21} should be reduced by 2; in both cases the new cycle time becomes 4. If one reduces A_{11} or A_{22} by this amount instead of A_{12} or A_{21}, then the cycle time remains $9/2$. The consequences of the four potential ways of reduction are expressed by

$$\begin{pmatrix} 1 & 7 \\ 2 & 4 \end{pmatrix} \begin{pmatrix} 2.5 \\ e \end{pmatrix} = 4.5 \begin{pmatrix} 2.5 \\ e \end{pmatrix} ; \quad \begin{pmatrix} 3 & 5 \\ 2 & 4 \end{pmatrix} \begin{pmatrix} 1 \\ e \end{pmatrix} = 4 \begin{pmatrix} 1 \\ e \end{pmatrix} ;$$

$$\begin{pmatrix} 3 & 7 \\ 0 & 4 \end{pmatrix} \begin{pmatrix} 3 \\ e \end{pmatrix} = 4 \begin{pmatrix} 3 \\ e \end{pmatrix} ; \quad \begin{pmatrix} 3 & 7 \\ 2 & 2 \end{pmatrix} \begin{pmatrix} 2.5 \\ e \end{pmatrix} = 4.5 \begin{pmatrix} 2.5 \\ e \end{pmatrix} .$$

To answer the question of which 'transportation line' to speed up for more general networks, application of the trial and error method as used above would become very laborious. Fortunately more elegant and more efficient methods exist. For those one needs the notion of a critical circuit, which is elaborated upon in Chapters 2 and 3. Without defining such a circuit in this section formally let us mention that, in Figure 1.1, this critical circuit consists of the arcs determined by A_{12} and A_{21}. Note that $(A_{12} + A_{21})/2 = \lambda = 9/2$, and this equality is not a coincidence.

Stochastic extensions are possible. Towards that end, consider

$$x(k+1) = A(k)x(k) \ ,$$

where the matrix A now depends on k in a stochastic way. Assume that $x \in \mathbb{R}^2$ and that for each k the matrix A is one of the following two matrices:

$$\begin{pmatrix} 3 & 7 \\ 2 & 4 \end{pmatrix}, \quad \begin{pmatrix} 3 & 5 \\ 2 & 4 \end{pmatrix}.$$

Both matrices occur with probability 1/2 and there is no correlation in time. A suitable definition of cycle time turns out to be

$$\lim_{k\to\infty} \mathbb{E}\left[x_i(k+1) - x_i(k)\right] \ ,$$

where $\mathbb{E}$ denotes mathematical expectation. Application of the theory presented in Chapters 7 and 8 shows that this cycle time is independent of i and is equal to 13/3.

Conventional linear systems with inputs and outputs are of the form (1.10), although (1.10) itself has the max-plus algebra interpretation. This equation is a representation of a linear system in the time domain. Its representation in the z-domain equals

$$Y(z) = C(zI - A)^{-1}BU(z) \ ,$$

where $Y(z)$ and $U(z)$ are defined by

$$Y(z) = \sum_{i=0}^{\infty} y(i)z^{-i} \ , \qquad U(z) = \sum_{i=0}^{\infty} u(i)z^{-i} \ ,$$

where it is tacitly assumed that the system was at rest for $t \leq 0$. The matrix $H(z) \stackrel{\text{def}}{=} C(zI - A)^{-1}B$ is called the transfer matrix of the system. Here I refers to the identity matrix in conventional algebra. The notion of transfer matrix is especially useful when subsystems are combined to build larger systems, by means of parallel, series and feedback connections.

In the max-plus algebra context, the z-transform also exists (see [72]), but here we will rather refer to the γ-transform where γ operates as z^{-1}. For instance, the γ-transform of u is defined as

$$U(\gamma) = \bigoplus_{i=0}^{\infty} u(i) \otimes \gamma^i \ ,$$

and $Y(\gamma)$ and $X(\gamma)$ are defined likewise. Multiplication of (1.12) by γ^k yields

$$\left.\begin{array}{rcl} \gamma^{-1}x(k+1)\gamma^{k+1} & = & A \otimes x(k)\gamma^k \oplus B \otimes u(k)\gamma^k \ , \\ y(k)\gamma^k & = & C \otimes x(k)\gamma^k \ . \end{array}\right\} \tag{1.43}$$

If these equations are summed with respect to $k = 0, 1, \ldots$, then we obtain

$$\left.\begin{array}{rcl} \gamma^{-1}X(\gamma) & = & A \otimes X(\gamma) \oplus B \otimes U(\gamma) \oplus \gamma^{-1}x_0 \ , \\ Y(\gamma) & = & C \otimes X(\gamma) \ . \end{array}\right\} \tag{1.44}$$

The first of these equations can be solved by first multiplying (max-plus algebra), equivalently adding (conventional) the left- and right-hand sides by γ and then repeatedly substituting the right-hand side for $X(\gamma)$ within this right-hand side. This results in

$$X(\gamma) = (\gamma A)^*(\gamma BU(\gamma) \oplus x_0) \ .$$

Thus we obtain $Y(\gamma) = H(\gamma)U(\gamma)$, provided that $x_0 = \varepsilon$, and where the transfer matrix $H(\gamma)$ is defined by

$$H(\gamma) = C \otimes (\gamma A)^* \otimes \gamma \otimes B = \gamma CB \oplus \gamma^2 CAB \oplus \gamma^3 CA^2B \oplus \cdots \ . \tag{1.45}$$

The transfer matrix is defined by means of an infinite series and the convergence depends on the value of γ. If the series is convergent for $\gamma = \gamma'$, then it is also convergent for all γ's which are smaller than γ'. If the series does not converge, it still has a meaning as a formal series.

Exactly as in conventional system theory, the product of two transfer matrices (in which it is tacitly assumed that the sizes of these matrices are such that the multiplication is possible), is a new transfer matrix which refers to a system which consists of the original systems connected in series. In the same way, the sum of two transfer matrices refers to two systems put in parallel. This section will be concluded by an example of such a parallel connection.

We are given two systems. The first one is given in (1.13), and is characterized by the 1×1 transfer matrix

$$H_1 = \varepsilon\gamma \oplus 11\gamma^2 \oplus 14\gamma^3 \oplus 20\gamma^4 \oplus 24\gamma^5 \oplus 29\gamma^6 \oplus \cdots \ .$$

It is easily shown that this series converges for $\gamma \leq -4.5$; the number 4.5 corresponds to the eigenvalue of A. The second system is given by

$$x(k+1) = \begin{pmatrix} e & \varepsilon & 4 \\ 1 & 1 & \varepsilon \\ \varepsilon & 6 & 3 \end{pmatrix} x(k) \oplus \begin{pmatrix} \varepsilon \\ 2 \\ e \end{pmatrix} u(k) \ ,$$

$$y(k) = \begin{pmatrix} 1 & 1 & 4 \end{pmatrix} x(k) \ ,$$

and its transfer matrix is

$$H_2 = 4\gamma \oplus 12\gamma^2 \oplus 15\gamma^3 \oplus 18\gamma^4 \oplus 23\gamma^5 \oplus 26\gamma^6 \oplus \cdots \ .$$

The transfer matrix of the two systems put in parallel has size 1×1 again (one can talk about a transfer function) and is obtained as

$$H_{\text{par}} = H_1 \oplus H_2 = 4\gamma \oplus 12\gamma^2 \oplus 15\gamma^3 \oplus 20\gamma^4 \oplus 24\gamma^5 \oplus 29\gamma^6 \oplus \cdots \ . \tag{1.46}$$

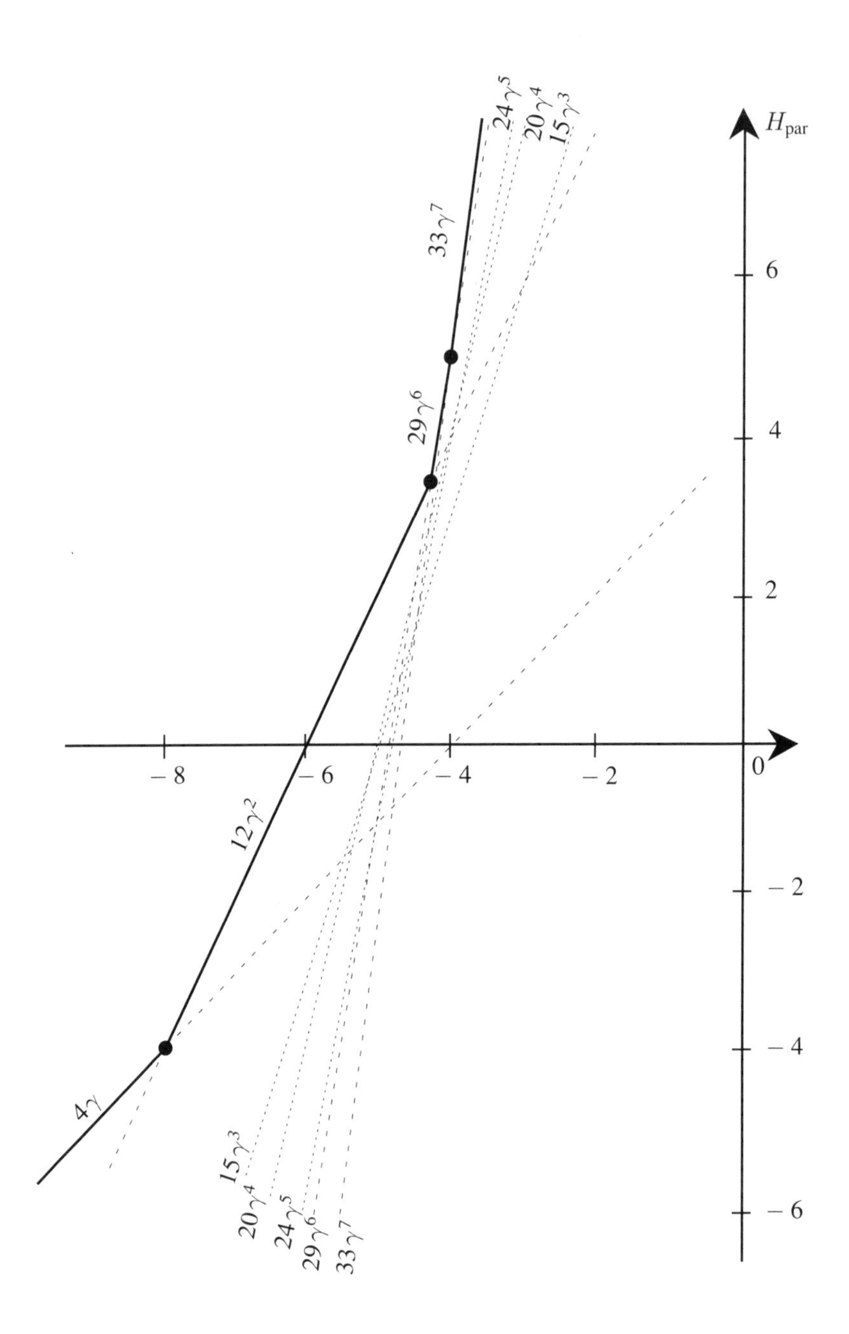

Figure 1.14: The transfer function H_{par} as a function of γ

A transfer function can easily be visualized. If $H(\gamma)$ is a scalar function, i.e. the system has one input and one output, then it is a continuous and piecewise linear function. As an example, the transfer function of the parallel connection considered above is pictured in Figure 1.14.

Above it was shown how to derive the transfer matrix of a system if the representation of the system in the 'time domain' is given. This time domain representation is characterized by the matrices A, B and C. Now one could pose the opposite question: How can we obtain a time domain representation, or equivalently, how can we find A, B and C if the transfer matrix is given? A partial answer to this question is given in [103]. For the example above, one would like to obtain a time domain representation of the two systems put in parallel starting from (1.46). This avenue will not be pursued now. Instead, one can always obtain such a representation by connecting the underlying networks of the two original systems in the appropriate ('parallel') way and then derive the state equations directly. In this way one gets for the above example,

$$x(k+1) = \begin{pmatrix} 3 & 7 & \varepsilon & \varepsilon & \varepsilon \\ 2 & 4 & \varepsilon & \varepsilon & \varepsilon \\ \varepsilon & \varepsilon & e & \varepsilon & 4 \\ \varepsilon & \varepsilon & 1 & 1 & \varepsilon \\ \varepsilon & \varepsilon & \varepsilon & 6 & 3 \end{pmatrix} x(k) \oplus \begin{pmatrix} \varepsilon \\ 1 \\ \varepsilon \\ 2 \\ e \end{pmatrix} u(k) \ ,$$

$$y(k) = \begin{pmatrix} 3 & \varepsilon & 1 & 1 & 4 \end{pmatrix} x(k) \ .$$

1.4 NOTES

A few times, reference has been made to (linear) system theory. Classic texts are [81], [32] or [72]. A few examples could be phrased in terms of dynamic programming. There are many elementary texts which explain the theory of dynamic programming. A more advanced text is [18]. Systems in the context of max-plus algebra were probably first described in [49], though most theory in this book is algebraically rather than system oriented. It was in [39] where the relation between system theory and max algebra was clearly shown. The shortest path problem is a standard example in dynamic programming texts. The Viterbi algorithm was found in [59]. The example on production was presented in [38], the examples of parallel computation and traffic can be found in [97]. Other simple examples can be found in [101]. A max-plus modeling of the Dutch intercity railway net is given in [27]. An application to chemical batch processing is given in [107]. A good introductory text to design methods of processors for parallel computation is [79]. The relation between eigenvalue and cycle time was developed in [39]. Stochastic extensions were given in [104]. The connection between transfer matrices and state equations in the max-plus algebra context was investigated in [103]; see also §9.2.3.

CHAPTER 2

Graph Theory and Petri Nets

2.1 INTRODUCTION

An overview of various results in the theories of graphs and Petri nets will be given. Several properties of DEDS can be expressed in terms of graph-theoretic notions such as strong connectedness and critical circuits.

A rich relationship exists between graphs and matrices, about which many books have been written. Here, we will emphasize some of these relationships between directed graphs and matrices, together with their consequences if such matrices and graphs are composed to build larger ones. The way to construct such compositions is by means of parallel and series connections.

Petri nets, which describe DEDS pictorially, can be viewed as bipartite graphs. An essential feature of Petri nets, not present in conventional graphs, is that they are dynamic systems. Tokens are used to reflect this dynamic behavior.

There is an equivalence between DEDS without concurrency and a subclass of Petri nets called 'event graphs'. For any timed event graph, we will show how to obtain a mathematical model in terms of recurrence equations. In the proper algebraic framework, these equations are linear and the model offers a strong analogy with conventional linear dynamic systems.

In the last part of this chapter, starting from the point of view of resources involved in DEDS, we propose a methodolgy to go from the specifications of a concrete system to its modeling by event graphs.

2.2 DIRECTED GRAPHS

A *directed graph* $\mathcal{G}$ is defined as a pair $(\mathcal{V}, \mathcal{E})$, where $\mathcal{V}$ is a set of elements called *nodes* and where $\mathcal{E}$ is a set the elements of which are ordered (not necessarily different) pairs of nodes, called *arcs*. The possibility of several arcs between two nodes exists (one then speaks about a multigraph); in this chapter, however, we almost exclusively deal with directed graphs in which there is at most one (i.e. zero or one) arc between any two nodes. One distinguishes graphs and directed graphs. The difference between the two is that in a graph the elements of $\mathcal{E}$ are not ordered while they are in a directed graph. Instead of nodes and arcs, one also speaks about vertices and edges, respectively. The origin of the symbols $\mathcal{V}$ and $\mathcal{E}$ in the definition of a (directed) graph is due to the first letters of the latter two names. Instead of directed graph one often uses the shorter word 'digraph', or even 'graph' if it is clear from the context that digraph is meant. In this chapter we will almost exclusively deal with digraphs (hence also called graphs).

Denote the number of nodes by n, and number the individual nodes $1, 2, \ldots, n$. If $(i, j) \in \mathcal{E}$, then i is called the initial node or the origin of the arc (i, j), and j the final node or the destination of the arc (i, j). Graphically, the nodes are represented by points, and the arc (i, j) is represented by an 'arrow' from i to j.

We now give a list of concepts of graph theory which will be used later on.

Predecessor, successor. If in a graph $(i, j) \in \mathcal{E}$, then i is called a predecessor of j and j is called a successor of i. The set of all predecessors of j is indicated by $\pi(j)$ and the set of all successors of i is indicated by $\sigma(i)$. A predecessor is also called an *upstream node* and a successor is also called a *downstream node*.

Source, sink. If $\pi(i) = \emptyset$, then node i is called a source; if $\sigma(i) = \emptyset$ then i is called a sink. Depending on the application, a source, respectively sink, is also called an *input(-node)*, respectively an *output(-node)* of the graph.

Path, circuit, loop, length. A *path* ρ is a sequence of nodes $(i_1, i_2, \ldots, i_p), p > 1$, such that $i_j \in \pi(i_{j+1})$, $j = 1, \ldots, p-1$. Node i_1 is the initial node and i_p is the final one of this path. Equivalently, one also says that a path is a sequence of arcs which connects a sequence of nodes. An *elementary path* is a path in which no node appears more than once. When the initial and the final nodes coincide, one speaks of a *circuit*. A circuit $(i_1, i_2, \ldots, i_p = i_1)$ is an *elementary circuit* if the path $(i_1, i_2, \ldots, i_{p-1})$ is elementary. A *loop* is a circuit (i, i), that is, a circuit composed of a single node which is initial and final. This definition assumes that $i \in \pi(i)$, that is, there does exist an arc from i to i. The *length* of a path or a circuit is equal to the sum of the lengths of the arcs of which it is composed, the lengths of the arcs being 1 unless otherwise specified. With this convention, the length of a loop is 1. The length of path ρ is denoted $|\rho|_1$. The subscript '1' here refers to the word 'length' (later on, another subscript 'w' will appear for a different concept). The set of all paths and circuits in a graph is denoted R. A digraph is said to be *acyclic* if R contains no circuits.

Descendant, ascendant. The set of descendants $\sigma^+(i)$ of node i consists of all nodes j such that a path exists from i to j. Similarly the set of ascendants $\pi^+(i)$ of node i is the set of all nodes j such that a path exists from j to i. One has, e.g., $\pi^+(i) = \pi(i) \cup \pi(\pi(i)) \cup \ldots$. The mapping $i \mapsto \pi^*(i) = \{i\} \cup \pi^+(i)$ is the transitive closure of π; the mapping $i \mapsto \sigma^*(i) = \{i\} \cup \sigma^+(i)$ is the transitive closure of σ.

Subgraph. Given a graph $\mathcal{G} = (\mathcal{V}, \mathcal{E})$, a graph $\mathcal{G}' = (\mathcal{V}', \mathcal{E}')$ is said to be a subgraph of $\mathcal{G}$ if $\mathcal{V}' \subset \mathcal{V}$ and if $\mathcal{E}'$ consists of the set of arcs of $\mathcal{G}$ which have their origins and destinations in $\mathcal{V}'$.

Chain, connected graph. A graph is called connected if for all pairs of nodes i and j there exists a chain joining i and j. A chain is a sequence of nodes

$(i_1, i_2, \ldots, i_p)$ such that between each pair of successive nodes either the arc (i_j, i_{j+1}) or the arc (i_{j+1}, i_j) exists. If one disregards the directions of the arcs in the definition of a path, one obtains a chain.

Strongly connected graph. A graph is called strongly connected if for any two different nodes i and j there exists a path from i to j. Equivalently, $i \in \sigma^*(j)$ for all $i, j \in \mathcal{V}$, with $i \neq j$. Note that, according to this definition, an isolated node, with or without a loop, is a strongly connected graph.

Bipartite graph. If the set of nodes $\mathcal{V}$ of a graph $\mathcal{G}$ can be partitioned into two disjoint subsets $\mathcal{V}_1$ and $\mathcal{V}_2$ such that every arc of $\mathcal{G}$ connects an element of $\mathcal{V}_1$ with one of $\mathcal{V}_2$ or the other way around, then $\mathcal{G}$ is called bipartite.

In §2.3, it will be useful to introduce the notion of an 'empty circuit' the length of which is equal to 0 by definition. An empty circuit contains no arcs. The circuit (i) is an empty circuit which should not be confused with the loop (i, i) of length 1 (the latter makes sense only if there exists an arc from node i to itself). Empty circuits are *not* included in the set R of paths.

To exemplify the various concepts introduced, consider the graph presented in Figure 2.1. It is a digraph since the arcs are indeed directed. The graph has seven nodes. Node 3 is a predecessor of node 6; $3 \in \pi(6)$. Similarly, $6 \in \sigma(3)$. The sequence of nodes 1, 3, 6, 4, 3, 2 is a nonelementary path. The arc $(1, 1)$ is a loop and the sequence of nodes 3, 6, 4, 3 is an elementary circuit of length 3. The sequence of nodes 2, 3, 6 is a chain. It should be clear that the graph of Figure 2.1 is connected.

Definition 2.1 (Equivalence relation $\mathcal{R}$) *Let $i, j \in \mathcal{V}$ be two nodes of a graph. We say that $i\mathcal{R}j$, if either $i = j$ or there exist paths from i to j and from j to i.*

Then $\mathcal{V}$ is split up into equivalence classes $\mathcal{V}_1, \ldots, \mathcal{V}_q$, with respect to the relation $\mathcal{R}$. Note that if node i belongs to $\mathcal{V}_\ell$, then $\mathcal{V}_\ell = \sigma^*(i) \cap \pi^*(i)$. To each equivalence class $\mathcal{V}_\ell$ corresponds a subgraph $\mathcal{G}_\ell = (\mathcal{V}_\ell, \mathcal{E}_\ell)$, where $\mathcal{E}_\ell$ is the restriction of $\mathcal{E}$ to $\mathcal{V}_\ell$, which is strongly connected.

Definition 2.2 (Maximal strongly connected subgraphs–m.s.c.s.) *The subgraphs $\mathcal{G}_i = (\mathcal{V}_i, \mathcal{E}_i)$ corresponding to the equivalence classes determined by $\mathcal{R}$ are the maximal strongly connected subgraphs of $\mathcal{G}$.*

Notation 2.3

- The subset of nodes of the m.s.c.s. containing node i (and possibly reduced to i) is denoted $[i]$.
- The subset of nodes $\bigcup_{j \in \pi^*(i)} [j]$ is denoted $[\leq i]$.
- The symbol $[< i]$ represents the subset of nodes $[\leq i] \setminus [i]$. ■

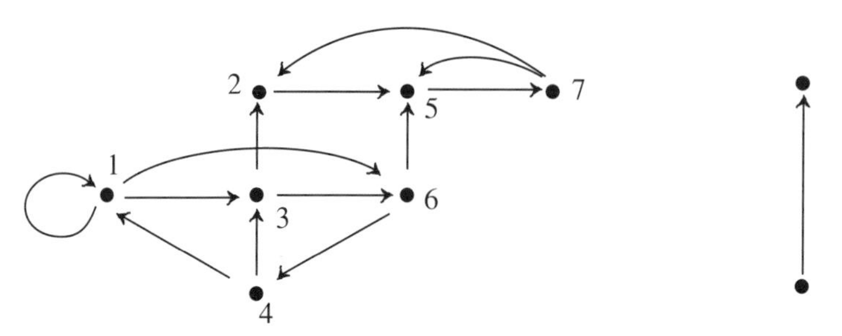

Figure 2.1: A digraph

Figure 2.2: The reduced graph

The graph of Figure 2.1 has two m.s.c.s.'s, namely the subgraphs consisting of the nodes 1, 3, 6, 4 and 2, 5, 7, respectively. If one 'lumps' the nodes of each m.s.c.s. into a single node, one obtains the so-called *reduced graph.*

Definition 2.4 (Reduced graph) *The reduced graph of $\mathcal{G}$ is the graph with nodes $\overline{\mathcal{V}} \stackrel{\text{def}}{=} \{1, \dots, q\}$ (one node per m.s.c.s.), and with arcs $\overline{\mathcal{E}}$, where $(i, j) \in \overline{\mathcal{E}}$ if $(k, l) \in \mathcal{E}$ for some node k of $\mathcal{V}_i$ and some node l of $\mathcal{V}_j$.*

Figure 2.2 shows the reduced graph of the graph in Figure 2.1.

Notation 2.5 A node i of the reduced graph corresponds to a collection of nodes of the original graph. Let x be a vector the entries of which are associated with nodes of the original graph. If we want to refer to the subvector associated with the nodes of m.s.c.s. i, we will use the notation $x_{(i)}$. Similarly, for a matrix A, $A_{(i)(j)}$ is the block extracted from A by keeping the rows associated with the nodes of m.s.c.s. i and the columns associated with the nodes of m.s.c.s. j. If node ℓ of the original graph belongs to m.s.c.s. i, the notation $x_{[\ell]}$ is equivalent to $x_{(i)}$. Similarly, $x_{(<i)}$, respectively $x_{(\leq i)}$, is equivalent to $x_{[<\ell]}$, respectively $x_{[\leq \ell]}$. ■

Lemma 2.6 *The reduced graph is acyclic.*

Proof If there is a path from $k \in \mathcal{V}_i$ to $l \in \mathcal{V}_j$, then there is no path from any node of $\mathcal{V}_j$ to any node of $\mathcal{V}_i$ (otherwise, k and l would be in the same m.s.c.s.). ■

Denote the existence of a path from one subgraph $\mathcal{G}_i$ to another one $\mathcal{G}_j$ by the binary relation $\mathcal{R}'$; $\mathcal{G}_i \mathcal{R}' \mathcal{G}_j$. Then these subgraphs $\mathcal{G}_1, \dots, \mathcal{G}_q$, together with the relation $\mathcal{R}'$ form a partially ordered set, see Chapter 4 and also [85].

2.3 GRAPHS AND MATRICES

In this section we consider matrices with entries belonging to an abstract alphabet $\mathcal{C}$ in which some algebraic operations will be defined in §2.3.1. Some relationships between these matrices and 'weighted graphs' will be introduced. Consider a graph

$\mathcal{G} = (\mathcal{V}, \mathcal{E})$ and associate an element $A_{ij} \in \mathcal{C}$ with each arc $(j, i) \in \mathcal{E}$: then $\mathcal{G}$ is called a *weighted graph*. The quantity A_{ij} is called the *weight* of arc (j, i). Note that the second subscript of A_{ij} refers to the initial (and not the final) node. The reason is that, in the algebraic context, we will work with column vectors (and not with row vectors) later on. In addition, we will also consider compositions of matrices and the resulting consequences for the corresponding graphs.

The alphabet $\mathcal{C}$ contains a special symbol ε the properties of which will be given in §2.3.1.

Definition 2.7 (Transition graph) *If an $n \times m$ matrix $A = (A_{ij})$ with entries in $\mathcal{C}$ is given, the transition graph of A is a weighted, bipartite graph with $n + m$ nodes, labeled $1, \ldots, m, m + 1, \ldots, m + n$, such that each row of A corresponds to one of the nodes $1, \ldots, m$; each column of A corresponds to one of the nodes $m + 1, \ldots, m + n$. An arc from j to $n + i$, $1 \leq i \leq m$, $1 \leq j \leq n$, is introduced with weight A_{ij} if $A_{ij} \neq \varepsilon$.*

As an example, consider the matrix

$$A = \begin{pmatrix} 3 & \varepsilon & \varepsilon & 7 & \varepsilon & \varepsilon & \varepsilon \\ \varepsilon & \varepsilon & 2 & \varepsilon & \varepsilon & \varepsilon & 1 \\ e & \varepsilon & \varepsilon & 2 & \varepsilon & \varepsilon & \varepsilon \\ \varepsilon & \varepsilon & \varepsilon & \varepsilon & \varepsilon & 5 & \varepsilon \\ \varepsilon & 4 & \varepsilon & \varepsilon & \varepsilon & 8 & 6 \\ 4 & \varepsilon & 1 & \varepsilon & \varepsilon & \varepsilon & \varepsilon \\ \varepsilon & \varepsilon & \varepsilon & \varepsilon & e & \varepsilon & \varepsilon \end{pmatrix}. \tag{2.1}$$

Its transition graph is depicted in Figure 2.3.

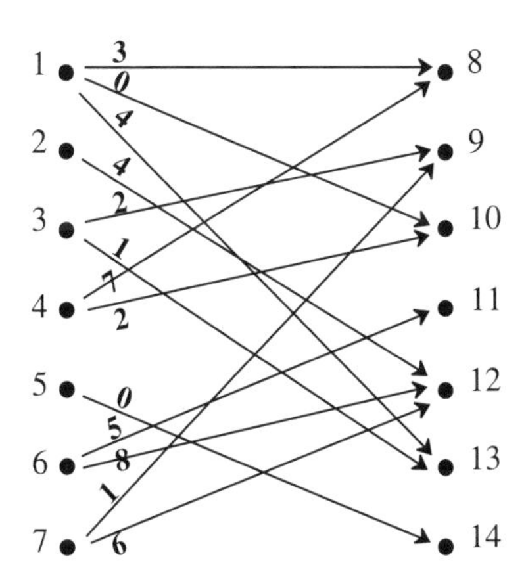

Figure 2.3: The transition graph of A

Definition 2.8 (Precedence graph) *The precedence graph of a square $n \times n$ matrix A with entries in $\mathcal{C}$ is a weighted digraph with n nodes and an arc (j, i) if $A_{ij} \neq \varepsilon$,* *in which case the weight of this arc receives the numerical value of A_{ij}. The precedence graph is denoted $\mathcal{G}(A)$.*

It is not difficult to see that any weighted digraph $\mathcal{G} = (\mathcal{V}, \mathcal{E})$ is the precedence graph of an appropriately defined square matrix. The weight A_{ij} of the arc from node j to node i defines the ij-th entry of a matrix A. If an arc does not exist, the corresponding entry of A is set to ε. The matrix A thus defined has $\mathcal{G}$ as its precedence graph.

The transition graph of a square $n \times n$ matrix, which has $2n$ nodes, can be transformed into a precedence graph of n nodes. Towards that end, one combines the nodes i and $n+i$ of the transition graph into one single node for the precedence graph, $i = 1, \ldots, n$. As an example, Figure 2.4 gives the precedence graph of the

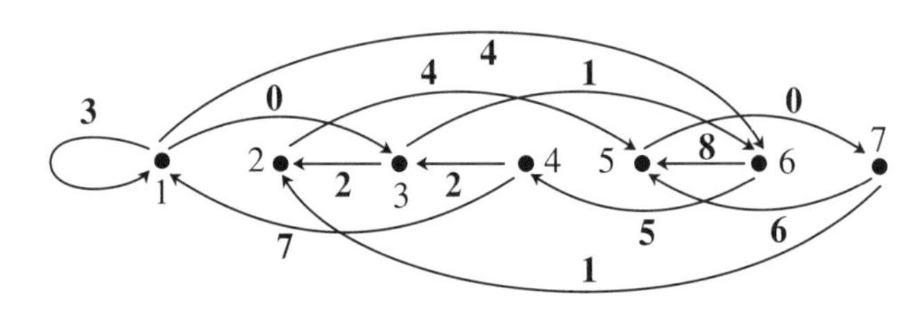

Figure 2.4: The precedence graph of A

matrix A defined in (2.1). One directly recognizes the relation of this graph with the transition graph in Figure 2.3. The latter graph has been 'folded' so as to obtain the precedence graph.

It may be convenient to consider that entries A_{ij} equal to ε define *dummy* arcs (which are not drawn) in the associated precedence or transition graph. Later, a path including a dummy arc will be called a dummy path. Dummy paths are not included in the set R of paths (which were taken into account in Definition 2.1; therefore these dummy paths are not involved in the definition of m.s.c.s.'s). The interest of the notion of dummy arcs is that these arcs may be considered as being of the same length as arcs associated with entries $A_{ij} \neq \varepsilon$ (generally this length is 1); hence arcs of the same length can be associated with all entries of a matrix.

Notation 2.9 The number of m.s.c.s.'s of $\mathcal{G}(A)$ is denoted N_A. ■

For later reference, the following definitions are given.

Definition 2.10 (Incidence matrix) *The incidence matrix $F = (F_{ij})$ of a graph $\mathcal{G} = (\mathcal{V}, \mathcal{E})$ is a matrix the number of columns of which equals the number of arcs and the number of rows of which equals the number of nodes of the graph. The entries of F can take the values 0, 1 or -1. If $l = (i, j) \in \mathcal{E}$, $i \neq j$, then $F_{il} = 1, F_{jl} = -1$ and the other entries of column l are 0. If $l = (i, i) \in \mathcal{E}$, then $F_{il} = 0$.*

Definition 2.11 (Adjacency matrix) *The adjacency matrix $G = (G_{ij})$ of a graph $\mathcal{G} = (\mathcal{V}, \mathcal{E})$ is a matrix the numbers of rows and columns of which are equal to* *the number of nodes of the graph. The entry G_{ij} is equal to 1 if $j \in \pi(i)$ and to 0 otherwise.*

Note that if $\mathcal{G} = \mathcal{G}(A)$, then $G_{ij} = 1$ if and only if $A_{ij} \neq \varepsilon$ (G describes the 'support' of A).

2.3.1 Composition of Matrices and Graphs

We now study two kinds of compositions of matrices, and the relation between the transition graph of these compositions and the original transition graphs. These compositions are, respectively, the *parallel composition*, denoted $\oplus$, and the *series composition*, denoted $\otimes$. These compositions will be defined by means of the corresponding composition operations of elements in the alphabet $\mathcal{C}$, for which the same symbols $\oplus$ and $\otimes$ will be used. The operation $\oplus$ is usually referred to as 'addition' or 'sum', and the operation $\otimes$ as 'multiplication' or 'product'. The alphabet $\mathcal{C}$ includes two special elements ε and e with specific properties to be defined in the following set of axioms:

Associativity of addition:

$$\forall a, b, c \in \mathcal{C} \ , \quad (a \oplus b) \oplus c = a \oplus (b \oplus c) \ .$$

Commutativity of addition:

$$\forall a, b \in \mathcal{C} \ , \quad a \oplus b = b \oplus a \ .$$

Associativity of multiplication:

$$\forall a, b, c \in \mathcal{C} \ , \quad (a \otimes b) \otimes c = a \otimes (b \otimes c) \ .$$

Right and left distributivity of multiplication over addition:

$$\forall a, b, c \in \mathcal{C} \ , \quad (a \oplus b) \otimes c = (a \otimes c) \oplus (b \otimes c) \ ,$$

$$\forall a, b, c \in \mathcal{C} \ , \quad c \otimes (a \oplus b) = (c \otimes a) \oplus (c \otimes b) \ .$$

Existence of a zero element:

$$\exists \varepsilon \in \mathcal{C} : \quad \forall a \in \mathcal{C} \ , \quad a \oplus \varepsilon = a \ .$$

Absorbing zero element:

$$\forall a \in \mathcal{C} \ , \quad a \otimes \varepsilon = \varepsilon \ .$$

Existence of an identity element:

$$\exists e \in \mathcal{C} : \quad \forall a \in \mathcal{C} \ , \quad a \otimes e = e \otimes a = a \ .$$

In Chapter 3 other related axioms will be discussed in detail. There the notion of a *semifield* will be introduced and its relation to axioms of this type will be made clear.

The *parallel composition* $\oplus$ of matrices is defined for matrices of the same size by the following rule: if $A = (A_{ij})$ and $B = (B_{ij})$ have the same size, then

$$(A \oplus B)_{ij} = A_{ij} \oplus B_{ij} \ .$$

A transition graph can of course be associated with the matrix $C = A \oplus B$. This transition graph has the same set of nodes as the transition graph of A (and therefore of B) and there exists a (nondummy) arc from node j to node i if and only if at least one of the transition graphs of A and B has a (nondummy) arc from j to i. This is a consequence of the axiom $A_{ij} \oplus \varepsilon = A_{ij}$. In general, this arc receives the weight $A_{ij} \oplus B_{ij}$. It may be viewed as the arc resulting from the merging of two parallel arcs (if both A_{ij} and B_{ij} are different from ε). The symbol ε is called the *zero element* of the operation $\oplus$. Two other axioms are that $\oplus$ is associative and commutative and they have obvious consequences for the parallel composition of transition graphs.

The *series composition* $\otimes$ of matrices A and B is defined only when the number of columns of A equals the number of rows of B (say, A is $m \times n$ and B is $n \times p$) by the following rule:

$$(A \otimes B)_{ij} = \bigoplus_{k=1}^{n} A_{ik} \otimes B_{kj} \ . \tag{2.2}$$

A transition graph can be associated with the matrix $C = A \otimes B$ of size $m \times p$. With the help of Figure 2.5, we explain how this graph is obtained. First, the

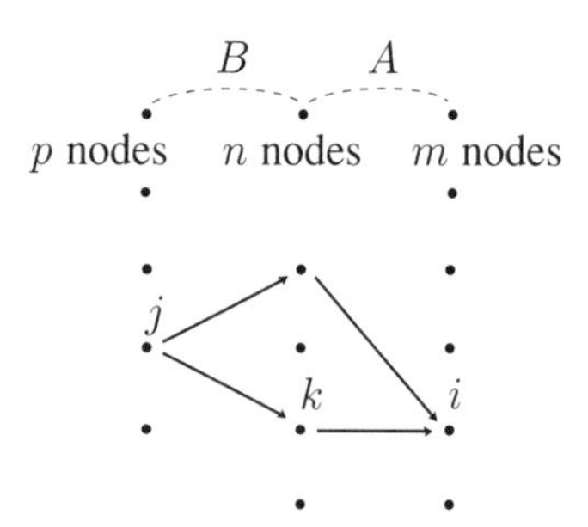

Figure 2.5: The series composition of two transition matrices

transition graphs of matrices B and A are concatenated. The graph so obtained is not a transition graph since it has n intermediate nodes in addition of its p input nodes and its m output nodes. These intermediate nodes are removed, and an arc from node j to node i exists in the transition graph of C if and only if there exists at least one (nondummy) path from node j to node i in the concatenated graph shown in Figure 2.5. This arc receives the weight indicated by Equation (2.2). In order to interpret this formula, let us first define the weight of a path $\rho = (i_1, \ldots, i_p)$ in a weighted graph as the product $A_{i_p,i_{p-1}} \otimes \cdots \otimes A_{i_2,i_1}$ of weights of arcs composing this path (observe the order). This weight is denoted $|\rho|_w$, where the subscript w refers to the word 'weight'. Note that a dummy path always has a weight ε thanks to the absorbing property of ε in products. Then, each term of the sum in (2.2) can be interpreted as the weight of some parallel path of length 2 from node j to node i, characterized by the intermediate node k it passes through (see Figure 2.5). The previous rule pertaining to the weights of parallel compositions of arcs is thus extended to parallel paths. Since ε is the zero element for addition, dummy paths do not contribute to (nonzero) weights in parallel compositions.

For the series composition of matrices, we have $(A \otimes B) \otimes C = A \otimes (B \otimes C)$. This associativity property of matrix multiplication is a direct consequence of the axioms given above, namely the associativity of $\otimes$ and the right and left distributivity of $\otimes$ over $\oplus$. It is easily seen that these right and left distributivities also hold for matrices of appropriate sizes.

The notion $(A^p)_{ji}, p = 1, 2, \ldots,$ where A is square, is defined by $(A^p)_{ji} = (A \otimes A^{p-1})_{ji}$. Observe that this involves the enumeration and the sums of weights of all paths of length p with initial node i and final node j. This definition makes sense for both the transition graph of A, concatenated p times, and the precedence graph of A. We get

$$(A^p)_{ji} = \bigoplus_{\{\rho \mid |\rho|_l = p; i_0 = i; i_p = j\}} A_{i_p,i_{p-1}} \otimes A_{i_{p-1},i_{p-2}} \otimes \cdots \otimes A_{i_1,i_0} \; .$$

Because we removed the intermediate nodes in the graph of Figure 2.5 in order to obtain the transition graph of $C = A \otimes B$, and similarly when one considers the transition graph of a matrix C equal to A^p, the information that weights of the graph of C have been obtained as weights of paths of length larger than 1 (namely 2 and p, respectively) has been lost. In order to keep track of this information, one may introduce the notion of *length of a transition graph*. This length is an integer number associated with the transition graph and hence also with all its individual arcs, including dummy arcs, and finally with the matrix associated with this graph. These considerations explain why the lengths of arcs may be taken greater than 1 and why dummy arcs may also have a nonzero length.

We now consider the transition graph corresponding to matrices A^0 where A is an $n \times n$ matrix. In the same way as A^p, for $p \geq 1$, describes the weights of paths of length p, A^0 should describe the weights of paths of length 0, that is, empty circuits (i) corresponding to 'no transition at all'. Pictorially, the corresponding transition graph of such matrices has the special form depicted in Figure 2.6a in which input and output nodes are not distinguishable. This transition graph must not be confused either with that of the 'zero matrix' (with all entries equal to ε) or with that of the 'identity matrix' (with diagonal entries equal to e and all off-diagonal entries equal to ε). The transition graph of the zero matrix is depicted in Figure 2.6b: all its arcs are dummy with length 1. The transition graph of the identity matrix is depicted in Figure 2.6c: weights are indicated in the figure and the length is 1 for all dummy and nondummy arcs. The length associated with all entries of A^0 is 0, that is, the series composition of the transition graph of A^0 with any other bipartite graph do not modify the lengths of paths. In the same way, we would like the weights not

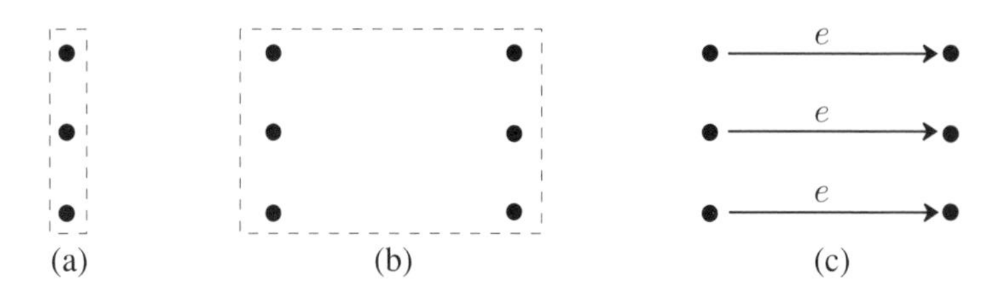

Figure 2.6: The transition graphs of A^0, ε and e

to be modified in that operation: this requires that the entries of A^0 be the same as those of the identity matrix e. But again the respective lengths and transition graphs of A^0 and e are different (see Figure 2.6a–c). Also, dummy loops (i,i) of the transition graph of the zero matrix should not be confused with empty circuits (i) (see Figure 2.6a–b) even if it is difficult to distinguish them with the help of precedence, rather than transition, graphs.

In Chapter 1 we already met several examples of $\oplus$ and $\otimes$ operations which satisfy the axioms just stated. A particular example is that $\mathcal{C}$ equals $\mathbb{R}$, $e = 0$, $\varepsilon = -\infty$, $\oplus$ equals maximization and $\otimes$ equals addition. Note that the max-operation is idempotent, i.e. $a \oplus a = a$, but this property is not (yet) assumed as an axiom.

Remark 2.12 An example of operations which do *not* satisfy some of the axioms is one where $\oplus$ is addition and $\otimes$ is minimization. The axioms of distributivity are not satisfied. Indeed,

$$\min(5, 3+6) \neq \min(5,3) + \min(5,6) \ .$$

As a consequence, associativity of multiplication with respect to matrices does not hold. If for instance,

$$A = \begin{pmatrix} 1 & 3 \\ 2 & 2 \end{pmatrix}, \quad B = \begin{pmatrix} 3 & 2 \\ 1 & 5 \end{pmatrix}, \quad C = \begin{pmatrix} 4 & 2 \\ 2 & 5 \end{pmatrix},$$

then

$$\begin{pmatrix} 4 & 6 \\ 5 & 6 \end{pmatrix} = (AB)C \neq A(BC) = \begin{pmatrix} 4 & 4 \\ 4 & 4 \end{pmatrix}.$$

A practical interpretation of a system with these operations is in terms of the calculation of the capacity of a network in which the arcs are pipes through which there is a continuous flow. The capacity of a pipe is assumed to be proportional to the diameter of this pipe. Then it is easily seen that the capacity of two pipes in parallel equals the sum of the two capacities. Similarly, the capacity of two pipes in series equals the minimum of their capacities. The reader should contemplate the physical consequences of the lack of associativity. ■

Definition 2.13 (Irreducibility) *The (square) matrix A is called irreducible if no permutation matrix P exists such that the matrix $\widetilde{A}$, defined by*

$$\widetilde{A} = P'AP \ ,$$

has an upper triangular block structure.

The reader should be aware of the fact that this definition is invariant with respect to the algebra used. Premultiplication of A by P' and postmultiplication by P simply refers to a renumbering of the nodes of the corresponding graph. Hence renumbering of the nodes of the same graph leads to different A-matrices. In an upper triangular block structure, diagonal blocks with non-ε entries are allowed. If one also wants the diagonal blocks to have ε-entries only, one should speak about a strictly upper triangular block structure.

Theorem 2.14 *A necessary and sufficient condition for the square matrix A to be irreducible is that its precedence graph be strongly connected.*

Proof Suppose that A is such that by an appropriate renumbering of the nodes, $\widetilde{A}$ has an upper triangular block structure. Call the diagonal blocks $A_1^d, \ldots, A_q^d$. If A_q^d has size $n_q \times n_q$, then there are no paths from any of the (renumbered) nodes $1, \ldots, n-n_q$, to any of the nodes $n-n_q+1, \ldots, n$. Hence this graph is not strongly connected.

On the other hand, if the graph is not strongly connected, determine its m.s.c.s.'s $\mathcal{G}_i, i = 1, \ldots, q$. These subgraphs form a partially ordered set. Number the individual nodes of $\mathcal{V}$ in such a way that if $\mathcal{G}_i \mathcal{R}' \mathcal{G}_j$, then the nodes of $\mathcal{V}_i$ have lower indices than those of $\mathcal{V}_j$ ($\mathcal{R}'$ was defined in §2.2). With this numbering of the nodes, the corresponding matrix A will be upper block triangular. ■

Definition 2.15 (Aperiodicity) *The irreducible square matrix A is aperiodic if there exists an integer N such that for all $n \geq N$ and for all i, j, $(A^n)_{ij} \neq \varepsilon$.*

Theorem 2.16 *An irreducible matrix A such that $A_{jj} \neq \varepsilon$ for all j is aperiodic.*

Proof From Theorem 2.14, the irreducibility assumption implies that for all i, j, there exists n such that $(A^n)_{ij} \neq \varepsilon$. This together with the assumption $A_{jj} \neq \varepsilon$ in turn implies that $(A^m)_{ij} \neq \varepsilon$ for all $m \geq n$. The assertion of the theorem follows immediately from this since the number of nodes is finite. ■

Definition 2.17 *A digraph is called a tree if there exists a single node such that there is a unique path from this node to any other node.*

In order to determine whether A is irreducible, one can calculate A^+ (see (1.18)). Matrix A is irreducible if and only if all entries of A^+ are different from ε. This algorithm for determining whether A is irreducible can be simplified by considering only Boolean variables. Replace A by the adjacency matrix G of its precedence graph (see Definition 2.11), except that 0 and 1 are replaced by ε and e, respectively,

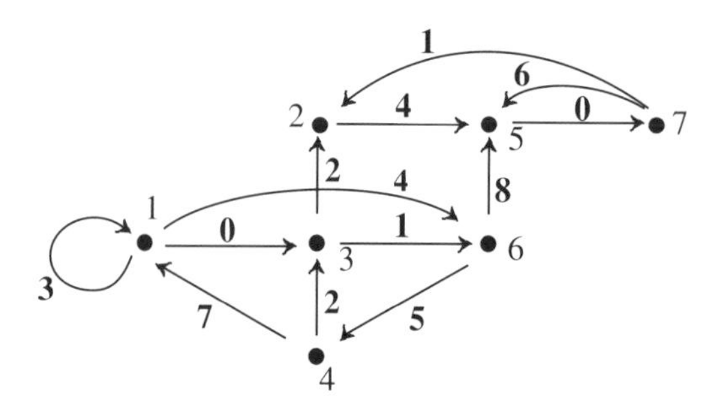

Figure 2.7: Weighted digraph consisting of two m.s.c.s.'s

in the present context. Then A is irreducible if and only if all entries of G^+ are identical to e. As an example, consider the matrix given in (2.1). The matrix G^+ becomes

$$\begin{pmatrix} e & \varepsilon & e & e & \varepsilon & e & \varepsilon \\ e & e & e & e & e & e & e \\ e & \varepsilon & e & e & \varepsilon & e & \varepsilon \\ e & \varepsilon & e & e & \varepsilon & e & \varepsilon \\ e & e & e & e & e & e & e \\ e & \varepsilon & e & e & \varepsilon & e & \varepsilon \\ e & e & e & e & e & e & e \end{pmatrix} . \tag{2.3}$$

Hence A is not irreducible. In fact it follows directly from (2.3) that the nodes $1, 3, 4$ and 6 form a m.s.c.s., as do the nodes $2, 5$ and 7. If one rearranges the nodes and arcs of Figure 2.4, one obtains Figure 2.7, in which the weights of the arcs have been indicated, and the two m.s.c.s.'s are clearly visible. Figure 2.7 is identical to Figure 2.1 apart from the fact that weights are given.

2.3.2 Maximum Cycle Mean

In this subsection the maximum cycle mean will be defined and some of its properties will be derived. The maximum cycle mean has a clear relation with eigenvalues of matrices within the context of the max algebra and with periodic regimes of systems described by linear equations, also within the same context. These relationships will be described in §3.

Let $\mathcal{G} = (\mathcal{V}, \mathcal{E})$ be a weighted digraph with n nodes. The weights are real numbers here and are given by means of the $n \times n$ matrix A. As discussed before, the numerical value of A_{ij} equals the weight of the arc from node j to node i. If no such arc exists, then $A_{ij} = \varepsilon$. It is known from Chapter 1 that the entry (i, j) of $A^k = A \otimes \cdots \otimes A$, considered within the algebraic structure $\mathbb{R}_{\max} \stackrel{\text{def}}{=} (\mathbb{R} \cup \{-\infty\}, \max, +)$, denotes the maximum weight with respect to all paths of length k which go from node j to node i. If no such path exists, then $(A^k)_{ij} = \varepsilon$. Within this algebraic structure, ε gets assigned the numerical value $-\infty$ and $e = 0$. *In this subsection we will confine ourselves to the algebraic structure* $\mathbb{R}_{\max}$.

Definition 2.18 (Cycle mean) *The mean weight of a path is defined as the sum of the weights of the individual arcs of this path, divided by the length of this path. If the path is denoted* ρ, *then the mean weight equals* $|\rho|_{\mathrm{w}}/|\rho|_1$. *If such a path is a circuit one talks about the mean weight of the circuit, or simply the cycle mean.*

We are interested in the maximum of these cycle means, where the maximum is taken over all circuits in the graph (empty circuits are not considered here). Consider an $n \times n$ matrix A with corresponding precedence graph $\mathcal{G} = (\mathcal{V}, \mathcal{E})$. The maximum weight of all circuits of length j which pass through node i of $\mathcal{G}$ can be written as $(A^j)_{ii}$. The maximum of these maximum weights over all nodes is $\bigoplus_{i=1}^n (A^j)_{ii}$ which can be written trace(A^j). The average weight is obtained by dividing this number by j in the conventional sense, but this can be written $(\text{trace}(A^j))^{1/j}$ in the max-plus algebra notation. Finally, we have to take the maximum with respect to the length j. It is not necessary to consider lengths larger than the number n of nodes since it is enough to limit ourselves to elementary circuits. It follows that a formula for the maximum cycle mean λ in the max-plus algebra notation is

$$\lambda = \bigoplus_{j=1}^{n} (\text{trace}(A^j))^{1/j} \ .$$

The following theorem provides an expression, due to R. Karp, for this maximum value. All circuits with a cycle mean equal to the maximum cycle mean are called critical circuits. Karp's theorem does not give the critical circuit(s).

Theorem 2.19 (Karp's theorem) *Given is an* $n \times n$ *matrix* A *with corresponding precedence graph* $\mathcal{G} = (\mathcal{V}, \mathcal{E})$. *The maximum cycle mean is given by*

$$\lambda = \max_{i=1,\dots,n} \ \min_{k=0,\dots,n-1} \frac{(A^n)_{ij} - (A^k)_{ij}}{n-k} , \quad \forall j \ . \tag{2.4}$$

In this equation, A^n *and* A^k *are to be evaluated in* $\mathbb{R}_{\max}$; *the other operations are conventional ones.*

Proof Note that the index j in (2.4) is arbitrary (it will be shown in this proof that one can take any $j \in \{1, \dots, n\}$). The resulting value of λ is independent of j.

Without loss of generality, we may assume that $\mathcal{G}$ is strongly connected. If it were not, we would consider each of its m.s.c.s.'s—since $\mathcal{G}$ is assumed to be finite, there are only a finite number of such m.s.c.s.'s—and determine the maximum cycle mean of each of them and then take the maximum one.

We first assume that the maximum cycle mean is 0. Then it must be shown that

$$\max_{i=1,\dots,n} \ \min_{k=0,\dots,n-1} \frac{(A^n)_{ij} - (A^k)_{ij}}{n-k} = 0 \ .$$

Since $\lambda = 0$ there exists a circuit of weight 0 and there exists no circuit with positive weight. Because there are no circuits (or loops) with positive weight, there is a maximum weight of all paths from node j to node i which is equal to

$$\chi_{ij} \stackrel{\text{def}}{=} \max \sum_{l=1}^{k} A_{i_l,i_{l-1}} \ , \quad \text{subject to} \quad i_0 = j \ , \quad i_k = i \ ,$$

where the maximum is taken with respect to all paths and all k. Since for $k \geq n$ the path would contain a circuit and since all circuits have nonpositive weight, we can restrict ourselves to $k < n$. Therefore we get

$$\chi_{ij} = \max_{k=0,\dots,n-1} (A^k)_{ij} \ .$$

Also, $(A^n)_{ij} \leq \chi_{ij}$, and hence

$$(A^n)_{ij} - \chi_{ij} = \min_{k=0,\dots,n-1} (A^n)_{ij} - (A^k)_{ij} \leq 0 \ .$$

Equivalently,

$$\min_{k=0,\dots,n-1} \frac{(A^n)_{ij} - (A^k)_{ij}}{n-k} \leq 0 \ . \tag{2.5}$$

Equality in (2.5) will only hold if $(A^n)_{ij} = \chi_{ij}$. It will be shown that indeed an index i exists such that this is true. Let ζ be a circuit of weight 0 and let l be a node of ζ. Let ρ_{lj} be a path from j to l with corresponding maximum weight $|\rho_{lj}|_{\mathrm{w}} = \chi_{lj}$. Now this path is extended by appending to it a number of repetitions of ζ such that the total length of this extended path, denoted ρ_{e}, becomes greater than or equal to n. This is again a path of maximum weight from j to l. Now consider the path consisting of the first n arcs of ρ_{e}; its initial node is j and denote its final node l'. Of course $l' \in \zeta$. Since any subpath of any path of maximum weight is of maximum weight itself, the path from j to l' is of maximum weight. Therefore $(A^n)_{l'j} = \chi_{l'j}$. Now choose $i = l'$ and we get

$$\max_{i=1,\dots,n} \left[\min_{k=0,\dots,n-1} \frac{(A^n)_{ij} - (A^k)_{ij}}{n-k} \right] = 0 \ .$$

This completes the part of the proof with $\lambda = 0$.

Now consider an arbitrary finite λ. A constant c is now subtracted from each weight A_{ij}. Then clearly λ will be reduced by c. Since $(A^k)_{ij}$ is reduced by kc, we get that

$$\frac{(A^n)_{ij} - (A^k)_{ij}}{n-k}$$

is reduced by c, for all i, j and k, and hence

$$\max_{i=1,\dots,n} \min_{k=0,\dots,n-1} \frac{(A^n)_{ij} - (A^k)_{ij}}{n-k}$$

is also reduced by c. Hence both sides of (2.4) are affected equally when all weights A_{ij} are reduced by the same amount. Now choose this amount such that λ becomes 0 and then we are back in the previous situation where $\lambda = 0$. ■

2.3.3 The Cayley-Hamilton Theorem

The Cayley-Hamilton theorem states that, in conventional algebra, a square matrix satisfies its own characteristic equation. In mathematical terms, let A be an $n \times n$ matrix and let

$$p_A(x) \stackrel{\text{def}}{=} \det(xI - A) = x^n + c_1 x^{n-1} + \cdots + c_{n-1}x + c_n x^0 \ , \tag{2.6}$$

where I is the identity matrix, be its characteristic polynomial. The term x^0 in the polynomial equals 1. Then $p_A(A) = 0$, where 0 is the zero matrix. The coefficients $c_i, i = 1, \ldots, n$, in (2.6) satisfy

$$c_k = (-1)^k \sum_{i_1 < i_2 < \cdots < i_k} \det \begin{pmatrix} A_{i_1,i_1} & \ldots & A_{i_1,i_k} \\ \vdots & & \vdots \\ A_{i_k,i_1} & \ldots & A_{i_k,i_k} \end{pmatrix} . \tag{2.7}$$

The reason for studying the Cayley-Hamilton theorem is the following. In conventional system theory, this theorem is used for the manipulation of different system descriptions and for analyzing such properties as controllability (see [72]). In the context of discrete event systems, a utilization of this theorem will be shown in §9.2.2.

In this section it will be shown that a Cayley-Hamilton theorem also exists in an algebraic structure defined by a set $\mathcal{C}$ of elements supplied with two operations denoted $\oplus$ and $\otimes$ which obey some of the axioms given in §2.3.1, namely

- associativity of addition,
- commutativity of addition,
- associativity of multiplication,
- both right and left distributivity of multiplication over addition,
- existence of an identity element,

provided we also have

Commutativity of multiplication:

$$\forall a, b \in \mathcal{C} \ , \quad a \otimes b = b \otimes a \ .$$

Note that the existence of a zero element (and its absorbing property) is not required in this subsection.

A *partial permutation* of $\{1, \ldots, n\}$ is a bijection ς of a subset of $\{1, \ldots, n\}$ onto itself. The domain of ς is denoted by $\text{dom}(\varsigma)$ and its cardinality is denoted $|\varsigma|_1$. A partial permutation ς for which $|\varsigma|_1 = n$ is called a complete permutation. The completion $\widehat{\varsigma}$ of a partial permutation ς is defined by

$$\widehat{\varsigma}(i) = \begin{cases} \varsigma(i) & \text{if } i \in \text{dom}(\varsigma) \ , \\ i & \text{if } i \in \{1, \ldots, n\} \setminus \text{dom}(\varsigma) \ . \end{cases}$$

The *signature* * of a partial permutation ς, denoted $\mathrm{sgn}^*(\varsigma)$, is defined by

$$\mathrm{sgn}^*(\varsigma) = \mathrm{sgn}(\widehat{\varsigma})(-1)^{|\varsigma|_1} \ ,$$

where $\mathrm{sgn}(\widehat{\varsigma})$, sometimes also written as $(-1)^{\widehat{\varsigma}}$, denotes the conventional signature of the permutation $\widehat{\varsigma}$, see [61].

Every (partial) permutation has a unique representation as a set of disjoint circuits. For example, the permutation

$$\begin{pmatrix} 1 & 2 & 3 & 4 & 5 & 6 \\ 4 & 6 & 3 & 5 & 1 & 2 \end{pmatrix}$$

has the circuit representation

$$\{(1,4,5),(3),(2,6)\} \ .$$

With the graph-theoretic interpretation of permutations in mind, these disjoint circuits correspond to m.s.c.s.'s. The unique partial permutation of cardinality 0 has the empty set as its circuit representation. If ς is a partial permutation with cardinality k consisting of a single circuit, then $\mathrm{sgn}^* = (-1)^{k-1}(-1)^k = -1$. It easily follows that for any partial permutation ς, $\mathrm{sgn}^* = (-1)^r$, where r is the number of circuits appearing in the circuit representation of ς.

Given an $n \times n$ matrix $A = (A_{ij})$, the weight of ς is defined by

$$|\varsigma|_{\mathrm{w}} = \bigotimes_{i \in \mathrm{dom}(\varsigma)} A_{\varsigma(i),i} \ .$$

The weight of the partial permutation with cardinality 0 equals e, in accordance with the theory presented in §2.3.1. Let $1 \le i,j \le n$ and let T^+_{ji} be the set of all pairs (ς, ρ) where ς is a partial permutation and where ρ is a path from i (the initial node) to j (the final node) in such a way that

$$|\varsigma|_1 + |\rho|_1 = n \ , \quad \mathrm{sgn}^*(\varsigma) = 1 \ .$$

The set T^-_{ji} is defined identically except for the fact that the condition $\mathrm{sgn}^*(\varsigma) = 1$ is replaced by $\mathrm{sgn}^*(\varsigma) = -1$.

Lemma 2.20 *For each pair (j,i), with $1 \le i,j \le n$, there is a bijection $\eta_{ji} : T^+_{ji} \to T^-_{ji}$ in such a way that $\eta_{ji}(\varsigma,\rho) = (\varsigma',\rho')$ implies $|\varsigma|_{\mathrm{w}} \otimes |\rho|_{\mathrm{w}} = |\varsigma'|_{\mathrm{w}} \otimes |\rho'|_{\mathrm{w}}$.*

Proof Each pair $(\varsigma,\rho) \in T^+_{ji} \cup T^-_{ji}$ is represented by a directed graph with nodes $\{1,\dots,n\}$. The set of arcs consists of two classes of arcs:

$$\mathcal{E}_\varsigma = \{(i,\varsigma(i)) \mid i \in \mathrm{dom}(\varsigma)\} \ , \quad \mathcal{E}_\rho = \{(l,k) \mid (l,k) \text{ is an arc of } \rho\} \ .$$

This graph will, in general, contain multiple arcs, since ρ may traverse the same arc more than once, or the same arc may appear in both $\mathcal{E}_\varsigma$ and $\mathcal{E}_\rho$. The expression

$|\varsigma|_{\rm w} \otimes |\rho|_{\rm w}$ is the series composition, with multiplicities taken into account, of all the A_{lk} for which (k,l) is an arc of the graph associated with (ς, ρ).

Let ρ be the path connecting the nodes $i_0, i_1, \ldots, i_q$, in this order. There is a smallest integer $v \geq 0$ such that either $i_u = i_v$ for some $u < v$, or $i_v \in \mathrm{dom}(\varsigma)$. If such a v did not exist, then ρ must have $|\rho|_1 + 1$ distinct nodes (because there are no u and v with $i_v = i_u$). But then there are at least $|\varsigma|_1 + |\rho|_1 + 1 = n + 1$ distinct nodes (because ς and ρ do not have any node in common), which is a contradiction. Furthermore, it is easily seen that this smallest integer v cannot have both properties. Hence either $i_u = i_v$ for some $u < v$ or $i_v \in \mathrm{dom}(\varsigma)$. An example of the first property is given in Figure 2.8 and an example of the second property is given in Figure 2.9. The dashed arcs refer to the set $\mathcal{E}_\varsigma$ and the solid arcs refer to the set $\mathcal{E}_\rho$. In Figure 2.8, v equals 3, and in Figure 2.9, v equals 1 (i.e. $i_1 = 2$).

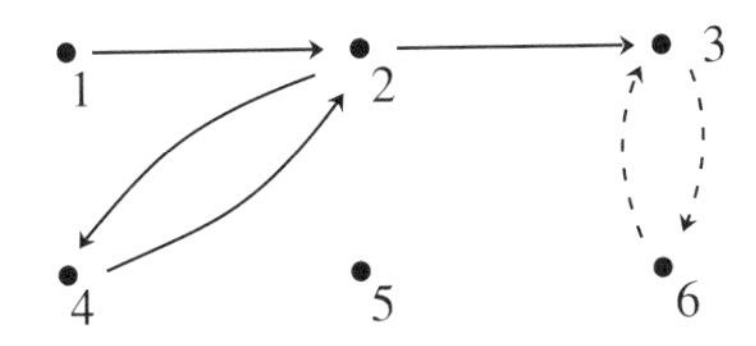

Figure 2.8: Example of a graph with the first property

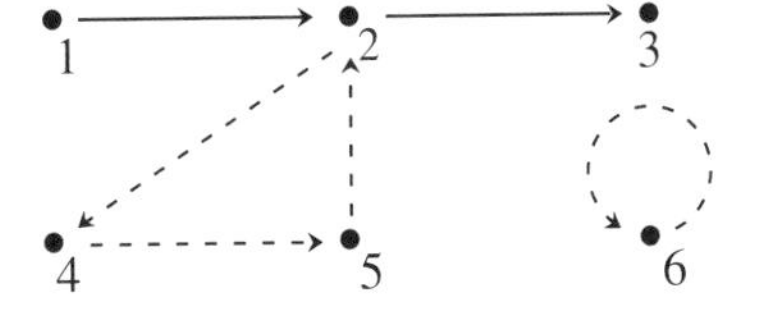

Figure 2.9: Example of a graph with the second property

Consider the situation with the first property such as pictured in Figure 2.8. We have $i_u = i_v$ for some u and v. The circuit passing through i_u is removed from ρ and adjoined as a new circuit to ς. The new path from i_0 to i_q is denoted ρ' and the new, longer, partial permutation is denoted ς'. The mapping η_{ji} is defined as $\eta_{ji}(\varsigma, \rho) = (\varsigma', \rho')$. Application of the mapping η_{31} to Figure 2.8 is given in Figure 2.10. Since the number of circuits of ς' and of ς differ by one, we have

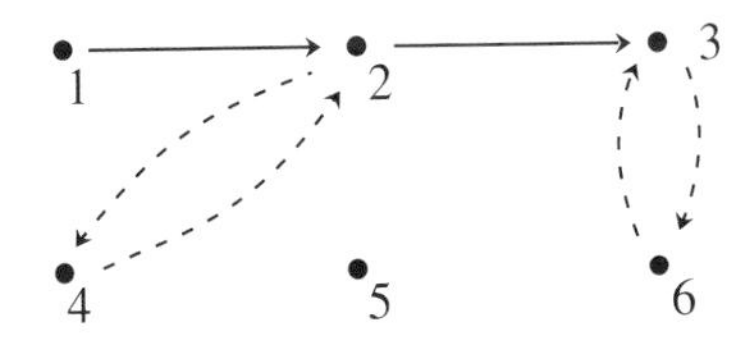

Figure 2.10: η_{31} applied to Figure 2.8

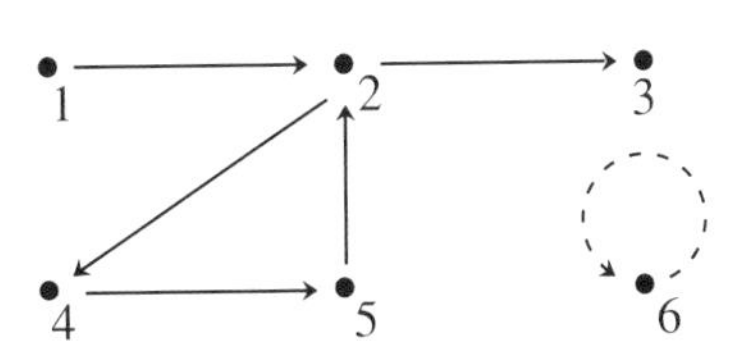

Figure 2.11: η_{31} applied to Figure 2.9

$\mathrm{sgn}^*(\varsigma') = -\mathrm{sgn}^*(\varsigma)$ and therefore η_{ji} maps T^+_{ji} into T^-_{ji} and vice versa.

Consider next the situation with the second property such as pictured in Figure 2.9. The mapping $\eta_{ji}(\varsigma, \rho) = (\varsigma', \rho')$ is obtained by removing the circuit containing i_v from ς and adjoining it to ρ. Application of η_{31} to Figure 2.9 is given in Figure 2.11. Also in this situation, the numbers of circuits of ς' and of ς differ by one, and again we have $\mathrm{sgn}^*(\varsigma') = -\mathrm{sgn}^*(\varsigma)$ which results in the fact that η_{ji} maps T^+_{ji} into T^-_{ji} and vice versa. In both situations $|\varsigma|_{\rm w} \otimes |\rho|_{\rm w} = |\varsigma'|_{\rm w} \otimes |\rho'|_{\rm w}$, since nothing has changed in the graph of (ς, ρ). It is in the derivation of this equality that the associativity and commutativity of multiplication, and the existence of an identity element, have been used. What remains to be shown is that the mapping η_{ji} is surjective. For this reason consider the iteration $\eta_{ji} \circ \eta_{ji}$. It easily follows that this mapping is the identity on $T_{ji}^+ \cup T_{ji}^-$, which is only possible if η_{ji} is surjective. ■

Definition 2.21 (Characteristic equation) *The characteristic equation is given by*

$$p_A^+(x) = p_A^-(x) \ , \tag{2.8}$$

where

$$p_A^+(x) = \bigoplus_{k=0}^{n} \left(\bigoplus_{|\varsigma|_1 = k, \mathrm{sgn}^*(\varsigma)=1} \left(\bigotimes_{i \in \mathrm{dom}(\varsigma)} A_{\varsigma(i),i} \right) \right) x^{n-k} \ ,$$

and

$$p_A^-(x) = \bigoplus_{k=0}^{n} \left(\bigoplus_{|\varsigma|_1 = k, \mathrm{sgn}^*(\varsigma)=-1} \left(\bigotimes_{i \in \mathrm{dom}(\varsigma)} A_{\varsigma(i),i} \right) \right) x^{n-k} \ .$$

It is easily verified that this characteristic equation, if considered the conventional algebra, coincides with the equation obtained by setting the characteristic polynomial (2.6) equal to zero. The crucial feature of (2.8) is that there are no terms with 'negative' coefficients (in contrast with the conventional characteristic equation, which can have negative coefficients). Since the inverse of $\oplus$ does not exist, the terms in (2.8) cannot freely be moved from one side to the other side of the equation.

Theorem 2.22 (The Cayley-Hamilton theorem) *The following identity holds true:*

$$p_A^+(A) = p_A^-(A) \ .$$

Proof For $k = 0, \ldots, n$,

$$(A^{n-k})_{ji} = \bigoplus_{\substack{|\rho|_1 = n-k \\ \text{initial node of } \rho \text{ equals } i \\ \text{final node of } \rho \text{ equals } j}} |\rho|_w \ .$$

It follows that

$$p_A^+(A)_{ji} = \bigoplus_{(\varsigma,\rho) \in T_{ji}^+} |\varsigma|_w \otimes |\rho|_w \ , \qquad p_A^-(A)_{ji} = \bigoplus_{(\varsigma,\rho) \in T_{ji}^-} |\varsigma|_w \otimes |\rho|_w \ .$$

Owing to Lemma 2.20, these two sums are identical. It is in these two equalities that the associativity and commutativity of addition, the distributivity and the existence of an identity element have been used. ■

Let us give an example. For the 3×3 matrix $A = (A_{ij})$, the characteristic equation is $p_A^+(x) = p_A^-(x)$, where

$$\begin{aligned} p_A^+(x) &= x^3 \oplus (A_{11}A_{22} \oplus A_{11}A_{33} \oplus A_{22}A_{33})x \oplus A_{13}A_{22}A_{31} \\ &\quad \oplus A_{12}A_{21}A_{33} \oplus A_{11}A_{32}A_{23} \ , \\ p_A^-(x) &= (A_{11} \oplus A_{22} \oplus A_{33})x^2 \oplus (A_{12}A_{21} \oplus A_{13}A_{31} \oplus A_{23}A_{32})x \\ &\quad \oplus A_{11}A_{22}A_{33} \oplus A_{12}A_{23}A_{31} \oplus A_{21}A_{13}A_{32} \ , \end{aligned}$$

where, as usual, the $\otimes$-symbols have been omitted. If we consider

$$A = \begin{pmatrix} 1 & 2 & 3 \\ 4 & 1 & \varepsilon \\ e & 5 & 3 \end{pmatrix}$$

in the algebraic structure $\mathbb{R}_{\max}$, then the characteristic equation becomes

$$x^3 \oplus 4x \oplus 9 = 3x^2 \oplus 6x \oplus 12 \ ,$$

which can be simplified to

$$x^3 = 3x^2 \oplus 6x \oplus 12 \ ,$$

since the omitted terms are dominated by the corresponding terms at the other side of the equality. A simple calculation shows that if one substitutes A in the latter equation, one obtains an identity indeed:

$$\begin{pmatrix} 12 & 11 & 9 \\ 10 & 12 & 10 \\ 12 & 11 & 12 \end{pmatrix} = \begin{pmatrix} 9 & 11 & 9 \\ 8 & 9 & 10 \\ 12 & 11 & 9 \end{pmatrix} \oplus \begin{pmatrix} 7 & 8 & 9 \\ 10 & 7 & \varepsilon \\ 6 & 11 & 9 \end{pmatrix} \oplus \begin{pmatrix} 12 & \varepsilon & \varepsilon \\ \varepsilon & 12 & \varepsilon \\ \varepsilon & \varepsilon & 12 \end{pmatrix} .$$

This section will be concluded with some remarks on minimal polynomial equations. The Cayley-Hamilton theorem shows that there exists at least one polynomial equation satisfied by a given $n \times n$ matrix A. This polynomial equation is of degree n. For example, if A is the 3×3 identity matrix, then

$$p_A^+(x) = x^3 \oplus x \ , \qquad p_A^-(x) = x^2 \oplus e \ ,$$

and A satisfies the equation $A^3 \oplus A = A^2 \oplus e$. There may exist equations of lower degree also satisfied by A. With the previous A, we also have $A = e$ and $x = e$ is a polynomial equation of degree 1 also satisfied by the identity matrix. A slightly less trivial example is obtained for

$$A = \begin{pmatrix} \varepsilon & 1 & \varepsilon \\ 1 & \varepsilon & \varepsilon \\ \varepsilon & \varepsilon & 1 \end{pmatrix} .$$

The characteristic equation is

$$x^3 \oplus 3 = 1x^2 \oplus 2x \ .$$

It is easily seen that A satisfies both $x^3 = 2x$ and $3 = 1x^2$. These equations have been obtained by a 'partitioning' of the characteristic equation; 'adding' these partitioned equations, one obtains the characteristic equation again. In this case, $3 = 1x^2$ is of degree 2. We may call a polynomial equation of least degree satisfied by a matrix, with the additional requirement that the coefficient of the highest power be equal to e, a *minimal polynomial equation* of this matrix. This is the counterpart of the notion of the minimal polynomial in conventional algebra and it is known that this minimal polynomial is a divisor of the characteristic polynomial [61]. In the present situation, it is not clear whether the minimal polynomial equation of a matrix is unique and how to extend the idea of division of polynomials to polynomial equations. In Chapter 3, a more detailed discussion on polynomials is given.

2.4 PETRI NETS

2.4.1 Definition

Petri nets are directed bipartite graphs. They are named after C.A. Petri, see [96]. The set of nodes $\mathcal{V}$ is partitioned into two disjoint subsets $\mathcal{P}$ and $\mathcal{Q}$. The elements of $\mathcal{P}$ are called places and those of $\mathcal{Q}$ are called transitions. Places will be denoted $p_i, i = 1, \ldots, |\mathcal{P}|$, and transitions, $q_j, j = 1, \ldots, |\mathcal{Q}|$. The directed arcs go from a place to a transition or vice versa. Since a Petri net is bipartite, there are no arcs from place to place or from transition to transition. In the graphical representation of Petri nets, places are drawn as circles and transitions as bars (the orientation of these bars can be anything). An example of a Petri net is given in Figure 2.12.

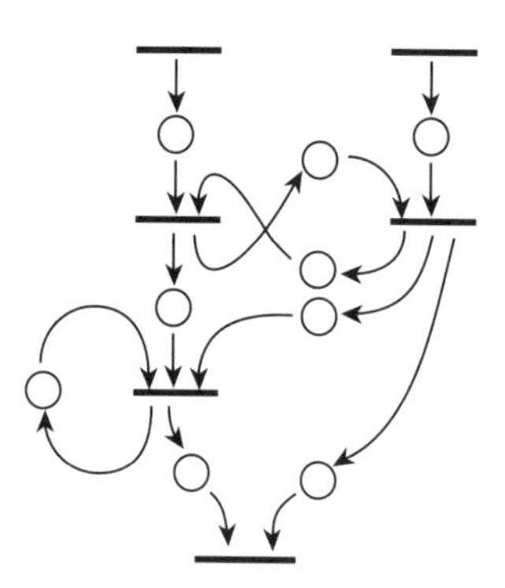

Figure 2.12: A Petri net with sources and sinks

In order to complete the formal definition of a Petri net, an initial marking must be introduced. The initial marking assigns a nonnegative integer μ_i to each place p_i. It is said that p_i is marked with μ_i initial tokens. Pictorially, μ_i dots (the tokens) are placed in the circle representing place p_i. The components μ_i form the vector μ, called the initial marking of the Petri net.

Definition 2.23 *A Petri net is a pair $(\mathcal{G}, \mu)$, where $\mathcal{G} = (\mathcal{E}, \mathcal{V})$ is a bipartite graph* *with a finite number of nodes (the set $\mathcal{V}$) which are partitioned into the disjoint sets $\mathcal{P}$ and $\mathcal{Q}$; $\mathcal{E}$ consists of pairs of the form (p_i, q_j) and (q_j, p_i), with $p_i \in \mathcal{P}$ and $q_j \in \mathcal{Q}$; the initial marking μ is a $|\mathcal{P}|$-vector of nonnegative integers.*

Notation 2.24 If $p_i \in \pi(q_j)$ (or equivalently $(p_i, q_j) \in \mathcal{E}$), then p_i is an upstream place for q_j. Downstream places are defined likewise. The following additional notation will also be used when we have to play with indices: if $p_i \in \pi(q_j)$, we write $i \in \pi^q(j), i = 1, \ldots, |\mathcal{P}|, j = 1, \ldots, |\mathcal{Q}|$; similarly, if $q_j \in \pi(p_i)$, we write $j \in \pi^p(i)$, with an analogous meaning for σ^p or σ^q. ■

Roughly speaking, places represent conditions and transitions represent events. A transition (i.e. an event) has a certain number of input and output places representing the pre-conditions and the post-conditions of the event, respectively. The presence of a token in a place is interpreted as the condition associated with that place being fulfilled. In another interpretation, μ_i tokens are put into a place to indicate that μ_i data items or resources are available. If a token represents data, then a typical example of a transition is a computation step for which these data are needed as an input. In Figure 2.13, the Petri net of the production example in §1.2.3 is given.

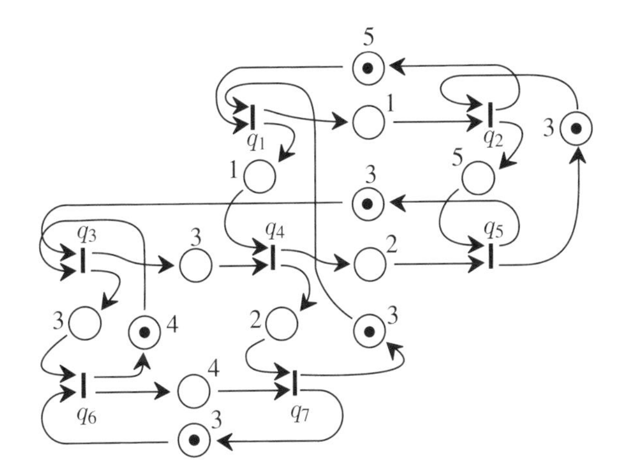

Figure 2.13: Petri net of the manufacturing system of §1.2.3

The tokens in this figure are located in such a way that machine M_1 starts working on product P_2 and M_2 on P_1. Note that M_1 cannot work on P_3 in Figure 2.13.

Within the classical Petri-net setting, the marking of the Petri net is identified with the state. Changes occur according to the following rules:

- A transition is said to be *enabled* if each upstream place contains at least one token.
- A *firing* of an enabled transition removes one token from each of its upstream places and adds one token to each of its downstream places.

Remark 2.25 The enabling rule given above is not the most general one. Sometimes integer valued 'weights' are attached to arcs. A transition is enabled if the upstream place contains at least the number of tokens given by the weight of the connecting arc. Similarly, after the firing of a transition, a downstream place receives the number of tokens given by the weight of the connecting arc. Instead of talking about such 'weights', one sometimes talks about multi-arcs; the weight equals the number of arcs between a transition and a place or between a place and a transition. In terms of 'modeling power', see [96] and [108] for a definition, this generalization is not more powerful than the rules which will be used here. The word 'weight' of an arc will be used in a different sense later on. ■

For q_j to be enabled, we need that

$$\mu_i \geq 1 \ , \quad \forall \ p_i \in \pi(q_j) \ .$$

If the enabled transition q_j fires, then a new marking $\widetilde{\mu}$ is obtained with

$$\widetilde{\mu}_i = \begin{cases} \mu_i - 1 & \text{if } p_i \in \pi(q_j) \ , \\ \mu_i + 1 & \text{if } \ p_i \in \sigma(q_j) \ , \\ \mu_i & \text{otherwise.} \end{cases}$$

In case both $p_i \in \pi(q_j)$ and $p_i \in \sigma(q_j)$ for the same place p_i, then $\widetilde{\mu}_i = \mu_i$.

In Figure 2.13, once M_1 has completed its work on P_2 and M_2 its work on P_1, then Figure 2.14 is obtained. The next transitions that are now enabled are described

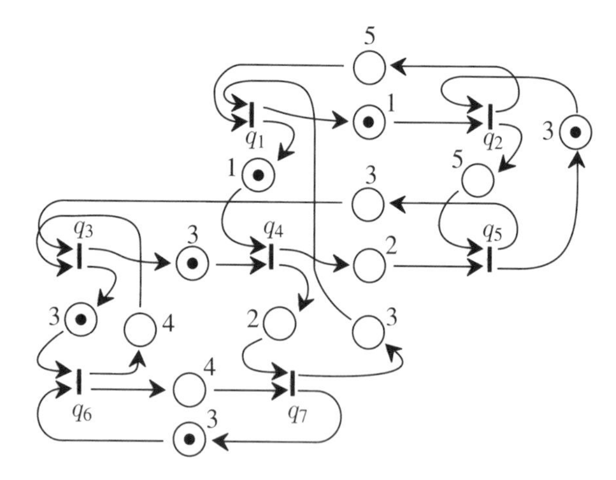

Figure 2.14: The tokens after the firing of q_1 and q_3

by the combinations (M_1, P_3), (M_2, P_2) and (M_3, P_1). Note that in general the total amount of tokens in the net is not left invariant by the firing of a transition, although this does not happen in Figure 2.13. If we have a 'join'-type transition (Figure 2.15), which is called an *and-convergence*, or a 'fork'-type transition (Figure 2.16), called an *and-divergence*, then clearly the number of tokens changes after a firing has taken

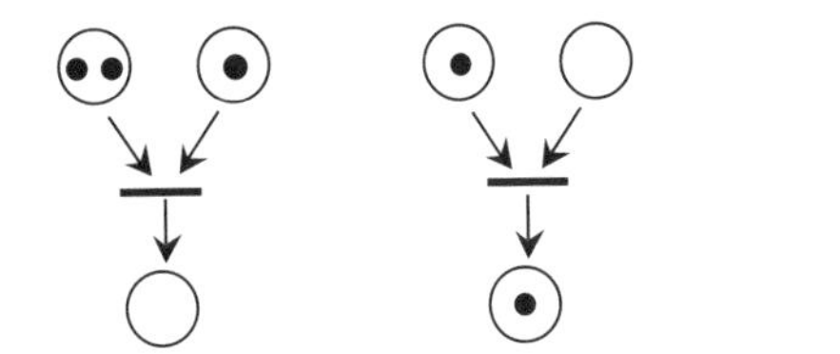

Figure 2.15: And-convergence before and after firing

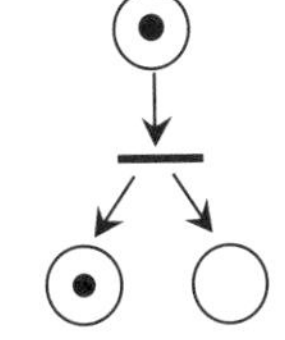

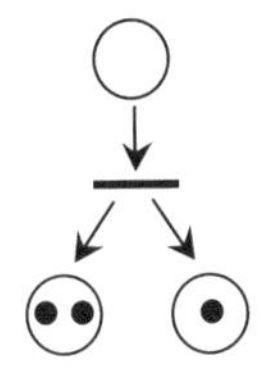

Figure 2.16: And-divergence before and after firing

place. In the same vein, an *or-convergence* refers to two or more arcs entering one place, and an *or-divergence* refers to two or more arcs originating from one place.

A transition without predecessor(s) is called a source transition or simply a source; it is enabled by the outside world. Similarly, a transition which does not have successor(s), is called a sink (or sink transition). Sink transitions deliver tokens to the outside world. In Figure 2.12 there are two transitions which are sources and there is one transition which is a sink. If there are no sources in the network, as in Figure 2.17, then we talk about an *autonomous* network. It is assumed that only transitions can be sources or sinks. This is no loss of generality, since one can always add a transition upstream or downstream of a place if necessary. A source transition is an input of the network, a sink transition is an output of the network.

The structure of a place p_i having two or more output transitions, as shown in Figure 2.18, is referred to as a *conflict*, since the transitions are competing for the

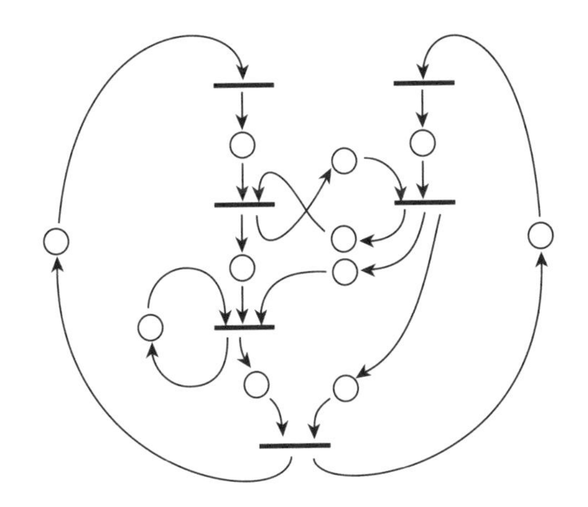

Figure 2.17: An autonomous Petri net

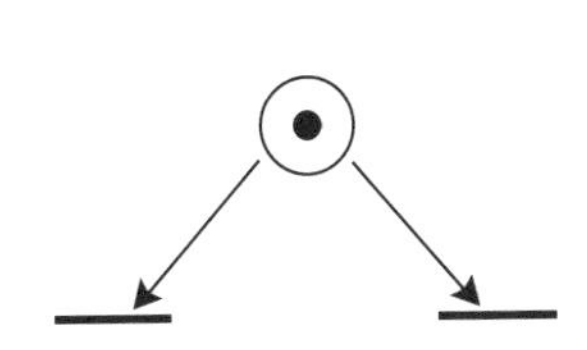

Figure 2.18: Part of a Petri net with a conflict

token in the place. The transitions concerned will be said to be *rivals*. There are no general rules as to which transition should fire first. One says that a Petri net with such an or-divergence exhibits nondeterminism. Depending on the application, one can also talk about a choice (between the transitions) or a decision.

The firing of an enabled transition will change the distribution of tokens. A sequence of firings will result in a sequence of markings. A marking $\overline{\mu}$ is said to be *reachable* from a marking μ if there exists a sequence of enabled firings that transforms μ into $\overline{\mu}$.

Definition 2.26 (Reachability tree) *The reachability tree of a Petri net $(\mathcal{G}, \mu)$ is a tree with nodes in $\mathbb{N}^{|\mathcal{P}|}$ which is obtained as follows: the initial marking μ is a node of this tree; for each q enabled in μ, the marking $\overline{\mu}$ obtained by firing q is a new node of the reachability tree; arcs connect nodes which are reachable from one another in one step; this process is applied recursively from each such $\overline{\mu}$.*

Definition 2.27 (Reachability graph) *The reachability graph is obtained from the reachability tree by merging all nodes corresponding to the same marking into a single node.*

Take as an example the Petri net depicted in Figure 2.19. The initial marking

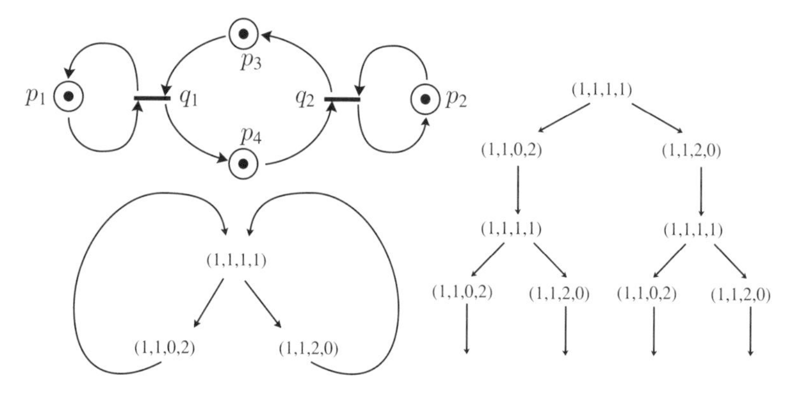

Figure 2.19: A Petri net with corresponding reachability graph and reachability tree

is $(1,1,1,1)$. Both transitions are enabled. If q_1 fires first, the next marking is $(1,1,0,2)$. If q_2 fires instead, the marking becomes $(1,1,2,0)$. From $(1,1,0,2)$, only the initial marking can be reached immediately by firing q_2; starting from $(1,1,2,0)$, only q_1 can fire, which also leads to the initial marking. Thus it has been shown that there are three different markings in the reachability graph of $(1,1,1,1)$.

Definition 2.28 *For a Petri net with n transitions and m places, the incidence matrix $G = (G_{ij})$ is an $n \times m$ matrix of integers -1, 0 and $+1$. The entry G_{ij} is defined by*

$$G_{ij} = G_{ij}^{\text{out}} - G_{ij}^{\text{in}} \ ,$$

where $G_{ij}^{\text{out}} = 1$ (0) if there is an (no) arc from q_i to p_j and $G_{ij}^{\text{in}} = 1$ (0) if there is an (no) arc from p_j to q_i. Matrices G^{out} and G^{in} are defined as $G^{\text{out}} = (G_{ij}^{\text{out}})$ and $G^{\text{in}} = (G_{ij}^{\text{in}})$, respectively.

Note that G does not uniquely define a Petri net since, if $G_{ii} = 0$, a path including exactly one place around the transition q_i is also possible. A circuit consisting of one transition and one place is called a *loop* in the context of Petri nets. If each place in the Petri net had only one upstream and one downstream transition, then the incidence matrix G would reduce to the well-known incidence matrix F introduced in Definition 2.10 by identifying each place p with the unique arc from $\pi(p)$ to $\sigma(p)$.

Transition q_j is enabled if and only if a marking μ is given such that

$$\mu \geq (G^{\text{in}})' e_j \ ,$$

where $e_j = (0, \ldots, 0, 1, 0, \ldots, 0)'$, with the 1 being the j-th component. If this transition fires, then the next marking $\widetilde{\mu}$ is given by

$$\widetilde{\mu} = \mu + G' e_j \ .$$

A destination marking $\overline{\mu}$ is reachable from μ if a firing sequence $e_{j_1}, \ldots, e_{j_d}$ exists such that

$$\overline{\mu} = \mu + G' \sum_{l=1}^{d} e_{j_l} \ .$$

Hence a necessary condition for $\overline{\mu}$ to be reachable from μ is that an n-vector x of nonnegative integers exists such that

$$G' x = \overline{\mu} - \mu \ . \tag{2.9}$$

The existence of such a vector x is not a sufficient condition; for a counterexample see for instance [108]. The vector x does not reflect the order in which the firings take place. In the next subsection a necessary and sufficient condition for reachability will be given for a subclass of Petri nets. An integer solution x to (2.9), with its components not necessarily nonnegative, exists if and only if $\overline{\mu}' y = \mu' y$, for any y that satisfies $Gy = 0$. The necessity of this statement easily follows if one takes the inner products of the left- and right-hand sides of (2.9) with respect to y. The sufficiency is easily shown if one assumes that x does not exist, i.e. $\text{rank}[G] < \text{rank}[G, \overline{\mu} - \mu]$—the notation $[G, \overline{\mu} - \mu]$ refers to the matrix consisting of G and the extra column $\overline{\mu} - \mu$. Then a vector y exists with $y'G' = 0$ and $y'(\overline{\mu} - \mu) \neq 0$, which is a contradiction.

2.4.2 Subclasses and Properties of Petri Nets

In this subsection we introduce some subclasses of Petri nets and analyze their basic properties. Not all of these properties are used later on; this subsection is also meant to give some background information on distinct features of Petri nets. The emphasis will be on event graphs.

Definition 2.29 (Event graph) *A Petri net is called an event graph if each place has exactly one upstream and one downstream transition.*

Definition 2.30 (State machine) *A Petri net is called a state machine if each transition has exactly one upstream and one downstream place.*

Event graphs have neither or-divergences nor or-convergences. In event graphs each place together with its incoming and outgoing arcs can be interpreted as an arc itself, connecting the upstream and downstream transition, directly.

In the literature, event graphs are sometimes also referred to as *marked graphs* or as *decision free Petri nets*. Figure 2.20 shows both a state machine which is not

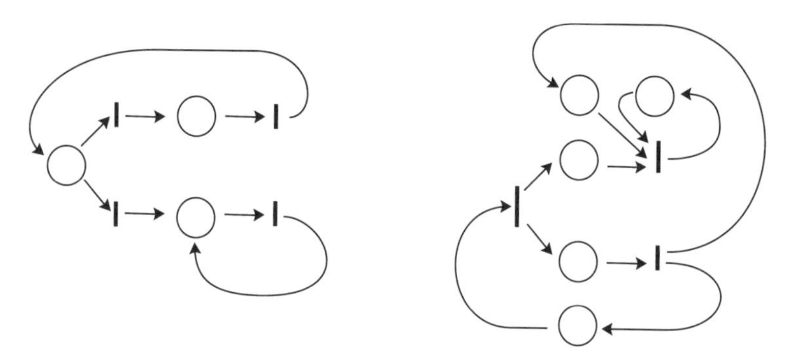

Figure 2.20: A state machine and an event graph

an event graph and an event graph which is not a state machine. An event graph does not allow and cannot model conflicts; a token in a place can be consumed by only one predetermined transition. In an event graph several places can precede a given transition. It is said that event graphs can model synchronization. State machines do not admit synchronization; however, they do allow competition. The number of tokens in an autonomous state machine never changes (Petri nets in which the number of tokens remains constant are called strictly conservative; a discussion follows later).

It can be shown [108] that state machines are equivalent to the finite state machines or automata in theoretical computer science. Each automaton can be rephrased as a state machine Petri net. This shows that Petri nets have more modeling power than automata.

Basic definitions and properties of Petri nets are now given.

Definition 2.31 (Bounded and safe nets) *A Petri net, with initial marking μ, is said to be k-bounded if the number of tokens in each place does not exceed a finite number k for any marking reachable from μ. Instead of 1-bounded ($k = 1$) Petri nets, one speaks of safe Petri nets.*

Concerning (practical) applications, it is important to know whether one deals with a bounded or safe Petri net, since one is then sure that there will be no overflows in the buffers or registers, no matter what the firing sequence will be.

Definition 2.32 (Live net) *A Petri net is said to be live for the initial marking μ if for each marking ν reachable from μ and for each transition q, there exists a marking o which is reachable from ν and such that q is enabled on o. A Petri net which is not live is called deadlocked.*

A Petri net is deadlocked if its reachability tree has a marking where a transition, or a set of transitions, can never fire whatever the firing sequences of the other transitions. For a live Petri net, whatever the finite initial sequence of firings, from that point onwards, any arbitrary transition can be fired an infinite number of times.

An example of a live net is a state machine the underlying graph of which is strongly connected and the initial marking of which has at least one token.

Definition 2.33 (Consistent net) *A Petri net is called consistent (weakly consistent) if there exists a firing sequence, characterized by the x-vector with positive (non-negative) integers as components, such that $G'x = 0$, where G is the incidence matrix.*

In a consistent Petri net we can choose a finite sequence of firings such that repeating this sequence results in a periodic behavior.

Definition 2.34 (Synchronous net) *A consistent net is called synchronous if the only solutions x of $G'x = 0$ are of the form $x = k(1, 1, \ldots, 1)'$.*

Definition 2.35 (Strictly conservative net) *A Petri net with initial marking μ is called strictly conservative if, for all reachable markings $\overline{\mu}$, we have $\sum_{p_i \in \mathcal{P}} \overline{\mu}_i = \sum_{p_i \in \mathcal{P}} \mu_i$.*

Definition 2.36 (Conservative net) *A Petri net with initial marking μ is called conservative if positive integers c_i exist such that, for all reachable markings $\overline{\mu}$, we have $\sum_{p_i \in \mathcal{P}} c_i \overline{\mu}_i = \sum_{p_i \in \mathcal{P}} c_i \mu_i$.*

Theorem 2.37 *The number of tokens in any circuit of an event graph is constant.*

Proof If a transition is part of a circuit, then exactly one of its incoming arcs and one of its outgoing arcs belong to the circuit. The firing of the transition removes one token from the upstream place connected to the incoming arc (it may remove tokens from other places as well, but they do not belong to the circuit) and it adds one token to the downstream place connected to the outgoing arc (it may add tokens to other downstream places as well, but they do not belong to the circuit). ■

An event graph is not necessarily strictly conservative. Consider an and-convergence, where two circuits merge. The firing of this transition removes one token from each of the upstream places and adds one token to the only downstream place. At an and-divergence, where two circuits split up, the firing of the transition removes one token from the only upstream place and adds one token to each of the downstream places.

Theorem 2.38 *An autonomous event graph is live if and only if every circuit contains at least one token with respect to the initial marking.*

Proof

Only if part: If there are no tokens in a circuit of the initial marking of an event graph, then this circuit will remain free of tokens and thus all transitions along this circuit never fire.

If part: If a transition is never enabled by any firing sequence, then by backtracking token-free places, one can find a token-free circuit. Indeed, if in an event graph a transition never fires, there is at least one upstream transition that never fires also (this statement cannot be made for general Petri nets). This backtracking is only possible if each place has a transition as predecessor and each transition has at least one place as predecessor. This holds for autonomous event graphs. Thus the theorem has been proved. ■

Theorem 2.39 *For a connected event graph, with initial marking μ, a firing sequence can lead back to μ if and only if it fires every transition an equal number of times.*

Proof In a connected event graph all transitions are either and-divergences, and-convergences or they are simple, i.e. they have one upstream place as well as one downstream place. These categories may overlap one another. If an and-divergence is enabled and it fires, then the number of tokens in all downstream places is increased by one. In order to dispose of these extra tokens, the downstream transitions in each of these places must fire also (in fact, they must fire as many times as the originally enabled transition fired in order to keep the number of tokens of the places in between constant). If an and-convergence wants to fire, then the upstream transitions of its upstream places must fire first in order that the number of tokens of the places in between do not change. Lastly, if a transition is simple and it can fire, both the unique downstream transition and upstream transition must fire the same number of times in order that the number of tokens in the places in between do not change. The reasoning above only fails for loops. Since loops are connected to the event graph also and since a firing of a transition in a loop does not change the number of tokens in the place in the loop, these loops can be disregarded in the above reasoning. ■

This theorem states that the equation $G'x = 0$ has only one positive independent solution $x = (k, \dots, k)'$. An immediate consequence is that every connected event graph is synchronous.

Theorem 2.40 *Consider an autonomous live event graph. It is safe if and only if the total number of tokens in each circuit equals one.*

Proof The if part of the proof is straightforward. Now consider the only if part. Assume that the graph is safe and that the total number of tokens in a circuit ζ_k, indicated by $\mu(\zeta_k)$, is not necessarily one. Consider all circuits $(\zeta_1, \zeta_2, \dots, \zeta_m)$ passing through a place p_i and its upstream transition t_j. Bring as many tokens as possible to each of the upstream places of t_j and subsequently fire t_j as many times as possible. It can be seen that the maximum number of tokens that can be brought in p_i is bounded from above by $\min\{\mu(\zeta_1), \mu(\zeta_2), \dots, \mu(\zeta_m)\}$. In particular, if this minimum equals one, then this maximum number of tokens is less than or equal to one. Since the event graph is live, t_j can be enabled, and therefore this maximum equals one. ■

The following theorem is stated in [95].

Theorem 2.41 *In a live event graph $\overline{\mu}$ is reachable from μ if and only if $\overline{\mu}'y = \mu'y$, for any y that satisfies $Gy = 0$.*

This last theorem sharpens the result mentioned at the end of §2.4.1, where the condition $\overline{\mu}'y = \mu'y$ was only a necessary condition.

2.5 TIMED EVENT GRAPHS

The original theory of Petri nets deals with the ordering of events, and questions pertaining to when events take place are not addressed. However, for questions related to performance evaluation (how fast can a network produce?) it is necessary to introduce time. This can be done in two basic ways by associating durations with either transition firings or with the sojourn of tokens in places.

Durations associated with firing times can be used to represent production times in a manufacturing environment, where transitions represent machines, the length of a code in a computer science setting, etc. We adopt the following definition.

Definition 2.42 (Firing time) *The firing time of a transition is the time that elapses between the starting and the completion of the firing of the transition.*

We also adopt the additional convention that the tokens to be consumed by a transition remain in the preceding places during the firing time; they are called *reserved tokens*.

Durations associated with places can be used to represent transportation or communication time. When a transition produces a token into a place, this token cannot immediately contribute to the enabling of the downstream transitions; it must first spend some *holding time* in that place, which actually represents the time it takes to transport this token from the initial transition to the place.

Definition 2.43 (Holding time) *The holding time of a place is the time a token must spend in the place before contributing to the enabling of the downstream transitions.*

Observe that there is a basic asymmetry between both types of durations: firing times represent the actual time it takes to fire a transition, while holding times can be viewed as the minimal time tokens have to spend in places (indeed it is not because a specific token has completed its holding time in a place that it can immediately be consumed by some transition; it may be that no transition capable of consuming this token is enabled at this time). In practical situations, both types of durations may be present. However, as we shall see later on, if one deals with event graphs, one can disregard durations associated with transitions without loss of generality.

Roughly speaking, a Petri net is said to be timed if such durations are given as new data associated with the network. A basic dichotomy arises depending on whether these durations are constant or variable. Throughout the book, only the dependence on the index of the firing (the firing of transition q of index k is the k-th to be initiated), or on the index of the token (the token of p of index k is the k-th token of p to contribute enabling $\sigma(p)$) will be considered. The other possible dependences, like for instance the dependence on time, or on some possibly changing environment, will not be addressed.

The timing of a Petri net will be said to be variable if the firing times of a transition depend on the index of the firing or if the holding times of tokens in a place depend on the index of the token. The timing is constant otherwise.

In the constant case, the first, the k-th and $(k+1)$-st firings of transition q take the same amount of time; this common firing time may however depend on q (see the examples below).

Remark 2.44 With our definitions, nothing prevents a transition from having several ongoing firings (indeed, a transition does not have to wait for the completion of an ongoing firing in order to initiate a new firing). If one wants to prevent such a phenomenon, one may add an extra place associated with this transition. This extra place should have the transition under consideration as unique predecessor and successor, and one token in the initial marking, as indicated in Figure 2.21. The addition of this loop models a mechanism that will be called a *recycling* of

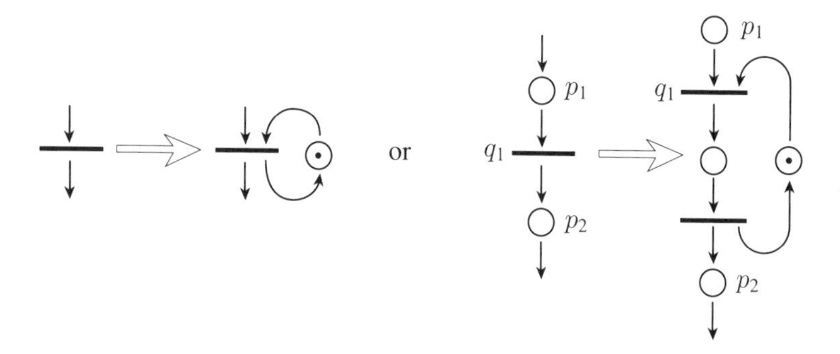

Figure 2.21: Recycling of a transition

the transition. Owing to this mechanism, the firings of the transition are properly serialized in the sense that its $(k+1)$-st firing can only start after the completion of the k-th firing. ■

In the rest of this chapter, unless otherwise specified, $\oplus$ will be maximization and $\otimes$ addition, so that $\varepsilon = -\infty$ and $e = 0$. We will also use the symbol $\oslash$ to denote subtraction.

2.5.1 Simple Examples

The global aim of the present section is to derive evolution equations for the variables $x_i(k), i = 1, \ldots, |\mathcal{Q}|, k \geq 0$, not counting the sources and sinks, and where $x_i(k)$ is defined as the epoch at which transition q_i starts firing for the k-th time. Both constant and variable timings will be considered. For general Petri nets, such equations are difficult to derive. This problem will only be addressed in Chapter 9. For the moment, we shall confine ourselves to event graphs. We start with some simple examples with constant timing before addressing the general case. The issue of the initial condition will not be addressed in these simple examples either (see §2.5.2.1). The general rule (to which we will return in the next subsection) is that transitions start firing as soon as they are enabled.

Example 2.45 (An autonomous timed event graph) The first example deals with the manufacturing system of §1.2.3, as depicted in Figure 1.5. The related Petri net, given in Figure 2.13, is the starting point for discussing the way the evolution equations are derived. The timing under consideration is limited to constant holding times on places, which are indicated in the figure. The firing times are all assumed to be 0. We have the following evolution equations:

$$\begin{aligned}
x_1(k+1) &= 5x_2(k) \oplus 3x_7(k) \ , \\
x_2(k+1) &= 1x_1(k+1) \oplus 3x_5(k) \ , \\
x_3(k+1) &= 3x_5(k) \oplus 4x_6(k) \ , \\
x_4(k+1) &= 1x_1(k+1) \oplus 3x_3(k+1) \ , \\
x_5(k+1) &= 5x_2(k+1) \oplus 2x_4(k+1) \ , \\
x_6(k+1) &= 3x_3(k+1) \oplus 3x_7(k) \ , \\
x_7(k+1) &= 2x_4(k+1) \oplus 4x_6(k+1) \ .
\end{aligned}$$

In order to get this set of equations, one must first observe that owing to our assumption that holding times are constant, overtaking of tokens in places is not possible: the k-th token to enter a place will be the k-th token to leave that place (at least if the initial marking in that place is 0). If one uses this observation, the equation for x_6 (for instance) is obtained as follows: q_6 is enabled for the $(k+1)$-st time at the latest of the two epochs when the $(k+1)$-st token to enter the place between q_3 and q_6 completes its holding time there, and when the k-th token to enter the place between q_7 and q_6 completes its holding time. The difference between the arguments k and $k+1$ comes from the fact that the place between q_7 and q_6 has one token in the initial marking. If one now uses the definition of holding times, it is easily seen that the first of these two epochs is $x_3(k+1)+3$, while the second one is $x_7(k)+3$, which concludes the proof. There are clearly some further problems to be addressed regarding the initial condition, but let us forget them for the moment. In matrix form this equation can be written as

$$x(k+1) = A_0 x(k+1) \oplus A_1 x(k) \ , \tag{2.10}$$

where

$$A_0 = \begin{pmatrix}
\varepsilon & \varepsilon & \varepsilon & \varepsilon & \varepsilon & \varepsilon & \varepsilon \\
1 & \varepsilon & \varepsilon & \varepsilon & \varepsilon & \varepsilon & \varepsilon \\
\varepsilon & \varepsilon & \varepsilon & \varepsilon & \varepsilon & \varepsilon & \varepsilon \\
1 & \varepsilon & 3 & \varepsilon & \varepsilon & \varepsilon & \varepsilon \\
\varepsilon & 5 & \varepsilon & 2 & \varepsilon & \varepsilon & \varepsilon \\
\varepsilon & \varepsilon & 3 & \varepsilon & \varepsilon & \varepsilon & \varepsilon \\
\varepsilon & \varepsilon & \varepsilon & 2 & \varepsilon & 4 & \varepsilon
\end{pmatrix} , \quad
A_1 = \begin{pmatrix}
\varepsilon & 5 & \varepsilon & \varepsilon & \varepsilon & \varepsilon & 3 \\
\varepsilon & \varepsilon & \varepsilon & \varepsilon & 3 & \varepsilon & \varepsilon \\
\varepsilon & \varepsilon & \varepsilon & \varepsilon & 3 & 4 & \varepsilon \\
\varepsilon & \varepsilon & \varepsilon & \varepsilon & \varepsilon & \varepsilon & \varepsilon \\
\varepsilon & \varepsilon & \varepsilon & \varepsilon & \varepsilon & \varepsilon & \varepsilon \\
\varepsilon & \varepsilon & \varepsilon & \varepsilon & \varepsilon & \varepsilon & 3 \\
\varepsilon & \varepsilon & \varepsilon & \varepsilon & \varepsilon & \varepsilon & \varepsilon
\end{pmatrix} .$$

The equation is written in a more convenient way as

$$x(k+1) = Ax(k) \ , \tag{2.11}$$

where $A = A_0^* A_1$ (see (1.22)), or, written out,

$$A = \begin{pmatrix} \varepsilon & 5 & \varepsilon & \varepsilon & \varepsilon & \varepsilon & 3 \\ \varepsilon & 6 & \varepsilon & \varepsilon & 3 & \varepsilon & 4 \\ \varepsilon & \varepsilon & \varepsilon & \varepsilon & 3 & 4 & \varepsilon \\ \varepsilon & 6 & \varepsilon & \varepsilon & 6 & 7 & 4 \\ \varepsilon & 11 & \varepsilon & \varepsilon & 8 & 9 & 9 \\ \varepsilon & \varepsilon & \varepsilon & \varepsilon & 6 & 7 & 3 \\ \varepsilon & 8 & \varepsilon & \varepsilon & 10 & 11 & 7 \end{pmatrix} .$$

■

Remark 2.46 Both equations (2.11) and (1.29) describe the evolution of the firing times, the first equation with respect to the state x, the second one with respect to the output y. Using the notation of §1.2.3, one can check that

$$y(k+1) = Cx(k+1) = CAx(k)$$

and that

$$y(k+1) = My(k) = MCx(k) \ ,$$

where A is defined above, and where CA equals MC. ■

Conversely, it is easy to derive a Petri net from (2.11); such a net has 7 transitions (the dimension of the state vector) and 22 places (the number of entries in A which are not equal to ε). Each of these places has a token in the initial marking. The holding time associated with the place connecting transition j to transition i is given by the appropriate A_{ij} entry. Thus at least two different Petri nets exist which both yield the same set of evolution equations. In this sense these Petri nets are equivalent.

Example 2.47 **(A nonautonomous timed event graph)** The starting point for the next example is Figure 2.22, which coincides with Figure 2.12 with an initial marking added. The timing is again limited to places. The evolution equations are given by

$$\begin{aligned} x(k+1) &= A_0 x(k+1) \oplus A_1 x(k) \oplus A_2 x(k-1) \oplus B_0 u(k+1) \oplus B_1 u(k) \ , \\ y(k) &= C_0 x(k) \oplus C_1 x(k-1) \ , \end{aligned}$$

where

$$A_0 = \begin{pmatrix} \varepsilon & \varepsilon & \varepsilon \\ 3 & \varepsilon & \varepsilon \\ 3 & 4 & \varepsilon \end{pmatrix}, \quad A_1 = \begin{pmatrix} \varepsilon & 4 & \varepsilon \\ \varepsilon & \varepsilon & \varepsilon \\ \varepsilon & \varepsilon & \varepsilon \end{pmatrix}, \quad A_2 = \begin{pmatrix} \varepsilon & \varepsilon & \varepsilon \\ \varepsilon & \varepsilon & \varepsilon \\ \varepsilon & \varepsilon & 2 \end{pmatrix},$$

$$B_0 = \begin{pmatrix} 1 & \varepsilon \\ \varepsilon & \varepsilon \\ \varepsilon & \varepsilon \end{pmatrix}, \quad B_1 = \begin{pmatrix} \varepsilon & \varepsilon \\ \varepsilon & 5 \\ \varepsilon & \varepsilon \end{pmatrix},$$

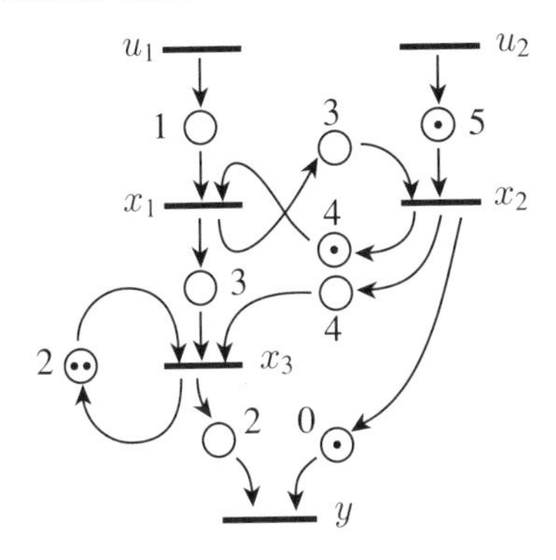

Figure 2.22: Petri net of Example 2.47

$$C_0 = \begin{pmatrix} \varepsilon & \varepsilon & 2 \end{pmatrix}, \quad C_1 = \begin{pmatrix} \varepsilon & e & \varepsilon \end{pmatrix} .$$

If one uses A_0^*, this system can be written as

$$\begin{aligned} x(k+1) &= \begin{pmatrix} \varepsilon & 4 & \varepsilon \\ \varepsilon & 7 & \varepsilon \\ \varepsilon & 11 & \varepsilon \end{pmatrix} x(k) \oplus \begin{pmatrix} \varepsilon & \varepsilon & \varepsilon \\ \varepsilon & \varepsilon & \varepsilon \\ \varepsilon & \varepsilon & 2 \end{pmatrix} x(k-1) \\ &\oplus \begin{pmatrix} 1 & \varepsilon \\ 4 & \varepsilon \\ 8 & \varepsilon \end{pmatrix} u(k+1) \oplus \begin{pmatrix} \varepsilon & \varepsilon \\ \varepsilon & 5 \\ \varepsilon & 9 \end{pmatrix} u(k) , \\ y(k) &= \begin{pmatrix} \varepsilon & \varepsilon & 2 \end{pmatrix} x(k) \oplus \begin{pmatrix} \varepsilon & e & \varepsilon \end{pmatrix} x(k-1) . \end{aligned}$$

This equation can be put into standard form by augmenting the state space: if one defines

$$\widetilde{x}(k) = (x_1(k), x_1(k-1), x_2(k), x_2(k-1), x_3(k), x_3(k-1), u_1(k), u_2(k))' ,$$

the system can be written as

$$\begin{aligned} \widetilde{x}(k+1) &= \widetilde{A}\widetilde{x}(k) \oplus \widetilde{B}u(k+1) , \\ y(k) &= \widetilde{C}\widetilde{x}(k) , \end{aligned} \tag{2.12}$$

where

$$\widetilde{A} = \begin{pmatrix} \varepsilon & \varepsilon & 4 & \varepsilon & \varepsilon & \varepsilon & \varepsilon & \varepsilon \\ e & \varepsilon & \varepsilon & \varepsilon & \varepsilon & \varepsilon & \varepsilon & \varepsilon \\ \varepsilon & \varepsilon & 7 & \varepsilon & \varepsilon & \varepsilon & \varepsilon & 5 \\ \varepsilon & \varepsilon & e & \varepsilon & \varepsilon & \varepsilon & \varepsilon & \varepsilon \\ \varepsilon & \varepsilon & 11 & \varepsilon & \varepsilon & 2 & \varepsilon & 9 \\ \varepsilon & \varepsilon & \varepsilon & \varepsilon & e & \varepsilon & \varepsilon & \varepsilon \\ \varepsilon & \varepsilon & \varepsilon & \varepsilon & \varepsilon & \varepsilon & \varepsilon & \varepsilon \\ \varepsilon & \varepsilon & \varepsilon & \varepsilon & \varepsilon & \varepsilon & \varepsilon & \varepsilon \end{pmatrix} , \quad B = \begin{pmatrix} 1 & \varepsilon \\ \varepsilon & \varepsilon \\ 4 & \varepsilon \\ \varepsilon & \varepsilon \\ 8 & \varepsilon \\ \varepsilon & \varepsilon \\ e & \varepsilon \\ \varepsilon & e \end{pmatrix} ,$$

and

$$\widetilde{C} = \begin{pmatrix} \varepsilon & \varepsilon & \varepsilon & e & 2 & \varepsilon & \varepsilon & \varepsilon \end{pmatrix} .$$

Further simplifications are possible in this equation. Since x_1 is not observable (see [72]) and since it does not influence the dynamics of x_2 or x_3 either, it can be discarded from the state, as can u_1. Thus a five-dimensional state vector suffices. Equation (2.12) is still not in the standard form since the argument of u is $k+1$ instead of k. If one insists on the 'precise' standard form, it can be shown that (2.12) and

$$\begin{aligned} \widehat{x}(k+1) &= \widetilde{A}\widehat{x}(k) \oplus \widetilde{A}\widetilde{B}u(k) , \\ y(k) &= \widetilde{C}\widehat{x}(k) \oplus \widetilde{C}\widetilde{B}u(k) , \end{aligned} \tag{2.13}$$

are identical in the sense that their γ-transforms (see Chapter 1) are identical. This is left as an exercise to the reader. The latter equation does have the standard form, though there is a direct throughput term (the input $u(k)$ has a direct influence on the output $y(k)$). ■

Example 2.48 (Discrete analogue of the system of §1.2.7) The purpose of this example is to play again with the 'counter' description already alluded to in Chapter 1, and to show that discrete event systems may obey similar equations as some continuous systems, up to the problem of 'quantization'. Figure 2.23 (left-hand side) represents a simple event graph: the three transitions are labelled u, x and y, holding

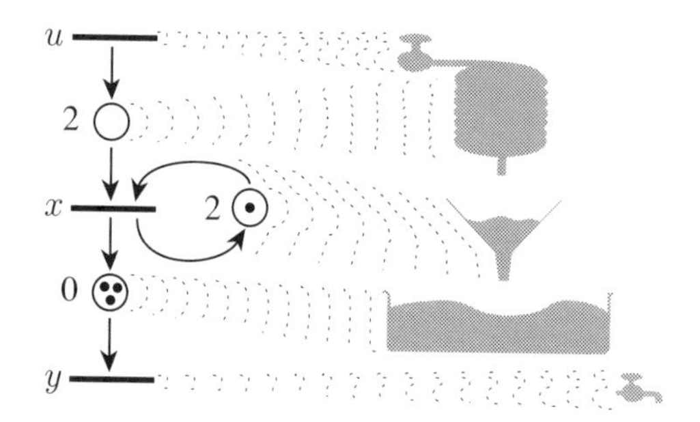

Figure 2.23: An event graph and its continuous analogue

times of places are indicated by numbers (firing times are all zero) and an initial marking is shown. With each transition, e.g. x, is associated a function of time t having the same name, e.g. $t \mapsto x(t)$, with the following meaning: $x(t)$ represents the number of firings of transition x up to time t, and it is assumed that $x(t) = e$ for $t < 0$. The following equations are obtained *in the min-plus algebra*:

$$x(t) = 1x(t-2) \oplus u(t-2) \; ; \qquad y(t) = 3x(t) ,$$

where the delays in time originate from the holding times and where the coefficients are connected with the initial number of tokens in the places. By successive substitutions, one obtains:

$$\begin{aligned} y(t) &= 3u(t-2) \oplus 4x(t-2) \\ &= 3u(t-2) \oplus 4u(t-4) \oplus 5x(t-4) \\ &\vdots \\ &= \bigoplus_{\substack{2 \le \tau \le t+2 \\ \tau \text{ even}}} h(\tau)u(t-\tau) \ , \end{aligned}$$

where h is the function defined by (1.38). Observe that the min-summation can be limited to $t+2$ because $u(-1) = u(-2) = \cdots = x(-1) = x(-2) = \cdots = e$ and the coefficient of $x(t)$ is larger than that of $u(t)$. Indeed, for the same reason, the min-summation can be extended to $\tau < +\infty$, and also to $-\infty < \tau$ because $h(\tau)$ remains equal to 3 for values of τ below 2, whereas $u(t-\tau)$ is nonincreasing with τ. Finally, one obtains that

$$y(t) = \bigoplus_{\substack{-\infty < \tau < +\infty \\ \tau \text{ even}}} h(\tau)u(t-\tau) \ ,$$

which compares with (1.39), except that now τ ranges in $2\mathbb{Z}$ instead of $\mathbb{R}$.

The right-hand side of Figure 2.23 suggests the correspondence of continuous elements (extracted from Figure 1.13) with their discrete counterparts.

Recalling the mixing operation explained in the last paragraph of §1.2.7, we see that the discrete analogous operation consists here in 'synchronizing' two event graphs similar to that of Figure 2.23 by a join at their output transition y. ■

2.5.2 The Basic Autonomous Equation

The event graphs of this section are assumed to be autonomous. The nonautonomous case will be considered in §2.5.5. We now turn to the derivation of a set of evolution equations for event graphs with variable timing, under the general assumptions that

- the transitions start firing as soon as they are enabled;
- the tokens of a place start enabling the transitions downstream as soon as they have completed their holding times.

The problems that arise with variable timing are slightly more complex than in the preceding examples. The main reason for this lies in the fact that tokens can then overtake one another when traversing places or transitions. As we will see in Chapter 9, this precludes a simple ordering of events to hold, and event graphs with overtaking lose the nice linearity property emphasized in the preceding examples. The discussion will hence be limited to the case of event graphs with First In First Out (FIFO) places and transitions, where the ordering of events preserves linearity.

2.5.2.1 Initial Condition

We start with a first discussion of the *initial condition* which is here understood as a set of initial delays attached to the tokens of the initial marking in a way that generalizes what is often done in queuing theory. This topic will be revisited in §5.4.4.1 and §5.4.4.2, and further considered from a system-theoretic point of view in §5.4.4.3.

Assume that one starts looking at the system evolution at time $t = 0$, and that the piecewise constant function $N_i(t)$ describing the evolution of the number of tokens present in p_i, $i = 1, \ldots, |\mathcal{P}|$, at time $t \in \mathbb{R}$, is right continuous. Let $N_i(0) = \mu_i$, where μ_i denotes the initial marking in place p_i.

The general idea behind the initial condition is as follows: the $N_i(0)$ $(= \mu_i)$ tokens visible at time $t = 0$ in p_i are assumed to have entered p_i before time 0; at time 0, each token is completing its holding time or it is being consumed by the transition (namely it is a reserved token), or it is ready to be consumed. We can equivalently define the initial condition through the entrance times of the initial tokens, or through the vector of $\mathbb{R}$-valued *lag times*, where

Definition 2.49 (Lag time) *The lag time of a token of the initial marking of p_i is the epoch when this token starts contributing to enabling $\sigma(p_i)$.*

However, these lag times should be compatible with the general rules that transitions fire as soon as they are enabled and that tokens start enabling the transition downstream as soon as they have completed their holding times. For instance

- if the lag time of an initial token exceeds its holding time, this token cannot have entered the place before time 0;
- if the lag times (which are possibly negative) are such that one of the transitions completes firing and consumes tokens of the initial marking before $t = 0$, these tokens cannot be part of the marking seen at time 0 since they must have left before time 0.

Definition 2.50 (Weakly compatible initial condition) *The initial condition of a timed event graph consists of an initial marking and a vector of lag times. This initial condition is weakly compatible if*

1. *the lag time of each initial token does not exceed its holding time;*
2. *the first epoch when a transition completes firing is nonnegative.*

2.5.2.2 FIFO Places and Transitions

A basic assumption that will be made throughout the chapter is that both places and transitions are First In First Out (FIFO) channels.

Definition 2.51 (FIFO place) *A place p_i is FIFO if the k-th token to enter this place is also the k-th which becomes available in this place.*

In view of the interpretation of holding times as communication or transportation times, this definition just means that the transportation or communication medium is overtake free. For instance, a place with constant holding times is FIFO.

Definition 2.52 (FIFO transition) *A transition q_j is FIFO if the k-th firing of q_j to start is also the k-th to complete.*

The interpretation is that tokens cannot overtake one another because of the firing mechanism, namely the tokens produced by the $(k+1)$-st firing of q_j to be initiated cannot enter the places of $\sigma(q_j)$ earlier than those of the k-th firing. For instance, a transition with constant firing times is always FIFO. If a transition is recycled, its $(k+1)$-st firing cannot start before the completion of the k-th one, so that a recycled transition is necessarily FIFO, regardless of the firing times.

Definition 2.53 (FIFO event graph) *An event graph is FIFO if all its places and transitions are FIFO.*

A typical example of a FIFO timed event graph is that of a system with constant holding times and recycled transitions with possibly variable firing times. An event graph with constant holding and firing times is always FIFO, even if its transitions are not recycled. Since the FIFO property is essential in order to establish the evolution equations of the present section, it is important to keep in mind that:

> The classes of timed event graphs considered throughout the book are those with
>
> 1. constant firing and holding times;
> 2. constant holding times and variable firing times, provided all transitions are recycled.

2.5.2.3 *Numbering of Events*

The following way of numbering the tokens that traverse a place and the firings of a transition will be adopted.

> By convention, the k-th token, $k \geq 1$, of place p_i is the k-th token to contribute enabling $\sigma(p_i)$ during the evolution of the event graph, including the tokens of the initial marking. The k-th firing, $k \geq 1$, of transition q_i is the k-th firing of q_i to be initiated, including the firings that consume initial tokens.

It may happen that two tokens in p_i contribute enabling $\sigma(p_i)$ at the same epoch, or that two firings of a transition are initiated at the same time (when the transition is not recycled). In this case, some ordering of these simultaneous events is chosen, keeping in mind that it should be compatible with the FIFO assumptions.

2.5.2.4 Dynamics

In what follows, the sequences of holding times $\alpha_i(k)$, $i = 1, \ldots, |\mathcal{P}|$, $k \in \mathbb{Z}$, and of firing times $\beta_j(k)$, $j = 1, \ldots, |\mathcal{Q}|$, $k \in \mathbb{Z}$, are assumed to be given nonnegative and finite real numbers. Initially, only the restriction of these sequences to $k \geq 1$ will be needed. However, we assume that these sequences can be continued to $k \leq 0$. Such a continuation is clear in the case of a constant timing, and we will see in due time how to define the continuation in more general circumstances (see §2.5.7).

We are now in a position to define the dynamics of the event graph more formally.

- The k-th token of place p_i incurs the holding time $\alpha_i(k)$.
- Once the k-th firing of transition q_j is enabled, the time for q_j to complete its k-th firing is the firing time $\beta_j(k)$. When this firing is completed, the reserved token is removed from each of the places of $\pi(q_j)$, and each place of $\sigma(q_j)$ receives one token.

We now state a few basic properties of the numbering in a FIFO event graph with a weakly compatible initial condition. For i such that $\mu_i \geq 1$, denote $w_i(1) \leq w_i(2) \leq \cdots \leq w_i(\mu_i) \in \mathbb{R}$, the lag times of the initial tokens of place p_i ordered in a nondecreasing way.

Lemma 2.54 *If the initial condition is weakly compatible and if the timed event graph is FIFO, then for all k such that $1 \leq k \leq \mu_i$ and $\mu_i \geq 1$, the initial token with lag time $w_i(k)$ is also the k-th token of place p_i (that is the k-th token to enable $\sigma(p_i)$).*

Proof If this last property does not hold for some place p_i, then a token which does not belong to the initial marking of p_i, and which hence enters p_i after time 0 (the initial condition is weakly compatible), contributes to enabling $\sigma(p_i)$ before one of the tokens of the initial marking does. Since the tokens of the initial marking enter p_i before time 0 (the initial condition is weakly compatible), this contradicts the assumption that p_i is FIFO. ■

Lemma 2.55 *The firing of q_j that consumes the k-th token of p_i (for all $p_i \in \pi(q_j)$) is the k-th firing of q_i.*

Proof Owing to the numbering convention, the set of k-th tokens of $p_i \in \pi(q_j)$ enables q_j before the set of $(k+1)$-st tokens. ■

Lemma 2.56 *The completion of the k-th firing of q_j, $k \geq 1$, produces the $(k+\mu_i)$-th token of p_i, for all $p_i \in \sigma(q_j)$.*

Proof The FIFO assumptions on transitions imply that the completion of the k-th firing of q_j produces the k-th token to enter the places that follow p_i. The property follows immediately from the FIFO assumption on places and from Lemma 2.54. ■

2.5.2.5 Evolution Equations

Definition 2.57 (State variables, daters) *The state variable $x_j(k)$, $j = 1, \ldots, |\mathcal{Q}|$, $k \geq 1$, of the event graph is the epoch when transition q_j starts firing for the k-th time, with the convention that for all q_i, $x_i(k) = \infty$ if q_i fires less than k times. These state variables will be called daters.*

These state variables are continued to negative values of k by the relation $x_j(k) = \varepsilon$, for all $k \leq 0$. Let

$$M = \max_{i=1,\ldots,|\mathcal{P}|} \mu_i \ . \tag{2.14}$$

In what follows, we will adopt the convention that the $\oplus$-sum over an empty set is ε. Define the $|\mathcal{Q}| \times |\mathcal{Q}|$ matrices $A(k,k), A(k,k-1), \ldots, A(k,k-M)$, by

$$A_{jl}(k,k-m) \stackrel{\text{def}}{=} \left(\bigoplus_{\{i \in \pi^q(j) | \pi^p(i)=l, \mu_i = m\}} \alpha_i(k) \right) \otimes \beta_l(k-m) \ , \tag{2.15}$$

and the $|\mathcal{Q}|$-dimensional vector $v(k), k = 1, \ldots, M$, by

$$v_j(k) \stackrel{\text{def}}{=} \bigoplus_{\{i \in \pi^q(j) | \mu_i \geq k\}} w_i(k) \ . \tag{2.16}$$

Theorem 2.58 *For a timed event graph with recycled transitions, the state vector $x(k) = (x_j(k))$ satisfies the evolution equations:*

$$\begin{aligned} x(k) &= A(k,k)x(k) \oplus A(k,k-1)x(k-1) \oplus \cdots \oplus A(k,k-M)x(k-M) \ , \\ & \quad k = M+1, M+2, \ldots , \end{aligned} \tag{2.17}$$

with the initial conditions

$$\begin{aligned} x(k) &= A(k,k)x(k) \oplus \cdots \oplus A(k,k-M)x(k-M) \oplus v(k) \ , \\ & \quad k = 1, 2, \ldots, M \ , \end{aligned} \tag{2.18}$$

where $x_j(k) \stackrel{\text{def}}{=} \varepsilon$ for all $k \leq 0$.

Proof We first prove that the variables $x_j(k), j = 1, \ldots, |\mathcal{Q}|$, satisfy the evolution equations:

$$\begin{aligned} x_j(k) &= \bigoplus_{\{i \in \pi^q(j) | k > \mu_i\}} \left(\alpha_i(k) \otimes \beta_{\pi^p(i)}(k - \mu_i) \otimes x_{\pi^p(i)}(k - \mu_i) \right) \\ & \quad \oplus \left(\bigoplus_{\{i \in \pi^q(j) | k \leq \mu_i\}} w_i(k) \right) , \quad k = 1, 2, \ldots . \end{aligned} \tag{2.19}$$

The k-th firing, $k \geq 1$, of transition q_j starts as soon as, for all $i \in \pi^q(j)$, the k-th token of p_i contributes to enabling q_j. In view of Lemmas 2.55 and 2.56, for

$k > \mu_i$, this k-th token is produced by the $(k - \mu_i)$-th firing of transition $\pi(p_i)$, so that the epoch when this token contributes enabling $\sigma(p_i)$ is $\alpha_i(k) \otimes \beta_{\pi^p(i)}(k - \mu_i) \otimes x_{\pi^p(i)}(k - \mu_i)$. For $k \leq \mu_i$, this event takes place at time $w_i(k)$, in view of Lemma 2.54, which completes the proof of (2.19).

We now use associativity and commutativity of $\oplus$, together with our convention on $\oplus$-sums over empty sets, to rewrite $x_j(k)$, $k > M$, as

$$\bigoplus_{m=0}^{M} \bigoplus_{l=1}^{|\mathcal{Q}|} \bigoplus_{\{i \in \pi^q(j) | \pi^p(i) = l, \ \mu_i = m\}} \alpha_i(k) \otimes \beta_l(k - m) \otimes x_l(k - m) \ .$$

The distributivity of $\otimes$ with respect to $\oplus$ implies in turn

$$x_j(k) = \bigoplus_{m=0}^{M} \bigoplus_{l=1}^{|\mathcal{Q}|} \left(\bigoplus_{\{i \in \pi^q(j) | \pi^p(i) = l, \ \mu_i = m\}} \alpha_i(k) \right) \otimes \beta_l(k - m) \otimes x_l(k - m) \ ,$$

which completes the proof of (2.17), in view of the definition of A.

The proof of (2.18) follows the same lines (using the continuation of the functions $x_j(k)$ to ε for $k \leq 0$). ■

Remark 2.59 Owing to the dynamics, the first transition to complete its firing is necessarily within the set of transitions q_j having at least one token in the initial marking of p_i for all $p_i \in \pi(q_j)$. Since the set of tokens with the smallest lag times is the first to be consumed, the second weak compatibility condition in Definition 2.50 can be translated into the requirement that

$$\beta_j(1) \otimes v_j(1) \geq e \ , \tag{2.20}$$

for all j such that $\mu_i \geq 1 \ \forall \ p_i \in \pi(q_j)$, which can be seen as a first set of linear constraints on the lag times in view of (2.16). Similarly, the first weak compatibility relation is translated into the following additional set of linear constraints:

$$w_i(k) \leq \alpha_i(k) \ , \quad i = 1, \ldots, |\mathcal{P}| \ , \quad 1 \leq k \leq \mu_i \ . \tag{2.21}$$

For instance, if $w_i(k) = \alpha_i(k)$ for all $i = 1, \ldots, |\mathcal{P}|, 1 \leq k \leq \mu_i$, then the initial condition is weakly compatible, provided the condition $w_i(1) \leq w_i(2) \leq \ldots \leq w_i(\mu_i)$ is satisfied. ■

Example 2.60 Consider the timed event graph of Figure 2.24. The firing times are assumed to be equal to e. The place connecting q_j to q_l is denoted p_{lj}. The holding times in this place are denoted $\alpha_{lj}(k)$, and the lag times of the initial tokens $w_{lj}(k)$. In order to have a FIFO event graph, places should be overtake free. This will always be true for p_{11} and p_{22}, regardless of the holding time sequences in these places (since there is always at most one token each of these places). A simple sufficient condition ensuring that the other places are overtake free is that the associated holding time sequences are non decreasing in k (for instance constant).

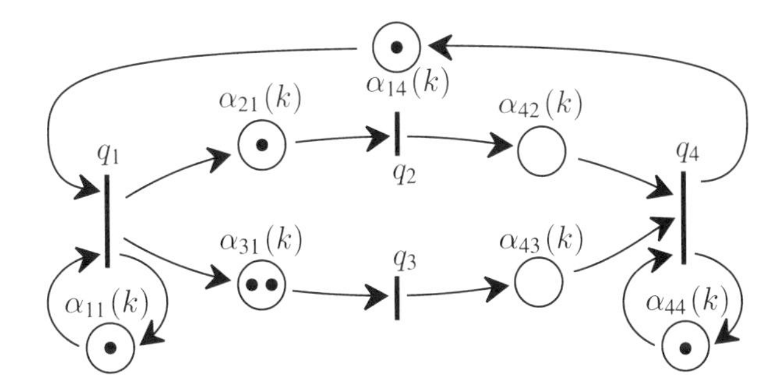

Figure 2.24: Event graph of Example 2.60

Under the assumtion that the event graph is FIFO, the matrices and vectors involved in the evolution equations are

$$A(k,k) = \begin{pmatrix} \varepsilon & \varepsilon & \varepsilon & \varepsilon \\ \varepsilon & \varepsilon & \varepsilon & \varepsilon \\ \varepsilon & \varepsilon & \varepsilon & \varepsilon \\ \varepsilon & \alpha_{42}(k) & \alpha_{43}(k) & \varepsilon \end{pmatrix}, \quad A(k,k-1) = \begin{pmatrix} \alpha_{11}(k) & \varepsilon & \varepsilon & \alpha_{14}(k) \\ \alpha_{21}(k) & \varepsilon & \varepsilon & \varepsilon \\ \varepsilon & \varepsilon & \varepsilon & \varepsilon \\ \varepsilon & \varepsilon & \varepsilon & \alpha_{44}(k) \end{pmatrix},$$

$$A(k,k-2) = \begin{pmatrix} \varepsilon & \varepsilon & \varepsilon & \varepsilon \\ \varepsilon & \varepsilon & \varepsilon & \varepsilon \\ \alpha_{31}(k) & \varepsilon & \varepsilon & \varepsilon \\ \varepsilon & \varepsilon & \varepsilon & \varepsilon \end{pmatrix},$$

and

$$v(1) = \begin{pmatrix} w_{11}(1) \oplus w_{14}(1) \\ w_{21}(1) \\ w_{31}(1) \\ w_{44}(1) \end{pmatrix}, \quad v(2) = \begin{pmatrix} \varepsilon \\ \varepsilon \\ w_{31}(2) \\ \varepsilon \end{pmatrix}.$$

The constraints (2.21) and (2.20) are translated into the bounds $w_{lj}(k) \leq \alpha_{lj}(k)$, and $w_{11}(1) \oplus w_{14}(1) \geq e$, $w_{21}(1) \geq e$, $w_{31}(1) \geq e$. ■

2.5.2.6 *Simplifications*

Firing Times The evolution equations (2.19) are unchanged if one sets all the firing times equal to e and if $\alpha_i(k)$ receives the value $\alpha_i(k) \otimes \beta_{\pi^p(i)}(k - \mu_i)$. Thus, one can always modify the holding times in order to get an 'equivalent' event graph with firing times of duration e, where equivalence means that the epochs when transitions fire are the same in both systems.

> It may be assumed without loss of generality that the firing times are equal to $e = 0$.

Observe that under this assumption the state variable $x_j(k)$ is also the epoch when transition q_j completes its k-th firing. Graphically, this is exemplified in Figure 2.25. If a firing time were assigned to a transition within an event graph, then this time can always be assigned to (i.e. added to) the holding times of all the upstream places. Consider Figure 2.25a, in which the transition has a firing time of 5 time units. The holding times of the places are 3, 4 and 6 as indicated in this figure. Figure 2.25b shows assignment of the firing time to the places. The firing of the transition in Figure 2.25b, which is instantaneous, corresponds to the completion of the firing of the transition in Figure 2.25a. Another, similar, solution is provided in Figure 2.25c, where the holding time has been assigned to all the downstream places. The firing of the transition in this figure corresponds to the initiation of the firing of the transition in Figure 2.25a.

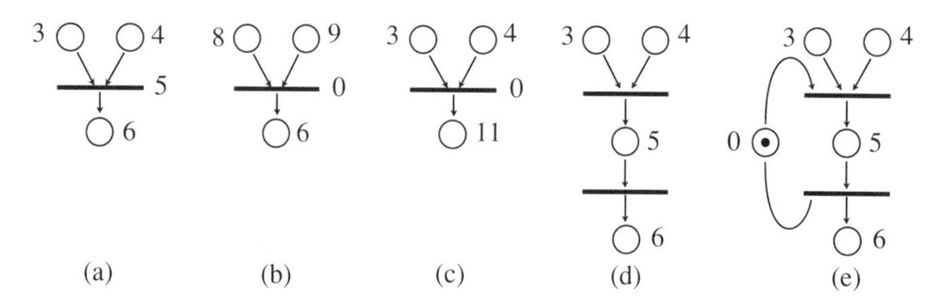

Figure 2.25: Firing times set to e

A different solution is provided in Figure 2.25d. In this figure, both the beginning and completion of the firing are now explicitly represented. In Figure 2.25e, the holding time at the transitions is also 0, but in contrast to the previous solutions the transitions cannot fire twice (or more times) within 5 time units. The transitions cannot be engaged in (partly) parallel activities.

In what follows, we shall therefore often assume that the firing times are zero. The practical implication of this mathematical simplification is clear in the constant firing and holding time case. In the variable case, one should however always keep in mind that the only meaningful initial situation is that of variable firing times (on recycled transitions) and constant holding times.

Initial Condition The end of this subsection is devoted to a discussion of the initial condition. It is shown that Equation (2.18), defining the initial condition, can be further simplified whenever the lag times satisfy certain additional and natural constraints.

For all i such that $\mu_i > 0$, denote by $y_i(k)$, $k \leq 0$, the *entrance time* function associated with place p_i, defined by the relation

$$y_i(k - \mu_i) \stackrel{\text{def}}{=} \begin{cases} w_i(k) \not{\otimes} \alpha_i(k) & \text{if } 1 \leq k \leq \mu_i \ ; \\ \varepsilon & \text{if } k > \mu_i \ , \end{cases} \tag{2.22}$$

where we recall that $\not{\otimes}$ denotes conventional subtraction.

The initial condition is said to be *compatible* if it is wealky compatible and if for any pair of places p_i and p_j which follow the same transition, the entrance times $y_i(k)$ and $y_j(k)$ coincide provided $k \geq \min(\mu_i, \mu_j)$.

Definition 2.61 (Compatible initial condition) *The initial condition is compatible if it is weakly compatible and if there exist functions $z_j(k), j = 1, \ldots, |\mathcal{Q}|$, $k \leq 0$, such that*

$$y_i(k) = z_{\pi^p(i)}(k) \ , \quad \forall i, k \quad \text{such that} \quad -\mu_i + 1 \leq k \leq 0 \ . \tag{2.23}$$

This condition is quite natural, should the initial condition result from a past evolution of the event graph: for instance, the last tokens of the initial marking to enter two places p_i and $p_{i'}$ that follow the same transition q_j, have then been produced at the same time $z_j(0)$ by a firing of q_j.

Let

$$M_j \stackrel{\text{def}}{=} \max_{i \in \sigma^q(j)} (\mu_i) \ .$$

Observe that the function $z_j(k)$ is only defined through (2.23) for $-M_j < k \leq 0$, provided $M_j \geq 1$. For other values of k, or if $M_j = 0$, we take $z_j(k) = \varepsilon$.

Instead of the former continuation of $x(k)$ to $k \leq 0$ (which consisted in taking $x_j(k) = \varepsilon$ for $k \leq 0$), we now take

$$x_j(k) = z_j(k) \ , \quad \forall k \leq 0 \ , \quad j = 1, \ldots, |\mathcal{Q}| \ . \tag{2.24}$$

Corollary 2.62 *For a FIFO timed event graph with a compatible initial condition, the state vector $x(k) = (x_j(k))$ satisfies the evolution equations:*

$$\begin{aligned} x(k) &= A(k,k)x(k) \oplus A(k,k-1)x(k-.1) \oplus \cdots \oplus A(k,k-M)x(k-M) \ , \\ &\quad k = 1, 2, \ldots, \end{aligned} \tag{2.25}$$

provided the continuation of $x(k)$ to negative values of k is the one defined by (2.24).

Proof By successively using (2.22) and (2.23), one gets

$$\begin{aligned} \bigoplus_{\{i \in \pi^q(j) | k \leq \mu_i\}} w_i(k) &= \bigoplus_{\{i \in \pi^q(j) | k \leq \mu_i\}} \alpha_i(k) \otimes y_i(k - \mu_i) \\ &= \bigoplus_{\{i \in \pi^q(j) | k \leq \mu_i\}} \alpha_i(k) \otimes z_{\pi^p(i)}(k - \mu_i) \ , \end{aligned}$$

for all $k = 1, 2, \ldots,$ so that one can rewrite (2.19) as

$$x_j(k) = \bigoplus_{\{i \in \pi^q(j)\}} \left(x_{\pi^p(i)}(k - \mu_i) \otimes \alpha_i(k) \right) \ , \quad k = 1, 2, \ldots, \tag{2.26}$$

when using the continuation of x proposed in (2.24). Equation (2.25) follows immediately from (2.26). ■

Remark 2.63 A simple example of compatible initial condition is obtained when choosing $w_i(k) = \alpha_i(k)$ for all $1 \leq k \leq \mu_i$. Practically speaking, this means that all the initial tokens enter at time 0. This corresponds to the continuation

$$x_j(k) = \begin{cases} e & \text{if } -M_j < k \leq 0 \ ; \\ \varepsilon & \text{if } k \leq M_j \ . \end{cases} \tag{2.27}$$

■

Example 2.64 (Example 2.60 continued) If the initial condition is compatible, let

$$\begin{aligned} z_1(0) &\stackrel{\text{def}}{=} w_{11}(1) \not{\circ} \alpha_{11}(1) = w_{21}(1) \not{\circ} \alpha_{21}(1) = w_{31}(2) \not{\circ} \alpha_{31}(2) \ , \\ z_1(-1) &\stackrel{\text{def}}{=} w_{31}(1) \not{\circ} \alpha_{31}(1) \ , \\ z_4(0) &\stackrel{\text{def}}{=} w_{14}(1) \not{\circ} \alpha_{14}(1) = w_{44}(1) \not{\circ} \alpha_{44}(1) \ . \end{aligned}$$

Define

$$x(0) = \begin{pmatrix} z_1(0) \\ \varepsilon \\ \varepsilon \\ z_4(0) \end{pmatrix} , \quad x(-1) = \begin{pmatrix} z_1(-1) \\ \varepsilon \\ \varepsilon \\ \varepsilon \end{pmatrix} .$$

It is easily checked that

$$v(2) = A(2,0)x(0)$$

and that

$$v(1) = A(1,0)x(0) \oplus A(1,-1)x(-1) \ .$$

Thus

$$x(k) = A(k,k)x(k) \oplus A(k,k-1)x(k-1) \oplus A(k,k-2)x(k-2), \ \ k = 1,2,\ldots .$$

■

2.5.3 Constructiveness of the Evolution Equations

The first natural question concerning Equation (2.17) is: is it implicit or constructive? The main result of this section establishes that the evolution equations (2.17) are not implicit and that they allow one to recursively define the value of $x_j(k)$ for all $j = 1, \ldots, |\mathcal{Q}|$, and $k \geq 1$, provided the event graph under consideration is live.

Lemma 2.65 *The event graph is live if and only if there exists a permutation P of the coordinates for which the matrix $P'A(k,k)P$ is strictly lower triangular for all k.*

Proof If the matrix $P'A(k,k)P$ is strictly lower triangular for some permutation P, then there is no circuit with 0 initial marking, in view of the definition of $A(k,k)$ (see (2.15)). Conversely, if the event graph is live, the matrix $A(k,k)$ has no circuit, and there exists a permutation of the coordinates that makes A strictly lower triangular. The proof is then concluded from Theorem 2.38. ■

Observe that the fact that P does not depend on k comes from the fact that the support of $A(k,k)$ does not depend on k (by 'the support' of matrix A we mean the matrix S with the same dimension as A defined by $S_{ij} = 1_{A_{ij} \neq \varepsilon}$).

If the matrix $P'A(k,k)P$ is strictly lower triangular, $A^n(k,k) = \varepsilon$ for $n \geq |\mathcal{Q}|$, and the matrix

$$A^*(k,k) \stackrel{\text{def}}{=} e \oplus A(k,k) \oplus A^2(k,k) \oplus \cdots$$

is finite. Let

$$\overline{A}(k,k-l) \stackrel{\text{def}}{=} A^*(k,k)A(k,k-l) \ , \quad k \in \mathbb{Z} \ , \quad l = 1,\ldots,M \ , \tag{2.28}$$

and

$$\overline{v}(k) \stackrel{\text{def}}{=} A^*(k,k)v(k) \ , \quad k \in \mathbb{Z} \ ,$$

with $v_j(k) \stackrel{\text{def}}{=} \varepsilon$ for $k \leq 0$ or $k > M$.

Theorem 2.66 *If the event graph is live, the evolution equations (2.17) and (2.18) can be rewritten as*

$$\begin{aligned} x(k) \quad = \quad & \overline{A}(k,k-1)x(k-1) \oplus \cdots \oplus \overline{A}(k,k-M)x(k-M) \oplus \overline{v}(k) \ , \\ & k = 1,2,\ldots, \end{aligned} \tag{2.29}$$

where $x_j(k) \stackrel{\text{def}}{=} \varepsilon$, for all $k \leq 0$.

Proof From (2.17) and (2.18), we obtain by induction on n that

$$\begin{aligned} x(k) = {} & A^{n+1}(k,k)x(k) \\ & \oplus \left(\bigoplus_{m=0}^{n} A^m(k,k) \right) \left(\bigoplus_{l=1}^{M} A(k,k-l)x(k-l) \oplus v(k) \right), \quad k = 1,2,\ldots . \end{aligned}$$

Equation (2.29) follows from the last relation by letting n go to ∞. ■

Remark 2.67 If the initial condition is compatible and if one now takes the continuation of $x(k)$ for $k \leq 0$, as defined in (2.24), the same type of arguments shows that (2.25) becomes

$$x(k) = \bigoplus_{l=1}^{M} \overline{A}(k,k-l)x(k-l) \ , \quad k = 1,2,\ldots .$$

■

Corollary 2.68 *If the event graph is live and if the holding times and the lag times are all finite, so are the state variables $x_j(k)$, $j = 1,\ldots,|\mathcal{Q}|$, $k \geq 1$.*

Proof The proof is by induction based on (2.29). ■

Remark 2.69 The matrix $\overline{A}(k, k-l)$, $l \geq 1$, has a simple graph-theoretic interpretation. Let $\mathcal{S}(j', j, l)$ be the set of paths in the graph $\mathcal{G}$ of the event graph, of length at least 2, with initial transition $q_{j'}$, with final transition q_j, and such that the first two transitions are connected by a place with initial marking equal to l, while the other transitions are connected by places with 0 initial marking. It is easily checked, using the results of §2.4, that $\overline{A}_{jj'}(k, k-l)$ is defined by the relation

$$\overline{A}_{jj'}(k, k-l) = \bigoplus_{\{\rho=(j_1,i_1,j_2,i_2\cdots,i_{h-1},j_h)\in\mathcal{S}(j',j,l)\}} \bigotimes_{n=1}^{h-1} \alpha_{i_n}(k) \ , \tag{2.30}$$

with the usual convention if the set $\mathcal{S}(j', j, l)$ is empty. The entry $\overline{A}_{jj'}(k, k-l)$ is hence simply the longest path in $\mathcal{S}(j', j, l)$. ■

Example 2.70 **(Example 2.60 continued)** We have

$$A^*(k,k) = \begin{pmatrix} e & \varepsilon & \varepsilon & \varepsilon \\ \varepsilon & e & \varepsilon & \varepsilon \\ \varepsilon & \varepsilon & e & \varepsilon \\ \varepsilon & \alpha_{42}(k) & \alpha_{43}(k) & e \end{pmatrix},$$

so that

$$\overline{A}(k,k-1) = \begin{pmatrix} \alpha_{11}(k) & \varepsilon & \varepsilon & \alpha_{14}(k) \\ \alpha_{21}(k) & \varepsilon & \varepsilon & \varepsilon \\ \varepsilon & \varepsilon & \varepsilon & \varepsilon \\ \alpha_{42}(k)\alpha_{21}(k) & \varepsilon & \varepsilon & \alpha_{44}(k) \end{pmatrix},$$

$$\overline{A}(k,k-2) = \begin{pmatrix} \varepsilon & \varepsilon & \varepsilon & \varepsilon \\ \varepsilon & \varepsilon & \varepsilon & \varepsilon \\ \alpha_{31}(k) & \varepsilon & \varepsilon & \varepsilon \\ \alpha_{43}(k)\alpha_{31}(k) & \varepsilon & \varepsilon & \varepsilon \end{pmatrix},$$

and

$$\overline{v}(1) = \begin{pmatrix} w_{11}(1) \oplus w_{14}(1) \\ w_{21}(1) \\ w_{31}(1) \\ w_{21}(1)\alpha_{42}(1) \oplus w_{31}(1)\alpha_{43}(1) \oplus w_{44}(1) \end{pmatrix}, \quad \overline{v}(2) = \begin{pmatrix} \varepsilon \\ \varepsilon \\ w_{31}(2) \\ w_{31}(2)\alpha_{43}(2) \end{pmatrix}.$$

■

Remark 2.71 **(Equivalent event graph with positive initial marking)** With the evolution equation (2.29), one can associate a *derived* event graph with the same set of transitions as the initial event graph, and where the initial marking is such that $\mu_i > 0$ for all places p_i. This event graph is equivalent to the initial one in the sense that corresponding transitions fire at the same times. The derived event graph associated with the event graph of Example 2.60 is given in Figure 2.26, left-hand side. This derived event graph can be defined from the original one by the following transformation rules:

1. take the same set of transitions as in the original event graph;
2. for each path of $\mathcal{S}(j', j, l)$, $l \geq 1$, in the original event graph, create a place connecting j' to j with l tokens and with the weight of the path as holding time. ■

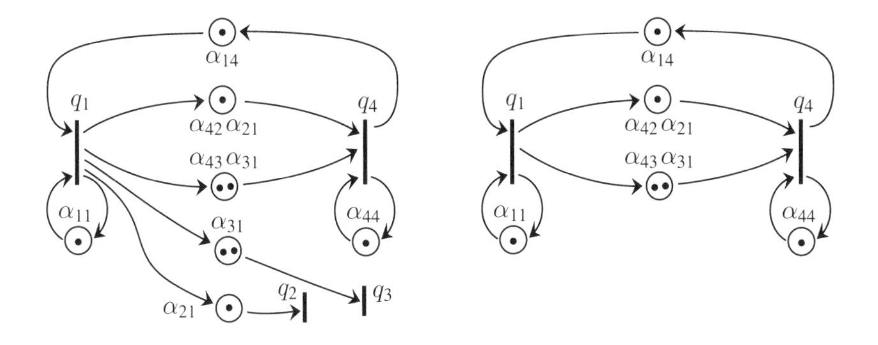

Figure 2.26: Illustration of Remarks 2.71 and 2.72

Remark 2.72 (Equivalent event graph with reduced state space) The dimension of the state space can be reduced using the following observation: the transitions followed by places all having a 0 initial marking, or equivalently the transitions q_j such that the entries of the j-th column of $A(k, k-l)$ are ε, are not present in the right-hand side of (2.29). Let $\mathcal{Q}'$ be the set of transitions followed by at least one place with a positive initial marking. One can take as reduced state variables $x_j(k), j \in \mathcal{Q}', k \geq 1$. The remaining variables are obtained from them via (2.29). ■

Example 2.73 (Example 2.60 continued) Here we have $\mathcal{Q}' = \{q_1, q_4\}$. With these new state variables, the evolution equations are reduced to

$$\begin{pmatrix} x_1(k) \\ x_4(k) \end{pmatrix} = \begin{pmatrix} \alpha_{11}(k) & \alpha_{14}(k) \\ \alpha_{42}(k)\alpha_{21}(k) & \alpha_{44}(k) \end{pmatrix} \begin{pmatrix} x_1(k-1) \\ x_4(k-1) \end{pmatrix} \oplus \begin{pmatrix} \varepsilon & \varepsilon \\ \alpha_{43}(k)\alpha_{31}(k) & \varepsilon \end{pmatrix} \begin{pmatrix} x_1(k-2) \\ x_4(k-2) \end{pmatrix} \oplus \begin{pmatrix} \overline{v}_1(k) \\ \overline{v}_4(k) \end{pmatrix}.$$

The event graph corresponding to these equations is depicted in Figure 2.26, right-hand side. It is obtained from the derived graph by deleting the transitions that do not belong to $\mathcal{Q}'$.

The other state variables are obtained from the reduced state variables by the relation

$$\begin{pmatrix} x_2(k) \\ x_3(k) \end{pmatrix} = \begin{pmatrix} \alpha_{21}(k) & \varepsilon \\ \varepsilon & \varepsilon \end{pmatrix} \begin{pmatrix} x_1(k-1) \\ x_4(k-1) \end{pmatrix} \oplus \begin{pmatrix} \varepsilon & \varepsilon \\ \alpha_{31}(k) & \varepsilon \end{pmatrix} \begin{pmatrix} x_1(k-2) \\ x_4(k-2) \end{pmatrix} \oplus \begin{pmatrix} \overline{v}_2(k) \\ \overline{v}_3(k) \end{pmatrix}.$$

The variables $(x_2(k), x_3(k))$ are output variables in the derived event graph. Equivalently, transitions q_2 and q_3 are sinks of the derived event graph. ■

2.5.4 Standard Autonomous Equations

The data of this section is a live event graph satisfying the evolution equations (2.29), with the reduction of the state space mentioned in Remark 2.72. We will assume that the transitions of $\mathcal{Q}'$ are numbered $1, \ldots, |\mathcal{Q}'|$, which introduces no loss of generality.

It may be desirable to replace the initial recurrence (2.29), which is of order M, by an equivalent recurrence of order 1. This is done by using the standard technique which consists in extending the state vector. As a new state vector, take the $\big(|\mathcal{Q}'| \times M\big)$-dimensional vector

$$\widetilde{x}(k) \stackrel{\text{def}}{=} \begin{pmatrix} x(k) \\ x(k-1) \\ \vdots \\ x(k+1-M) \end{pmatrix} .$$

Let $\widetilde{A}(k),\ k \in \mathbb{Z}$, be the $(|\mathcal{Q}'| \times M) \times (|\mathcal{Q}'| \times M)$ matrix defined by the relation

$$\widetilde{A}(k) = \begin{pmatrix} \overline{A}(k+1,k) & \overline{A}(k+1,k-1) & \ldots & \ldots & \overline{A}(k+1,k+1-M) \\ e & \varepsilon & \ldots & \varepsilon & \varepsilon \\ \varepsilon & e & \ddots & \vdots & \varepsilon \\ \vdots & \ddots & e & \varepsilon & \varepsilon \\ \varepsilon & \ldots & \varepsilon & e & \varepsilon \end{pmatrix} ,$$

where e and ε denote the $|\mathcal{Q}'| \times |\mathcal{Q}'|$ identity and zero matrices, respectively, and let $\widetilde{v}(k)$ be the $\big(|\mathcal{Q}'| \times M\big)$-dimensional vector

$$\widetilde{v}(k) \stackrel{\text{def}}{=} \begin{pmatrix} \overline{v}(k+1) \\ \varepsilon \\ \vdots \\ \varepsilon \end{pmatrix} ,$$

where ε represents here the $|\mathcal{Q}'|$-dimensional zero vector.

Adopting the convention that $x_j(k)$ and $v_j(k)$ are equal to ε for $k \leq 0$, it should be clear that Equation (2.31) in the following corollary is a mere rewriting of the evolution equations (2.29).

Corollary 2.74 *The extended state space vector $\widetilde{x}(k)$ satisfies the $\big(M \times |\mathcal{Q}'|\big)$-dimensional recurrence relation of order* 1

$$\widetilde{x}(k+1) = \widetilde{A}(k)\widetilde{x}(k) \oplus \widetilde{v}(k) \ , \quad k = 1, 2, \ldots . \tag{2.31}$$

Equation (2.31) will be referred to as the *standard form* of the evolution equations of an autonomous timed event graph satisfying (2.29).

Remark 2.75 In the particular case of a compatible initial condition, these equations read

$$\widetilde{x}(k+1) = \widetilde{A}(k)\widetilde{x}(k) \ , \quad k = 1, 2, \ldots , \tag{2.32}$$

provided the continuation of $x_j(k)$ for $k \leq 0$, is that of Equation (2.24). Whenever the entrance times of the tokens of the initial marking are all equal to e (see Remark 2.63), it is easily checked from (2.27) that in this case

$$\widetilde{x}_{l|Q'|+j}(0) = \begin{cases} e & \text{if } 0 \leq l < M_j \ ; \\ \varepsilon & \text{for } l \geq M_j \ , \end{cases} \tag{2.33}$$

for $l = 0, \ldots, M-1; j = 1, \ldots, |Q'|$. ■

Example 2.76 (Example 2.60 continued) Here we have

$$\widetilde{x}(k) = \begin{pmatrix} x_1(k) \\ x_4(k) \\ x_1(k-1) \\ x_4(k-1) \end{pmatrix} , \quad \widetilde{v}(k) = \begin{pmatrix} \overline{v}_1(k+1) \\ \overline{v}_4(k+1) \\ \varepsilon \\ \varepsilon \end{pmatrix} ,$$

and

$$\widetilde{A}(k) = \begin{pmatrix} \alpha_{11}(k+1) & \alpha_{14}(k+1) & \varepsilon & \varepsilon \\ \alpha_{42}(k+1)\alpha_{21}(k+1) & \alpha_{44}(k+1) & \alpha_{43}(k+1)\alpha_{31}(k+1) & \varepsilon \\ e & \varepsilon & \varepsilon & \varepsilon \\ \varepsilon & e & \varepsilon & \varepsilon \end{pmatrix} .$$

In the special case mentioned at the end of the preceding remark, we have $\widetilde{x}(0) = (e, e, e, \varepsilon)'$. ■

Remark 2.77 (Equivalent net with at most one token in the initial marking) One can associate a timed event graph with the evolution equations (2.31). The interesting property of this event graph is that its initial marking is such that $M = 1$ (more precisely, each μ_i in this event graph is 1).

In view of Corollary 2.74, one can hence state that for any timed event graph, one can construct another 'equivalent' event graph with initial marking equal to 1 everywhere. The equivalence means here that one can find a bijective mapping from the set of transitions of the initial event graph to a subset of the transitions of the second one, such that two corresponding transitions fire at the same time. In particular, any observation of these transitions will be identical.

For instance, in the case of a compatible initial condition, this event graph can be obtained from the original one by first transforming it into an event graph with positive initial marking (as it was done in Remark 2.71), and by then applying the following transformation rules:

1. for each transition q_j of $\mathcal{Q}'$ in the original event graph, create M transitions $q_{jl}, l = 0, \ldots, M-1$;

2. for each q_{jl}, $l = 0, \ldots, M-2$, create a place that connects q_{jl} to $q_{i,l+1}$, with 0 holding times. Put one token in its initial marking, with initial lag time $z_j(-l)$;

3. for each place connecting $q_i \in \mathcal{Q}'$ to q_j, and with $l+1$ initial tokens, $l \geq 0$, in the original system, create a place with one token with initial lag time $z_i(-l)$, and with the same holding times sequence as the original place. This new place has q_{il} as input transition, and q_{j0} as output transition.

For Example 2.60, the corresponding event graph is given in Figure 2.27. The behavior of q_{10} in this graph is the same as that of q_1 in Figure 2.24. The same property holds for q_{40} and q_4 respectively. ■

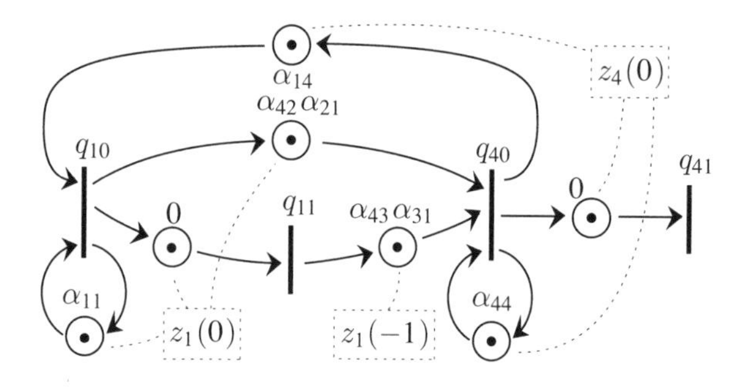

Figure 2.27: Illustration of Remark 2.77

2.5.5 The Nonautonomous Case

This subsection focuses on FIFO timed event graphs with external inputs. The firing times can be taken equal to 0, without loss of generality.

To the framework of the preceding sections, we add a new class of transitions called *input transitions*. This set of transitions will be denoted $\mathcal{I}$.

Definition 2.78 (Input transition) *An input transition consists of a source transition and of a nondecreasing real valued sequence, called the input sequence .*

The input sequence associated with transition $q_j \in \mathcal{I}$ will be denoted $u_j(k)$, $k \geq 1$; the interpretation of this sequence is that $u_j(k)$, $k \geq 1$, gives the epoch when q_j fires for the k-th time, due to some external trigger action. The input sequences are assumed to be given.

Definition 2.79 (Weakly compatible input sequence) *The input sequence $u_j(k)$, $k \in \mathbb{Z}$, is weakly compatible if $u_j(1) \geq 0$.*

In what follows, all input sequences will be assumed to be weakly compatible. As in the autonomous case, the initial condition is said to be weakly compatible if in addition

- the entrance times of the tokens of the initial marking are nonpositive;
- the firings which consume tokens of the initial marking complete at nonnegative epochs (the assumption that the input sequences are weakly compatible may contribute ensuring that this holds).

The definition of compatibility is the same as in the autonomous case. For instance, if the lag times of a nonautonomous event graph are compatible, one can continue the input sequence $\{u_j(k)\}$ (with $j \in \mathrm{I}$) to a nondecreasing sequence $\{u_j(k)\}_{k \in \mathbb{Z}}$, with $u_j(0) \leq 0$, such that for all $p_i \in \sigma(q_j)$ with $\mu_i \geq 1$,

$$w_i(k) = \alpha_i(k) \otimes u_j(k - \mu_i) \ , \quad \forall k : 1 \leq k \leq \mu_i \ . \tag{2.34}$$

2.5.5.1 *Basic Nonautonomous Equations*

The derivation of the evolution equations is based on the same type of assumptions as in the autonomous case, namely the event graph is FIFO and the initial condition is weakly compatible.

Define the $|\mathrm{Q}| \times |\mathrm{I}|$ matrices $B(k,k), \ldots, B(k, k-M)$ by

$$B_{jl}(k, k-m) \stackrel{\text{def}}{=} \bigoplus_{\{i \in \pi^q(j) | \pi^p(i) = l, \mu_i = m\}} \alpha_i(k) \ , \tag{2.35}$$

and the $|\mathrm{I}|$-dimensional vector $u(k) = (u_1(k), \ldots, u_{|\mathrm{I}|}(k)), k = 1, 2, \ldots$. Using the same arguments as in Theorem 2.58, we get the following result.

Theorem 2.80 *Under the foregoing assumptions, the state vector* $x(k) = (x_1(k), \ldots, x_{|\mathrm{Q}|}(k))'$ *satisfies the evolution equations:*

$$\begin{aligned} x(k) = A(k,k)x(k) \oplus \cdots \oplus A(k, k-M)x(k-M) \oplus B(k,k)u(k) \\ \oplus \cdots \oplus B(k, k-M)u(k-M) \oplus v(k) \ , \qquad k = 1, 2, \ldots, \end{aligned} \tag{2.36}$$

where $x_j(k) \stackrel{\text{def}}{=} \varepsilon$ *and* $u_j(k) = \varepsilon$ *for all* $k \leq 0$*;* $v_j(k)$ *is defined as in (2.16) for* $1 \leq k \leq M$ *and it is equal to* ε *otherwise.*

If the initial lag times and the input sequences are both compatible, this equation can be simplified by using the same arguments as in Corollary 2.62, which leads to the equation

$$\begin{aligned} x(k) = A(k,k)x(k) \oplus \cdots \oplus A(k, k-M)x(k-M) \oplus B(k,k)u(k) \\ \oplus \cdots \oplus B(k, k-M)u(k-M) \ , \qquad k = 1, 2, \ldots, \end{aligned} \tag{2.37}$$

where the continuations that are taken for $x(k)$ and $u(k)$, $k \leq 0$, are now those defined in Corollary 2.62 and Equation 2.34, respectively.

In what follows, we will say that the nonautonomous event graph is live if the associated autonomous event graph (namely the one associated with the equation $x(k) = A(k,k)x(k) \oplus \cdots \oplus A(k,k-M)x(k-M)$) is live. Let

$$\overline{B}(k,k-l) \stackrel{\text{def}}{=} A^*(k,k)B(k,k-m) \ , \quad k \in \mathbb{Z} \ , \quad l = 0,\dots,M \ .$$

The following theorem is proved like Theorem 2.66.

Theorem 2.81 *If the event graph is live, the evolution equations (2.36) can be rewritten as*

$$\begin{aligned} x(k) = {} & \overline{A}(k,k-1)x(k-1) \oplus \cdots \oplus \overline{A}(k,k-M)x(k-M) \\ & \oplus \overline{B}(k,k)u(k) \oplus \cdots \oplus \overline{B}(k,k-M)u(k-M) \oplus \overline{v}(k) \ , \\ & k = 1,2,\dots , \end{aligned} \tag{2.38}$$

with the same simplification as in Corollary 2.62, provided the initial lag times and the input sequences are compatible.

The graph theoretic interpretation of $\overline{B}_{jj'}(k,k-l)$ is again the longest path in $\mathcal{S}(j',j,l)$ (see Remark 2.69).

2.5.5.2 Standard Nonautonomous Equations

Define the $\big(M \times |\mathrm{I}|\big)$-dimensional vector

$$\widetilde{u}(k) \stackrel{\text{def}}{=} \begin{pmatrix} u(k+1) \\ u(k) \\ \vdots \\ u(k+2-M) \end{pmatrix} ,$$

and the $(|\mathrm{I}| \times M) \times (|\mathrm{Q}'| \times M)$ matrix

$$\widetilde{B}(k) = \begin{pmatrix} \overline{B}(k+1,k+1) & \overline{B}(k+1,k) & \dots & \overline{B}(k+1,k+2-M) \\ \varepsilon & \varepsilon & \dots & \varepsilon \\ \vdots & \vdots & \vdots & \vdots \\ \varepsilon & \varepsilon & \dots & \varepsilon \end{pmatrix} .$$

Corollary 2.82 (Standard nonautonomous equation) *The extended state space vector $\widetilde{x}(k)$ satisfies the $\big(M \times |\mathrm{Q}'|\big)$-dimensional recurrence of order 1:*

$$\widetilde{x}(k+1) = \widetilde{A}(k)\widetilde{x}(k) \oplus \widetilde{B}(k)\widetilde{u}(k) \oplus \widetilde{v}(k) \ , \quad k = 1,2,\dots , \tag{2.39}$$

with the same simplification as in Corollary 2.62, provided the initial lag times and the input sequences are compatible.

Remark 2.83 Formally, the autonomous equation (2.17) can also be seen as that of a nonautonomous event graph with a set of input transitions $I = \{q'_1, q'_2, \ldots\}$ of the same cardinality as the set Q', with input sequence vector $v(k)$, and with $B(k,k) = e$, $B(k,k-l) = \varepsilon$ for $l = 1, \ldots, M$ (the input transition q'_j is connected to the internal transition q_j by a single place with 0 initial marking).

However, the requirement that an input sequence should be nondecreasing contradicts our foregoing assumption on $v(k)$ (with our definitions, $v_j(k)$ eventually becomes ε for k large, as it can be seen from (2.16)). However, when using the fact that the sequences $x_j(k)$ are nondecreasing, it is easy to check that one can take the input sequence $u_j(k)$ defined by the function

$$u_j(k) \stackrel{\text{def}}{=} \begin{cases} v_j(k) & \text{if } 1 \leq k \leq M_j \ ; \\ v_j(M_j) & \text{if } k \geq M_j \ , \end{cases}$$

instead of $v(k)$, without altering the values of $x(k)$.

This representation of an autonomous event graph as a nonautonomous one, where all initial lag times can be taken equal to ε, is exemplified on the event graph of Example 2.60 in Figure 2.28. ■

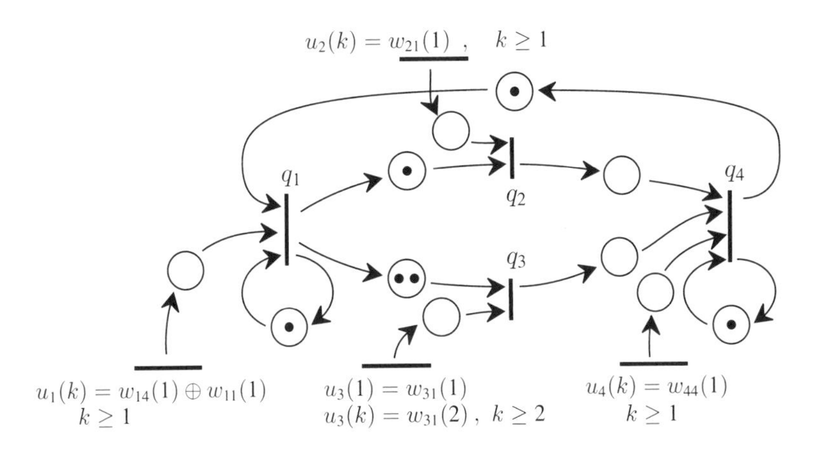

Figure 2.28: Illustration of Remark 2.83

2.5.6 Construction of the Marking

This subsection is concerned with the construction of the marking from the state variables. We will limit ourselves to the construction of the marking at certain epochs for which their expression is quite simple. However, the formulæ that are obtained are nonlinear in the max-plus algebra. We will therefore return to classical algebra, at least in this subsection. For the sake of simplicity, it is assumed that the initial condition is compatible. Pick some place p_i in $\mathfrak{P}$, and let $q_j = \pi(p_i)$,

$q_l = \sigma(p_i)$. Let $N_i(t)$ be the number of tokens in place p_i at time $t, t \geq 0$, with the convention that this piecewise constant function is right-continuous.

Let

$$N_i^-(k) \stackrel{\text{def}}{=} N_i\left(x_{\pi^p(i)}(k)\right), \quad k \geq 1 , \tag{2.40}$$

$$N_i^+(k) \stackrel{\text{def}}{=} N_i\left(x_{\sigma^p(i)}(k)\right), \quad k \geq 1 . \tag{2.41}$$

Owing to our definitions, $N_i^-(k)$ is the number of tokens in p_i just after the k-th token entrance into p_i after $t = 0$, while $N_i^+(k)$ is the number of tokens p_i just after the departure of the k-th token to leave p_i after $t = 0$.

Lemma 2.84 *Under the foregoing assumptions,*

$$N_i^-(k) = \sum_{h=1}^{k+\mu_i} 1_{\{x_l(h) > x_j(k)\}} , \quad k = 1, 2, \ldots , \tag{2.42}$$

$$N_i^+(k) = \sum_{h=k+1-\mu_i}^{\infty} 1_{\{x_l(k) \geq x_j(h)\}} , \quad k = 1, 2, \ldots . \tag{2.43}$$

Proof The tokens present in p_i at time (just after) $x_j(k)$ are those that arrived into p_i no later than $x_j(k)$ and which are still in p_i at time $x_j(k)$. The tokens that arrived no later than $x_j(k)$ are those with index h with respect to this place, with $1 \leq h \leq k + \mu_i$. Among these tokens, those which satisfy the relation $x_l(h) > x_j(k)$ are still in p_i at time $x_j(k)$. Similarly, the only tokens that can be present in place p_i just after time $x_l(k)$ are those with index $h > k$ with respect to p_i. The token of index h with respect to p_i is the token produced by transition j at time $x_j(h - \mu_i)$ (where the continuation of $x_j(k)$ to $k \leq 0$ is the one defined in §2.5.2.6). Among these tokens, those which entered p_i no later than time $x_l(k)$ are in p_i at time $x_l(k)$. ∎

2.5.7 Stochastic Event Graphs

Definition 2.85 (Stochastic event graph) *A timed event graph is a stochastic event graph if the holding times, the firing times and the lag times are all random variables defined on a common probability space.*

Different levels of generality can be considered. The most general situation that will be considered in Chapters 7 and 8 is the case when the sequences $\{\alpha_i(k)\}_{k\in\mathbb{Z}}$, $i = 1, \ldots, |\mathcal{P}|$, and $\{\beta_j(k)\}_{k\in\mathbb{Z}}$, $j = 1, \ldots, |\mathcal{Q}|$, are jointly stationary and ergodic sequences of nonnegative and integrable random variables defined on a common probability space $(\Omega, \mathbb{F}, \mathbb{P})$. Similarly, the lag times $w_i(k - \mu_l)$, $1 \leq k \leq \mu_i$, are assumed to be finite and integrable random variables defined on $(\Omega, \mathbb{F}, \mathbb{P})$.

More specific situations will also be considered, like for instance the case when the sequences $\{\alpha_l(k)\}_{k\in\mathbb{N}}$, $l = 1, \ldots, |\mathcal{P}|$, and $\{\beta_i(k)\}_{k\in\mathbb{N}}$, $i = 1, \ldots, |\mathcal{Q}|$, are mutually independent sequences of independent and identically distributed (i.i.d.)

random variables. For instance, all these variables could be exponentially distributed, with a parameter that depends on l or i, namely

$$\mathbb{P}[\alpha_l(k) \leq x] = \begin{cases} 1 - \exp(a_l x) & \text{if } x \geq 0 \ ; \\ 0 & \text{otherwise,} \end{cases} \tag{2.44}$$

and

$$\mathbb{P}[\beta_i(k) \leq x] = \begin{cases} 1 - \exp(b_i x) & \text{if } x \geq 0 \ ; \\ 0 & \text{otherwise,} \end{cases} \tag{2.45}$$

where $a_j > 0$ and $b_j > 0$. Another particular case arises when all the sequences are constant and deterministic, and the case with *constant timing* is thus a special (and degenerate) case of the i.i.d. situation.

2.6 MODELING ISSUES

In this section some issues related to the modeling of Petri nets will be described briefly.

2.6.1 Multigraphs

Multigraphs are graphs in which more than one arc between two nodes is allowed. The fact that in this chapter no such multigraphs have been considered is twofold.

The first reason is that the modeling power of Petri nets with multiple arcs and Petri nets with single arcs is the same [108]. This 'modeling power' is defined in terms of 'reachability', as discussed in §2.4.1. Petri nets with multiple arcs can straightforwardly be represented by Petri nets with single arcs, as shown in Figure 2.29. Note, however, that in the second of these figures a conflict situation

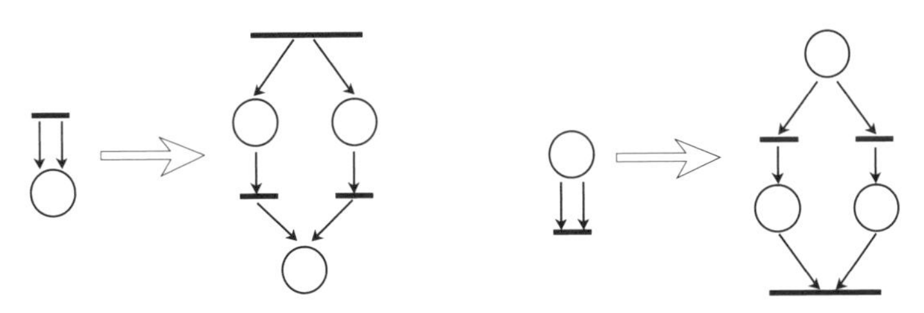

Figure 2.29: The change of multiple arcs into single arcs

has arisen. In order that this single arc representation of the originally multiple arc from place to transition behaves in the same way as this multiple arc, it is necessary that the two rival transitions receive tokens alternately.

The second reason is that it is not at all clear how to obtain equations for timed Petri nets in which there are multiple arcs. Observe that such nets are not event graphs, as is directly seen from Figure 2.29. It seems that some transitions fire in the long run twice as often as some other transitions; the $2k$-th enabling of such a 'fast' transition is caused by the k-th firing (approximately) of such a 'slow' transition.

2.6.2 Places with Finite Capacity

For the rule for enabling transitions it has tacitly been assumed that each place can accommodate an unlimited number of tokens. For the modeling of many physical systems it is natural to consider an upper limit for the number of tokens that each place can hold. Such a Petri net is referred to as a finite capacity net. In such a net, each place has an associated capacity K_i, being the maximum number of tokens that p_i can hold at any time. For a transition in a finite capacity net to be enabled, there is the additional condition that the number of tokens in each $p_i \in \sigma(q_j)$ cannot exceed its capacity after the firing of q_j.

In the discussion to come we confine ourselves to event graphs, although the extension to Petri nets is quite straightforward, see [96]. Suppose that place p_i has a capacity constraint K_i, then the finite capacity net will be 'remodeled' as another event graph, without capacity constraints. If $\pi(p_i) \cap \sigma(p_i) \neq \emptyset$, then there is a loop. The number of tokens in a loop before and after the firing is the same and hence the capacity constraint is never violated (provided the initial number of tokens was admissible). Assume now that $\pi(p_i) \cap \sigma(p_i) = \emptyset$. Add another place $p_{i'}$ to the net. This new place will have $\mu_{i'} = K_i - \mu_i$ tokens. Add an arc from $\sigma(p_i)$ to $p_{i'}$ and an arc from $p_{i'}$ to $\pi(p_i)$. The number of tokens in this new circuit is constant according to Theorem 2.37. The liveness of the event graph is not influenced by the addition of such a new circuit, see Theorem 2.38. An example is provided in Figure 2.30 where $K_i = 3$. It is easily verified that this newly constructed Petri net, without

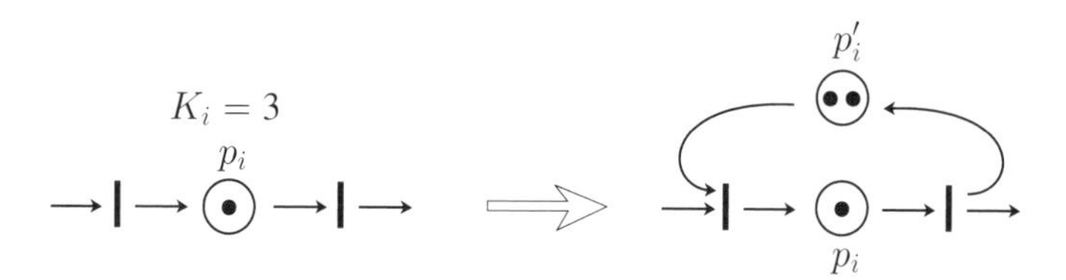

Figure 2.30: Node with and without a capacity constraint

capacity constraints, behaves in exactly the same way in terms of possible firing sequences as the original finite capacity net. In this sense the nets are 'equivalent'.

2.6.3 Synthesis of Event Graphs from Interacting Resources

This aim of this subsection is to show how, starting from the physical understanding of 'resources', and some assumptions, one can build up event graphs in a somewhat systematic way. This approach leads to a subclass of event graphs (essentially restricted by the kind of initial marking they can receive) for which the issue of initial conditions is transparent.

2.6.3.1 General Observations

The exercise of modeling is more an art than an activity which obeys rigid and precise rules. For example, the degree of details retained in a model is a matter of appraisal with respect to future uses of the model. Models may be equivalent in some respects but they may differ in the physical insights they provide. These observations are classical in conventional system theory: it is well known that different state space realizations, even with different dimensions, may yield the same input-output behavior, but some of these realizations may capture a physical meaning of the state variables whereas others may not.

To be more specific, a clear identification of what corresponds to 'resources' in an abstract Petri net model may not be crucial if available resources are given once and for all and if the problem only consists in evaluating the performance of a specified system, but it may become a fundamental issue when resource quantities enter the decision variables, e.g. in optimal resource sizing at the design stage. It has already been noticed that, in an event graph, although the number of tokens in each circuit is invariant during the evolution of the system, this is not generally the case of the *total number of tokens* in the system, even if the graph is strongly connected and autonomous. In this case, it is unclear how tokens and physical resources are related to each other.

Figure 2.31 shows the simplest example of this type for which there are only two distinct situations for the distribution of tokens, the total number of tokens being either one or two. Indeed, if we redraw this event graph as in Figure 2.32, it becomes possible to interpret the two tokens as two resources circulating in the

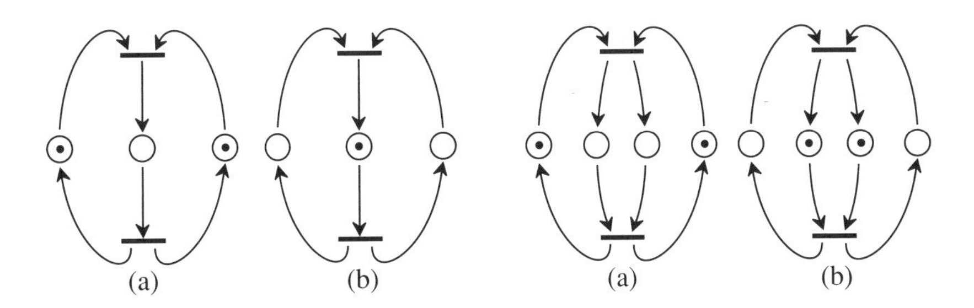

Figure 2.31: Merging two resources

Figure 2.32: An alternative model

system sometimes alone (case (a)) and sometimes jointly (case (b)). The problem with the former model of Figure 2.31 is that, when they stay together (case (b)), the two resources are represented by a single token. Obviously, these two models are equivalent if one is only interested in the number of transition firings within a certain time (assuming that some holding times have been defined properly in corresponding places), but the difference becomes relevant if one is willing to play with individual resource quantities in order to 'optimize' the design.

As long as modeling is an art, what is described hereafter should be considered as a set of practical guidelines—rather than as a rigorous theory—which should help in trying to construct models which capture as much physical meaning as possible. The dynamic systems we have in mind are supposed to involve the combined evolution of several interacting resources designed to achieve some specific overall task or service. Our approach is in three stages:

1. we first describe the evolution of each type of resource individually;
2. then we describe the mechanism of interaction between resources;
3. finally, we discuss the problem of initialization.

2.6.3.2 *State Evolution of Individual Resources*

The word 'resource' should be understood in the broadest sense: a machine, a tool, a part, a channel, a transportation link, a position in a storage or in a buffer, etc. are all resources. Of course, in practice, it is worthwhile modeling a resource explicitly as long as it is a 'scarce' resource, that is a resource the limited availability of which may have some influence on the evolution of the system at some time. The capacity of a buffer may be large enough for the buffer to behave as if it were infinite. Again, it is a matter of feeling to decide whether an explicit model of the finite capacity is a priori needed or not.

The evolution of each resource in the system is modeled by a sequence of 'stages'. What is considered a 'stage' is a matter of judgment since several *consecutive* stages may be aggregated into a single stage. For example, considering the evolution of a part in a workshop, traveling to the next machine and then waiting for the temperature to be cool enough before entering this next machine may be considered a single stage. We assume that

- the nature and the order of the stages experienced by each type of resource are known in advance;
- for a given resource, every stage is preceded (followed)—if it is not the first (last) stage—by a *single* other stage.

Consequently, for a given type of resource, if we draw a graph such that nodes (drawn as places in a Petri net) represent stages, and arcs indicate precedence of the upstream stages over the downstream stages, then this graph is a path (i.e. it has neither circuits nor branches) and it reflects the total order of stages.

Obviously, in a workshop, this assumes that a 'deterministic scheduling' of operations has been defined a priori. For example, a machine is supposed to work on a specified sequence of parts in a specified order, each operation being represented by a particular stage. If the stage 'idle' is possible between two such operations (i.e. the machine may deliver the part it holds without receiving its next part—hence it stays alone for a while), then additional stages should be introduced in between. Notice that

- the stage 'idle' is *not* represented by a *unique* node, but that a node of this type is introduced if necessary between any two successive operations;
- a machine may be idle, while still holding the part it was working on, because there is no storage available downstream, and the next operation of this machine cannot be undertaken before the downstream machine can absorb the present part. In this case, it is not necessary to introduce an additional stage, that is, the stage 'working' and the stage 'waiting for downstream delivery' need not be distinguished.

For a part, it is also assumed that the sequence of machines visited is specified in advance, but notice that storages visited should be considered as an operation of the same type as machines visited. Therefore, a storage position, if it is common to several types of parts, must know in advance the order in which it will be visited by these different parts, which is certainly a restrictive constraint in the modeling phase, but this is the price to pay for remaining eventually in the class of event graphs.

Finally, the evolution of a resource is simply a path (in other words, this evolution is represented by a serial automaton). For convenience, each arc will be 'cut' by a bar resembling a transition, which will serve later on for synchronization purposes. Each stage receives a sequence of holding times representing *minimal* times spent in this stage by the successive resources. For example, waiting in a storage should involve a minimal time equal to 0. At any time, the present stage of a given resource is represented by a token marking the corresponding place. Transitions represent changes of stages and they are instantaneous.

Resources may enter the system and leave it after a while (nonreusable resources) or they may revisit the same stages indefinitely because they are 'recycled' (reusable resources). For example, raw materials come in a workshop and leave after a transformation, whereas machines may indefinitely resume their work on the same repetitive sequences of parts. Sometimes, nonreusable resources are tightly associated with reusable resources so that it is only important to model these reusable resources: for example, parts may be fixed on pallets that are recycled after the parts leave the workshop. For reusable resources, we introduce an additional stage called the 'recycling stage', and we put an arc from the last stage (of an elementary sequence) to this recycling stage and another arc from the recycling stage to the first stage. Hence we obtain a circuit. Physically, the recycling stage might represent a certain 'reconditioning' operation (possibly involving other resources too), and therefore it might receive a nonzero holding time (transportation time, set-up time, etc.). However, it will be preferable to suppose that the recycling stage of any resource corresponds to an abstract operation which involves only this resource, and which is immediate (holding time 0). The positioning of this recycling stage with respect to the true reconditioning operation (before or after it) is left to the appraisal of the user in each specific situation. Indeed, each stage along the circuit of the reusable resource where this resource stands alone may be a candidate to play the role of the recycling stage, a remark which should be kept in mind when we speak of canonical initialization later on (this is related to the issue of the point at which one starts a periodic behavior).

2.6.3.3 Synchronization Mechanism

Any elementary operation may involve only one particular type of resource, or it may also involve several different resources. For example, a part waiting in a storage involves both the part and the storage position (whereas an idle machine involves only this machine). It should be realized that, so far, the same physical operation, as long as it involves n different resources simultaneously, has been represented by n different places.

If two stages belonging to two distinct resource paths (or circuits) correspond to the same operation, we must express that these stages are entered and left simultaneously by the two resources (moreover, the holding times of these stages should be the same). This may be achieved by putting two synchronization circuits, one connecting the 'beginning' transitions, the other connecting the 'end' transitions, as indicated in Figure 2.33. In order to comply with the standard event graph representation, we have put new places—represented here in grey color—over the arcs of these circuits. However, these grey places do not represent 'stages' as other places do. They are never marked with tokens and they have holding times equal to 0. Then, it is realized that these circuits involving no tokens and having a total holding time equal to 0 express simultaneity of events, that is entering or leaving the considered stage by anyone of the two resources precedes the same type of event achieved by the other resource and vice versa.

Remark 2.86 What we have just done here is nonstandard from the point of view of Petri net theory, although it is mathematically correct for the purpose of expressing simultaneity. Indeed, having a *live* Petri net which includes circuits with no tokens seems to contradict Theorem 2.38. More specifically, every transition having a 'grey place' upstream (which will never receive tokens) will never fire if we stick to the general rule about how transition firings are enabled. We propose the following (tricky) adaptation of this rule to get out of this contradiction: '*in a timed Petri net, a transition may "borrow" tokens to enable its own firing during a duration of* 0 *time units, that is, provided that it can "return" the same amount of tokens immediately*'. This condition is satisfied for transitions preceded and followed by (the same number of) 'grey places' since tokens may be 'borrowed' in upstream 'grey places' only at the epoch of firing (since those tokens are then immediately available—holding time 0), and, for the same reason, tokens produced in downstream 'grey places' can immediately be 'returned' after firing (they are immediately available when produced).

Mathematically, consider the pair of transitions at the left-hand side of Figure 2.33 and let $x_1(k)$ and $x_2(k)$ denote their respective daters (see Definition 2.57). The 'grey circuit' translates into the following two inequalities:

$$x_1(k) \geq x_2(k) \quad \text{and} \quad x_2(k) \geq x_1(k) \ ,$$

which imply equality: this is exactly what we want and everything is consistent.

On the contrary, the reader should think of what happens if one of the holding times put on 'grey places' is strictly positive and there are still no tokens in the initial marking of these places. ∎

To avoid this discussion, two alternative solutions can be adopted. The first one consists in merging the simultaneous transitions as shown by Figure 2.34, which removes the synchronization circuits and the 'grey places'. We then come up with a representation similar to that of Figure 2.32. A further step towards simplification would be to merge the places and arcs in between so as to arrive at a representation similar to that of Figure 2.31, but then the resource interpretation would be obscured.

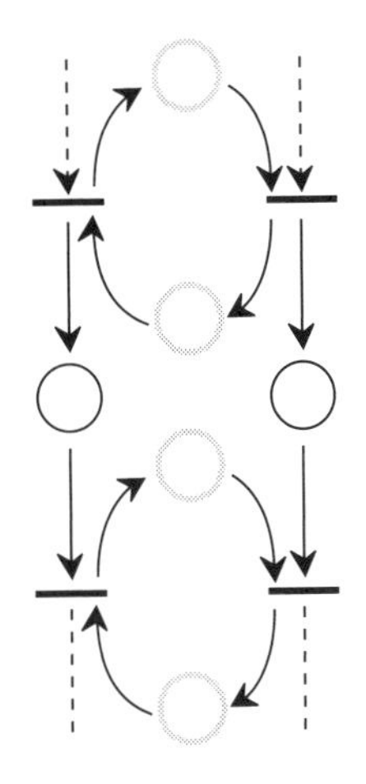

Figure 2.33: Synchronization mechanism

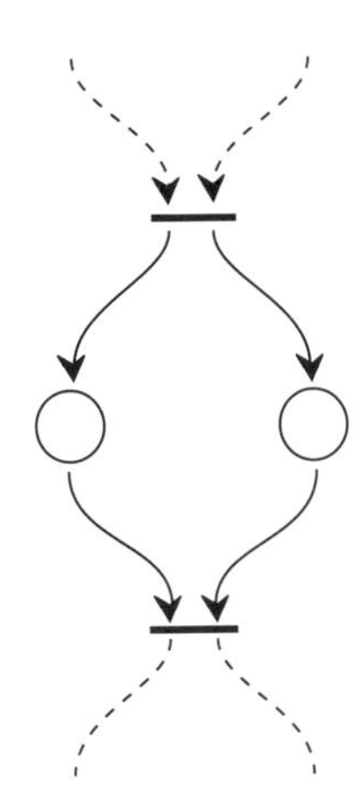

Figure 2.34: Alternative representation of synchronization

The second solution involves the introduction of fake transitions x'_1 and x'_2 upstream the real transitions to be synchronized. This mechanism is explained by Figure 2.35 with the equations proving that the firing times of x_1 and x_2 (denoted after the name of the transitions) are equal.

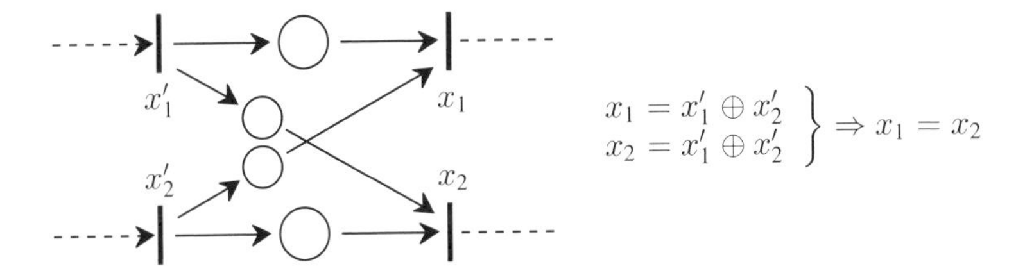

Figure 2.35: Another synchronization mechanism

Notice that the holding times of two synchronized stages, which were equal by construction, can now be different—e.g. one of the two may be taken to be 0—since only the greater holding time is important. All the above considerations extend to the case of n, rather than 2, resources simultaneously. It is realized that, with this approach, every transition has as many output arcs as it has input arcs. A problem arises with resources which are physically merged into a single product (assembly) or split up into two or more products (disassembly). That is, the end nodes of a resource path do not correspond to inlet or outlet nodes of the system. There is a choice of considering that a cluster of two resources that enter the system separately is still made of two resources traveling together (which amounts to doubling all arcs and places after the assembly operation) or to accept that the number of output arcs at some transition (representing the beginning of an assembly operation) be less than the number of input arcs (which means that a new type of resource starts its life at that point of the system). Dual considerations apply to the problem of disassembly.

2.6.3.4 Initialization

As already indicated, during the evolution of the system the present stage of each resource will be marked on the graph by a token in a certain place along the path or the circuit of that resource. Two resources of the same type, e.g. two machines performing exactly the same sequence of operations, may use the same path or circuit as long as they never need to be distinguished. For example, a storage with n positions, and which is dedicated to a single type of stored resource, will be represented by a circuit with two stages, 'occupied' and 'available' (the only distinguishable stages of a single position), and with n tokens the distribution of which in the circuit indicates how many positions are available at any time. For a storage accommodating several types of resources, we refer the reader to the example at the end of this section.

Epochs at which resources move from one stage to the next one will be given by the dater attached to the transition in between. We now define a canonical initial condition. For reusable resources, it corresponds to all tokens put at the corresponding recycling stages. As discussed earlier, these recycling stages are supposed to involve a single type of resource each, and a holding time equal to 0 (therefore, it is irrelevant to know when the tokens had been put there). For any nonreusable resource, since it passes through the system, we first complete its path by adding an inlet transition (upstream from the first place) and an outlet transition (downstream from the last place) so as to attach the epochs of inputs and outputs to these transitions (unless one of these transitions is already represented because the resource participates in an assembly operation, or because it is issued from a disassembly operation). The canonical initial condition for nonreusable resources corresponds to their paths being *empty*: all tokens must be introduced at the inlet transitions *after* the origin of time. Observe that the canonical initial condition is compatible in the sense of Definition 2.61.

From this given canonical initial condition, and given a sequence of epochs at all input transitions at which tokens are introduced into the system (see Definition 2.78), tokens will evolve within the system (whereas other tokens will leave) according to the general rules of timed event graphs, and they will reach some positions at some other given epoch. Obviously, all situations thus obtained, which we may call 'reachable conditions', are also acceptable 'initial conditions' by changing the origin of time to the present time. Such a candidate to play the role of an initial condition obviously fulfills some constraints (see hereafter), and it is not defined only by the positions of tokens (called the 'initial marking'): the time already spent by each token of the initial marking in this position before the (new) origin of time is also part of the definition of the 'initial condition' (at least for places where the holding time is nonzero). Alternatively, the first epoch at which each token of the initial marking may be consumed must be given: this corresponds to the notion of lag time introduced in Definition 2.49. Necessary, but maybe not sufficient, requirements of reachability from the canonical initial condition can be stated for all places included between any pair of synchronized (Figure 2.33) or merged (Figure 2.34) transitions. Say there are n such places, then the same number of tokens must mark all these places, say p tokens per place, and there must exist p n-tuples of tokens (with one token per place) with the same difference between their holding times and their lag times. This assumption that there must exist exactly the same number p of tokens in each of the n places makes this notion of an 'initial condition reachable from the canonical initial condition' even more restrictive than the notion of a 'compatible initial condition' of Definition 2.61.

For example, if we return to the graph of Figure 2.32, with the interpretation of two reusable resources, position (a) corresponds to the canonical initial condition, position (b) is another acceptable initial marking. Suppose now that we add an additional exemplary of the resource represented by the left-hand circuit. Figure 2.36a represents an initial marking which we can interpret, but Figure 2.36b does not, although it is perfectly correct from the abstract point of view of event graphs. Moreover, it may be noted that the graph of Figure 2.36b is not reducible

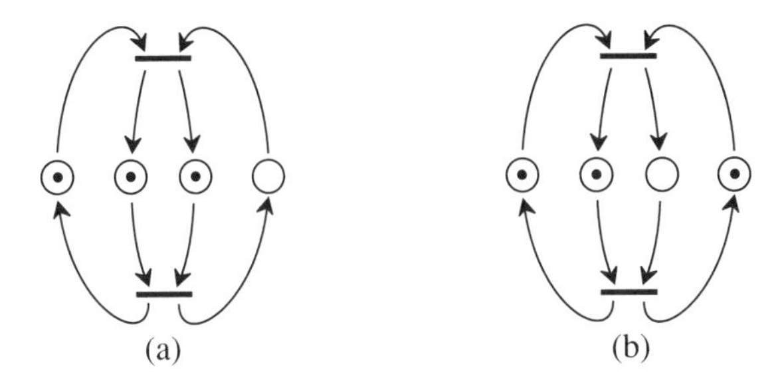

Figure 2.36: Two plus one resources

to something similar to Figure 2.31.

2.6.3.5 An Example

We consider two types of parts, say ♠ and ♡, which can wait in separate buffers of large capacities before being heated individually by a furnace. The furnace deals with parts ♠ and ♡ alternately. The parts can then wait in a stove having the room for three parts, each one of either type, the purpose being to maintain the temperature of the parts until they are assembled in pairs (♠, ♡) by a machine. Finally, parts leave the workshop. A part must stay in the furnace if it cannot enter the stove. In the same way, parts can leave the stove in pairs (♠, ♡) only when the assembly machine can handle them.

Figure 2.37 represents the stage sequences of parts ♠ and ♡ as vertical paths on the left-hand and right-hand sides, respectively. In the middle, from top to bottom, the three circuits represent the stage sequences of the furnace, of the stove and of the machine. Each stage is labeled by a letter and Table 2.1 gives the interpretation of these stages together with their holding times.

Table 2.1: An example: stage interpretation

Stage label	Stage interpretation	Holding time
A	Part ♠ waits in buffer	0
B	Part ♠ stays in furnace	α_1
C	Part ♠ waits in stove	0
D	Part ♠ is assembled with part ♡	α_2
E	Part ♡ waits in buffer	0
F	Part ♡ stays in furnace	α_3
G	Part ♡ waits in stove	0
H	Part ♡ is assembled with part ♠	α_2
I	Furnace waits for part of type ♠	0
J	Furnace holds part of type ♠	α_1
K	Furnace waits for part of type ♡	0
L	Furnace holds part of type ♡	α_3
M	One position in stove waits for part of type ♠	0
N	One position in stove holds part of type ♠	0
O	One position in stove waits for part of type ♡	0
P	One position in stove holds part of type ♡	0
Q	Machine assembles a pair (♠, ♡)	α_2
R	Machine waits for a pair (♠, ♡)	0

The transitions which are synchronized are connected by a dotted line. The initial marking assumes that the furnace starts with a part ♠ and that two positions out of the three in the stove accept a part ♠ first.

Figure 2.38 shows the reduced event graph obtained by merging the synchronized transitions and the stages representing the same operation. However, we keep multiple labels when appropriate so that the order of multiplicity of places between synchronized transitions is still apparent. Single labels on places in circuits indicate possible recycling stages. Observe that, for any transition, the total number of labels of places upstream balances the total number of labels of places downstream. Indeed, the Petri net of Figure 2.37 (with transitions connected by dotted lines merged together) is strictly conservative (Definition 2.35) whereas that of Figure 2.38 is conservative (Definition 2.36) with weights (used in that definition) indicated by the number of labels.

2.7 NOTES

Graph theory is a standard course in many mathematical curricula. The terminology varies, however. Here we followed the terminology as used in [67] as much as possible. Karp's algorithm, the subject of §2.3.2, was first published in [73]. Since the 1970s, a lot of attention has been paid to the theory of Petri nets. This theory has the potential of being suitable as an excellent modeling aid in many fields of application. Its modeling power is larger than the one

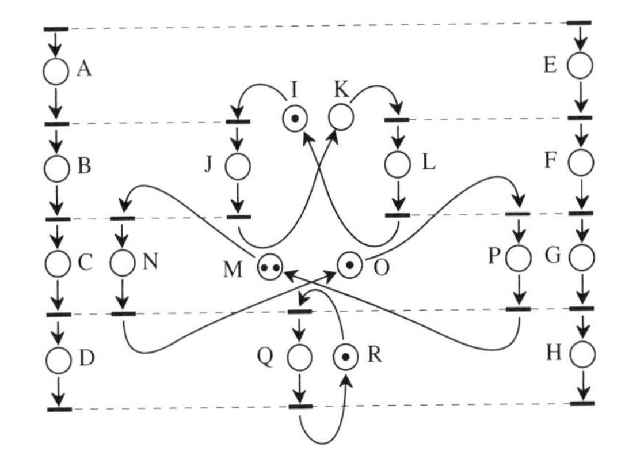

Figure 2.37: Expanded model

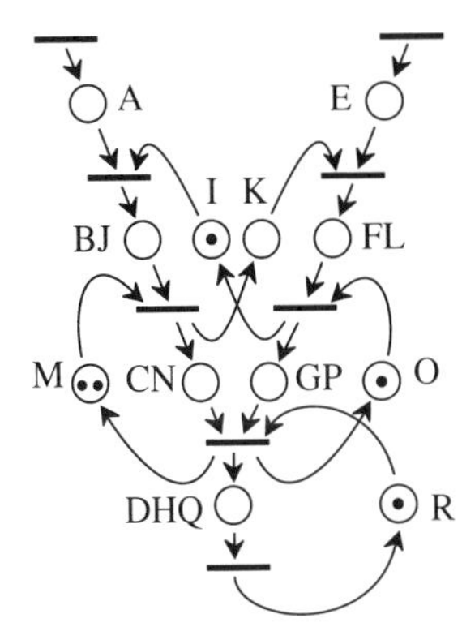

Figure 2.38: Reduced model

of automata. If one were to allow inhibitor arcs, see [1], then the modeling power is the same as that of Turing machines. A 'system theory' including maximization, addition as well as negation (which is what an inhibitor arc represents), has not yet been developed. An excellent overview of the theory of Petri nets is given in [96], where also many other references can be found. The section on timed Petri nets, however, is rather brief in [96]. Some material on cycle times can be found in [47] and [115], which is also discussed in Chapter 9. A good, though somewhat dated, introduction to Petri nets is [108]. Other sources which contain a lot of material on Petri nets are [30] and [29]. For a recent discussion on modeling power related to Petri nets, see [82]. Section 2.5 on equations for timed Petri nets is mainly based on [39] for the constant timing case and on [11] for the general case. Equivalence of systems represented by different graphs is also discussed in [84]. This reference, however, deals with systolic systems, in which there is a fixed clock frequency. Some of the results obtained there also seem plausible within the timed event graph setting. Relationships between graphs and binary dynamic systems are described in [25].

In [92] a novel scheme, called *kanban*, is described and analyzed for the coordination of tokens in an event graph. The essence is that the introduction of a new circuit with enough tokens regulates the 'speed' of the original event graph and it controls the number of tokens in various places.

For a recent development in continuous Petri nets, where the the number of tokens is real rather than integer valued, see [53] and [99]. In the latter reference some results given in Chapter 3 for 'discrete' Petri nets are directly extended to continuous ones.

PART II

Algebra

CHAPTER 3

Max-Plus Algebra

3.1 INTRODUCTION

In this chapter we systematically revisit the classical algebraic structures used in conventional calculus and we substitute the idempotent semifield $\mathbb{R}_{\max}$ (the set of real numbers endowed with the operations max and plus) for the field of scalars. The purpose is to provide the mathematical tools needed to study linear dynamical systems in $(\mathbb{R}_{\max})^n$.

This chapter is divided in three main parts. In the first part we study linear systems of equations and polynomial functions in $\mathbb{R}_{\max}$. In the second part we consider a more advanced topic which can be skipped in a first reading. This topic is the problem of the linear closure of $\mathbb{R}_{\max}$ and its consequences for solving systems of linear and polynomial equations (the linear closure is the extension of a set in such a way that any nondegenerated linear equation has one and only one solution). The third part is concerned with a max-plus extension of the Perron-Frobenius theory. It gives conditions under which event graphs reach a periodic behavior and it characterizes their periodicities. It can be seen as a more advanced topic of the spectral theory of max-plus matrices given in the first part of this chapter.

We first introduce the algebraic structure $\mathbb{R}_{\max}$ and we study its basic properties.

3.1.1 Definitions

Definition 3.1 (Semifield) *A semifield $\mathcal{K}$ is a set endowed with two operations $\oplus$ and $\otimes$ such that:*

- *the operation $\oplus$ is associative, commutative and has a* zero element *ε;*
- *the operation $\otimes$ defines a group on $\mathcal{K}_* \stackrel{\text{def}}{=} \mathcal{K} \setminus \{\varepsilon\}$, it is distributive with respect to $\oplus$ and its* identity element *e satisfies $\varepsilon \otimes e = e \otimes \varepsilon = \varepsilon$.*

We say that the semifield is

- idempotent *if the first operation is idempotent, that is, if $a \oplus a = a$, $\forall a \in \mathcal{K}$;*
- commutative *if the group is commutative.*

Theorem 3.2 *The zero element ε of an idempotent semifield is* absorbing *for the second operation, that is $\varepsilon \otimes a = a \otimes \varepsilon = \varepsilon, \forall a \in \mathcal{K}$.*

Proof We have that

$$\varepsilon = \varepsilon e = \varepsilon(\varepsilon \oplus e) = \varepsilon^2 \oplus \varepsilon = \varepsilon^2 \ ,$$

and then,

$$\forall a \in \mathcal{K}_* \ , \quad \varepsilon = \varepsilon e = \varepsilon a^{-1} a = \varepsilon(a^{-1} \oplus \varepsilon)a = \varepsilon a^{-1} a \oplus \varepsilon^2 a = \varepsilon^2 a = \varepsilon a \ .$$

■

Definition 3.3 (The algebraic structure $\mathbb{R}_{\max}$) *The symbol $\mathbb{R}_{\max}$ denotes the set $\mathbb{R} \cup \{-\infty\}$ with* max *and* $+$ *as the two binary operations* $\oplus$ *and* $\otimes$*, respectively.*

We call this structure the max-plus algebra. Sometimes this is also called an ordered group. We remark that the natural order on $\mathbb{R}_{\max}$ may be defined using the $\oplus$ operation

$$a \leq b \quad \text{if} \quad a \oplus b = b \ .$$

Definition 3.4 (The algebraic structure $\overline{\mathbb{R}}_{\max}$) *The set $\mathbb{R} \cup \{-\infty\} \cup \{+\infty\}$ endowed with the operations* max *and* $+$ *as* $\oplus$ *and* $\otimes$ *and with the convention that $(-\infty) + \infty = -\infty$ is denoted $\overline{\mathbb{R}}_{\max}$. The element $+\infty$ is denoted $\top$.*

Theorem 3.5 *The algebraic structure $\mathbb{R}_{\max}$ is an idempotent commutative semifield.*

The proof is straightforward. If we compare the properties of $\oplus$ and $\otimes$ with those of $+$ and $\times$, we see that:

- we have lost the symmetry of addition (for a given a, an element b does not exist such that $\max(b, a) = -\infty$ whenever $a \neq -\infty$);
- we have gained the idempotency of addition;
- there are no zero divisors in $\mathbb{R}_{\max}$ ($a \oplus b = -\infty \Rightarrow a = -\infty$ or $b = -\infty$).

If we try to make some algebraic calculations in this structure, we soon realize that idempotency is as useful as the existence of a symmetric element in the simplification of formulæ. For example, the analogue of the binomial formula

$$(a+b)^n = \begin{pmatrix} n \\ 0 \end{pmatrix} a^n + \begin{pmatrix} n \\ 1 \end{pmatrix} a^{n-1}b + \cdots + \begin{pmatrix} n \\ n-1 \end{pmatrix} ab^{n-1} + \begin{pmatrix} n \\ 0 \end{pmatrix} b^n$$

is $n \max(a, b) = \max(na, nb)$, which is much simpler. On the other hand, we now face the difficulty that the max operation is no longer cancellative, e.g. $\max(a, b) = b$ does not imply that $a = -\infty$.

3.1.2 Notation

First of all, to emphasize the analogy with conventional calculus, max has been denoted $\oplus$, and $+$ has been denoted $\otimes$. We also introduce the symbol $\not{o}$ for the conventional $-$ (the inverse operation of $+$ which plays the role of multiplication, that is, the 'division'). Hence $a \not{o} b$ means $a - b$. Another notation for $a \not{o} b$ is the two-dimensional display notation

$$\frac{a}{b} \; .$$

We will omit the sign $\otimes$ if this does not lead to confusion. To prevent mistakes, we use ε and e for the 'zero' and the 'one', that is, the neutral elements of $\oplus$ and $\otimes$, respectively, namely $-\infty$ and 0. To get the reader acquainted with this new notation, we propose the following table.

$\mathbb{R}_{\max}$ notation	Conventional notation	$=$
$2 \oplus 3$	$\max(2,3)$	3
$1 \oplus 2 \oplus 3 \oplus 4 \oplus 5$	$\max(1,2,3,4,5)$	5
$2 \otimes 3 = 5$	$2 + 3$	5
$2 \oplus \varepsilon$	$\max(2, -\infty)$	2
$\varepsilon = \varepsilon \otimes 2$	$-\infty + 2$	$-\infty$
$(-1) \otimes 3$	$-1 + 3$	2
$e \otimes 3$	$0 + 3$	3
$3^2 = 2^3 = 3 \otimes 3 = 2 \otimes 2 \otimes 2$	$3 \times 2 = 2 \times 3 = 3 + 3 = 2 + 2 + 2$	6
$e = e^2 = 2^0$	$0 \times 2 = 2 \times 0$	0
$(2 \otimes 3) \not{o} (2 \oplus 3)$	$(2 + 3) - \max(2,3)$	2
$(2 \oplus 3)^3 = 2^3 \oplus 3^3$	$3 \times \max(2,3) = \max(3 \times 2, 3 \times 3)$	9
$6 \not{o} e$	$6 - 0$	6
$e \not{o} 3$	$0 - 3$	-3
$\sqrt[2]{8}$	$8/2$	4
$\sqrt[5]{15}$	$15/5$	3

There is no distinction, hence there are risks of confusion, between the two systems of notation as far as the power operation is concerned. As a general rule, a formula is written in one system of notation. Therefore if, in a formula, an operator of the max-plus algebra appears explicitly, then usually all the operators of this formula are max-plus operators.

3.1.3 The min Operation in the Max-Plus Algebra

It is possible to derive the min operation from the two operations $\otimes$ and $\oplus$ as follows:

$$\min(a,b) = \frac{ab}{a \oplus b} \; .$$

Let us now prove the classical properties of the min by pure rational calculations in the max-plus algebra.

- $-\min(a,b) = \max(-a,-b)$:

$$\frac{e}{\frac{ab}{a \oplus b}} = \frac{a \oplus b}{ab} = \frac{e}{b} \oplus \frac{e}{a} \; ;$$

- $-\max(a,b) = \min(-a,-b)$:

$$\frac{e}{a \oplus b} = \frac{\frac{e}{ab}}{\frac{a \oplus b}{ab}} = \frac{\frac{e}{a} \otimes \frac{e}{b}}{\frac{e}{a} \oplus \frac{e}{b}} \; ;$$

- $\min(a, \min(b,c)) = \min(\min(a,b),c)$:

$$\frac{a \frac{bc}{b \oplus c}}{a \oplus \frac{bc}{b \oplus c}} = \frac{abc}{ab \oplus ac \oplus bc}$$

  and the symmetry of the formula with respect to a, b and c proves the result;

- $\max(c, \min(a,b)) = \min(\max(c,a), \max(c,b))$:

$$\left\{ c \oplus \frac{ab}{a \oplus b} = \frac{(c \oplus a)(c \oplus b)}{(c \oplus a) \oplus (c \oplus b)} \right\} \Leftrightarrow \left\{ \frac{ca \oplus cb \oplus ab}{a \oplus b} = \frac{(c \oplus a)(c \oplus b)}{a \oplus b \oplus c} \right\}$$

$$\Leftrightarrow \{(ca \oplus cb \oplus ab)(a \oplus b \oplus c) = (c \oplus a)(c \oplus b)(a \oplus b)\} \; .$$

  To check the last identity, we consider the expressions in both sides as polynomials in c and we first remark that the coefficient of c^2, namely $a \oplus b$, is the same in both sides. The coefficient of c^0, namely $ab(a \oplus b)$, also is the same in both sides. Now, considering the coefficient of c, it is equal to $(a \oplus b)^2 \oplus ab$ in the left-hand side, and to $(a \oplus b)^2$ in the right-hand side: these two expressions are clearly always equal.

- $\min(c, \max(a,b)) = \max(\min(c,a), \min(c,b))$:

$$\left\{ \frac{c(a \oplus b)}{c \oplus a \oplus b} = \frac{ca}{c \oplus a} \oplus \frac{cb}{c \oplus b} \right\} \Leftrightarrow \left\{ \frac{a \oplus b}{c \oplus a \oplus b} = \frac{a}{c \oplus a} \oplus \frac{b}{c \oplus b} \right\} \; .$$

  The latter identity is amenable to the same verification as earlier.

3.2 MATRICES IN $\mathbb{R}_{\max}$

In this section we are mainly concerned with systems of linear equations. There are two kinds of linear systems in $\mathbb{R}_{\max}$ for which we are able to compute solutions: $x = Ax \oplus b$ and $Ax = b$ (the general system being $Ax \oplus b = Cx \oplus d$). We also study the spectral theory of matrices. There exist good notions of eigenvalue and eigenvector but there is often only one eigenvalue: this occurs when the precedence graph associated with the matrix is strongly connected (see Theorem 2.14).

3.2.1 Linear and Affine Scalar Functions

Definition 3.6 (Linear function) *The function $f : \mathbb{R}_{\max} \to \mathbb{R}_{\max}$ is linear if it satisfies*

$$f(c) = c \otimes f(e) \ , \quad \forall c \in \mathbb{R}_{\max} \ .$$

Thus any linear function is of the form $y = f(c) = a \otimes c$, where $a = f(e)$. The graph of such a function consists of a straight line with slope equal to one and which intersects the y-axis at a (see Figure 3.1).

Definition 3.7 (Affine function) *The function $f : \mathbb{R}_{\max} \to \mathbb{R}_{\max}$, $f(c) = ac \oplus b, a \in \mathbb{R}_{\max}, b \in \mathbb{R}_{\max}$ is called affine.*

Observe that, as usual, $b = f(\varepsilon)$ and $a = \lim_{c \to \infty} f(c) \not/ c$, but here the limit is reached for a finite value of x (see Figure 3.2).

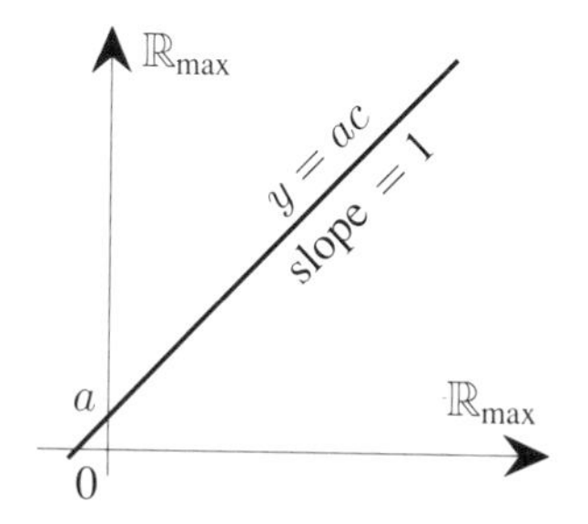

Figure 3.1: A linear function

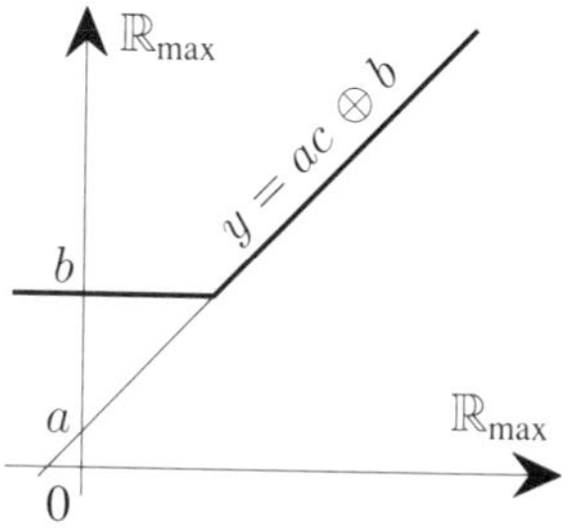

Figure 3.2: An affine function

In the primary school, the first algebraic problem that we have to solve is to find the solution of a scalar affine equation.

Definition 3.8 (Affine equation) *The general scalar affine equation is*

$$ax \oplus b = a'x \oplus b' \ . \tag{3.1}$$

Indeed, since $\oplus$ has no inverse, Equation (3.1) cannot be reduced to the usual form $ax \oplus b = \varepsilon$, which motivates the definition above.

Theorem 3.9 *The solution of the general scalar affine equation is obtained as follows:*

- *if*

$$((a' < a) \ \text{and} \ (b < b')) \quad \text{or} \quad ((a < a') \ \text{and} \ (b' < b)) \tag{3.2}$$

  hold true, then the solution is unique and it is given by $x = (b \oplus b') \not/ (a \oplus a')$;

- *if $a \neq a'$, $b \neq b'$, and (3.2) does not hold, no solutions exist in $\mathbb{R}_{\max}$;*

- *if* $a = a'$ *and* $b \neq b'$*, the solution is nonunique and all solutions are given by* $x \geq (b \oplus b') \oslash a$*;*
- *if* $a \neq a'$ *and* $b = b'$*, the solution is nonunique and all solutions are given by* $x \leq b \oslash (a \oplus a')$*;*
- *if* $a = a'$ *and* $b = b'$*, all* $x \in \mathbb{R}$ *are solutions.*

The proof is straightforward from the geometric interpretation of affine equations as depicted in Figure 3.3.

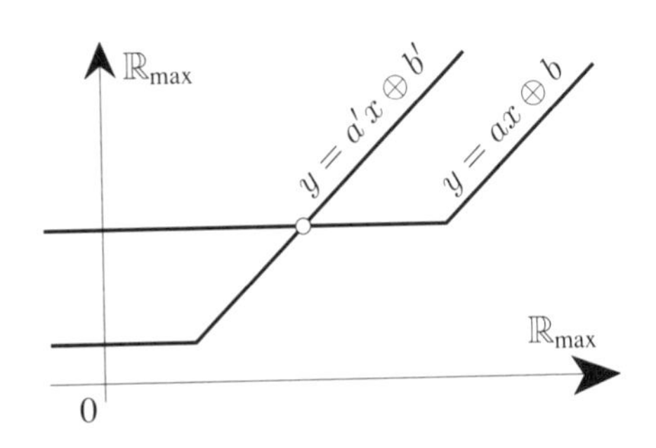

Figure 3.3: An affine equation

In practice, it is better to simplify (3.1) before solving it. For example, if $a > a'$ and $b' > b$, then $ax \oplus b = a'x \oplus b' \Leftrightarrow ax = b'$. Let us give all different kinds of simplified equations that may come up.

Definition 3.10 (Canonical form of an affine equation) *An affine equation is in canonical form if it is in one of the simplified forms:*

- $ax = b$*;*
- $ax \oplus b = \varepsilon$*;*
- $ax \oplus b = ax$*;*
- $ax \oplus b = b$*.*

3.2.2 Structures

The moduloid structure is a kind of module structure which, in turn, is a kind of vector space, that is, a set of vectors with an internal operation and an external operation defined over an idempotent semifield.

Definition 3.11 (Moduloid) *A moduloid* $\mathcal{M}$ *over an idempotent semifield* $\mathcal{K}$ *(with operations* $\oplus$ *and* $\otimes$*, zero element* ε *and identity element* e*) is a set endowed with*

- *an internal operation also denoted* $\oplus$ *with a zero element also denoted* ε*;*

- *an external operation defined on $\mathcal{K} \times \mathcal{M}$ with values in $\mathcal{M}$ indicated by the simple juxtaposition of the scalar and vector symbols;*

which satisfies the following properties:

- *$\oplus$ is associative, commutative;*
- *$\alpha(x \oplus y) = \alpha x \oplus \alpha y$;*
- *$(\alpha \oplus \beta)x = \alpha x \oplus \beta x$;*
- *$\alpha(\beta x) = (\alpha\beta)x$;*
- *$ex = x$;*
- *$\varepsilon x = \varepsilon$;*

for all $\alpha, \beta \in \mathcal{K}$ and all $x, y \in \mathcal{M}$.

We will only be concerned with some special cases of such a structure.

Example 3.12 $(\mathbb{R}_{\max})^n$ is a moduloid over $\mathbb{R}_{\max}$. Its zero element is $(\varepsilon, \ldots, \varepsilon)'$. ■

Other examples of moduloids will be presented later on.

Definition 3.13 (Idempotent algebra) *A moduloid with an additional internal operation also denoted $\otimes$ is called an idempotent algebra if $\otimes$ is associative, if it has an identity element also denoted e, and if it is distributive with respect to $\oplus$.*

This idempotent algebra is the main structure in which the forthcoming system theory is going to be developed.

Example 3.14 Let $(\mathbb{R}_{\max})^{n\times n}$ be the set of $n \times n$ matrices with coefficients in $\mathbb{R}_{\max}$ endowed with the following two internal operations:

- the componentwise addition denoted $\oplus$;
- the matrix multiplication already used in Chapters 1 and 2 denoted $\otimes$:

$$(A \otimes B)_{ij} = \bigoplus_{k=1}^{n} A_{ik} \otimes B_{kj} \ ;$$

and the external operation:

- $\forall \alpha \in \mathbb{R}_{\max}, \forall A \in (\mathbb{R}_{\max})^{n\times n}, \alpha A = (\alpha A_{ij})$.

The set $(\mathbb{R}_{\max})^{n\times n}$ is an idempotent algebra with

- the zero matrix, again denoted ε, which has all its entries equal to ε;
- the identity matrix, again denoted e, which has the diagonal entries equal to e and the other entries equal to ε. ■

3.2.3 Systems of Linear Equations in $(\mathbb{R}_{\max})^n$

In this subsection we are mainly interested in systems of linear equations. To define such systems, we use matrix notation. A linear mapping $(\mathbb{R}_{\max})^n \to (\mathbb{R}_{\max})^n$ will be represented by a matrix as it was done in the previous chapters. With the max-plus algebra, the general system of equations is

$$Ax \oplus b = Cx \oplus d \ ,$$

where A and C are $n \times n$ matrices and b and d are n-vectors. This system can be put in canonical form in the same way as we have done in the scalar case.

Definition 3.15 (Canonical form of a system of affine equations) *The system $Ax \oplus b = Cx \oplus d$ is said to be in canonical form if A, C, b, and d satisfy*

- $C_{ij} = \varepsilon$ *if* $A_{ij} > C_{ij}$, *and* $A_{ij} = \varepsilon$ *if* $A_{ij} < C_{ij}$;
- $d_i = \varepsilon$ *if* $b_i > d_i$, *and* $b_i = \varepsilon$ *if* $b_i < d_i$.

Example 3.16 Consider the system

$$\begin{pmatrix} 3 & 2 \\ \varepsilon & 2 \end{pmatrix} \begin{pmatrix} x_1 \\ x_2 \end{pmatrix} \oplus \begin{pmatrix} 1 \\ 2 \end{pmatrix} = \begin{pmatrix} 4 & 1 \\ 1 & 1 \end{pmatrix} \begin{pmatrix} x_1 \\ x_2 \end{pmatrix} \oplus \begin{pmatrix} e \\ 3 \end{pmatrix} ,$$

which can be simplified as follows:

$$\begin{pmatrix} \varepsilon & 2 \\ \varepsilon & 2 \end{pmatrix} \begin{pmatrix} x_1 \\ x_2 \end{pmatrix} \oplus \begin{pmatrix} 1 \\ \varepsilon \end{pmatrix} = \begin{pmatrix} 4 & \varepsilon \\ 1 & \varepsilon \end{pmatrix} \begin{pmatrix} x_1 \\ x_2 \end{pmatrix} \oplus \begin{pmatrix} \varepsilon \\ 3 \end{pmatrix} ,$$

which implies

$$\left.\begin{array}{ll} 2x_2 \oplus 1 & = 4x_1 \\ 2x_2 & = 1x_1 \oplus 3 \end{array}\right\} \Rightarrow 4x_1 = 1x_1 \oplus 3 \Rightarrow 4x_1 = 3 \Rightarrow x_1 = -1 \Rightarrow x_2 = 1 \ .$$

This system has a solution. In general, a linear system may or may not have a solution. Moreover, even if a solution exists, it may be nonunique. ■

There are two classes of linear systems for which we have a satisfactory theory, namely,

- $x = Ax \oplus b$;
- $Ax = b$.

Let us study the former case first.

3.2.3.1 Solution of $x = Ax \oplus b$

Theorem 3.17 *If there are only circuits of nonpositive weight in $\mathcal{G}(A)$, there is a solution to $x = Ax \oplus b$ which is given by $x = A^*b$. Moreover, if the circuit weights are negative, the solution is unique.*

The reader should recall the definition of A^* as given by (1.22).

Proof If A^*b does exist, it is a solution; as a matter of fact,

$$A(A^*b) \oplus b = (e \oplus AA^*)b = A^*b \ .$$

Existence of A^*b. The meaning of $(A^*)_{ij}$ is the maximum weight of all paths of any length from j to i. Thus, a necessary and sufficient condition for the existence of $(A^*)_{ij}$ is that no strongly connected components of $\mathcal{G}(A)$ have a circuit with positive weight. Otherwise, there would exist a path from j to i of arbitrarily large weight for all j and i belonging to the strongly connected component which includes the circuit of positive weight (by traversing this circuit a sufficient number of times).

Uniqueness of the solution. Suppose that x is a solution of $x = Ax \oplus b$. Then x satisfies

$$\begin{aligned} x &= b \oplus Ab \oplus A^2x \ , \\ x &= b \oplus Ab \oplus \cdots \oplus A^{k-1}b \oplus A^kx \ , \end{aligned} \tag{3.3}$$

and thus $x \geq A^*b$. Moreover, if all the circuits of the graph have negative weights, then $A^k \to \varepsilon$ when $k \to \infty$. Indeed, the entries of A^k are the weights of the paths of length k which necessarily traverse some circuits of A a number of times going to ∞ with k, but the weights of these circuits are all negative. Using this property in Equation (3.3) for k large enough, we obtain that $x = A^*b$. ■

Remark 3.18 If the maximum circuit weight is zero, a solution does exist, but there is no uniqueness anymore. For example, the equation $x = x \oplus b$ admits the solution $x = a^*b = b$ but all $x > b$ are solutions too. ■

Example 3.19 Consider the two-dimensional equation

$$x = \begin{pmatrix} -1 & 2 \\ -3 & -1 \end{pmatrix} x \oplus \begin{pmatrix} e \\ 2 \end{pmatrix} .$$

Then,

$$b = \begin{pmatrix} e \\ 2 \end{pmatrix}, \quad Ab = \begin{pmatrix} 4 \\ 1 \end{pmatrix}, \quad A^2b = \begin{pmatrix} 3 \\ 1 \end{pmatrix}, \quad A^3b = \begin{pmatrix} 3 \\ e \end{pmatrix}, \quad A^4b = \begin{pmatrix} 2 \\ e \end{pmatrix} \ldots .$$

Thus,

$$x = \begin{pmatrix} e \\ 2 \end{pmatrix} \oplus \begin{pmatrix} 4 \\ 1 \end{pmatrix} \oplus \begin{pmatrix} 3 \\ e \end{pmatrix} \oplus \begin{pmatrix} 2 \\ e \end{pmatrix} \oplus \cdots = \begin{pmatrix} 4 \\ 2 \end{pmatrix} .$$

This is the unique solution. ■

Theorem 3.20 *If $\mathcal{G}(A)$ has no circuit with positive weight, then*

$$A^* = e \oplus A \oplus \cdots \oplus A^{n-1} ,$$

where n is the dimension of matrix A.

Proof All the paths of length greater than or equal to n are necessarily made up of a circuit and of a path with length strictly less than n. Therefore, because the weights of circuits are nonpositive by assumption, we have

$$\forall m \geq n , \quad A^m \leq e \oplus \cdots \oplus A^{n-1} .$$

■

Returning to Example 3.19, we remark that $A^2 b, A^3 b, \ldots$ are less than $b \oplus Ab$.

3.2.3.2 Solution of $Ax = b$

The second class of linear systems for which we can obtain a general result consists of the systems $Ax = b$. However, we must first consider the problem in $\overline{\mathbb{R}}_{\max}$ rather than in $\mathbb{R}_{\max}$, and second, we must somewhat weaken the notion of 'solution'. A *subsolution* of $Ax = b$ is an x which satisfies $Ax \leq b$, where the order relation on the vectors can also be defined by $x \leq y$ if $x \oplus y = y$.

Theorem 3.21 *Given an $n \times n$ matrix A and an n-vector b with entries in $\overline{\mathbb{R}}_{\max}$, the greatest subsolution of $Ax = b$ exists and is given by*

$$-x_j = \max_i (-b_i + A_{ij}) .$$

For reasons that will become apparent in §4.4.4 and §4.6.2, the vector form of this formula can be written $e \phi x = (e \phi b) A$.

Proof We have that

$$\begin{aligned}
\{Ax \leq b\} \quad &\Leftrightarrow \quad \left\{ \bigoplus_j A_{ij} x_j \leq b_i , \quad \forall i \right\} \\
&\Leftrightarrow \quad \{ x_j \leq b_i - A_{ij} , \quad \forall i, j \} \\
&\Leftrightarrow \quad \left\{ x_j \leq \min_i \left(b_i - A_{ij} \right) , \quad \forall j \right\} \\
&\Leftrightarrow \quad \left\{ -x_j \geq \max_i \left(-b_i + A_{ij} \right) , \quad \forall j \right\} .
\end{aligned}$$

Conversely, it can be checked similarly that the vector x defined by $-x_j = \max_i \left(-b_i + A_{ij} \right), \forall j$, is a subsolution. Therefore, it is the greatest one. ■

As a consequence, in order to attempt to solve the system $Ax = b$, we may first compute its greatest subsolution and then check by inspection whether it satisfies the equality.

Example 3.22 Let us compute the greatest subsolution of the following equality:

$$\begin{pmatrix} 2 & 3 \\ 4 & 5 \end{pmatrix} \begin{pmatrix} x_1 \\ x_2 \end{pmatrix} = \begin{pmatrix} 6 \\ 7 \end{pmatrix} .$$

According to the preceding considerations, let us first compute

$$(e \not/ b)A = \begin{pmatrix} -6 & -7 \end{pmatrix} \begin{pmatrix} 2 & 3 \\ 4 & 5 \end{pmatrix} = \begin{pmatrix} -3 & -2 \end{pmatrix} .$$

Then the greatest subsolution is $(x_1, x_2) = (3, 2)$; indeed,

$$\begin{pmatrix} 2 & 3 \\ 4 & 5 \end{pmatrix} \begin{pmatrix} 3 \\ 2 \end{pmatrix} = \begin{pmatrix} 5 \\ 7 \end{pmatrix} \leq \begin{pmatrix} 6 \\ 7 \end{pmatrix} .$$

It is easily verified that the second inequality would not be satisfied if we increase x_1 and/or x_2. Therefore, the first inequality cannot be reduced to an equality. ■

3.2.4 Spectral Theory of Matrices

Given a matrix A with entries in $\mathbb{R}_{\max}$, we consider the problem of existence of eigenvalues and eigenvectors, that is, the existence of (nonzero) λ and x such that

$$Ax = \lambda x \ . \tag{3.4}$$

The main result is as follows.

Theorem 3.23 *If A is irreducible, or equivalently if $\mathcal{G}(A)$ is strongly connected, there exists one and only one eigenvalue (but possibly several eigenvectors). This eigenvalue is equal to the maximum cycle mean of the graph (see § 2.3.2):*

$$\lambda = \max_{\zeta} \frac{|\zeta|_{\mathrm{w}}}{|\zeta|_{\mathrm{l}}} ,$$

where ζ ranges over the set of circuits of $\mathcal{G}(A)$.

Proof

Existence of x and λ. Consider matrix $B = A \not/ \lambda \stackrel{\text{def}}{=} (e \not/ \lambda)A$, where $\lambda = \max_{\zeta} |\zeta|_{\mathrm{w}}/|\zeta|_{\mathrm{l}}$. The maximum circuit weight of $\mathcal{G}(B)$ is e. Hence B^* and $B^+ = BB^*$ exist. Matrix B^+ has some columns with diagonal entries equal to e. To prove this claim, pick a node k of a circuit ξ such that $\xi \in \arg\max_{\zeta} |\zeta|_{\mathrm{w}}/|\zeta|_{\mathrm{l}}$. The maximum weight of paths from k to k is e. Therefore we have $e = B^+_{kk}$. Let $B_{\cdot k}$ denote the k-th column of B. Then, since, generally speaking, $B^+ = BB^*$ and $B^* = e \oplus B^+$ (e the identity matrix), for that k,

$$B^+_{\cdot k} = B^*_{\cdot k} \Rightarrow BB^*_{\cdot k} = B^+_{\cdot k} = B^*_{\cdot k} \Rightarrow AB^*_{\cdot k} = \lambda B^*_{\cdot k} \ .$$

Hence $x = B^*_{\cdot k} = B^+_{\cdot k}$ is an eigenvector of A corresponding to the eigenvalue λ. The set of nodes of $\mathcal{G}(A)$ corresponding to nonzero entries of x is called the *support* of x.

Graph interpretation of λ. If λ satisfies Equation (3.4), there exists a nonzero component of x, say x_{i_1}. Then we have $(Ax)_{i_1} = \lambda x_{i_1}$ and there exists an index i_2 such that $A_{i_1 i_2} x_{i_2} = \lambda x_{i_1}$. Hence $x_{i_2} \neq \varepsilon$ and $A_{1 i_2} \neq \varepsilon$. We can repeat this argument and obtain a sequence $\{i_j\}$ such that $A_{i_{j-1} i_j} x_{i_j} = \lambda x_{i_{j-1}}$, $x_{i_j} \neq \varepsilon$ and $A_{i_{j-1} i_j} \neq \varepsilon$. At some stage we must reach an index i_l already encountered in the sequence since the number of nodes is finite. Therefore, we obtain a circuit $\beta = (i_l, i_m, \ldots, i_{l+1}, i_l)$. By multiplication along this circuit, we obtain

$$A_{i_l i_{l+1}} A_{i_{l+1} i_{l+2}} \ldots A_{i_m i_l} x_{i_{l+1}} x_{i_{l+2}} \ldots x_{i_m} x_{i_l} = \lambda^{m-l+1} x_{i_l} x_{i_{l+1}} \ldots x_{i_m} \ .$$

Since $x_{i_j} \neq \varepsilon$ for all i_j, we may simplify the equation above which shows that λ^{m-l+1} is the weight of the circuit of length $m - l + 1$, or, otherwise stated, λ is the average weight of circuit β. Observe that this part of the proof did not use the irreducibility assumption.

If A is irreducible, all the components of x are different from ε. Suppose that the support of x does not cover the whole graph. Then, there are arcs going from the support of x to other nodes because the graph $\mathcal{G}(A)$ has only one strongly connected component. Therefore, the support of Ax is larger than the support of x, which contradicts Equation (3.4).

Uniqueness in the irreducible case. Consider any circuit $\gamma = (i_1, \ldots, i_p, i_1)$ such that its nodes belong to the support of x (here any node of $\mathcal{G}(A)$). We have

$$A_{i_2 i_1} x_{i_1} \leq \lambda x_{i_2} \ , \quad \ldots \quad , \quad A_{i_p i_{p-1}} x_{i_{p-1}} \leq \lambda x_{i_p} \ , \quad A_{i_1 i_p} x_{i_p} \leq \lambda x_{i_1} \ .$$

Hence, by the same argument as in the paragraph on the graph interpretation of λ, we see that λ is greater than the average weight of γ. Therefore λ is the maximum cycle mean and thus it is unique. ■

It is important to understand the role of the support of x in the previous proof. If $\mathcal{G}(A)$ is not strongly connected, the support of x is not necessarily the whole set of nodes and, in general, there is no unique eigenvalue (see Example 3.26 below).

Remark 3.24 The part of the proof on the graph interpretation of λ indeed showed that, for a general matrix A, any eigenvalue is equal to some cycle mean. Therefore the maximum cycle mean is equal to the maximum eigenvalue of the matrix. ■

Example 3.25 (Nonunique eigenvector) With the only assumption of Theorem 3.23 on irreducibility, the uniqueness of the eigenvector is not guaranteed as is shown by the following example:

$$\begin{pmatrix} 1 & e \\ e & 1 \end{pmatrix} \begin{pmatrix} e \\ -1 \end{pmatrix} = \begin{pmatrix} 1 \\ e \end{pmatrix} = 1 \begin{pmatrix} e \\ -1 \end{pmatrix} ,$$

and

$$\begin{pmatrix} 1 & e \\ e & 1 \end{pmatrix} \begin{pmatrix} -1 \\ e \end{pmatrix} = \begin{pmatrix} e \\ 1 \end{pmatrix} = 1 \begin{pmatrix} -1 \\ e \end{pmatrix} .$$

The two eigenvectors are obviously not 'proportional'. ■

Example 3.26 **(A not irreducible)**

- The following example is a trivial counterexample to the uniqueness of the eigenvalue when $\mathcal{G}(A)$ is not connected:

$$\begin{pmatrix} 1 & \varepsilon \\ \varepsilon & 2 \end{pmatrix} \begin{pmatrix} e \\ \varepsilon \end{pmatrix} = 1 \begin{pmatrix} e \\ \varepsilon \end{pmatrix}, \quad \begin{pmatrix} 1 & \varepsilon \\ \varepsilon & 2 \end{pmatrix} \begin{pmatrix} \varepsilon \\ e \end{pmatrix} = 2 \begin{pmatrix} \varepsilon \\ e \end{pmatrix}.$$

- In the following example $\mathcal{G}(A)$ is connected but not strongly connected. Nevertheless there is only one eigenvalue:

$$\begin{pmatrix} 1 & e \\ \varepsilon & e \end{pmatrix} \begin{pmatrix} e \\ \varepsilon \end{pmatrix} = 1 \begin{pmatrix} e \\ \varepsilon \end{pmatrix},$$

  but

$$\begin{pmatrix} 1 & e \\ \varepsilon & e \end{pmatrix} \begin{pmatrix} a \\ e \end{pmatrix} = \lambda \begin{pmatrix} a \\ e \end{pmatrix}$$

  has no solutions because the second equation implies $\lambda = e$, and then the first equation has no solutions for the unknown a.

- In the following example $\mathcal{G}(A)$ is connected but not strongly connected and there are two eigenvalues:

$$\begin{pmatrix} e & e \\ \varepsilon & 1 \end{pmatrix} \begin{pmatrix} e \\ \varepsilon \end{pmatrix} = e \begin{pmatrix} e \\ \varepsilon \end{pmatrix}, \quad \begin{pmatrix} e & e \\ \varepsilon & 1 \end{pmatrix} \begin{pmatrix} e \\ 1 \end{pmatrix} = 1 \begin{pmatrix} e \\ 1 \end{pmatrix}.$$

■

More generally, consider the block triangular matrix

$$F = \begin{pmatrix} A & \varepsilon \\ B & C \end{pmatrix},$$

where $\mathcal{G}(A)$ and $\mathcal{G}(C)$ are strongly connected, and $\mathcal{G}(C)$ is downstream of $\mathcal{G}(A)$. Let λ_A and λ_C, be the eigenvalues of blocks A and C, respectively, and let x_A and x_C be the corresponding eigenvectors. Observe that $\begin{pmatrix} \varepsilon & x_C \end{pmatrix}'$ is an eigenvector of F for the eigenvalue λ_C. In addition, if $\lambda_A > \lambda_C$, the expression $(C \not\!/ \lambda_A)^*$ is well-defined. The vector

$$\begin{pmatrix} x_A \\ (C \not\!/ \lambda_A)^* (B \not\!/ \lambda_A) x_A \end{pmatrix}$$

is an eigenvector of F for the eigenvalue λ_A. In conclusion, F has two eigenvalues if the upstream m.s.c.s. is 'slower' than the downstream one. Clearly this kind of result can be generalized to a decomposition into an arbitrary number of blocks. This generalization will not be considered here (see [62]).

3.2.5 Application to Event Graphs

Consider an autonomous event graph with n transitions, that is, an event graph without sources, with constant holding times and zero firing times. In §2.5.2 we saw that it can be modeled by

$$x(k) = \bigoplus_{i=0}^{M} A(i)x(k-i) \ , \tag{3.5}$$

where the $A(i)$ are $n \times n$ matrices with entries in $\mathbb{R}_{\max}$. We assume that the event graph (in which transitions are viewed as the nodes and places as the arcs) is strongly connected. In §2.5.4, it was shown that an equation in the standard form

$$\overline{x}(k+1) = \overline{A}\,\overline{x}(k) \tag{3.6}$$

can also describe the same physical system. A new event graph (with a different number of transitions in general) can be associated with (3.6). In this new event graph, each place has exactly one token in the initial marking. Therefore, in this graph, the length of a circuit or path can either be defined as the total number of arcs, or as the total number of tokens in the initial marking along this circuit or path.

We refer the reader to the transformations explained in §2.5.2 to §2.5.4 to see that some transitions in the two graphs can be identified to each other and that the circuits are in one-to-one correspondence. Since the original event graph is assumed to be strongly connected, the new event graph can also be assumed to be strongly connected, provided unnecessary transitions (not involved in circuits) be canceled. Then $\overline{A}$ is irreducible. Hence there exists a unique eigenvalue λ, and at least one eigenvector $\overline{x}$. By starting the recurrence in (3.6) with the initial value $\overline{x}(0) = \overline{x}$, we obtain that $\overline{x}(k) = \lambda^k \overline{x}$ for all k in $\mathbb{N}$. Therefore, a token leaves each transition every λ units of time, or, otherwise stated, the throughput of each transition is $1/\lambda$.

It was shown that λ can be evaluated as the maximum cycle mean of $\mathcal{G}\left(\overline{A}\right)$, that is, as the maximum ratio 'weight divided by length' over all the circuits of $\mathcal{G}\left(\overline{A}\right)$. The purpose of the following theorem is to show that λ can also be evaluated as the same maximum ratio over the circuits of the *original* graph, provided the length of an arc be understood as the number of tokens in the initial marking of the corresponding place (the weight is still defined as the holding time).

Let us return to (3.5). The graph $\mathcal{G}(A(i))$ describes the subgraph of the original graph obtained by retaining only the arcs corresponding to places marked with i tokens. Since $A(i)$ is an $n \times n$ matrix, all original nodes are retained in this subgraph which is however not necessarily connected. Consider the following $n \times n$ matrix:

$$B(\lambda) \stackrel{\text{def}}{=} \bigoplus_{i=0}^{M} \lambda^{-i} A(i) \ ,$$

where λ is any real number.

Remark 3.27 For a given value of λ, a circuit of $B(\lambda)$ can be defined by a sequence of nodes (in which the last node equals the first). Indeed, once this sequence of nodes is given, the arc between a pair of successive nodes (a, b) is selected by the argument $i_\lambda(a, b)$ of the $\max_i$ which is implicit in the expression of $(B(\lambda))_{ba}$. If this $i_\lambda(a, b)$ is not unique, it does not matter which one is selected since any choice leads to the same weight. Therefore, the set of circuits of $\mathcal{G}(B(\lambda))$ is a subset of the set of circuits of the original graph (a circuit of the original graph can be specified by a given sequence of nodes *and* by a mapping $(a, b) \mapsto i(a, b)$ in order to specify one of the possible parallel arcs between nodes a and b). The set of circuits of $\mathcal{G}(B(\lambda))$ is thus changing with λ. However, for any given value of λ, if we are only interested in the maximum circuit weight of $\mathcal{G}(B(\lambda))$ (or in the maximum cycle mean, assuming that the length of a circuit is defined as the number of arcs in $\mathcal{G}(B(\lambda))$), the maximum can be taken over the whole set of circuits of the original graph (this set is independent of λ). Indeed, the additional circuits thus considered do not contribute to the maximum since they correspond to choices of $i(a, b)$ which are not optimal. ■

Theorem 3.28 *We assume that*

1. $\mathcal{G}(B(e))$ *is strongly connected;*
2. $\mathcal{G}(A(0))$ *has circuits of negative weight only;*
3. *there exists at least one circuit of* $\mathcal{G}(B(\lambda))$ *containing one token.*

Then, there exist a vector x *and a unique scalar* λ *satisfying* $x = B(\lambda)x$. *The graph interpretation of* λ *is*

$$\lambda = \max_\zeta \frac{|\zeta|_w}{|\zeta|_t} , \tag{3.7}$$

where ζ *ranges over the set of circuits of the original event graph and* $|\zeta|_t$ *denotes the number of tokens in circuit* ζ.

Proof To solve the equation $x = B(\lambda)x$, we must find λ and x such that e is an eigenvalue of the matrix $B(\lambda)$. The graph $\mathcal{G}(B(e))$ being strongly connected, $\mathcal{G}(B(\lambda))$ is also strongly connected for any real value of λ and therefore $B(\lambda)$ admits a unique eigenvalue $\Lambda(\lambda)$. Owing to the graph interpretation of $\Lambda(\lambda)$, $\Lambda(\lambda) = \max_\zeta |\zeta|_w / |\zeta|_l$, where ζ ranges over the set of circuits of $\mathcal{G}(B(\lambda))$. However, Remark 3.27 showed that we can as well consider that ζ ranges over the set of circuits of the original graph. Hence, in conventional notation, we have

$$\begin{aligned}
\Lambda(\lambda) &= \max_\zeta \left(\frac{1}{|\zeta|_l} \sum_{(a,b)\in\zeta} \left(B(\lambda)\right)_{ba} \right) \\
&= \max_\zeta \left(\frac{1}{|\zeta|_l} \sum_{(a,b)\in\zeta} \max_{i\in\{1,\dots,M\}} \left((A(i))_{ba} - i \times \lambda\right) \right) \\
&= \max_\zeta \left(\frac{1}{|\zeta|_l} \max_{i(a,b)} \sum_{(a,b)\in\zeta} \left((A(i(a,b))_{ba} - i(a,b) \times \lambda\right) \right) .
\end{aligned} \tag{3.8}$$

If we assume that there exists a λ such that $\Lambda(\lambda) = e = 0$, then, for any circuit ζ of the original graph, and thus for any mapping $i(\cdot,\cdot)$ which completes the specification of the circuit, we have

$$\lambda \geq \frac{\sum_{(a,b)\in\zeta}(A(i(a,b)))_{ba}}{\sum_{(a,b)\in\zeta} i(a,b)} ,$$

and the equality is obtained for some circuit ζ. This justifies the interpretation (3.7) of λ.

Let us now prove that $\Lambda(\lambda) = e$ has a unique solution. Because of (3.8), and since, according to Remark 3.27, the mappings $i(\cdot,\cdot)$ can be viewed as ranging in a set independent of λ, $\Lambda(\cdot)$ is the upper hull of a collection of affine functions (of λ). Each affine function has a nonpositive slope which, in absolute value, equals the number of tokens in a circuit divided by the number of arcs in the circuit. Therefore, Λ is a nonincreasing function of λ. Moreover, due to the third assumption of the theorem, there is at least one strictly negative slope. Hence $\lim_{\lambda\to-\infty}\Lambda(\lambda) = +\infty$. On the other hand, owing to the second assumption, and since the affine functions with zero slope stem necessarily from the circuits of $A(0)$, $\lim_{\lambda\to+\infty}\Lambda(\lambda) < 0$. Finally Λ is a convex nonincreasing function which decreases from $+\infty$ to a strictly negative value, and thus its graph crosses the x-axis at a single point. ■

It is easy to see that if we start the recurrence (3.5) with $x(0) = x, x(1) = \lambda \otimes x, \ldots, x(M) = \lambda^M \otimes x$, then $x(k) = \lambda^k \otimes x$ for all k. Hence $1/\lambda$ is the throughput of the system at the periodic regime. At the end of this chapter, we will give conditions under which this regime is asymptotically reached, and conditions under which it is reached after a finite time, whatever the initial condition is.

3.3 SCALAR FUNCTIONS IN $\mathbb{R}_{\max}$

In this section we discuss nonlinear real-valued functions of one real variable, considered as mappings from $\mathbb{R}_{\max}$ into $\mathbb{R}_{\max}$. We classify them in polynomial, rational and algebraic functions in the max-plus algebra sense.

3.3.1 Polynomial Functions $\mathcal{P}(\mathbb{R}_{\max})$

Polynomial functions are a subset of piecewise linear functions (in the conventional sense) for which we have the analogue of the fundamental theorem of algebra. This set is not isomorphic to the set of formal polynomials of $\mathbb{R}_{\max}$, that is, the set of finite sequences endowed with a product which is the sup-convolution of sequences.

3.3.1.1 Formal Polynomials and Polynomial Functions

Definition 3.29 (Formal polynomials) *We consider the set of finite real sequences of any length*

$$p = (p(k), \ldots, p(i) \ldots p(n)) , \quad k, i, n \in \mathbb{N} , \quad p(i) \in \mathbb{R}_{\max} .$$

If the extreme values k and n are such that $p(k)$ and $p(n)$ are different from ε, then $\mathrm{val}(p) \stackrel{\mathrm{def}}{=} k$ *is called the* valuation *of p, and* $\deg(p) \stackrel{\mathrm{def}}{=} n$ *is called the* degree *of p. This set is endowed with the following two internal operations:*

- *componentwise addition $\oplus$;*
- *sup-convolution $\otimes$ of sequences, that is,*

$$(p \otimes q)(l) \stackrel{\mathrm{def}}{=} \bigoplus_{\substack{i+j=l \\ \mathrm{val}(p) \leq i \leq \deg(p) \\ \mathrm{val}(q) \leq j \leq \deg(q)}} p(i)q(j) \ ,$$

and with the following external operation involving scalars in $\mathbb{R}_{\max}$:

- *multiplication of all the elements of the sequence p by the same scalar of $\mathbb{R}_{\max}$,*

We thus define $\mathbb{R}_{\max}[\gamma]$ which is called the set of formal polynomials.

Note that if the polynomial γ is defined by

$$\gamma(k) = \begin{cases} e & \text{if } k = 1 \ ; \\ \varepsilon & \text{otherwise,} \end{cases}$$

then any polynomial of $\mathbb{R}_{\max}[\gamma]$ can be written as $p = \bigoplus_{l=k}^{n} p(l)\gamma^l$.

Let us give a list of definitions related to the notion of formal polynomials.

Definition 3.30

Polynomial functions: *associated with a formal polynomial p, we define the polynomial function by*

$$\widehat{p} : \mathbb{R}_{\max} \to \mathbb{R}_{\max} \ , \quad c \mapsto \widehat{p}(c) = p(k)c^k \oplus \cdots \oplus p(n)c^n \ .$$

The set of polynomial functions is denoted $\mathfrak{P}(\mathbb{R}_{\max})$.

Support: *the support* $\mathrm{supp}(p)$ *is the set of indices of the nonzero elements of p, that is,* $\mathrm{supp}(p) = \{i \mid k \leq i \leq n, p(i) \neq \varepsilon\}$.

Monomial: *a formal polynomial reduced to a sequence of one element is called a monomial*[1].

Head monomial: *the monomial of highest degree one can extract from a polynomial p, that is $p(n)\gamma^n$, is called the head monomial.*

Tail monomial: *the monomial of lowest degree out of p, that is, $p(k)\gamma^k$, is called the tail monomial.*

[1] We do not to make the distinction between formal monomials and monomial functions because, unlike polynomials—see Remark 3.34 below—a formal monomial is in one-to-one correspondence with its associated function.

Full Support: *we say that a formal polynomial has a full support if*

$$p(i) \neq \varepsilon \ , \quad \forall i : k \leq i \leq n \ .$$

The following two theorems are obvious.

Theorem 3.31 *The set of formal polynomials $\mathbb{R}_{\max}[\gamma]$ is an idempotent algebra.*

Remark 3.32 Because we can identify scalars with monomials of degree 0, this idempotent algebra can be viewed as the idempotent semiring obtained by considering the two internal operations only, since the external multiplication by a scalar is amenable to the internal multiplication by a monomial. ■

Theorem 3.33 *The set of polynomial functions $\mathcal{P}(\mathbb{R}_{\max})$ endowed with the two internal operations:*

- *pointwise addition denoted $\oplus$, that is, $(\widehat{p} \oplus \widehat{q})(c) \stackrel{\text{def}}{=} \widehat{p}(c) \oplus \widehat{q}(c)$;*
- *pointwise multiplication denoted $\otimes$, that is, $(\widehat{p} \otimes \widehat{q})(c) \stackrel{\text{def}}{=} \widehat{p}(c) \otimes \widehat{q}(c)$,*

and the external operation over $\mathbb{R}_{\max} \times \mathcal{P}(\mathbb{R}_{\max})$, namely,

- $(b\widehat{p})(c) \stackrel{\text{def}}{=} b \otimes \widehat{p}(c)$,

is an idempotent algebra.

The same remark as above applies to this idempotent algebra too.

Polynomial functions are convex piecewise linear integer-sloped nondecreasing functions (see Figure 3.4). Indeed the monomial $p(i)c^i$ is nothing but the conventional affine function $ic + p(i)$. Owing to the meaning of addition of monomials, polynomial functions are thus upper hulls of such affine functions.

Remark 3.34 There is no one-to-one correspondence between formal polynomials and polynomial functions. For example,

$$\forall c \ , \quad c^2 \oplus 2 = (c \oplus 1)^2 = c^2 \oplus 1c \oplus 2 \ .$$

The monomial $1c$ is dominated by $c^2 \oplus 2$. In other words, $1c$ does not contribute to the graph of $c^2 \oplus 1c \oplus 2$ (see Figure 3.5), and thus, two different formal polynomials are associated with the same function. ■

The following lemma should be obvious.

Lemma 3.35 (Evaluation homomorphism) *Consider the mapping $\mathcal{F} : \mathbb{R}_{\max}[\gamma] \to \mathcal{P}(\mathbb{R}_{\max}), p \mapsto \widehat{p}$. Then, $\mathcal{F}$ is a homomorphism between the algebraic structures defined in Theorems 3.31 and 3.33. It will be referred to as the* evaluation homomorphism.

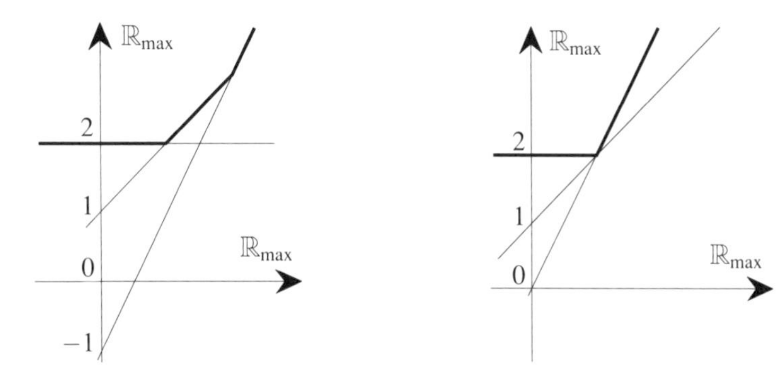

Figure 3.4: The graph of $y = (-1)c^2 \oplus 1c \oplus 2$ **Figure 3.5**: The graph of $y = (c \oplus 1)^2$

Unlike in conventional algebra, the evaluation homomorphism is not a one-to-one correspondence, as shown in Remark 3.34. More precisely, it is surjective but not injective. This is why it is important to distinguish between formal objects and their associated numerical functions.

Remark 3.36 The mapping $\mathcal{F}$ is in fact closely related to the Fenchel transform [58]. In convexity theory [119], with a numerical function f over a Hilbert space V, the Fenchel transform $\mathcal{F}_e$ associates a numerical function $\mathcal{F}_e(f)$ over the dual Hilbert space V^* as follows:

$$\forall c \in V^* , \quad \left[\mathcal{F}_e(f)\right](c) = \sup_{z \in V} \left(\langle c, z\rangle - f(z)\right) ,$$

where $\langle\cdot,\cdot\rangle$ denotes the duality product over $V^* \times V$. If we consider the formal polynomial p as a function from $\mathbb{N}$ into $\mathbb{R}_{max}$, $p : l \mapsto p(l)$ (the domain of which can be extended to the whole $\mathbb{N}$ by setting $p(l) = \varepsilon$ if $l < \text{val}(p)$ or $l > \deg(p)$), then,

$$\widehat{p}(c) = \max_{l \in \mathbb{N}}(lc + p(l)) = \mathcal{F}_e(-p)(c) . \tag{3.9}$$

■

Before studying the properties of $\mathcal{F}$ and the equivalence relation it induces in $\mathbb{R}_{max}[\gamma]$, let us first give some terminology related to convexity that we will use later on.

Definition 3.37

Convex set in a disconnected domain: *we say that a subset $F \subset E = \mathbb{N} \times \mathbb{R}$ is convex if, for all $\mu \in [0,1]$ and for all $x, y \in F$ for which $\mu x + (1-\mu)y \in E$, it follows that $\mu x + (1-\mu)y \in F$.*

Hypograph and epigraph of a real function $f : X \to \mathbb{R}$: *these are the sets defined respectively by*

$$\text{hypo}(f) \stackrel{\text{def}}{=} \{(x,y) \mid x \in X, y \in \mathbb{R}, y \leq f(x)\} ,$$

$$\mathrm{epi}(f) \stackrel{\mathrm{def}}{=} \{(x, y) \mid x \in X, y \in \mathbb{R}, y \geq f(x)\} \ .$$

Convex (concave) mapping: *a function $f : X \to \mathbb{R}$ is convex (respectively concave) if its epigraph (respectively its hypograph) is convex.*

Extremal point of a convex set F: *it is a point which is not a convex combination of two other distinct points of F.*

Theorem 3.38 *Consider the equivalence relation*

$$\forall p, q \in \mathbb{R}_{\max}[\gamma] \ , \quad p \stackrel{\mathcal{F}}{\equiv} q \Leftrightarrow \widehat{p} = \widehat{q} \Leftrightarrow q \in \mathfrak{F}^{-1}(\widehat{p}) \ .$$

1. *With a given $p \in \mathbb{R}_{\max}[\gamma]$, we associate two other elements of $\mathbb{R}_{\max}[\gamma]$ denoted $p_1^\sharp$ and $p_2^\sharp$, such that, for all $l \in \mathbb{N}$,*

$$\begin{aligned} p_1^\sharp(l) &\stackrel{\mathrm{def}}{=} \max_{\substack{0 \leq \mu \leq 1 \\ i,j \in \mathbb{N}}} (\mu p(i) + (1-\mu)p(j)) \ , \\ \textit{subject to} \quad & l = \mu i + (1-\mu) j \ , \end{aligned} \tag{3.10}$$

$$p_2^\sharp(l) \stackrel{\mathrm{def}}{=} \min_{c \in \mathbb{R}_{\max}} \left(\widehat{p}(c) - lc\right) \ . \tag{3.11}$$

Then $p_1^\sharp = p_2^\sharp$ (denoted simply $p^\sharp$) and $p^\sharp$ belongs to the same equivalence class as p of which it is the maximum element. The mapping $l \mapsto p^\sharp(l)$ is the concave upper hull of $l \mapsto p(l)$. Hence hypo $(p^\sharp)$ *is convex.*

2. *Let now $p^\flat \in \mathbb{R}_{\max}[\gamma]$ be obtained from p by canceling the monomials of p which do not correspond to extremal points of* hypo $(p^\sharp)$. *Then $p^\flat$ belongs to the same equivalence class as p of which it is the minimum element.*

3. *Two members p and q of the same equivalence class have the same degree and valuation. Moreover $p^\sharp = q^\sharp$ and $p^\flat = q^\flat$.*

Proof

1. Using (3.10) with the particular values $\mu = 1$, hence $l = i$, we first prove that $p_1^\sharp \geq p$ for the pointwise conventional order (which is also the natural order associated with the addition in $\mathbb{R}_{\max}[\gamma]$). Combining (3.9) (written for $p_1^\sharp$) with (3.10), we obtain

$$\begin{aligned} \widehat{p_1^\sharp}(c) &= \max_l \left(lc + \max_{0 \leq \mu \leq 1,\ i,j \in \mathbb{N}} (\mu p(i) + (1-\mu)p(j)) \right) \\ & \quad \text{subject to} \quad l = \mu i + (1-\mu) j \\ &= \max_{0 \leq \mu \leq 1} \left(\mu \max_i \left(ic + p(i)\right) + (1-\mu) \max_j \left(jc + p(j)\right) \right) \\ &= \max_{0 \leq \mu \leq 1} \left(\mu \widehat{p}(c) + (1-\mu)\widehat{p}(c)\right) \\ &= \widehat{p}(c) \ . \end{aligned}$$

This shows that $p_1^\sharp$ belongs to the same equivalence class as p and that it is greater than any such p, hence it is the maximum element in this equivalence class.

On the other hand, combining (3.11) with (3.9), we get

$$p_2^\sharp(l) = \min_{c \in \mathbb{R}_{\max}} \left(\max_{m \in \mathbb{N}} (mc + p(m)) - lc \right) .$$

By choosing particular value $m = l$, it is shown that $p_2^\sharp \geq p$. Since $\mathcal{F}$ is a homomorphism, it preserves the order, and thus $\widehat{p_2^\sharp} \geq \widehat{p}$. But, if we combine (3.9) (written for $p_2^\sharp$) with (3.11), we get

$$\widehat{p_2^\sharp}(c) = \max_{l \in \mathbb{N}} \left(lc + \min_{c' \in \mathbb{R}_{\max}} (\widehat{p}(c') - lc') \right) .$$

By picking the particular value $c' = c$, it is shown that $\widehat{p_2^\sharp} \leq \widehat{p}$. Hence, we have proved that $\widehat{p_2^\sharp} = \widehat{p}$ and that $p_2^\sharp \geq p$. Therefore $p_2^\sharp$ is the maximum element in the equivalence class of p. Since the maximum element is unique (see § 4.3.1), it follows that $p_1^\sharp = p_2^\sharp$.

From (3.11), it is apparent that $p_2^\sharp$ is concave as the lower hull of a family of affine functions. Hence, since it is greater than p, it is greater than its concave upper hull, but (3.10) shows that indeed it coincides with this hull.

2. It is now clear that the equivalence class of p can be characterized by $p^\sharp$, or equivalently by its hypograph which is a convex set. Since a convex set is fully characterized by the collection of its extreme points, this collection is another characterization of the class. Since $p^\flat$ has precisely been defined from this collection of extreme points, it is clearly an element of the same class and the minimum one (dropping any further monomial would change the collection of extreme points and thus the equivalence class).

3. In particular, the head and tail monomials of a given p correspond to members of the collection of extreme points. Therefore, all elements of an equivalence class have the same degree and valuation. ∎

Definition 3.39 (Canonical forms of polynomial functions) *According to the previous theorem, we may call $p^\sharp$ and $p^\flat$ the* concavified polynomial *and the* skeleton *of p, respectively.*

- *The skeleton $p^\flat$ is also called the* minimum canonical form *of the polynomial function $\widehat{p}$;*

- *the concavified polynomial $p^\sharp$ is also called the* maximum canonical form *of $\widehat{p}$.*

Figure 3.6 illustrates these notions. It should be clear that necessarily $p^\sharp$ has full support.

Example 3.40 For $p = \gamma^2 \oplus \gamma \oplus 2$, we have $p^\flat = \gamma^2 \oplus 2$. For $p = \gamma^3 \oplus 3$, we have $p^\sharp = \gamma^3 \oplus 1\gamma^2 \oplus 2\gamma \oplus 3$. ■

Lemma 3.41 *A formal polynomial p of valuation k and degree n is a maximal canonical form of a polynomial function $\widehat{p}$, that is, $p = p^\sharp$, if and only if p has full support (hence $p(l) \neq \varepsilon$, for $l = k, \ldots, n$), and*

$$\frac{p(n-1)}{p(n)} \geq \frac{p(n-2)}{p(n-1)} \geq \cdots \geq \frac{p(k)}{p(k+1)} \ . \tag{3.12}$$

Proof The fact that a maximal canonical form must have full support has been established earlier and this ensures that the 'ratios' in (3.12) are well defined. In the proof of Theorem 3.38 it has also been shown that $p^\sharp$ is the concave upper hull of the function $l \mapsto p(l)$ (concavity in the sense of Definition 3.37). Conversely, if p is concave, it is equal to its own concave upper hull, and thus $p = p^\sharp$. Now, (3.12) simply expresses that the slopes of the lines defined by the successive pairs of points $\big((l-1, p(l-1)), (l, p(l))\big)$ are decreasing with l, which is obviously a necessary and sufficient condition for p to be concave. ■

3.3.1.2 Factorization of Polynomials

Let us now show that polynomial functions and concave formal polynomials can be factored into a product of linear factors.

Definition 3.42 (Corners, multiplicity) *The nonzero corners of a polynomial function $\widehat{p}$ are the abscissæ of the extremal points of the epigraph of $\widehat{p}$. Since $\widehat{p}$ is convex, at such a corner, the (integer) slope increases by some integer which is called the multiplicity of the corner. A corner is called multiple if the multiplicity is larger than one.*

The zero corner exists if the least slope appearing in the graph of $\widehat{p}$ is nonzero: the multiplicity of this zero corner is then equal to that (integer) slope.

Figure 3.7 shows a nonzero corner of multiplicity 2.

Theorem 3.43 (Fundamental theorem of algebra)

1. *Any formal polynomial of the form $p = p(n) \bigotimes_{i=1}^{n} (\gamma \oplus c_i)$ (where some c_i may be equal to ε) satisfies $p = p^\sharp$.*

2. *Conversely, if a formal polynomial $p = p(k)\gamma^k \oplus p(k+1)\gamma^{k+1} \oplus \cdots \oplus p(n)\gamma^n$ is such that $p = p^\sharp$, then it has full support, the numbers $c_i \in \mathbb{R}_{\max}$ defined by*

$$c_i \stackrel{\text{def}}{=} \begin{cases} p(n-i) \not/ p(n-i+1) & \text{for } 1 \leq i \leq n-k \ ; \\ \varepsilon & \text{for } n-k < i \leq n \ , \end{cases} \tag{3.13}$$

are such that

$$c_1 \geq c_2 \geq \cdots \geq c_n \ , \tag{3.14}$$

and p can be factored as $p = p(n)\left(\bigotimes_{i=1}^{n} (\gamma \oplus c_i)\right)$.

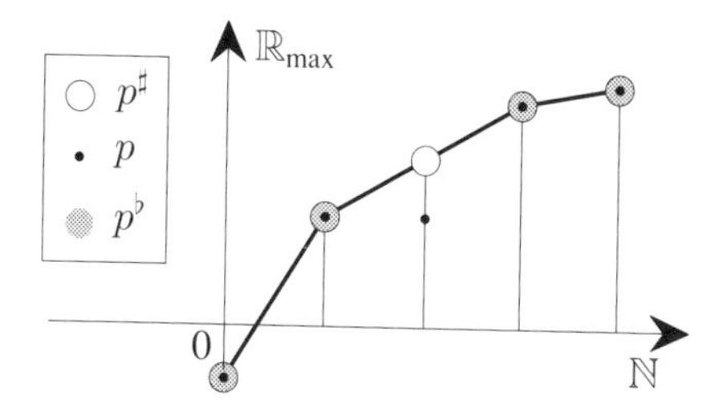

Figure 3.6: The functions p, $p^\sharp$ and $p^\flat$

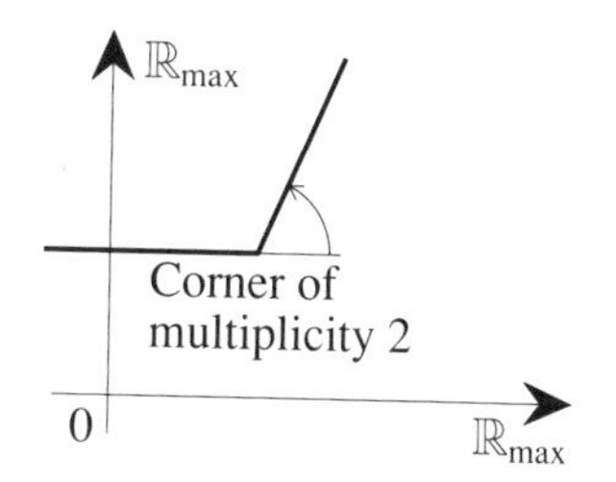

Figure 3.7: Corner of a polynomial function

3. *Any polynomial function $\widehat{p}$ can be factored as*

$$\widehat{p}(c) = p(n)\left(\bigotimes_{i=1}^{n}(c \oplus c_i)\right) ,$$

where the c_i are the zero and/or nonzero corners of the polynomial function $\widehat{p}$ repeated with their order of multiplicity. These corners are obtained by Equation (3.13) using the maximum canonical form.

Proof

1. Let $p = \bigotimes_{i=1}^{n} p(n)(\gamma \oplus c_i)$ and assume without loss of generality that the c_i have been numbered in such a way that (3.14) holds. We consider the nontrivial case $p(n) \neq \varepsilon$. By direct calculation, it can be seen that $p(n-k) = p(n)\sigma_k$, where σ_k is the k-th symmetric product of the corners c_i, that is,

$$\sigma_1 = \bigoplus_{i=1}^{n} c_i \ , \quad \sigma_2 = \bigoplus_{i \neq j = 1} c_i c_j \ , \ldots .$$

Owing to our assumption on the ordering of the c_i, it is clear that $\sigma_k = \bigotimes_{i=1}^{k} c_j$. Therefore,

$$\frac{p_{n-k}}{p_{n-k+1}} = \frac{\sigma_k}{\sigma_{k-1}} = c_k \leq c_{k-1} = \frac{p_{n-k+1}}{p_{n-k+2}} .$$

Thus $p = p^\sharp$ by Lemma 3.41.

2. If $p = p^\sharp$, the c_i can be defined by (3.13) unambiguously and then (3.14) follows from Lemma 3.41. The fact that p can be factored as indicated is checked as previously by direct calculation.

3. From the preceding considerations, provided that we represent $\widehat{p}$ with the help of its maximum canonical form, the factored form can be obtained if we define the c_i with Equation (3.13). To complete the proof, it must be shown that any

such c_i is a corner in the sense of Definition 3.42 and that, if a c_i appears k_i times, then the slope jumps by k_i at c_i. To see this, rewrite the factored form in conventional notation, which yields

$$\widehat{p}(c) = \sum_{i=1}^{n} \max(c, c_i) \ .$$

Each elementary term $c \oplus c_i$ has a graph represented in Figure 3.2 with a slope discontinuity equal to one at c_i. If a c_i appears k_i times, the term $k_i \times \max(c, c_i)$ causes a slope discontinuity equal to k_i. All other terms with $c_j \neq c_i$ do not cause any slope discontinuity at c_i. ■

Example 3.44 The formal polynomial $\gamma^2 \oplus 3\gamma \oplus 2$ is a maximum canonical form because $c_1 = 3 \not/ e = 3 \geq c_2 = 2 \not/ 3 = -1$, and therefore it can be factored into $(\gamma \oplus 3)(\gamma \oplus (-1))$. The formal polynomial $(\gamma^2 \oplus 2)^\sharp = \gamma^2 \oplus 1\gamma \oplus 2$ can be factored into $(\gamma \oplus 1)^2$. ■

3.3.2 Rational Functions

In this subsection we study rational functions in the $\mathbb{R}_{\max}$ algebra. These functions are continuous, piecewise linear, integer-sloped functions. We give the multiplicative form of such functions, which completely defines the points where the slope changes. Moreover, we show that the Euclidean division and the decomposition into simple elements is not always possible.

3.3.2.1 Definitions

Definition 3.45 *Given* $p(0), \ldots, p(n)$, $q(0), \ldots, q(m) \in \mathbb{R}_{\max}$, $p(n)$ *and* $q(m) \neq \varepsilon$, *the rational function* $\widehat{r}$, *associated with these coefficients, is given by*

$$\widehat{r} : \mathbb{R}_{\max} \to \mathbb{R}_{\max} \ , \quad c \mapsto \widehat{r}(c) = \frac{p(0) \oplus \cdots \oplus p(n)c^n}{q(0) \oplus \cdots \oplus q(m)c^m} \ .$$

Such a function is equal to the difference of two polynomial functions (see Figure 3.8): hence it is still continuous, piecewise linear, integer-sloped, but it is neither convex nor increasing anymore.

Definition 3.46 *The corners of the numerator are called* zero corners *or* root corners, *and the corners of the denominator are called* pole corners.

Using the fundamental theorem of algebra, we can write any rational function $\widehat{r}$ as

$$\widehat{r}(c) = a \frac{\bigotimes_{i=1}^{n} (c \oplus c_i)}{\bigotimes_{j=1}^{m} (c \oplus d_j)} \ ,$$

where the zero and pole corners are possibly repeated with their order of multiplicity. At a zero corner of multiplicity k_i, the change of slope is k_i (unless this zero corner coincides with some pole corner). At a pole corner of multiplicity l_j, the change of slope is $-l_j$ (see Figure 3.9).

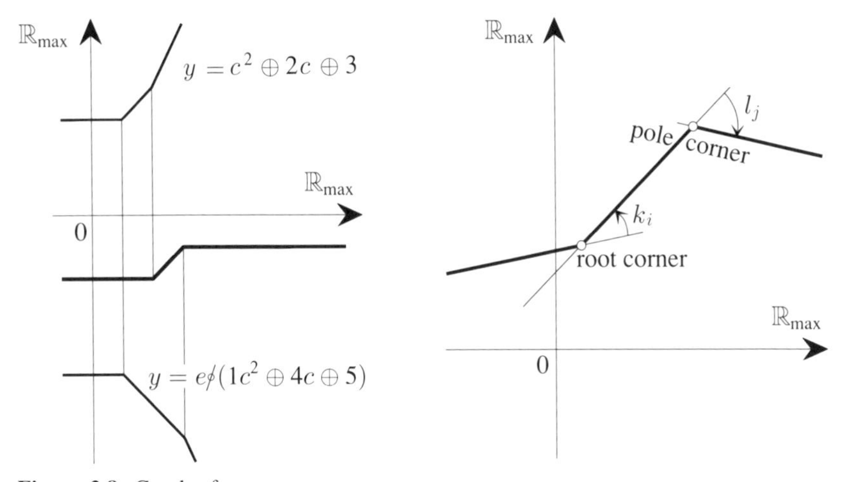

Figure 3.8: Graph of $y = (c^2 \oplus 2c \oplus 3) \oint (1c^2 \oplus 4c \oplus 5)$

Figure 3.9: Root and pole corners

3.3.2.2 *Euclidean Division*

In general, a polynomial $\widehat{p}$ cannot be expressed as $\widehat{b}\widehat{q} \oplus \widehat{r}$ with $\deg(\widehat{r}) < \deg(\widehat{b})$ for some given polynomial $\widehat{b}$ as shown by Example 3.47. Nevertheless, sometimes we can obtain such a decomposition, as shown by Example 3.48.

Example 3.47 The equation $c^2 \oplus e = \widehat{q}(c \oplus 1) \oplus \widehat{r}$ has no solutions. Indeed, $\widehat{q}$ must be of degree 1 and $\widehat{r}$ of degree 0; thus $q = q(1)\gamma \oplus q(0)$, $r = r(0)$, with $q(0), q(1), r(0) \in \mathbb{R}_{\max}$. By identifying the coefficients of degree 2 in both sides, we must have $q(1) = e$. Now, since the maximal canonical form in the left-hand side is $c^2 \oplus c \oplus e$, by considering the coefficient of degree 1 in the right-hand side, we must have $q(0) \oplus 1q(1) \leq e$, which contradicts $q(1) = e$. ■

Example 3.48 For $\widehat{p}(c) = c^2 \oplus 3$ and $\widehat{b} = c \oplus 1$, we have

$$c^2 \oplus 3 = (c \oplus 1)^2 \oplus 3 = (c \oplus 1)(c \oplus 1.5) \oplus 3 \ .$$

■

As in conventional algebra, this issue of the Euclidean division leads to solving a triangular system of linear equations. However, in $\mathbb{R}_{\max}$, the difficulty dwells in the fact that a triangular system of linear equations with nonzero diagonal elements may have no solutions.

Example 3.49 The system $(x_1 = 1, x_1 \oplus x_2 = e)$ has no solutions in $\mathbb{R}_{\max}$. ■

3.3.2.3 Decomposition of a Rational Function into Simple Elements

Definition 3.50 *A* proper rational function $\widehat{r} \not\!\!/ \widehat{q}$ *is a rational function which satisfies* $\deg(\widehat{r}) < \deg(\widehat{q})$.

In general, it is not possible to express a rational function $\widehat{p} \not\!\!/ \widehat{q}$ as $\widehat{s} \oplus \widehat{r} \not\!\!/ \widehat{q}$, where $\widehat{r} \not\!\!/ \widehat{q}$ is proper and $\widehat{s}$ is a polynomial function. Nevertheless, given a proper rational function, we may attempt to decompose it into simple elements.

Definition 3.51 *A proper rational function* $\widehat{r}$ *is* decomposable into simple elements *if it can be written as*

$$\widehat{r} = \bigoplus_{i=1}^{n} \bigoplus_{k=1}^{K_i} a_{ik} \not\!\!/ (c \oplus c_i)^k \ ,$$

where the a_{ik} *are constants.*

Such a decomposition is not always possible.

Example 3.52 We first consider a rational function for which the decomposition into simple elements is possible:

$$\frac{c \oplus 1}{(c \oplus e)^2} = \frac{c \oplus e}{(c \oplus e)^2} \oplus \frac{1}{(c \oplus e)^2} = \frac{e}{c \oplus e} \oplus \frac{1}{(c \oplus e)^2} \ .$$

The rational function $(c \oplus e) \not\!\!/ (c \oplus 1)^2$, however, cannot be decomposed. Indeed, if such a decomposition exists, we would have

$$\frac{c \oplus e}{(c \oplus 1)^2} = \frac{a}{c \oplus 1} \oplus \frac{b}{(c \oplus 1)^2} \ .$$

Then $a(c \oplus 1) \oplus b = c \oplus e$, hence $a = e$, and also $a1 \oplus b = 1 \oplus b = e$, which is impossible. ■

The graph of a proper rational function which can be decomposed into simple elements is necessarily nonincreasing because it is the upper hull of nonincreasing functions. But a rational function with the degree of the numerator lower than the degree of the denominator is not always nonincreasing: this depends on the relative ordering of its pole and zero corners.

Example 3.53 The function $y = 2(c \oplus e) \not\!\!/ (c \oplus 1)^2$ is proper but not monotonic (see Figure 3.10). ■

However, being nonincreasing is a necessary but not a sufficient condition to be decomposable into simple elements as shown by the following example.

Example 3.54 The function $\widehat{r}(c) = e \not\!\!/ (c^2 \oplus c)$, the graph of which is displayed in Figure 3.11, cannot be decomposed into simple elements. Indeed,

$$\left\{ \widehat{r}(c) = \frac{a}{c} \oplus \frac{b}{c \oplus e} \right\} \Rightarrow \{a(c \oplus e) \oplus bc = e\} \Rightarrow \left\{ \begin{array}{rcl} a \oplus b & = & \varepsilon \\ a & = & e \end{array} \right\},$$

which is impossible in $\mathbb{R}_{\max}$. ■

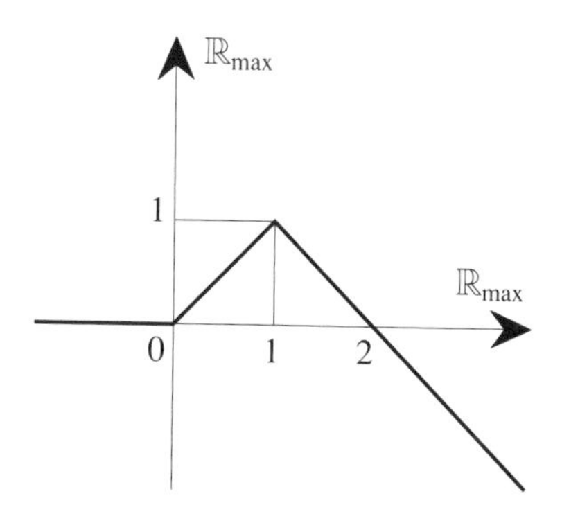

Figure 3.10: Graph of $y = 2(c \oplus e) \not/ (c \oplus 1)^2$

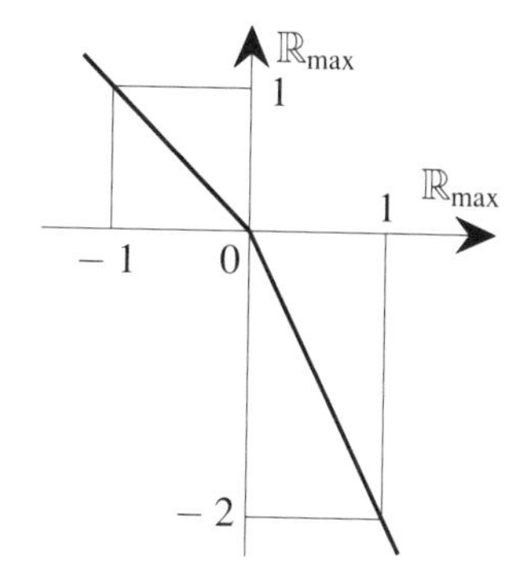

Figure 3.11: Graph of $y = e \not/ (c^2 \oplus c)$

Theorem 3.55 *A proper rational function has a simple element decomposition if and only if it has general root corners and special pole corners which are at the intersection of a zero-sloped line and a negative-integer-sloped line (Figure 3.12).*

Proof If $\widehat{r}$ is decomposable, it can be written as $\widehat{r} = \bigoplus r_i$ with

$$\widehat{r}_i(c) = \bigoplus_{k=1}^{K_i} a_{ik} \not/ (c \oplus c_i)^k \ .$$

Reducing the right-hand side to a common denominator $(c \oplus c_i)^{K_i}$, we obtain $\widehat{r}_i = \widehat{p}(c \oplus c_i) \not/ (c \oplus c_i)^{K_i}$, where $\widehat{p}$ is a polynomial function of degree $K_i - 1$. The polynomial function $\widehat{p}$ is characterized by the fact that the abscissæ of its corners are greater than c_i. Therefore $\widehat{r}_i(c)$ is constant on the left-hand side of c_i, has a pole corner of order K_i at c_i, and a root corner on the right-hand side of c_i. Conversely, a function having this shape can easily be realized by an $\widehat{r}_i$. The proof is completed by considering the fact that $\widehat{r}$ is the supremum of a finite number of such $\widehat{r}_i$. ■

3.3.3 Algebraic Equations

Definition 3.56 (Polynomial equation) *Given two polynomial functions $\widehat{p}$ and $\widehat{q}$ of degree n and m, respectively, the equality $\widehat{p}(c) = \widehat{q}(c)$ is called a polynomial equation. Solutions to this equation are called* roots. *The* degree *of the polynomial equation is the integer* $\max(n, m)$.

Some polynomial equations have roots, some do not. For example, $c^n = a$ has the root $c = \sqrt[n]{a}$, that is, $c = a/n$ in conventional algebra. On the other hand, the equation $\widehat{p}(c) = \varepsilon$ has no roots when p is a general polynomial of degree $n \geq 1$.

3.3.3.1 Canonical Form of an Equation

Before studying equations, it is useful to write them in their simplest form. An equation $\widehat{p}(c) = \widehat{q}(c)$ can generally be simplified even if it is in the form $\widehat{p^\flat}(c) = \widehat{q^\flat}(c)$.

Indeed, if two monomials $p^\flat(k)\gamma^k$ and $q^\flat(k)\gamma^k$ of the same degree appear simultaneously, and if $p^\flat(k) < q^\flat(k)$, we can further simplify the equation by canceling the monomial $p^\flat(k)\gamma^k$.

Example 3.57 The equation $c^2 \oplus 3c \oplus 2 = 3c^2 \oplus 2c \oplus e$ can be reduced to $3c \oplus 2 = 3c^2$ (see Figure 3.13). ■

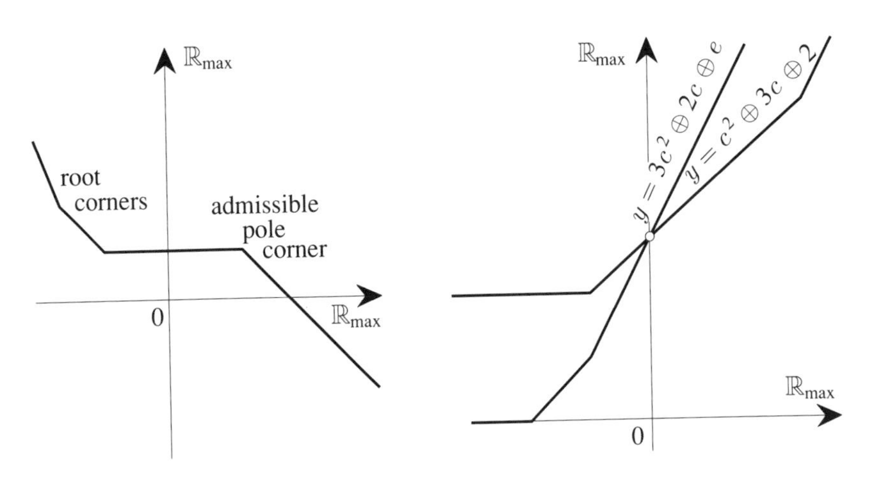

Figure 3.12: A rational function decomposable into simple elements

Figure 3.13: Equation $c^2 \oplus 3c \oplus 2 = 3c^2 \oplus 2c \oplus e$

Definition 3.58 *If there exist two identical monomials on both sides of the equation, we say that the equation is* degenerated. *The nondegenerated equation $\widehat{p}(c) = \widehat{q}(c)$ is in* minimal canonical form *if $(p \oplus q)^\flat = p \oplus q$.*

In the case of a degenerated equation, there is a segment of solutions.

3.3.3.2 Solving Polynomial Equations

Theorem 3.59 *Any root of the nondegenerated equation $\widehat{p}(c) = \widehat{q}(c)$ is a corner of $\widehat{p \oplus q}$.*

Proof Let us take the canonical form of the equation. Any root of $\widehat{p}(c) = \widehat{q}(c)$ is the solution of $p(k)c^k = q(l)c^l$ for some different k and l and thus it is a corner of $\widehat{p \oplus q}$ because the equation is in minimal canonical form. ■

The converse of this theorem is not true, i.e. a corner is not always a root of the equation $\widehat{p}(c) = \widehat{q}(c)$.

Example 3.60 The polynomial equation $3c \oplus 2 = 3c^2$ has the corner $c = -1$ which is not a root $(3(-1) \oplus 2 = 2,\ 3(-1)^2 = 1)$. ■

Now we give a characterization of the situation where the polynomial equation of degree n has exactly n roots.

Theorem 3.61 *Suppose that n is even (the case when n is odd is similar). Let*

$$\widehat{p}(c) = p(0) \oplus p(2)c^2 \oplus \cdots \oplus p(2k)c^{2k} \ ,$$

$$\widehat{q}(c) = p(1)c \oplus p(3)c^3 \oplus \cdots \oplus p(2k-1)c^{2k-1} \ ,$$

and suppose that $\widehat{p}(c) = \widehat{q}(c)$ is a nondegenerated equation in canonical form, then $c_i \stackrel{\text{def}}{=} p(i-1)\not{/}p(i)$, $i = 1, \ldots, n$, are the n roots of the equation.

Proof No corner of $\widehat{p \oplus q}$ is multiple. Hence each one is obtained as the intersection of two monomials of consecutive degrees. Therefore, these monomials belong to different sides of the equation. Thus the corrresponding corners are roots of the equation.

Conversely if the equation has n roots, $\widehat{p \oplus q}$ has n corners which therefore are distinct. But each of these corners is a root, hence these corners are characterized by the intersection of two monomial functions of consecutive degrees, the monomials being in different sides of the equation. ■

3.4 SYMMETRIZATION OF THE MAX-PLUS ALGEBRA

We have seen that the theory of linear system of equations in the max-plus algebra is not satisfactory at all, not even in the scalar case. We now extend $\mathbb{R}_{\max}$ to a larger set $\mathbb{S}$ for which $\mathbb{R}_{\max}$ can be viewed as the positive part. The construction is similar to the construction of $\mathbb{Z}$ as an extension of $\mathbb{N}$ in conventional algebra, but we cannot avoid some complications coming from the idempotency of the max operator. With this new set, we can generalize the notion of an equation that we call a balance. Then all nondegenerated scalar linear balances have a unique solution. Thus a linear closure of $\mathbb{R}_{\max}$ has been achieved.

3.4.1 The Algebraic Structure $\mathbb{S}$

A natural approach to our problem is to embed the max-plus algebra into a structure in which every nontrivial scalar linear equation has at least one solution. In particular we would like to have a solution to the equation $a \oplus x = \varepsilon$, that is, we would like to find a symmetric element to a. But this not possible, as shown by the following theorem.

Theorem 3.62 *Every idempotent group is reduced to the zero element.*

Proof Assume that the group $(\mathcal{G}, \oplus)$ is idempotent with zero element ε. Let b be the symmetric element of $a \in \mathcal{G}$. Then

$$a = a \oplus \varepsilon = a \oplus (a \oplus b) = (a \oplus a) \oplus b = a \oplus b = \varepsilon \ .$$

■

Nevertheless we can adapt the idea of the construction of $\mathbb{Z}$ from $\mathbb{N}$ to build a 'balancing' element rather than a symmetric one. This is the purpose of the following subsections.

3.4.1.1 *The Algebra of Pairs*

Let us consider the set of pairs $\mathbb{R}^2_{\max}$ endowed with the natural idempotent semifield structure:

$$(x', x'') \oplus (y', y'') = (x' \oplus y', x'' \oplus y'') \ ,$$

$$(x', x'') \otimes (y', y'') = (x'y' \oplus x''y'', x'y'' \oplus x''y') \ ,$$

with $(\varepsilon, \varepsilon)$ as the zero element and (e, ε) as the identity element.

Let $x = (x', x'')$ and define the *minus* sign as $\ominus x = (x'', x')$, the *absolute value* of x as $|x| = x' \oplus x''$ and the *balance operator* as $x^\bullet = x \ominus x = (|x|, |x|)$. Clearly, these operators have the following properties:

1. $a^\bullet = (\ominus a)^\bullet$;
2. $a^{\bullet\bullet} = a^\bullet$;
3. $ab^\bullet = (ab)^\bullet$;
4. $\ominus(\ominus a) = a$;
5. $\ominus(a \oplus b) = (\ominus a) \oplus (\ominus b)$;
6. $\ominus(a \otimes b) = (\ominus a) \otimes b$.

These properties allow us to write $a \oplus (\ominus b) = a \ominus b$ as usual.

3.4.1.2 *Quotient Structure*

Definition 3.63 (Balance relation) *Let $x = (x', x'')$ and $y = (y', y'')$. We say that x balances y (which is denoted $x \nabla y$) if $x' \oplus y'' = x'' \oplus y'$.*

It is fundamental to notice that ∇ is *not* transitive and thus is not an equivalence relation. For instance, consider $(e, 1) \nabla (1, 1)$, $(1, 1) \nabla (1, e)$, but $(e, 1) \not\nabla (1, e)$! Since ∇ cannot be an equivalence relation, it is not possible to define the quotient structure of $\mathbb{R}^2_{\max}$ by means of ∇ (unlike in conventional algebra in which $\mathbb{N}^2 / \nabla \simeq \mathbb{Z}$). However, we can introduce the equivalence relation $\mathcal{R}$ on $\mathbb{R}^2_{\max}$ closely related to the balance relation, namely,

$$(x', x'') \mathcal{R} (y', y'') \Leftrightarrow \begin{cases} x' \oplus y'' = x'' \oplus y' & \text{if } x' \neq x'', y' \neq y'' \ , \\ (x', x'') = (y', y'') & \text{otherwise.} \end{cases}$$

It is easy to check that $\mathcal{R}$ is compatible with the addition and multiplication of $\mathbb{R}^2_{\max}$, the balance relation ∇, and the $\ominus$, $|\cdot|$ and $^\bullet$ operators.

Definition 3.64 *The* symmetrized algebra $\mathbb{R}^2_{\max} / \mathcal{R}$ *of* $\mathbb{R}_{\max}$ *is called* $\mathbb{S}$.

We distinguish three kinds of equivalence classes:

$$\begin{aligned}
&\overline{(t,-\infty)} = \{(t,x'') \mid x'' < t\} \ , && \text{called positive elements;} \\
&\overline{(-\infty,t)} = \{(x',t) \mid x' < t\} \ , && \text{called negative elements;} \\
&\overline{(t,t)} = \{(t,t)\} \ , && \text{called balanced elements.}
\end{aligned}$$

By associating $\overline{(t,-\infty)}$ with $t \in \mathbb{R}_{\max}$, we can identify $\mathbb{R}_{\max}$ with the semifield of positive or zero classes denoted $\mathbb{S}^{\oplus}$. The set of negative or zero classes (of the form $\ominus x$ for $x \in \mathbb{S}^{\oplus}$) will be denoted $\mathbb{S}^{\ominus}$. This set is not stable by multiplication and thus it is not a semifield. The set of balanced classes (of the form $x^{\bullet}$) is denoted $\mathbb{S}^{\bullet}$; it is also isomorphic to $\mathbb{R}_{\max}$. This yields the decomposition

$$\mathbb{S} = \mathbb{S}^{\oplus} \cup \mathbb{S}^{\ominus} \cup \mathbb{S}^{\bullet} \ . \tag{3.15}$$

The element ε is the *only* element common to $\mathbb{S}^{\oplus}$ and $\mathbb{S}^{\ominus}$ and $\mathbb{S}^{\bullet}$. This decomposition of $\mathbb{S}$ should be compared with $\mathbb{Z} = \mathbb{N}^{+} \cup \mathbb{N}^{-}$. This notation allows us to write $3 \ominus 2$ instead of $\overline{(3,-\infty)} \oplus \overline{(-\infty,2)}$. We thus have $3 \ominus 2 = \overline{(3,2)} = \overline{(3,-\infty)} = 3$. More generally, calculations in $\mathbb{S}$ can be summarized as follows:

$$\begin{aligned}
&a \ominus b = a \ , && \text{if } a > b \ ; \\
&b \ominus a = \ominus a \ , && \text{if } a > b \ ; \\
&a \ominus a = a^{\bullet} \ .
\end{aligned} \tag{3.16}$$

Because of its importance, we introduce the notation $\mathbb{S}^{\vee}$ for the set $\mathbb{S}^{\oplus} \cup \mathbb{S}^{\ominus}$ and $\mathbb{S}^{\vee}_{\star} = \mathbb{S}^{\vee} \setminus \{\varepsilon\}$. The elements of $\mathbb{S}^{\vee}$ are called *signed* elements. They are either positive, negative or zero.

Theorem 3.65 *The set $\mathbb{S}^{\vee}_{\star} = \mathbb{S} \setminus \mathbb{S}^{\bullet}$ is the set of all invertible elements of $\mathbb{S}$.*

Proof The obvious identity $t \otimes (-t) = (\ominus t) \otimes (\ominus -t) = e$ for $t \in \mathbb{R}_{\max} \setminus \{\varepsilon\}$ implies that every nonzero element of $\mathbb{S}^{\vee}$ is invertible. Moreover, the absorbing properties of the balance operator show that $\mathbb{S}^{\vee}$ is absorbing for the product. Thus, $x^{\bullet} y \neq e$ for all $y \in \mathbb{S}$ since $e \notin \mathbb{S}^{\bullet}$. ∎

Remark 3.66 Thus, in $\mathbb{S}$, with each element $a \in \mathbb{S}^{\vee}$, we can associate an element $\ominus a$ such that $b = a \ominus a \in \mathbb{S}^{\bullet}$ but in general $b \neq \varepsilon$. This is the main difference with the usual symmetrization. Here the whole set $\mathbb{S}^{\bullet}$ plays the role of the usual zero element. ∎

3.4.2 Linear Balances

Before solving general linear balances, we need to understand the meaning of the generalization of equations in $\mathbb{R}_{\max}$ by balances in $\mathbb{S}$. This can be done by studying the properties of balances.

Theorem 3.67 *The relation ∇ satisfies the following properties:*

1. *$a \nabla a$;*

2. $a \nabla b \Leftrightarrow b \nabla a$;

3. $a \nabla b \Leftrightarrow a \ominus b \nabla \varepsilon$;

4. $\{a \nabla b, c \nabla d\} \Rightarrow a \oplus c \nabla b \oplus d$;

5. $a \nabla b \Rightarrow ac \nabla bc$.

Proof Let us prove Property 5. Obviously, $a \nabla b \Leftrightarrow a \ominus b \in \mathbb{S}^{\bullet}$ and, since $\mathbb{S}^{\bullet}$ is absorbing, $(a \ominus b)c = ac \ominus bc \in \mathbb{S}^{\bullet}$, i.e. $ac \nabla bc$. ■

Although ∇ is not transitive, when some variables are signed, we can manipulate balances in the same way as we manipulate equations.

Theorem 3.68

1. Weak substitution *If $x \nabla a$, $cx \nabla b$ and $x \in \mathbb{S}^{\vee}$, we have $ca \nabla b$.*

2. Weak transitivity *If $a \nabla x$, $x \nabla b$ and $x \in \mathbb{S}^{\vee}$, we have $a \nabla b$.*

3. Reduction of balances *If $x \nabla y$ and $x, y \in \mathbb{S}^{\vee}$, we have $x = y$.*

Proof

1. We have either $x \in \mathbb{S}^{\oplus}$ or $x \in \mathbb{S}^{\ominus}$. Assume for instance that $x \in \mathbb{S}^{\oplus}$, that is, $x = \overline{(x', \varepsilon)}$. With the usual notation, $x' \oplus a'' = a'$ and $c'x' \oplus b'' = c''x' \oplus b'$. Adding $c'a'' \oplus c''a''$ to the last equality, we get

$$c'x' \oplus c'a'' \oplus c''a'' \oplus b'' = c''x' \oplus c'a'' \oplus c''a'' \oplus b' ,$$

which yields, by using $x' \oplus a'' = a'$,

$$c'a' \oplus c''a'' \oplus b'' = c''a' \oplus c'a'' \oplus b' ,$$

that is, $ca \nabla b$.

2. This a consequence of the weak substitution for $c = e$.

3. This point is trivial but is important in order to derive equalities from balances. ■

The introduction of these new notions is justified by the fact that any linear balance (which is not degenerated) has one and only one solution in $\mathbb{S}^{\vee}$.

Theorem 3.69 *Let $a \in \mathbb{S}^{\vee}_{\star}$ and $b \in \mathbb{S}^{\vee}$, then $x = \ominus a^{-1} b$ is the unique solution of the balance*

$$ax \oplus b \nabla \varepsilon , \tag{3.17}$$

which belongs to $\mathbb{S}^{\vee}$.

Proof From the properties of balances it follows that $ax \oplus b \nabla \varepsilon \Leftrightarrow x \nabla \ominus a^{-1}b$. Then using the reduction property and the fact that $\ominus a^{-1}b \in \mathbb{S}^{\vee}$, we obtain $x = \ominus a^{-1}b$. ■

Remark 3.70 If $b \notin \mathbb{S}^{\vee}$, we lose the uniqueness of signed solutions. Every x such that $|ax| \leq |b|$ (i.e. $|x| \leq |a^{-1}b|$) is a solution of the balance (3.17). If $a \notin \mathbb{S}^{\vee}$, we again lose uniqueness. Assume $b \in \mathbb{S}^{\vee}$ (otherwise, the balance holds for all values of x), then every x such that $|ax| \geq |b|$ is a solution. ■

Remark 3.71 We can describe all the solutions of (3.17). For all $t \in \mathbb{R}_{\max}$, we obviously have $at^{\bullet} \nabla \varepsilon$. Adding this balance to $ax \oplus b \nabla \varepsilon$, where x is the unique signed solution, we obtain $a(x \oplus t^{\bullet}) \oplus b \nabla \varepsilon$. Thus,

$$x_t = x \oplus t^{\bullet} \tag{3.18}$$

is a solution of (3.17). If $t \geq |x|$, then $x_t = t^{\bullet}$ is balanced. Conversely, it can be checked that every solution of (3.17) may be written as in (3.18). Finally the *unique* signed solution x is also the least solution. ■

Remark 3.72 Nontrivial linear balances (with data in $\mathbb{S}^{\vee}$) always have solutions in $\mathbb{S}$; this is why $\mathbb{S}$ may be considered as a *linear closure* of $\mathbb{R}_{\max}$. ■

3.5 LINEAR SYSTEMS IN $\mathbb{S}$

It is straightforward to extend balances to the vector case. Theorems 3.67 and 3.68 still hold when a, b, x, y and c are matrices with appropriate dimensions, provided we replace 'belongs to $\mathbb{S}^{\vee}$' by 'every entry belongs to $\mathbb{S}^{\vee}$'. Therefore, we say that a vector or a matrix is *signed* if all its entries are signed.

We now consider a solution $x \in \mathbb{R}_{\max}$ of the equation

$$Ax \oplus b = Cx \oplus d \ . \tag{3.19}$$

Then the definition of the balance relation implies that

$$(A \ominus C)x \oplus (b \ominus d) \nabla \varepsilon \ . \tag{3.20}$$

Conversely, assuming that x is a positive solution of (3.20), we obtain

$$Ax \oplus b \nabla Cx \oplus d \ ,$$

with $Ax \oplus b$ and $Cx \oplus d \in \mathbb{S}^{\oplus}$. Using Theorem 3.68, we obtain

$$Ax \oplus b = Cx \oplus d \ .$$

Therefore we have the following theorem.

Theorem 3.73 *The set of solutions of the general system of linear equations (3.19) in* $\mathbb{R}_{\max}$ *and the set of* positive *solutions of the associated linear balance (3.20) in* $\mathbb{S}$ *coincide.*

Hence, studying Equation (3.19) is reduced to solving linear balances in $\mathbb{S}$.

Remark 3.74 The case when a solution x of (3.20) has some negative and some positive entries is also of interest. We write $x = x^+ \ominus x^-$ with $x^+, x^- \in (\mathbb{S}^\oplus)^n$. Partitioning the columns of A and C according to the sign of the entries of x, we obtain $A = A^+ \oplus A^-$, $C = C^+ \oplus C^-$, so that $Ax = A^+x^+ \ominus A^-x^-$ and $Cx = C^+x^+ \ominus C^-x^-$. We can thus claim the existence of a solution in $\mathbb{R}_{\max}$ to the new problem

$$A^+x^+ \oplus C^-x^- \oplus b = A^-x^- \oplus C^+x^+ \oplus d \ .$$

The solution of nondegenerated problems is not unique, but the set of solutions forms a single class of $\mathbb{R}^2_{\max}$ (for the equivalence relation $\mathcal{R}$). ■

3.5.1 Determinant

Before dealing with general systems, we need to extend the determinant machinery to the $\mathbb{S}$-context. We define the signature of a permutation σ by

$$\operatorname{sgn}(\sigma) = \begin{cases} e & \text{if } \sigma \text{ is even;} \\ \ominus e & \text{otherwise.} \end{cases}$$

Then the determinant of an $n \times n$ matrix $A = (A_{ij})$ is given (as usual) by

$$\bigoplus_{\sigma} \operatorname{sgn}(\sigma) \bigotimes_{i=1}^{n} A_{i\sigma(i)} \ ,$$

and is denoted either $|A|$ or $\det(A)$. The transpose of the matrix of cofactors is denoted $A^\natural$ ($\left(A^\natural\right)_{ij} \stackrel{\text{def}}{=} \operatorname{cof}_{ji}(A)$). The classical properties of the determinant are still true.

Theorem 3.75 *The determinant has the following properties:*

linearity:

$$|(u_1, \ldots, \lambda u_i \oplus \mu v_i, \ldots, u_n)| = \lambda|(u_1, \ldots, u_i, \ldots, u_n)| \oplus \mu|(u_1, \ldots, v_i, \ldots, u_n)| \ ;$$

antisymmetry:

$$|(u_{\sigma(1)}, \ldots, u_{\sigma(n)})| = \operatorname{sgn}(\sigma)|(u_1, \ldots, u_n)| \ ;$$

and consequently

$$|(u_1, \ldots, v, \ldots, v, \ldots, u_n)| \ \nabla\ \varepsilon \ ;$$

expansion with respect to a row:

$$|A| = \bigoplus_{k=1}^{n} a_{ik} \operatorname{cof}_{ik}(A) \ ;$$

transposition: $|A| = |A'|$.

direct consequence is that some classical proofs lead to classical identities in this ew setting. Sometimes weak substitution limits the scope of this approach.

heorem 3.76 *For an $n \times n$ matrix A with entries in $\mathbb{S}$, we have*

ramer formula: *$AA^\natural \nabla |A|e$, and if $|A|$ is signed, then the diagonal of $AA^\natural$ is signed;*

ecursive computation of the determinant:

$$|A| = \left| \begin{pmatrix} F & G \\ H & a_{nn} \end{pmatrix} \right| = |F|a_{nn} \ominus HF^\natural G$$

for a partition of matrix A where a_{nn} is a scalar;

ayley-Hamilton theorem: *p being the characteristic polynomial of matrix A, i.e. $p(\lambda) \stackrel{\text{def}}{=} |A \ominus \lambda e|$, we have $p(A) \nabla \varepsilon$.*

emark 3.77 We define the positive determinant of a matrix A, denoted $|A|^+$, by he sum of terms $\bigotimes_i A_{i\sigma(i)}$ where the sum is limited to even permutations, and a egative determinant, denoted $|A|^-$, by the same sum limited to odd permutations. he matrix of positive cofactors is defined by

$$\left(A^{\natural+}\right)_{ij} = \begin{cases} |A^{ji}|^+ & \text{if } i+j \text{ is even,} \\ |A^{ji}|^- & \text{if } i+j \text{ is odd,} \end{cases}$$

where A^{ji} denotes the matrix derived from A by deleting row j and column i. The natrix of negative cofactors $A^{\natural-}$ is defined similarly. With this notation, Theo-em 3.76 can be rewritten as follows:

$$AA^{\natural+} \oplus |A|^- e = AA^{\natural-} \oplus |A|^+ e \ .$$

This formula does not use the $\ominus$ sign and is valid in any semiring. The symmetrized lgebra appears to be a natural way of handling such identities (and giving proofs n an algebraic way). ■

3.5.2 Solving Systems of Linear Balances by the Cramer Rule

In this subsection we study solutions of systems of linear equations with entries in $\mathbb{S}$. We only consider the solutions belonging to $(\mathbb{S}^\vee)^n$, that is, we only consider *signed* solutions. Indeed, in a more general setting we cannot hope to have a result of uniqueness; see Remark 3.70. We can now state the fundamental result for the existence and uniqueness of signed solutions of linear systems.

Theorem 3.78 (Cramer system) *Let A be an $n \times n$ matrix with entries in $\mathbb{S}$, $|A| \in \mathbb{S}^\vee_\star$, $b \in \mathbb{S}^n$ and $A^\natural b \in (\mathbb{S}^\vee)^n$. Then, in $\mathbb{S}^\vee$ there exists a unique solution of*

$$Ax \nabla b \ , \tag{3.21}$$

and it satisfies

$$x \nabla A^\natural b \oslash |A| \ . \tag{3.22}$$

Proof By right-multiplying the identity $AA^\natural \nabla |A|e$ by $|A|^{-1}b$, we see that x is a solution. Let us now prove uniqueness. The proof is by induction on the size of the matrix. It is based on Gauss elimination in which we manage the balances using weak substitution. Let us prove (3.22) for the last row, i.e. $|A|x_n \nabla (A^\natural b)_n$. Developing $|A|$ with respect to the last column, $|A| = \bigoplus_{k=1}^{n} a_{kn}\mathrm{cof}_{kn}(A)$, we see that at least one term is invertible, say $a_{1n}\mathrm{cof}_{1n}(A)$. We now partition A, b and x in such a way that the scalar a_{1n} becomes a block:

$$A = \begin{pmatrix} H & a_{1n} \\ F & G \end{pmatrix}, \quad b = \begin{pmatrix} b_1 \\ B \end{pmatrix}, \quad x = \begin{pmatrix} X \\ x_n \end{pmatrix}.$$

Then $Ax \nabla b$ can be written as

$$HX \oplus a_{1n}x_n \nabla b_1 \ , \tag{3.23}$$

$$FX \oplus Gx_n \nabla B \ . \tag{3.24}$$

Since $|F| = (\ominus e)^{n+1}\mathrm{cof}_{1n}(A)$ is invertible, we can apply the induction hypothesis to (3.24). This implies that

$$X \nabla |F|^{-1}F^\natural(B \ominus Gx_n) \ .$$

Using the weak substitution property, we can replace $X \in (\mathbb{S}^\vee)^{n-1}$ in Equation (3.23) to obtain

$$|F|^{-1}HF^\natural(B \ominus Gx_n) \oplus a_{1n}x_n \nabla b_1 \ ,$$

that is,

$$x_n(|F|a_{1n} \ominus HF^\natural G) \nabla |F|b_1 \ominus HF^\natural B \ .$$

Here we recognize the developments of $|A|$ and $(A^\natural b)_n$, therefore

$$x_n \nabla (A^\natural b)_n \not/ |A| \ .$$

Since the same reasoning can be applied to the entries of x other than n, this concludes the proof. ■

Remark 3.79 Let us write D_i for the determinant of the matrix obtained by replacing the i-th column of A by the column vector b; then $D_i = (A^\natural b)_i$. Assume that $D \stackrel{\text{def}}{=} |A|$ is invertible, then Equation (3.22) is equivalent to

$$x_i \nabla D^{-1}D_i \ , \quad \forall i \ .$$

If $A^\natural b \in (\mathbb{S}^\vee)^n$, then by using the reduction of balances (see Theorem 3.68), we obtain

$$x_i = D^{-1}D_i \ ,$$

which is exactly the classical Cramer formula. ■

Example 3.80 The $\mathbb{R}_{\max}$ equation

$$\begin{pmatrix} e & -4 \\ 3 & 2 \end{pmatrix} \otimes \begin{pmatrix} x_1 \\ x_2 \end{pmatrix} \oplus \begin{pmatrix} 1 \\ -5 \end{pmatrix} = \begin{pmatrix} -1 & 1 \\ \varepsilon & 2 \end{pmatrix} \otimes \begin{pmatrix} x_1 \\ x_2 \end{pmatrix} \oplus \begin{pmatrix} 2 \\ 7 \end{pmatrix}, \tag{3.25}$$

corresponds to the balance

$$\begin{pmatrix} e & \ominus 1 \\ 3 & 2^\bullet \end{pmatrix} \begin{pmatrix} x_1 \\ x_2 \end{pmatrix} \nabla \begin{pmatrix} 2 \\ 7 \end{pmatrix} . \tag{3.26}$$

Its determinant is $D = 4$.

$$D_1 = \left| \begin{pmatrix} 2 & \ominus 1 \\ 7 & 2^\bullet \end{pmatrix} \right| = 8 \ , \quad D_2 = \left| \begin{pmatrix} e & 2 \\ 3 & 7 \end{pmatrix} \right| = 7 \ ,$$

$$A^\natural b = \begin{pmatrix} D_1 \\ D_2 \end{pmatrix} = \begin{pmatrix} 8 \\ 7 \end{pmatrix} \in (\mathbb{S}^\vee)^2 \ .$$

The system is invertible and has a unique solution. Thus $(x_1 = D_1 \not{/} D = 8 - 4 = 4, x_2 = D_2 \not{/} D = 7 - 4 = 3)$ is the unique positive solution in $\mathbb{S}$ of the balance (3.26). Hence it is the unique solution in $\mathbb{R}_{\max}$ of Equation (3.25). ■

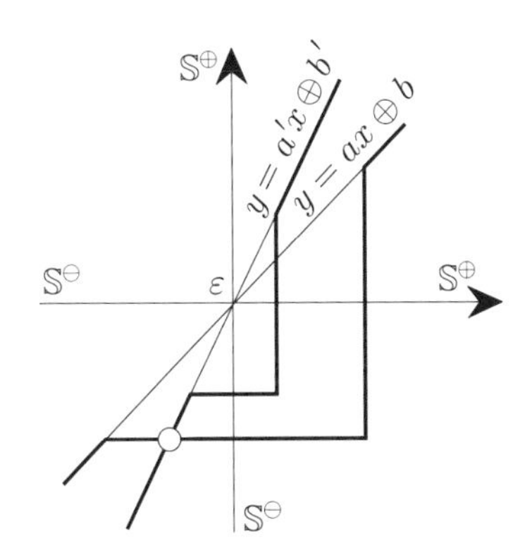

Figure 3.14: A $(2, 2)$ linear system of equations

Example 3.81 In the two-dimensional case the condition $A^\natural b \in (\mathbb{S}^\vee)^n$ has a very clear geometric interpretation (see Figure 3.14) as the intersection of straight lines in $\mathbb{S}$. First we can choose an exponential scaling of the x_1 and x_2 axes. The exponential maps $\mathbb{S}^\oplus$ to $\mathbb{R}^+$ and $\mathbb{S}^\ominus$ to $\mathbb{R}^-$, if we identify $\mathbb{S}^\ominus$ with $i\pi + \mathbb{R} \cup \{-\infty\}$. We do not represent the balance axis in this representation. Therefore the straight line $ax_2 \oplus bx_1 \oplus c \nabla \varepsilon$ is a broken line (in the usual sense) composed of four segments:

- two of them are symmetric with respect of the origin; they correspond to the contribution of $ax_2 \oplus bx_1$ to the balance (they belong to the conventional line $ax_2 + bx_1 = 0$);
- the horizontal segment corresponds to the contribution of $ax_2 \oplus c$ to the balance;
- the vertical segment corresponds to the contribution of $bx_1 \oplus c$ to the balance.

Then it is easy to see that two such lines have one and only one point of intersection, or there exists a complete segment of solutions. This latter case is called singular and is not further considered here. ∎

Remark 3.82 The invertibility of $|A|$ is *not* a necessary condition for the existence of a signed solution to the system $Ax \nabla b$ for some value of b. Let us consider

$$A = \begin{pmatrix} e & e & \varepsilon \\ e & e & \varepsilon \\ e & e & \varepsilon \end{pmatrix} .$$

Because $|A| = \varepsilon$, the matrix A is not invertible. Let $t \in \mathbb{S}^{\vee}$ be such that $|b_i| \leq |t|$ for all i, and let $x = \begin{pmatrix} t & \ominus t & \varepsilon \end{pmatrix}'$. Then $Ax \nabla b$. But in this case, the signed solution is not unique. ∎

Remark 3.83 As already noticed, in [65] for example, determinants have a natural interpretation in terms of assignment problems. So the Cramer calculations have the same complexity as $n+1$ assignment problems, which can be solved using flow algorithms. ∎

3.6 POLYNOMIALS WITH COEFFICIENTS IN $\mathbb{S}$

We have built the linear closure $\mathbb{S}$ of $\mathbb{R}_{\max}$. Therefore any linear equation in $\mathbb{S}$ has in general a solution in this set. The purpose of this section is to show that $\mathbb{S}$ is almost algebraically closed, that is, any 'closed' polynomial equation of degree n in $\mathbb{S}$ has n solutions. The term 'closed' refers to the fact that the class of formal polynomials having the same polynomial function defined on $\mathbb{S}$ is not always closed (in a topological sense that will be made precise later). We will see that $\mathbb{S}$ is algebraically closed only for the subset of 'closed' polynomial equations. Moreover, we will see that any polynomial function can be transformed into a closed function by modifying it in a finite number of points.

3.6.1 Some Polynomial Functions

We can generalize the notions of formal polynomials and of polynomial functions to the set $\mathbb{S}$. We restrict our study to polynomial functions in $\mathcal{P}(\mathbb{S}^{\vee}) \stackrel{\text{def}}{=} \mathcal{F}(\mathbb{S}^{\vee}[\gamma])$, where $\mathbb{S}^{\vee}[\gamma]$ denotes the class of formal polynomials with coefficients in $\mathbb{S}^{\vee}$ and $\mathcal{F}$ is the straightforward extension of the evaluation homomorphism introduced in Lemma 3.35. We will see that such functions assume values in $\mathbb{S}^{\vee}$ when their argument ranges in $\mathbb{S}^{\vee}$, except perhaps at a finite number of points. For the more general class of polynomial functions with coefficients in $\mathbb{S}$, denoted $\mathcal{P}(\mathbb{S})$, it may happen that the function takes a balanced value on a continuous set of points of $\mathbb{S}^{\vee}$ (in which case we say that we have a balanced facet). Because we are mainly interested in the analogy with conventional polynomials, we do not deal with this latter situation.

To get a better understanding of polynomial functions on $\mathbb{S}$ we study some particular cases. Let us start by plotting the graphs of polynomial functions of degree one, two and three (see Figure 3.15). We must study the graphs over each

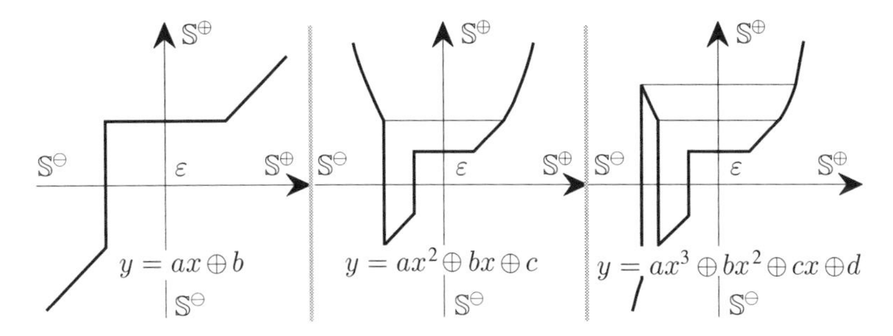

Figure 3.15: Polynomial functions of degree one, two and three

of the three components $\mathbb{S}^{\oplus}$, $\mathbb{S}^{\ominus}$ and $\mathbb{S}^{\bullet}$. The value of the function itself, however, is an element of $\mathbb{S}$ which also belongs to one of these components. Therefore the plot is quite complicated. In order to simplify the study, only the plot over $\mathbb{S}^{\vee}$ is considered. Moreover we will use the exponential system of coordinates that we have discussed in the previous section. Figure 3.15 shows the points of discontinuity of the three component functions. These discontinuities always appear at abscissæ which are symmetric with respect to corner abscissæ. At these discontinuities the polynomial functions take balanced values (in the graph, we see sign changes). They correspond to what we call 'roots' of polynomial functions. If a polynomial is of degree n, it has in general n corners and n points in $\mathbb{S}^{\vee}$ where the polynomial function takes balanced values.

Let us study the polynomial functions of degree two in more detail. We consider

$$\widehat{p}(c) = p(0) \oplus p(1)c \oplus p(2)c^2 \ , \quad p(i) \in \mathbb{S}^{\vee} \ , \quad i = 0, 1, 2 \ .$$

The polynomial function $|\widehat{p}| \in \mathcal{P}(\mathbb{R}_{\max})$ is defined as $\widehat{p}(c) = |p(0)| \oplus |p(1)|c \oplus |p(2)|c^2$. We will later prove, but it should be clear, that if c is a root of $\widehat{p}$, then $|c|$ is a corner of $|p|$. Owing to Theorem 3.43, $|\widehat{p}|$ can be factored into the product of two linear polynomials. We have the following four possibilities:

1. If $|p(1) \not/ p(2)| > |p(0) \not/ p(1)|$, then $|\widehat{p}|$ has two distinct corners $c_1 = |p(1) \not/ p(2)|$ and $c_2 = |p(0) \not/ p(1)|$. It can be checked that $\widehat{p} = p(2)(c \oplus p(1) \not/ p(2))(c \oplus p(0) \not/ p(1))$ and that this factorization is unique. In addition, the moduli of the roots are c_1 and c_2.

2. If $|p(1) \not/ p(2)| = |p(0) \not/ p(1)|$, then $|\widehat{p}|$ has a corner $c_1 = |p(0) \not/ p(2)|^{1/2}$ of multiplicity 2 and the roots of $\widehat{p} = p(2)(c \oplus p(1) \not/ p(2))(c \oplus p(0) \not/ p(1))$ have modulus c_1.

3. If $|p(1) \not/ p(2)| < |p(0) \not/ p(1)|$ and $p(2)p(0) \in \mathbb{S}^{\oplus}$, then $|\widehat{p}|$ has a corner $c_1 = |p(0) \not/ p(2)|^{1/2}$ of multiplicity 2 and $\widehat{p}$ cannot be factored. Indeed at c_1 and $\ominus c_1$, $\widehat{p}$ has signed values. We have $\widehat{p} = p(2)(c^2 \oplus p(0) \not/ p(2))$.

4. If $|p(1)\not/p(2)| < |p(0)\not/p(1)|$ and $p(2)p(0) \in \mathbb{S}^{\ominus}$, then the polynomial $|\widehat{p}|$ has a corner $c_1 = |p(0)\not/p(2)|^{1/2}$ of multiplicity 2. We have $\widehat{p} = p(2)(c \oplus c_1)(c \ominus c_1)$ and thus $\widehat{p}$ has been factored.

This discussion suggests that if the corners of $|\widehat{p}|$ (now a polynomial function of degree n) are distinct, then there are n roots; but if $|\widehat{p}|$ has multiple corners, then we are not guaranteed to have a factorization in linear factors. We now study the situation in more detail.

3.6.2 Factorization of Polynomial Functions

We can consider formal polynomials $\mathbb{S}[\gamma]$ and polynomial functions $\mathcal{P}(\mathbb{S})$ with coefficients in $\mathbb{S}$ as we have done for polynomials with coefficients in $\mathbb{R}_{\max}$. They define algebras. The following mapping from $\mathcal{P}(\mathbb{S})$ into $\mathcal{P}(\mathbb{R}_{\max})$

$$p = p(k)\gamma^k \oplus \cdots \oplus p(n)\gamma^n \mapsto |p| = |p(k)|\gamma^k \oplus \cdots \oplus |p(n)|\gamma^n$$

is a surjective morphism.

Definition 3.84 (Root) *The root of a polynomial function $\widehat{p} \in \mathcal{P}(\mathbb{S})$ is an element $c \in \mathbb{S}^{\vee}$ such that $\widehat{p}(c) \nabla \varepsilon$.*

Remark 3.85 It should be clear that the computation of the linear factors of a polynomial yields its roots. Indeed we have

$$\bigotimes_i (c \ominus c_i) \nabla \varepsilon \Leftrightarrow c \ominus c_i \nabla \varepsilon \ .$$

■

Lemma 3.86 *If, for a polynomial $\widehat{p} \in \mathcal{P}(\mathbb{S}^{\vee})$, c is a root of $\widehat{p}(c)$, then $|c|$ is a corner of $\widehat{|p|}$.*

Proof If c is a root, then $\widehat{p}(c)$ is balanced and hence $|\widehat{p}(c)| = |\widehat{p}(|c|)|$ is a quantity which is achieved by two distinct monomials of $\widehat{p}$ since, by assumption, the coefficients of $\widehat{p}$ belong to $\mathbb{S}^{\vee}$. Thus $|c|$ is a corner of $|p|$. ■

For the same reason as in the $\mathbb{R}_{\max}$ case, the mapping

$$\mathcal{F} : \mathbb{S}[\gamma] \rightarrow \mathcal{P}(\mathbb{S}) \ , \quad p \mapsto \widehat{p} \ ,$$

is not injective. In order to study the set valued-inverse mapping $\mathcal{F}^{-1}$, we introduce the notion of a closed polynomial function.

Definition 3.87 (Closed polynomial function) *We say that the polynomial function $\widehat{p} \in \mathcal{P}(\mathbb{S}^{\vee})$ is closed if $\mathcal{F}^{-1}(\widehat{p})$ admits a maximum element denoted $p^{\sharp} \in \mathcal{P}(\mathbb{S})$. This element is called the maximum representative of $\widehat{p}$.*

Example 3.88 The polynomial function $c^2 \ominus e$ is closed because it has the same graph as $c^2 \oplus e^\bullet c \ominus e$. The polynomial function $c^2 \oplus e$ is not closed because its graph is different from that of any polynomial $\widehat{p}_a = c^2 \oplus ac \oplus e$ with $a \in \{e, \ominus e, e^\bullet\}$ although it is the same for all $a < e$. ∎

The notion of closed polynomial function is relevant because the inverse of the evaluation homomorphism is simple for this class, and because this class is exactly the set of polynomial functions which can be factored into linear factors.

Theorem 3.89 *A polynomial function $\widehat{p} \in \mathcal{P}(\mathbb{S}^\vee)$ can be factored in linear factors if and only if it is closed.*

Proof

Any polynomial which can be factored is closed. Let us suppose that the degree of $\widehat{p}$ is n, its valuation 0, and the coefficient of its head monomial is e (the proof can be generalized to avoid these assumptions). Let us suppose that the degree of $\widehat{p}$ is n, its valuation 0, and the coefficient of its head monomial is e (the proof can be adapted to the general case). Let $c_i, i = 1, \ldots, n$, denote the roots of $\widehat{p}$ numbered according to the decreasing order of their modulus. If the c_i are all distinct, we have

$$\widehat{p}(c) = \bigoplus_{i=0}^{n} m_i \quad \text{with} \quad m_i \stackrel{\text{def}}{=} \left(\bigotimes_{j=1}^{i} c_j \right) c^{n-i} \ .$$

Because

$$m_i > m_j \ , \quad \forall j > i \ , \quad \forall c : c_{i+1} < c < c_i \ ,$$

we cannot increase the coefficient of a monomial without changing the graph of $\widehat{p}$ and therefore it is closed. The situation is more complicated when there is a multiple root because then at least three monomials take the same value in modulus at this root. To understand what happens at such a multiple root, let us consider the case when $\widehat{p}$ has only the roots e or $\ominus e$, that is, $\widehat{p}(c) = (c \oplus e)^{n-m}(c \ominus e)^m$. The expansion of this polynomial gives five kinds of polynomials:

- $c^n \oplus c^{n-1} \oplus \cdots \oplus e$;
- $c^n \ominus c^{n-1} \oplus c^{n-2} \cdots \oplus e$;
- $c^n \ominus c^{n-1} \oplus c^{n-2} \cdots \ominus e$;
- $c^n \oplus e^\bullet c^{n-1} \oplus e^\bullet c^{n-2} \cdots \oplus e$;
- $c^n \oplus e^\bullet c^{n-1} \oplus e^\bullet c^{n-2} \cdots \ominus e$.

By inspection, we verify that we cannot increase any coefficient of these polynomials without changing their graphs. Therefore, they are closed.

We remark that some polynomials considered in this enumeration do not belong to $\mathcal{P}(\mathbb{S}^\vee)$. For example, it is the case of $(c \ominus e)^2(c \oplus e)$. Some other polynomials have their coefficients in $\mathbb{S}^\bullet$ and do not seem to belong to the class that we study here, but they have other representatives in $\mathcal{P}(\mathbb{S}^\vee)$. For example, we have

$$(c \ominus e)(c \oplus e) = c^2 \oplus e^\bullet c \ominus e = c^2 \ominus e \ .$$

Any closed polynomial can be factored. If $\widehat{p}$ is closed, let p denote its maximum representative. Its coefficients $c_i = p(n-i) \not{/} p(n-i+1)$ are nonincreasing with i in modulus. Indeed, if that were not the case, there would exist i and k such that $|c_{i-k}| > |c_{i+1}| > |c_i|$. Then it would be possible to increase p_{n-i} while preserving the inequalities $|c_{i-k}| > |c_{i+1}| > |c_i|$. Because this operation would not change the graph of $\widehat{p}$, we would have contradicted the maximality of p.

Now, if the $|c_i|$ are *strictly* decreasing with i, we directly verify by expansion that $\widehat{p}(c) = \bigotimes_i (c \oplus c_i)$. If the c_i are simply nonincreasing, we will only examine the particular case when the c_i have their modulus equal to e. There are four subcases:

1. $\widehat{p}(e) \not\nabla \varepsilon$ and $\widehat{p}(\ominus e) \not\nabla \varepsilon$;
2. $\widehat{p}(e) \not\nabla \varepsilon$ and $\widehat{p}(\ominus e) \nabla \varepsilon$;
3. $\widehat{p}(e) \nabla \varepsilon$ and $\widehat{p}(\ominus e) \not\nabla \varepsilon$;
4. $\widehat{p}(e) \nabla \varepsilon$ and $\widehat{p}(\ominus e) \nabla \varepsilon$.

The first case can appear only if n is even and

$$\widehat{p}(c) = c^n \oplus c^{n-2} \oplus \cdots \oplus e \ ,$$

which contradicts the fact the c_i are nonincreasing. The other cases correspond to a factorization studied in the first part of the proof. ■

Corollary 3.90 *A sufficient condition for $\widehat{p} \in \mathcal{P}(\mathbb{S}^\vee)$ to be closed is that $\widehat{|p|}$ has distinct corners.*

Example 3.91

- $c^2 \oplus e$ is always positive and therefore cannot be factored;
- $c^2 \oplus c \oplus e$ is closed and can be factored into $(c \oplus e)^2$;
- $(\gamma \ominus e)(\gamma \oplus e)^2 = (\gamma \oplus e)(\gamma \ominus e)^2 = \gamma^3 \oplus e^\bullet \gamma^2 \oplus e^\bullet \gamma \oplus e$ is a maximum representative of a closed polynomial;
- $\gamma^3 \oplus e^\bullet \gamma^2 \oplus e\gamma \oplus e$ is not a maximum representative of a closed polynomial.

■

In the following theorem the form of the inverse function of $\mathcal{F}$, which is a set-valued mapping, is made precise.

Theorem 3.92 *The set $\mathcal{F}^{-1}(\widehat{p})$, with $\widehat{p} \in \mathcal{P}(\mathbb{S}^\vee)$, admits the minimum element $p^\flat \in \mathbb{S}^\vee[\gamma]$ and the maximum element $p^\sharp \in \mathbb{S}[\gamma]$ which satisfy*

- $p^\flat \le \mathcal{F}^{-1}(\widehat{p}) \le p^\sharp$;
- $p^\flat, p^\sharp \in \mathcal{F}^{-1}(\widehat{p})$;
- $|p^\flat| = |p|^\flat$;
- $|p^\sharp| = |p|^\sharp$.

Proof The proof follows from the epimorphism property of $p \mapsto |p|$, from the corresponding result in $\mathbb{R}_{\max}$, and from the previous result on closed polynomial functions. ■

Example 3.93 For $\widehat{p} = c^2 \oplus 1c \oplus e$ we have $p^\flat = p^\sharp = \gamma^2 \oplus 1\gamma \oplus e$. For $\widehat{p} = c^2 \oplus (-1)c \oplus e$ we have $p^\flat = \gamma^2 \oplus e$, but $p^\sharp$ does not exist in this case. ■

3.7 ASYMPTOTIC BEHAVIOR OF A^k

In this section we prove a max-plus algebra analogue of the Perron-Frobenius theorem, that is, we study the asymptotic behavior of the mapping $k \mapsto A^k$, where A is an $n \times n$ matrix with entries in $\mathbb{R}_{\max}$. We can restrict ourselves to the case when $\mathcal{G}(A)$ contains at least a circuit since, otherwise, A is nilpotent, that is, $A^k = \varepsilon$ for k sufficiently large.

We suppose that the maximum cycle mean is equal to e. If this is not the case, the matrix A is normalized by dividing all its entries by the maximum cycle mean λ. Then the behavior of the general recurrent equation is easily derived from the formula $A^k = \lambda^k(\lambda^{-1}A)^k$. Therefore, in this section all circuits have nonpositive weights and some do have a weight equal to $e = 0$ which is also the maximum cycle mean. We recall that, in this situation, e is the maximum eigenvalue of A (see Remark 3.24).

3.7.1 Critical Graph of a Matrix A

Definition 3.94 *For an $n \times n$ normalized matrix A, the following notions are defined:*

Critical circuit: *a circuit ζ of the precedence graph $\mathcal{G}(A)$ is called critical if it has maximum weight, that is, $|\zeta|_{\mathrm{w}} = e$.*

Critical graph: *the critical graph $\mathcal{G}^{\mathrm{c}}(A)$ consists of those nodes and arcs of $\mathcal{G}(A)$ which belong to a critical circuit of $\mathcal{G}(A)$. Its nodes constitute the set $\mathcal{V}^{\mathrm{c}}$.*

Saturation graph: *given an eigenvector y associated with the eigenvalue e, the saturation graph $\mathcal{S}(A, y)$ consists of those nodes and arcs of $\mathcal{G}(A)$ such that $A_{ij}y_j = y_i$ for some i and j with $y_i, y_j \neq \varepsilon$.*

Cyclicity of a graph: *the cyclicity of a m.s.c.s. is the* gcd *(greatest common divisor) of the lengths of all its circuits. The cyclicity $c(\mathcal{G})$ of a graph $\mathcal{G}$ is the* lcm *(least common multiple) of the cyclicities of all its m.s.c.s.'s.*

Example 3.95 Consider the matrix

$$A = \begin{pmatrix} e & e & \varepsilon & \varepsilon \\ -1 & -2 & \varepsilon & \varepsilon \\ \varepsilon & -1 & -1 & e \\ \varepsilon & \varepsilon & e & e \end{pmatrix} .$$

- Its precedence graph $\mathcal{G}(A)$ has three critical circuits $\{1\}$, $\{3,4\}$,$\{4\}$.
- Its critical graph is the precedence graph of the matrix

$$C = \begin{pmatrix} e & \varepsilon & \varepsilon & \varepsilon \\ \varepsilon & \varepsilon & \varepsilon & \varepsilon \\ \varepsilon & \varepsilon & \varepsilon & e \\ \varepsilon & \varepsilon & e & e \end{pmatrix} .$$

- Matrix A has the eigenvector $\begin{pmatrix} e & -1 & -2 & -2 \end{pmatrix}'$ associated with the eigenvalue e. The corresponding saturation graph is the precedence graph of the matrix

$$S = \begin{pmatrix} e & \varepsilon & \varepsilon & \varepsilon \\ -1 & \varepsilon & \varepsilon & \varepsilon \\ \varepsilon & -1 & \varepsilon & e \\ \varepsilon & \varepsilon & e & e \end{pmatrix} .$$

- The cyclicity of the critical graph is 1. Indeed, the critical graph has two m.s.c.s.'s with nodes $\{1\}$ and $\{3,4\}$, respectively. The second one has two critical circuits, $\{4\}$ and $\{3,4\}$, of length 1 and 2, respectively. The cyclicity of the first m.s.c.s. is 1, the cyclicity of the second m.s.c.s. is $\gcd(1,2) = 1$. Therefore the cyclicity of $\mathcal{G}^c(A)$ is $\operatorname{lcm}(1,1) = 1$. ■

Let us give now some simple results about these graphs which will be useful in the following subsections.

Theorem 3.96 *Every circuit of $\mathcal{G}^c(A)$ is critical.*

Proof If this were not the case, we could find a circuit ζ, composed of subpaths ζ_i of critical circuits γ_i, with a weight different from e. If this circuit had a weight greater than e, it would contradict the assumption that the maximum circuit weight of $\mathcal{G}(A)$ is e. If the weight of ζ were less than e, the circuit ζ' composed of the union of the complements of ζ_i in γ_i would be a circuit of weight greater than e and this would also be a contradiction. ■

Corollary 3.97 *Given a pair of nodes (i,j) in $\mathcal{G}^c(A)$, all paths connecting i to j in $\mathcal{G}^c(A)$ have the same weight.*

Proof If there exists a path p from i to j in $\mathcal{G}^c(A)$, it can be completed by a path p' from j to i also in $\mathcal{G}^c(A)$ to form a critical circuit. If there exists another path p'' from i to j in $\mathcal{G}^c(A)$, the concatenations of p and p' on the one hand, and of p'' and p' on the other hand, form two critical circuits with the same weight. Hence p and p'' must have the same weight. ■

Theorem 3.98 *For each node in a saturation graph, there exists a circuit upstream in this graph. The circuits of any saturation graph belong to* $\mathcal{G}^c(A)$.

Proof Indeed, if i is one of its nodes, there exists another node j, upstream with respect to i, such that $y_i = A_{ij} y_j; y_i, y_j \neq \varepsilon$. The same reasoning shows that there exists another node upstream with respect to j, etc. Because the number of nodes of $\mathcal{S}(A, y)$ is finite, the path $(i, j, \ldots)$ obtained by this construction contains a circuit. A circuit $(i_0, i_1, \ldots, i_k, i_0)$ of a saturation graph $\mathcal{S}(A, y)$ satisfies

$$y_{i_1} = A_{i_1 i_0} y_{i_0} \ , \quad \ldots \ , \quad y_{i_0} = A_{i_0 i_k} y_{i_k} \ .$$

The multiplication of all these equalities shows that the weight of the circuit $(i, i_1, \ldots, i_k, i)$ is e. ■

Example 3.99 In Example 3.95, node 2 has the critical circuit (indeed a loop) $\{1\}$ upstream. ■

3.7.2 Eigenspace Associated with the Maximum Eigenvalue

In this subsection we describe the set of eigenvectors of matrix A associated with the eigenvalue e. Clearly, this set is a moduloid. We characterize a nonredundant set of generators of this moduloid as a subset of the columns of A^+.

> Here the word 'eigenvector' must be understood as 'eigenvector associated with the eigenvalue e'. We will use the notation A^p_{ij} for $(A^p)_{ij}$ and A^+_{ij} for $(A^+)_{ij}$.

Theorem 3.100 *If y is an eigenvector of A, it is also an eigenvector of A^+. It is the linear combination of the columns $A^+_{\cdot i}, i \in \mathcal{V}^c$. More precisely,*

$$y = \bigoplus_{i \in \mathcal{V}^c} y_i A^+_{\cdot i} \ . \tag{3.27}$$

Proof The first part of the theorem is trivial. Let us prove Formula (3.27). Consider two nodes i and j in the same m.s.c.s. of the saturation graph $\mathcal{S}(A, y)$. There exists a path $(i, i_1, \ldots, i_k, j)$ which satisfies

$$y_{i_1} = A_{i_1 i} y_i \ , \quad \ldots, \quad y_j = A_{j i_k} y_{i_k} \ .$$

Therefore, $y_j = w y_i$ with

$$w = A_{j i_k} \cdots A_{i_1 i} \leq A^+_{ji} \ ,$$

and we have

$$A^+_{lj} y_j = A^+_{lj} w y_i \leq A^+_{lj} A^+_{ji} y_i \leq A^+_{li} y_i \ , \quad \forall l \ . \tag{3.28}$$

We could have chosen i in a circuit of the saturation graph according to Theorem 3.98. This i will be called $i(j)$ in the following. We have

$$\begin{aligned} y_l &= \bigoplus_{j \in \mathcal{S}(A,y)} A^+_{lj} y_j && \text{(by definition of } \mathcal{S}(A,y)) \\ &\leq \bigoplus_{j \in \mathcal{S}(A,y)} A^+_{li(j)} y_{i(j)} && \text{(by (3.28))} \\ &\leq \bigoplus_{i \in \mathcal{V}^c} A^+_{li} y_i \ , \quad \forall l \ , \end{aligned}$$

where the last inequality stems from the fact that $i(j)$, belonging to a circuit of a saturation graph, belongs also to $\mathcal{V}^c$ by Theorem 3.98. The reverse inequality is derived immediately from the fact that y is an eigenvector of A^+. ∎

Theorem 3.101 *Given a matrix A with maximum circuit weight e, any eigenvector associated with the eigenvalue e is obtained by a linear combination of N^c_A columns of A^+, where N^c_A denotes the number of m.s.c.s.'s of $\mathcal{G}^c(A)$. More precisely we have*

1. *the columns $A^+_{\cdot i}, i \in \mathcal{V}^c$, are eigenvectors;*
2. *if nodes i and j belong to the same m.s.c.s. of $\mathcal{G}^c(A)$, then $A^+_{\cdot i}$ and $A^+_{\cdot j}$ are 'proportional';*
3. *no $A^+_{\cdot i}$ can be expressed as a linear combination of columns $A^+_{\cdot j}$ which only makes use of nodes j belonging to m.s.c.s.'s of $\mathcal{G}^c(A)$ distinct from $[i]$.*

Proof The first statement has already been proved in Theorem 3.23. Consider now the second statement: since $A^+A^+ \leq A^+$ and $A^+_{ij}A^+_{ji} = e$, if nodes i and j belong to the same m.s.c.s. of $\mathcal{G}^c(A)$, hence to the same critical circuit by Theorem 3.96, we have

$$A^+_{lj}A^+_{ji} \leq A^+_{li} = A^+_{li}A^+_{ij}A^+_{ji} \leq A^+_{lj}A^+_{ji} \ , \quad \forall l \ ,$$

which shows that

$$A^+_{lj}A^+_{ji} = A^+_{li} \ , \quad \forall l \ .$$

This result, together with (3.27), show that N^c_A columns of A^+ are sufficient to generate all eigenvectors.

The third statement of the theorem claims that we cannot further reduce this number of columns. Otherwise, one column of A^+, say i, could be expressed as a linear combination of other columns of A^+ selected in other m.s.c.s.'s of $\mathcal{G}^c(A)$. Let K denote the set of columns involved in this linear combination. Then, construct a matrix B as follows. Let $J = K \cup \{i\}$. Matrix B is obtained from A^+ by deleting all rows and columns with indices out of J. By construction, $B_{kk} = e, \forall k$, and the weights of all circuits of $\mathcal{G}(B)$ are less than or equal to e. The linear combination of the columns of A^+ is preserved when restricting this combination to matrix B. Owing to the multilinearity and antisymmetry of the determinant, and to the decomposition of any permutation into circular permutations, $\det B = e^{\bullet}$. Since $\bigotimes_k B_{kk} = e$, there must exist another permutation ς such that $\bigotimes_k B_{k\varsigma(k)} = e$. Stated differently, there must exist a critical circuit connecting some m.s.c.s.'s of $\mathcal{G}^c(A)$ and this yields a contradiction. ■

Example 3.102 If we return to Example 3.95, matrix A^+ is equal to

$$\begin{pmatrix} e & e & \varepsilon & \varepsilon \\ -1 & -1 & \varepsilon & \varepsilon \\ -2 & -1 & e & e \\ -2 & -1 & e & e \end{pmatrix} .$$

Nodes 1, 3, 4 belong to $\mathcal{G}^c(A)$ which has two m.s.c.s.'s, namely $\{1\}$ and $\{3,4\}$. Columns 1 and 3 are independent eigenvectors. Column 4 is equal to column 3. ■

3.7.3 Spectral Projector

In this subsection we build the spectral projector on the eigenspace (invariant moduloid) associated with the eigenvalue e.

Definition 3.103 (Spectral projector) *A matrix Q satisfying $AQ = QA = Q^2$ is called a spectral projector of A associated with the eigenvalue e.*

Theorem 3.104 *The matrices*

$$Q_i \stackrel{\text{def}}{=} A^+_{\cdot i} A^+_{i \cdot} \ , \quad i \in \mathcal{V}^c \ , \tag{3.29}$$

are spectral projectors of A.

Proof The properties $AQ_i = Q_i A = Q_i$ follow from the fact that the columns $A^+_{\cdot i}, i \in \mathcal{V}^c$, are eigenvectors of A (from which we deduce by transposition that the rows of A^+ such that $A^+_{ii} = e$ are left eigenvectors). Let us prove that $Q_i^2 = Q_i$, that is,

$$A^+_{\cdot i} A^+_{i \cdot} = \bigoplus_k A^+_{\cdot i} A^+_{ik} A^+_{ki} A^+_{i \cdot} \ .$$

This relation is true because $A^+_{ii} = e$ implies $\bigoplus_k A^+_{ik} A^+_{ki} = e$. ■

Theorem 3.105 *The matrix $Q \stackrel{\text{def}}{=} \bigoplus_{i \in \mathcal{V}^c} Q_i$, where the matrices Q_i are defined by (3.29), is a spectral projector.*

Proof The only nontrivial fact to prove is that $Q^2 = Q$. This relation will be proved if we prove that $Q_i Q_j \leq Q_i \oplus Q_j$. This last inequality is true because it means that the greatest weight of the paths connecting a pair of nodes and traversing i and j is less than the maximum weight of the paths connecting the same pair of nodes and traversing either i or j. ■

Example 3.106 Continuing with Example 3.95, we obtain two elementary spectral projectors:

$$Q_1 = \begin{pmatrix} e \\ -1 \\ -2 \\ -2 \end{pmatrix} \begin{pmatrix} e & e & \varepsilon & \varepsilon \end{pmatrix} = \begin{pmatrix} e & e & \varepsilon & \varepsilon \\ -1 & -1 & \varepsilon & \varepsilon \\ -2 & -2 & \varepsilon & \varepsilon \\ -2 & -2 & \varepsilon & \varepsilon \end{pmatrix},$$

$$Q_2 = \begin{pmatrix} \varepsilon \\ \varepsilon \\ e \\ e \end{pmatrix} \begin{pmatrix} -2 & -1 & e & e \end{pmatrix} = \begin{pmatrix} \varepsilon & \varepsilon & \varepsilon & \varepsilon \\ \varepsilon & \varepsilon & \varepsilon & \varepsilon \\ -2 & -1 & e & e \\ -2 & -1 & e & e \end{pmatrix},$$

$$Q = Q_1 \oplus Q_2 = \begin{pmatrix} e & e & \varepsilon & \varepsilon \\ -1 & -1 & \varepsilon & \varepsilon \\ -2 & -1 & e & e \\ -2 & -1 & e & e \end{pmatrix}.$$

■

3.7.4 Convergence of A^k with k

In this subsection we give a necessary and sufficient condition for the convergence of the powers of matrix A. To achieve this goal, we equip $\mathbb{R}_{\max}$ with the topology: when $n \to +\infty$,

$$x_n \to x \Leftrightarrow |x_n - x|_e \stackrel{\text{def}}{=} |\exp(x_n) - \exp(x)| \to 0 .$$

The purpose of this topology is to simplify the study of the convergence towards ε. Indeed, because $\exp(\varepsilon) = e$, $\lim_n |x_n - \varepsilon|_e = 0$ and $\lim_n |x'_n - \varepsilon|_e = 0$ imply that $\lim_n |x_n - x'_n|_e = 0$. This property, which is not true with respect to the usual absolute value in $\mathbb{R}$, is useful for the asymptotic cyclicity notion that we will introduce later on. We first recall the following result on the diophantine linear equation (see [33]).

Lemma 3.107 *For all p and n which are coprime and for all $q \geq (p-1)(n-1)$, there exist two integers $a(q)$ and $b(q)$ such that $q = a(q)p + b(q)n$.*

Theorem 3.108 *A necessary and sufficient condition to have $\lim_{k\to\infty} A^k = Q$ is that the cyclicity—see Definition 3.94—of each m.s.c.s. of $\mathcal{G}^c(A)$ is equal to 1.*

Proof Let us prove the sufficient condition first. Consider a node i in $\mathcal{G}^c(A)$. For any other node j, there exists a path from i to j of maximum weight (possibly equal to ε). If there happens to be more than one such path, we take the one with the least length and call this length $p(i, j)$ which is less than $n - 1$ if A is an $n \times n$ matrix. If the maximum weight is ε, we consider that the length is equal to 1. By Lemma 3.107 and the assumption on cyclicity, there exists some integer $M(i)$ such that, for all m greater than $M(i)$, there exists a critical circuit of length m in

$[i]_{\mathcal{G}^c(A)}$ (the m.s.c.s. of $\mathcal{G}^c(A)$ to which i belongs). Therefore, because the maximum weight of the circuits is e, any maximum-weight path from i to j of length q greater than $M(i)+p$ is composed of a critical circuit of length $q-p$ traversing i and a maximum-weight path from i to j (of any length). Therefore $A^q_{ji} = A^+_{ji}$ for all q greater than $p+M(i)$ and this holds for all i in $\mathcal{G}^c(A)$. Since $A^+_{ii} = e$, we also have $A^q_{ji} = A^+_{ji}A^+_{ii}$.

Consider now another node l which does not belong to $\mathcal{G}^c(A)$. Let i be a node in $\mathcal{G}^c(A)$, let q be large enough and let $p \le n$ be such that $A^p_{il} = A^+_{il}$ (such a p exists because circuits have weights less than e, hence the lengths of maximum-weight paths between any two nodes do not need to exceed n). We have

$$A^q_{jl} \ge A^{q-p}_{ji} A^p_{il} = A^+_{ji} A^+_{il} \ ,$$

where the inequality is a consequence of the matrix product in $\mathbb{R}_{\max}$ whereas the equality arises from the previous part of the proof and from the property of p. If we have a strict inequality, it means that the paths with maximum weight from j to l do not traverse i, and since this is true for any i in $\mathcal{G}^c(A)$, these paths do not traverse $\mathcal{G}^c(A)$. On the other hand, for q large enough, they must traverse some circuits which therefore have a strictly negative weight. When q increases, these paths have weights arbitrarily close to ε. Finally, this situation is possible only if there is no node of $\mathcal{G}^c(A)$ located downstream of l in $\mathcal{G}(A)$. In this case $A^+_{il} = \varepsilon$ for all i in $\mathcal{G}^c(A)$ and therefore

$$\lim_{q\to\infty} A^q_{jl} = \varepsilon = \bigoplus_{i\in\mathcal{V}^c} A^+_{ji} A^+_{il} \ .$$

By collecting all the above results, we have proved that

$$\lim_{q\to\infty} A^q_{jl} = \bigoplus_{i\in\mathcal{V}^c} A^+_{ji} A^+_{il} \ , \quad \forall j,l \in \mathcal{G}(A) \ .$$

Conversely, suppose that the above limit property holds true and that at the same time the cyclicity of $\mathcal{G}^c(A)$ is strictly greater than 1. Let us consider a node $i \in \mathcal{G}^c(A)$ (hence $A^+_{ii} = e$). We have

$$\exp(A^{k\times d}_{ii}) = \exp(e) = \exp(A^{k\times d+1}_{ii}) + \eta \ ,$$

where η can be arbitrarily small because of the assumed limit. But $A^{k\times d+1}_{ii} = A^p_{ii}$ for $0 \le p \le n$ (again because circuits have nonpositive weights). Therefore, $A^{k\times d+1}_{ii}$ can assume values out of a finite set. From the relation above, it should be clear that $A^{k\times d+1}_{ii} = e$. This means that there exists a circuit of length $k\times d+1$. But the gcd of kd and $k\times d+1$ is 1, which is a contradiction. ∎

Theorem 3.109 *Suppose that $\mathcal{G}(A)$ is strongly connected. Then there exists K such that*

$$\forall k \ge K \ , \quad A^k = Q \ ,$$

if and only if the cyclicity of each m.s.c.s. of $\mathcal{G}^c(A)$ is equal to 1.

Proof The proof is similar to the previous one. The only difference lies in the second part of the 'if' part. Under the assumption that $\mathcal{G}(A)$ is strongly connected, a path of maximum weight from l to j with length large enough necessarily crosses $\mathcal{G}^c(A)$. Therefore, for q large enough we have

$$A^q_{jl} = A^+_{ji} A^+_{il} ,$$

where i belongs to $\mathcal{G}^c(A)$. ∎

Example 3.110

- Using Example 3.95 once more, we have

$$A^2 = \begin{pmatrix} e & e & \varepsilon & \varepsilon \\ -1 & -1 & \varepsilon & \varepsilon \\ -2 & -2 & e & e \\ \varepsilon & -1 & e & e \end{pmatrix},$$

$$A^3 = A^4 = \cdots = \begin{pmatrix} e & e & \varepsilon & \varepsilon \\ -1 & -1 & \varepsilon & \varepsilon \\ -2 & -1 & e & e \\ -2 & -1 & e & e \end{pmatrix}.$$

  Therefore A^n, $n \geq 3$, is equal to $Q_1 \oplus Q_2$ given in Example 3.106.

- In the previous example the periodic regime is reached after a finite number of steps. This is true for the submatrix associated with the nodes of the critical graph but it is not true in general for the complete matrix. To show this, take the example

$$A = \begin{pmatrix} -1 & \varepsilon \\ \varepsilon & e \end{pmatrix}.$$

- In the previous example there is an entry which goes to ε. When all entries converge to a finite number, the periodic regime can be reached also, but the time needed may be arbitrarily long. Consider the matrix

$$A = \begin{pmatrix} -\eta & -1 \\ e & e \end{pmatrix}.$$

  The matrix A^k converges to the matrix

$$\begin{pmatrix} -1 & -1 \\ e & e \end{pmatrix}$$

  if η is a small positive number. But we have to wait for a power of order $1/\eta$ to reach the asymptote. ∎

3.7.5 Cyclic Matrices

In this subsection we use the previous theorem to describe the general behavior of the successive powers of matrix A which turns out to be essentially cyclic.

Definition 3.111

Cyclicity of a matrix: *a matrix A is said to be cyclic if there exist d and M such that $\forall m \geq M, A^{m+d} = A^m$. The least such d is called the cyclicity of matrix A and A is said to be d-cyclic.*

Asymptotic cyclicity: *a matrix A is said to be asymptotically cyclic if there exists d such that, for all $\eta > 0$, there exists M such that, for all $m \geq M$, $\sup_{ij} |(A^{m+d})_{ij} - (A^m)_{ij}|_{\mathrm{e}} \leq \eta$. The least such d is called the asymptotic cyclicity of matrix A and A is said to be d-asymptotically cyclic.*

Theorem 3.112 *Any matrix is asymptotically cyclic. The asymptotic cyclicity d of matrix A is equal to the cyclicity ρ of $\mathcal{G}^{\mathrm{c}}(A)$. Moreover if $\mathcal{G}(A)$ and $\mathcal{G}(A^\rho)$ are connected, the matrix is ρ-cyclic.*

Proof This result is already proved in the case $\rho = 1$. For any matrix A, if we consider $B = A^\rho$, then the asympotic cyclicity ρ' of B is equal to 1. Indeed, the nodes of the critical graph of B are a subset of $\mathcal{G}^{\mathrm{c}}(A)$, and around each such node there exists a loop. The necessary and sufficient conditions of convergence of the powers of a matrix can be applied to B (see Theorems 3.108 and 3.109). They show the convergence (possibly in a finite number of stages) of $B^k = A^{k \times \rho}$ to the spectral projector Q associated with B. Because any m can be written $h + k \times \rho$, $A^m = A^{h+k\times\rho} = A^h B^k$ converges to $A^h Q$ when k goes to infinity. This is equivalent to saying that matrix A is d-asymptotically-cyclic (or d-cyclic in the case of finite convergence), with $d \leq \rho$.

Let us prove that the asymptotic cyclicity d of matrix A is greater than or equal to ρ (hence it is equal to ρ). The proof when the matrix is not only asymptotically cyclic but cyclic is similar. Consider a node of the m.s.c.s. $l = [i]_{\mathcal{G}^{\mathrm{c}}(A)}$ of $\mathcal{G}^{\mathrm{c}}(A)$ and let ρ_l denote its cyclicity. By definition of d, we have

$$\exp\left(A_{ii}^{k\times\rho_l}\right) = \exp(e) = \exp(A_{ii}^{k\times\rho_l+d}) + \eta \ ,$$

for η arbitrarily small and for k large enough. Therefore $A_{ii}^{k\times\rho_l} = A_{ii}^{k\times\rho_l+d}$ and there is a circuit of length d in the m.s.c.s. l of $\mathcal{G}^{\mathrm{c}}(A)$. Therefore ρ_l divides d. But this is true for all m.s.c.s.'s of $\mathcal{G}^{\mathrm{c}}(A)$ and therefore d is divided by the lcm of all the ρ_l which is ρ. ■

Example 3.113 The matrix

$$A = \begin{pmatrix} -1 & \varepsilon & \varepsilon \\ \varepsilon & \varepsilon & e \\ \varepsilon & e & \varepsilon \end{pmatrix}$$

has cyclicity 2. Indeed,

$$A^{2n} = \begin{pmatrix} -2n & \varepsilon & \varepsilon \\ \varepsilon & e & \varepsilon \\ \varepsilon & \varepsilon & e \end{pmatrix}, \qquad A^{2n+1} = \begin{pmatrix} -(2n+1) & \varepsilon & \varepsilon \\ \varepsilon & \varepsilon & e \\ \varepsilon & e & \varepsilon \end{pmatrix}.$$

■

3.8 NOTES

The max-plus algebra is a special case of a more general structure which is called a dioid structure. This is the topic of the next chapter. Nevertheless the max-plus algebra, and the algebras of vector objects built up on it, are important examples of dioids for this book because they are perfectly tailored to describe synchronization mechanisms. They were also used to compute paths of maximum length in a graph in operations research. This is why they are sometimes called 'path algebra' [67].

Linear systems of equations in the max-plus algebra were systematically studied in [49]. Some other very interesting references on this topic are [67], [130]. In these references the Gauss elimination algorithm can be found. It was not discussed here. Linear dependence has been studied in [49], [65], [93], [126] and [62]. Several points of view exist but none of them is completely satisfactory. Moreover the geometry of linear manifolds in the max-plus algebra is not well understood (on this aspect, see [126]).

The only paper that we know on a systematic study of polynomial and rational functions in the max-plus algebra is [51]. In this paper one can find some results on rational functions not detailed in this book.

The symmetrization of the max-plus algebra was discussed earlier in [109] and [110]. The presentation given here is based on these references. This symmetrization is more deeply studied in [62]. The reference [65] has been an important source of ideas even though symmetrization has been avoided in this paper. The proof of the Cramer formula is mainly due to S. Gaubert and M. Akian. Relevant references are [118], [124], [93].

The attempt made here to discuss polynomials in $\mathbb{S}$ is new. It could give a new insight into the eigenvalue problem. Because of the lack of space this discussion has not been continued here.

The section on the max-plus Perron-Frobenius theorem is a new version of the report [37]. The proof is mainly due to M. Viot. Some other relevant references are [64], [49], [127].

CHAPTER 4

Dioids

4.1 INTRODUCTION

In previous chapters, the set $\mathbb{R} \cup \{-\infty\}$ (respectively $\mathbb{R} \cup \{+\infty\}$) endowed with the max (respectively the min) operation as addition and the usual addition as multiplication has appeared as a suitable algebraic structure for obtaining 'linear' models of some discrete event systems. In Chapter 5 it will be shown that another slightly more complex structure is also appropriate for the same class of systems.

All these algebraic structures share some common features that will be studied in the present chapter. However, there is yet no universal name nor a definite set of axioms everybody agrees upon in this general field. We refer the reader to the notes at the end of this chapter where some related works are briefly discussed. Here we adopt the following point of view: we introduce a first 'minimal' set of axioms according to what seems to be the intersection of axioms generally retained in the works alluded to above, and also according to what seems to be appropriate for the linear system theory we are going to develop in Chapter 5. Starting from this minimal set of axioms, we derive some basic results. To obtain further results we may need to introduce some additional assumptions or properties, which we do only when necessary: it is then clear where this added structure is really needed.

We use the word 'dioid' as the generic name for the algebraic structure studied in this chapter. The linguistic roots of this name and its introduction in the literature are discussed in the notes section.

Dioids are structures that lie somewhere between conventional linear algebra and semilattices endowed with an internal operation generally called multiplication. With the former, it shares combinatorial properties such as associativity and commutativity of addition, associativity of multiplication, distributivity of multiplication with respect to addition, and of course the existence of zero and identity elements. With the latter, it shares the features of an ordered structure (adding is then simply taking the upper bound) endowed with another 'compatible' operation. Therefore one may expect that the results of linear algebra which depend only on combinatorial properties will generalize to dioids. A typical case is the Cayley-Hamilton theorem. On the other hand, since neither addition nor multiplication are invertible in general dioids (in this respect the max-plus algebra is special since the structure associated with $+$ is a group), one appeals to the classical theory of residuation in lattice structures to provide alternative notions of inversion of the basic operations and of other order-preserving mappings. This yields a way to 'solve' some equations in a certain sense related to the order structure even if there is no solution in a more classical sense.

A section of this chapter is devoted to rational calculus, in the sense in which this expression is used in automata and formal languages theory. The motivation in terms of system theory and especially in terms of realization theory should be clear and this will be illustrated by Chapter 5. However, the problem of *minimal* realization is yet unsolved in the present framework (see also Chapters 6 and 9).

In this chapter we will also be interested in constructing more elaborate dioids from given basic dioids. This is generally done by considering the quotient of simple dioids by certain 'congruences' (equivalence relations which are compatible with the original dioid structure). Particular congruences will be considered for their usefulness regarding the developments in Chapter 5. As a motivation, the reader may think of this quotient operation yielding a 'coarser' dioid as a way to 'filter' elements of the original dioid that are 'undesirable' for the system theory one is concerned with. For example, if trajectories of discrete event systems are the basic objects, one is often interested in *nondecreasing* trajectories whereas the basic dioid may also contain nonmonotonic trajectories. Nonmonotonic trajectories are then mapped to nondecreasing ones in a canonical way by special congruences.

4.2 BASIC DEFINITIONS AND EXAMPLES

4.2.1 Axiomatics

Definition 4.1 (Dioid) *A dioid is a set $\mathcal{D}$ endowed with two operations denoted $\oplus$ and $\otimes$ (called 'sum' or 'addition', and 'product' or 'multiplication') obeying the following axioms:*

Axiom 4.2 (Associativity of addition)

$$\forall a, b, c \in \mathcal{D} \ , \quad (a \oplus b) \oplus c = a \oplus (b \oplus c) \ .$$

Axiom 4.3 (Commutativity of addition)

$$\forall a, b \in \mathcal{D} \ , \quad a \oplus b = b \oplus a \ .$$

Axiom 4.4 (Associativity of multiplication)

$$\forall a, b, c \in \mathcal{D} \ , \quad (a \otimes b) \otimes c = a \otimes (b \otimes c) \ .$$

Axiom 4.5 (Distributivity of multiplication with respect to addition)

$$\forall a, b, c \in \mathcal{D} \ , \quad \begin{aligned} (a \oplus b) \otimes c &= (a \otimes c) \oplus (b \otimes c) \ , \\ c \otimes (a \oplus b) &= c \otimes a \oplus c \otimes b \ . \end{aligned}$$

This is right, respectively left, distributivity of product with respect to sum. One statement does not follow from the other since multiplication is not assumed to be commutative.

Axiom 4.6 (Existence of a zero element)

$$\exists \varepsilon \in \mathcal{D} : \forall a \in \mathcal{D} \ , \quad a \oplus \varepsilon = a \ .$$

Axiom 4.7 (Absorbing zero element)

$$\forall a \in \mathcal{D} \ , \quad a \otimes \varepsilon = \varepsilon \otimes a = \varepsilon \ .$$

Axiom 4.8 (Existence of an identity element)

$$\exists e \in \mathcal{D} : \forall a \in \mathcal{D} \ , \quad a \otimes e = e \otimes a = a \ .$$

Axiom 4.9 (Idempotency of addition)

$$\forall a \in \mathcal{D} \ , \quad a \oplus a = a \ .$$

Definition 4.10 (Commutative dioid) *A dioid is commutative if multiplication is commutative.*

> Most of the time, the symbol '$\otimes$' is omitted as is the case in conventional algebra. Moreover, a^k, $k \in \mathbb{N}$, will of course denote $\underbrace{a \otimes \cdots \otimes a}_{k \text{ times}}$ and $a^0 = e$.

With the noticeable exception of Axiom 4.9, most of the axioms of dioids are required for rings too. Indeed, Axiom 4.9 is the most distinguishing feature of dioids. Because of this axiom, addition cannot be cancellative, that is, $a \oplus b = a \oplus c$ does not imply $b = c$ in general, for otherwise $\mathcal{D}$ would be reduced to ε (see Theorem 3.61). In fact, Axiom 4.9 is at the basis of the introduction of an order relation; as mentioned in the introduction, this is the other aspect of dioids, their lattice structure. This aspect is dealt with in §4.3.

Multiplication is not necessarily cancellative either (of course, because of Axiom 4.7, cancellation would anyway only apply to elements different from ε). We refer the reader to Example 4.15 below. A weaker requirement would be that the dioid be 'entire'.

Definition 4.11 (Entire dioid) *A dioid is entire if*

$$ab = \varepsilon \Rightarrow a = \varepsilon \text{ or } b = \varepsilon \ .$$

If $a \neq \varepsilon, b \neq \varepsilon$, and $ab = \varepsilon$, then a and b are called *zero divisors*. Hence, an entire dioid is a dioid which does not contain zero divisors. Not every dioid is entire (see Example 4.81 below). If multiplication is cancellative, the dioid is entire. As a matter of fact, $ab = \varepsilon \Rightarrow ab = a\varepsilon \Rightarrow b = \varepsilon$ if $a \neq \varepsilon$ by cancellation of a.

4.2.2 Some Examples

For the following examples of dioids, we let the reader check the axioms and define what ε and e should be. All of them are commutative dioids.

Example 4.12 The first example of a dioid encountered in this book was $\mathbb{R} \cup \{-\infty\}$ with max as $\oplus$ and $+$ as $\otimes$. It was denoted $\mathbb{R}_{\max}$. ■

Example 4.13 $(\mathbb{R} \cup \{+\infty\}, \min, +)$ is another dioid which is isomorphic—this terminology is precisely defined later on—to the previous one by the compatible bijection: $x \mapsto -x$. It will be denoted $\mathbb{R}_{\text{min}}$. ■

Example 4.14 Using the bijection $x \mapsto \exp(x)$, $\mathbb{R} \cup \{-\infty\}$ is mapped onto $\mathbb{R}^+$. For this bijection to preserve the dioid structure of $\mathbb{R}_{\text{max}}$, one has to define $\oplus$ in $\mathbb{R}^+$ as max again and $\otimes$ as $\times$ (the conventional product). This yields the dioid $(\mathbb{R}^+, \max, \times)$. ■

Example 4.15 Consider the set $\mathbb{R} \cup \{-\infty\} \cup \{+\infty\}$ and define $\oplus$ as max and $\otimes$ as min. ■

Example 4.16 In the previous example, replace the set by $\{0, 1\}$ and keep the same operations: this is the Boole algebra and also the unique dioid (up to an isomorphism) reduced to $\{\varepsilon, e\}$. ■

Example 4.17 Let $2^{\mathbb{R}^2}$ denote the set of all subsets of the $\mathbb{R}^2$ plane, including $\emptyset$ and the whole $\mathbb{R}^2$ itself. Then define $\oplus$ as $\cup$ and $\otimes$ as $+$, that is, the 'vector sum' of subsets

$$\forall A, B \subseteq \mathbb{R}^2 \ , \quad A \otimes B = A + B = \left\{ x \in \mathbb{R}^2 \mid x = y + z, y \in A, z \in B \right\} \ .$$

■

Example 4.18 A similar example in dimension 1 is provided by considering the subset of $2^{\mathbb{R}}$ consisting only of half-lines infinite to the left, that is, intervals $(-\infty, x]$ for all $x \in \mathbb{R}$, including $\emptyset$ but not $\mathbb{R}$ itself, with again $\cup$ as $\oplus$ and $+$ as $\otimes$. Observe that this subset of half-lines is closed—see below—for these two operations. This dioid is isomorphic to $\mathbb{R}_{\text{max}}$ by the bijection $x \in \mathbb{R} \mapsto (-\infty, x] \in 2^{\mathbb{R}}$ and $\varepsilon = -\infty \mapsto \emptyset$. ■

In all the examples above, except Examples 4.15 and 4.17, $\otimes$ induces a group structure on $\mathcal{D} \setminus \{\varepsilon\}$ ($\mathcal{D}$ minus ε). This implies of course that $\otimes$ is cancellative. Obviously $\otimes$ is not cancellative in Example 4.15. This is also true for Example 4.17: this fact follows from Lemma 4.35 below. However, in both cases the dioid is entire.

4.2.3 Subdioids

Definition 4.19 (Subdioid) *A subset $\mathcal{C}$ of a dioid is called a subdioid of $\mathcal{D}$ if*

- $\varepsilon \in \mathcal{C}$ *and* $e \in \mathcal{C}$*;*
- $\mathcal{C}$ *is closed for* $\oplus$ *and* $\otimes$*.*

The second statement means that $\forall a, b \in \mathcal{C}, a \oplus b \in \mathcal{C}$ and $a \otimes b \in \mathcal{C}$. We emphasize the first condition. For example, the dioid in Example 4.16 (Boole algebra) is *not* a subdioid of the one in Example 4.15. The dioid $(\mathbb{N} \cup \{-\infty\}, \max, +)$ is a subdioid of $\mathbb{R}_{\text{max}}$.

4.2.4 Homomorphisms, Isomorphisms and Congruences

Most of the material in this subsection is not very specific to the dioid structure and can be found in elementary textbooks on algebra. Here, we reconsider this material in the framework of dioids.

Definition 4.20 (Homomorphism) *A mapping Π from a dioid $\mathcal{D}$ into another dioid $\mathcal{C}$ is a homomorphism if*

$$\forall a, b \in \mathcal{D} \ , \qquad \Pi(a \oplus b) = \Pi(a) \oplus \Pi(b) \quad \text{and} \quad \Pi(\varepsilon) = \varepsilon \ , \qquad (4.1)$$

$$\Pi(a \otimes b) = \Pi(a) \otimes \Pi(b) \quad \text{and} \quad \Pi(e) = e \ . \qquad (4.2)$$

Of course the operations and neutral elements on the left-hand (respectively right-hand) side are those of $\mathcal{D}$ (respectively $\mathcal{C}$). If Π is surjective, it is clear that the former part of (4.1) (respectively (4.2)) implies the latter part which is thus redundant.

A mapping having only property (4.1) will be called 'a $\oplus$-morphism', and a mapping having property (4.2) will be called 'a $\otimes$-morphism'.

Definition 4.21 (Isomorphism) *A mapping Π from a dioid $\mathcal{D}$ into another dioid $\mathcal{C}$ is an isomorphism if Π^{-1} is defined over $\mathcal{C}$ and Π and Π^{-1} are homomorphisms.*

Lemma 4.22 *If Π is a homomorphism from $\mathcal{D}$ to $\mathcal{C}$ and if it is a bijection, then it is an isomorphism.*

Proof It suffices to prove that Π^{-1} satisfies (4.1)–(4.2). Applying (4.1) to $a = \Pi^{-1}(x), x \in \mathcal{C}$ and $b = \Pi^{-1}(y), y \in \mathcal{C}$, we get

$$\Pi\left(\Pi^{-1}(x) \oplus \Pi^{-1}(y)\right) = \Pi\left(\Pi^{-1}(x)\right) \oplus \Pi\left(\Pi^{-1}(y)\right) = x \oplus y \ ,$$

and therefore

$$\Pi^{-1}(x) \oplus \Pi^{-1}(y) = \Pi^{-1}(x \oplus y) \ .$$

Also

$$\Pi(\varepsilon) = \varepsilon \Rightarrow \varepsilon = \Pi^{-1}(\varepsilon) \ ,$$

which proves that Π^{-1} is a $\oplus$-morphism. The same reasoning can be applied to (4.2). ■

Definition 4.23 (Congruence) *A congruence in a dioid $\mathcal{D}$ is an equivalence relation (denoted $\equiv$) in $\mathcal{D}$ which is compatible with $\oplus$ and $\otimes$, that is,*

$$\forall a, b, c \in \mathcal{D} \ , \quad a \equiv b \Rightarrow a \oplus c \equiv b \oplus c \ ,$$

and the same for $\otimes$.

Lemma 4.24 *The quotient of $\mathcal{D}$ by a congruence (that is, the set of equivalence classes) is a dioid for the addition and multiplication induced by those of $\mathcal{D}$.*

Proof The main difficulty here is to show how $\oplus$ and $\otimes$ can be properly defined in the quotient. Let $[a]$ denote the equivalence class of a ($b \in [a] \Leftrightarrow b \equiv a \Leftrightarrow [b] = [a]$). Then define $[a] \oplus [b]$ by $[a \oplus b]$. This definition is correct because if $a' \in [a]$ and $b' \in [b]$, then $[a' \oplus b'] = [a \oplus b]$ from the compatibility of $\equiv$ with $\oplus$, that is, $[a \oplus b]$ only depends on $[a]$ and $[b]$, not on particular representatives of these classes. The same considerations apply to $\otimes$ too. ∎

Example 4.25 One special instance of a congruence is the following. Let Π be a homomorphism from a dioid $\mathcal{D}$ to another dioid $\mathcal{C}$. We can define an equivalence relation in $\mathcal{D}$ as follows:

$$\forall a, b \in \mathcal{D} \ , \quad a \overset{\Pi}{\equiv} b \Leftrightarrow \Pi(a) = \Pi(b) \ . \tag{4.3}$$

∎

Corollary 4.26 *If Π is a homomorphism, $\overset{\Pi}{\equiv}$ is a congruence. Therefore, the quotient set denoted $\mathcal{D}/\Pi$ is a dioid; it is isomorphic to $\Pi(\mathcal{D})$.*

The proof is straightforward. Of course, $\mathcal{D}/\Pi$ is isomorphic to $\mathcal{D}$ if Π is injective.

4.3 LATTICE PROPERTIES OF DIOIDS

4.3.1 Basic Notions in Lattice Theory

Hereafter, we list a few basic notions from lattice theory, the main purpose being to make our vocabulary more precise, especially when there are some variations with respect to other authors. The interested reader may refer to [22] or [57]. In a set, we adopt the following definitions.

Order relation: a binary relation (denoted $\geq$) which is reflexive, transitive and antisymmetric.

Total (partial) order: the order is *total* if for each pair of elements (a, b), the order relation holds true either for (a, b) or for (b, a), or otherwise stated, if a and b are always 'comparable'; otherwise, the order is *partial*.

Ordered set: a set endowed with an order relation; it is sometimes useful to represent an ordered set by an undirected graph the nodes of which are the elements of the set; two nodes are connected by an arc if the corresponding elements are comparable, the greater one being higher in the diagram; the minimal number of arcs is represented, the other possible comparisons being derived by transitivity. Figure 4.1 below gives an example of such a graph called a *'Hasse diagram'*.

Chain: a totally ordered set; its Hasse diagram is 'linear'.

Please note that the following elements do not necessarily exist.

Top element (of an ordered set): an element which is greater than any other element of the set (elsewhere also called 'universal').

Bottom element (of an ordered set): similar definition (elsewhere also called 'zero', but we keep this terminology for the neutral element of addition, although, as it will be seen hereafter, both notions coincide in a dioid).

Maximum element (of a subset): an element *of the subset* which is greater than any other element of the subset; if it exists, it is unique; it coincides with the top element if the subset is equal to the whole set.

Minimum element (of a subset): similar definition.

Maximal element (of a subset): an element *of the subset* which is not less than any other element of the subset; Figure 4.1 shows the difference between a maximum and a maximal element; if the subset has a maximum element, it is the unique maximal element.

Majorant (of a subset): an element *not necessarily belonging to the subset* which is greater than any other element of the subset (elsewhere also called 'upper bound' but we keep this terminology for a notion introduced below); if a majorant belongs to the subset, it is the maximum element.

Minorant (of a subset): similar definition (elsewhere also called 'lower bound', but we reserve this for a more specific notion).

Upper bound (of a subset): the least majorant, that is, the minimum element of the subset of majorants (elsewhere, when 'majorant' is called 'upper bound', this notion is called 'least upper bound').

Lower bound (of a subset): similar definition (elsewhere also called 'greatest lower bound').

The following items introduce more specific ordered sets and mappings between these sets.

Sup-semilattice: an ordered set such that there exists an upper bound for each pair of elements.

Inf-semilattice: similar definition.

Lattice: an ordered set which is both a sup- and an inf-semilattice.

Complete sup-semilattice: an ordered set such that there exists an upper bound for each finite or infinite subset.

Complete inf-semilattice: similar definition.

Complete lattice: obvious definition.

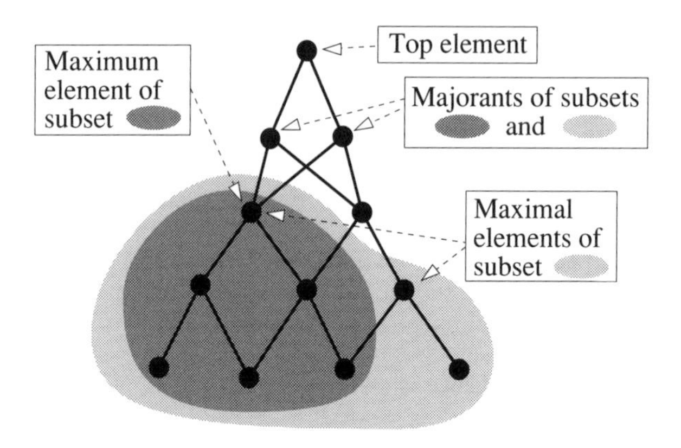

Figure 4.1: Top, maximum, majorants and maximal elements

Distributive lattice: let $a \vee b$ (respectively, $a \wedge b$) denote the upper (respectively, lower) bound of a and b in a lattice; then the lattice is *distributive* if

$$\forall a, b, c \ , \quad a \vee (b \wedge c) = (a \vee b) \wedge (a \vee c) \ ;$$

in fact, as shown in [57, p. 188], if this equality holds true, the same equality with $\vee$ and $\wedge$ interchanged also holds true, and conversely.

Isotone mapping: a mapping Π from an ordered set $\mathcal{D}$ into an ordered set $\mathcal{C}$ such that

$$\forall a, b \in \mathcal{D} \ , \quad a \geq b \Rightarrow \Pi(a) \geq \Pi(b) \ .$$

We conclude this brief enumeration (more facts from lattice theory will be recalled later on) by mentioning a fundamental result [57, pp. 175–176].

Theorem 4.27 *A complete sup-semilattice having a bottom element is a complete lattice.*

Proof Let $\mathcal{C}$ be a subset of a complete sup-semilattice $\mathcal{D}$; we must prove that it admits a lower bound. Consider the subset $\mathcal{T}$ of minorants of $\mathcal{C}$. This subset is nonempty since it contains at least the bottom element. Let c be the upper bound of $\mathcal{T}$, which exists since $\mathcal{D}$ is a complete sup-semilattice. Let us check whether c obeys the definition of the lower bound of $\mathcal{C}$. First, c itself is below $\mathcal{C}$ (that is, it belongs to $\mathcal{T}$—it is thus the maximum element of $\mathcal{T}$). As a matter of fact, $\mathcal{T}$ is bounded from above by all $b \in \mathcal{C}$ (by definition). Since c is less than, or equal to, every element greater than $\mathcal{T}$ (by definition of the upper bound), $c \leq b, \forall b \in \mathcal{C}$, hence $c \in \mathcal{T}$. Therefore, c is less than, or equal to, all elements of $\mathcal{C}$ and greater than every other element below $\mathcal{C}$, namely the elements in $\mathcal{T}$. Hence it is the lower bound of $\mathcal{C}$. ■

4.3.2 Order Structure of Dioids

Theorem 4.28 (Order relation) *In a dioid $\mathcal{D}$, one has the following equivalence:*

$$\forall a, b : a = a \oplus b \Leftrightarrow \exists c : a = b \oplus c \ .$$

Moreover these equivalent statements define a (partial) order relation denoted $\geq$ as follows:

$$a \geq b \Leftrightarrow a = a \oplus b \ .$$

This order relation is compatible with addition, namely

$$a \geq b \Rightarrow \{\forall c \ , \quad a \oplus c \geq b \oplus c\} \ ,$$

and multiplication, that is,

$$a \geq b \Rightarrow \{\forall c \ , \quad ac \geq bc\}$$

(the same for the left product). Two elements a and b in $\mathcal{D}$ always have an upper bound, namely $a \oplus b$, and ε is the bottom element of $\mathcal{D}$.

Proof

- Clearly, if $a = a \oplus b$, then $\exists c : a = b \oplus c$, namely $c = a$. Conversely, if $a = b \oplus c$, then adding b on both sides of this equality yields $a \oplus b = b \oplus (b \oplus c) = b \oplus c = a$.

- The relation $\geq$ is reflexive ($a = a \oplus a$ from Axiom 4.9), antisymmetric ($a = a \oplus b$ and $b = b \oplus a$ implies $a = b$), and transitive since

$$\begin{Bmatrix} a = a \oplus b \\ b = b \oplus c \end{Bmatrix} \Rightarrow \begin{Bmatrix} a \oplus c = a \oplus b \oplus c \\ b = b \oplus c \\ a = a \oplus b \end{Bmatrix} \Rightarrow \begin{Bmatrix} a \oplus c = a \oplus b \\ a = a \oplus b \end{Bmatrix} \Rightarrow \{a \oplus c = a\} .$$

  Therefore $\geq$ is an order relation.

- The compatibility of $\geq$ with addition is a straightforward consequence of Axioms 4.2 and 4.3. The compatibility of multiplication involves Axiom 4.5. The expression 'the (left or right) multiplication is isotone' is also used for this property. But, as will be discussed in §4.4.1, the mapping $x \mapsto ax$ is more than simply isotone: it is a $\oplus$-morphism.

- Obviously, $a \oplus b$ is greater than a and b. Moreover, if $c \geq a$ and $c \geq b$, then $c = c \oplus c \geq a \oplus b$. Hence $a \oplus b$ is the upper bound of a and b.

- Finally,

$$\{\forall a \ , \quad a = a \oplus \varepsilon\} \Leftrightarrow \{\forall a \ , \quad a \geq \varepsilon\} \ ,$$

  which means that ε is the bottom element of $\mathcal{D}$. ∎

Notation 4.29 As usual, we may use $a \leq b$ as an equivalent statement for $b \geq a$, and $b > a$ (or $a < b$) as an equivalent statement for $[b \geq a \text{ and } b \neq a]$. ■

The following lemma deals with the problem of whether the order relation induced by $\oplus$ is total or only partial.

Lemma 4.30 (Total order) *The order relation defined in Theorem 4.28 is total if and only if*

$$\forall a, b \in \mathcal{D} \ , \quad a \oplus b = \text{either } a \text{ or } b \ .$$

Proof It is just a matter of rewriting the claim 'either $a \geq b$ or $b \geq a$' using $\oplus$ and the very definition of $\geq$. ■

Let us revisit the previous examples of dioids and discover what is the order relation associated with $\oplus$. It is the conventional order of numbers for Examples 4.12, 4.14, 4.15 and 4.16. However, in Example 4.13 it is the *reversed* order: $2 \geq 3$ in this dioid since $2 = 2 \oplus 3$. As for Example 4.17 and 4.18, $\geq$ is simply $\supseteq$. All these dioids are chains except for Example 4.17.

Theorem 4.28 essentially shows that an idempotent addition in $\mathcal{D}$ induces a structure of sup-semilattice over $\mathcal{D}$. But we could have done it the other way around: considering a sup-semilattice, we can define the result of the addition of two elements as their upper bound; this obviously defines an idempotent addition. The sup-semilattice has then to be endowed with another operation called $\otimes$. This multiplication should be assumed not only isotone but also distributive, except that isotony is sufficient if $\mathcal{D}$ is a chain (see §4.4.1). We now present a counterexample to the statement that isotony of multiplication implies distributivity (and therefore of the statement that isotony of a mapping would imply that this mapping is a $\oplus$-morphism).

Example 4.31 Consider Example 4.17 again but change addition to $\cap$ instead of $\cup$. Now $A \geq B$ means $A \subseteq B$ and it is true that this implies $A \otimes C \geq B \otimes C$ or equivalently $A + C \subseteq B + C$. Since $\otimes = +$ is isotone, we do have, as a translation of (4.12),

$$(B \cap C) + D \subseteq (B + D) \cap (C + D) \tag{4.4}$$

(because here $\geq$ is $\subseteq$, not $\supseteq$!), but equality does not hold in general, as shown by the particular case: B is the subset reduced to the point $(1, 0) \in \mathbb{R}^2$, C is similarly reduced to the point $(0, 1)$, whereas D is the square $[-1, 1] \times [-1, 1]$. Clearly, the left-hand side of (4.4) is equal to $\emptyset$, whereas the right-hand side is the square $[0, 1] \times [0, 1]$ (see Figure 4.2). In conclusion, $\left(2^{\mathbb{R}^2}, \cap, +\right)$ is not a dioid. ■

4.3.3 Complete Dioids, Archimedian Dioids

In accordance with the definition of complete sup-semilattices, we adopt the following definition.

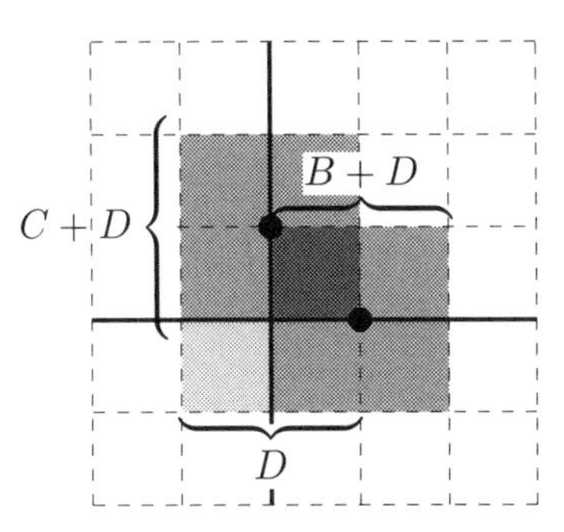

Figure 4.2: $+$ is not distributive with respect to $\cap$

Definition 4.32 (Complete dioid) *A dioid is complete if it is closed for infinite sums and Axiom 4.5 extends to infinite sums.*

With the former requirement, the upper bound of any subset is simply the sum of all its elements. The latter requirement may be viewed as a property of 'lower-semicontinuity' of multiplication.

In a complete dioid the top element of the dioid, denoted $\top$, exists and is equal to the sum of all elements in $\mathcal{D}$. The top element is always absorbing for addition since obviously $\forall a, \top \oplus a = \top$. Also

$$\top \otimes \varepsilon = \varepsilon \ , \tag{4.5}$$

because of Axiom 4.7.

If we consider our previous examples again, Examples 4.12, 4.13, 4.14 and 4.18 are not complete dioids, whereas Examples 4.15, 4.16 and 4.17 are. For $\mathbb{R}_{\max}$ to be complete, we should add the top element $\top = +\infty$ with the rule $-\infty + \infty = -\infty$ which is a translation of (4.5). This completed dioid is called $\overline{\mathbb{R}}_{\max}$ and $\overline{\mathbb{R}}$ denotes $\mathbb{R} \cup \{-\infty\} \cup \{+\infty\}$. Similarly, the dioid in Example 4.18 is not complete but can be completed by adding the 'half-line' $\mathbb{R}$ itself to the considered set. However something is lost when doing this completion since multiplication does not remain cancellative (see Lemma 4.35 below). Of course, a subdioid of a complete dioid may not be complete. For example $\left(\mathbb{Q} \cup \{-\infty\} \cup \{+\infty\}, \max, +\right)$ is a subdioid of $\overline{\mathbb{R}}_{\max}$ which is not complete.

The question arises whether $\top$ is in general absorbing for multiplication, that is,

$$\forall a \in \mathcal{D} \ , \quad a \neq \varepsilon \ , \quad \top \otimes a = a \otimes \top = \top \ ? \tag{4.6}$$

Property (4.6) can be proved for 'Archimedian' dioids.

Definition 4.33 (Archimedian dioid) *A dioid is Archimedian if*

$$\forall a \neq \varepsilon \ , \quad \forall b \in \mathcal{D} \ , \quad \exists c \text{ and } d \in \mathcal{D} : ac \geq b \text{ and } da \geq b \ .$$

Theorem 4.34 *In a complete Archimedian dioid, the absorbing property (4.6) holds true.*

Proof We give the proof only for right multiplication by $\top$. From Definition 4.33, given a, for all b, there exists c_b such that $ac_b \geq b$. One has that

$$a\top = a\left(\bigoplus_{b\in\mathcal{D}} b\right) \geq a\left(\bigoplus_{b\in\mathcal{D}} c_b\right) = \bigoplus_{b\in\mathcal{D}} ac_b \geq \bigoplus_{b\in\mathcal{D}} b = \top \ .$$

■

Among our previous examples, all dioids, except for the one in Example 4.15, are Archimedian, but only Examples 4.16 and 4.17 correspond to complete dioids for which (4.6) holds true. Example 4.15 is a case of a dioid which is complete but not Archimedian, and (4.6) fails to be true.

Theorem 4.35 *If a dioid is complete, Archimedian, and if it has a cancellative multiplication, then it is isomorphic to the Boole algebra.*

Proof Since (4.6) holds true and since $\otimes$ is cancellative, it is realized that every element different from ε is equal to $\top$. Hence, the dioid is reduced to $\{\varepsilon, \top\}$. ■

4.3.4 Lower Bound

Since a complete dioid is a complete sup-semilattice, and since there is also a bottom element ε, the lower bound can be constructed for any subset $\mathcal{C}$ of elements of $\mathcal{D}$ and the semilattice becomes then a complete lattice (Theorem 4.27). If $\mathcal{C} = \{x, y, z, \ldots\}$, its lower bound is denoted $x \wedge y \wedge z \wedge \ldots$. In general, we use the notation $\bigwedge_{x\in\mathcal{C}} x$. One has the following equivalences:

$$a \geq b \Leftrightarrow a = a \oplus b \Leftrightarrow b = a \wedge b \ . \tag{4.7}$$

This operation $\wedge$ is also associative, commutative, idempotent and has $\top$ as neutral element ($\top \wedge a = a, \forall a$). The following property, called 'absorption law', holds true [57, p. 184]:

$$\forall a, b \in \mathcal{D} \ , \quad a \wedge (a \oplus b) = a \oplus (a \wedge b) = a \ .$$

Returning to our examples, the reader should apply the formal construction of the lower bound recalled in Theorem 4.27 to Example 4.17 (a complete dioid) and prove that $\wedge$ is simply $\cap$ in this case. As for the other examples, since all of them are chains, and even when the dioid is not complete, a simpler definition of $a \wedge b$ can be adopted: indeed, owing to Lemma 4.30, (4.7) may serve as a definition. Moreover, in the case of a chain, since a lower bound can be defined anyway, and because there exists a bottom element ε, the dioid is a complete inf-semilattice even if it is not a complete sup-semilattice.

Equivalences (4.7) may leave the impression that $\oplus$ and $\wedge$ play symmetric roles in a complete dioid. This is true from the lattice point of view, but this is not true when considering the behavior with respect to the other operation of the dioid, namely $\otimes$. Since multiplication is isotone, from Lemma 4.42 in §4.4.1 below it follows that

$$(a \wedge b)c \leq (ac) \wedge (bc) \tag{4.8}$$

(similarly for left multiplication) what we may call 'subdistributivity' of $\otimes$ with respect to $\wedge$. The same lemma shows that distributivity holds true for chains. But this is not true in general for partially-ordered dioids. A counterexample is provided by Example 4.17 ($\oplus$ is $\cup$, $\otimes$ is $+$ and $\wedge$ is $\cap$). In Example 4.31 we showed that $+$ is not distributive with respect to $\cap$. There are, however, situations in which distributivity of $\otimes$ with respect to $\wedge$ occurs for certain elements. Here is such a case.

Lemma 4.36 *If a admits a left inverse b and a right inverse c, then*

- *$b = c$ and this unique inverse is denoted a^{-1};*
- *moreover, $\forall x, y, a(x \wedge y) = ax \wedge ay$.*

The same holds true for right multiplication by a, and also for right and left multiplication by a^{-1}.

Proof

- One has $b = b(ac) = (ba)c = c$, proving uniqueness of a right and left inverse.
- Then, $\forall x, y$, define ξ and η according to $(\xi = ax, \eta = ay)$, which is equivalent to $(x = a^{-1}\xi, y = a^{-1}\eta)$. One has

$$\xi \wedge \eta = aa^{-1}(\xi \wedge \eta) \leq a[a^{-1}\xi \wedge a^{-1}\eta] = a[x \wedge y] \leq ax \wedge ay = \xi \wedge \eta \ .$$

Hence equality holds throughout. ■

4.3.5 Distributive Dioids

Once the lower bound has been introduced, this raises the issue of the mutual behavior of $\oplus$ and $\wedge$. In fact, $\wedge$ is not necessarily distributive with respect to $\oplus$ and conversely, except again for chains. The following inequalities are again consequences of Lemma 4.42 in §4.4.1 below, and of the fact that $x \mapsto x \oplus c$ and $x \mapsto x \wedge c$ are isotone:

$$\forall a, b, c \in \mathcal{D} \ , \qquad \begin{array}{l} (a \wedge b) \oplus c \leq (a \oplus c) \wedge (b \oplus c) \ , \\ (a \oplus b) \wedge c \geq (a \wedge c) \oplus (b \wedge c) \ , \end{array}$$

which means that $\oplus$ is *subdistributive* with respect to $\wedge$, and $\wedge$ is *superdistributive* with respect to $\oplus$.

As already defined, a lattice is distributive when equality holds true in the two inequalities above.

Example 4.37 Here is an example of a complete lattice which is not distributive. Consider all the intervals of $\mathbb{R}$ (including $\emptyset$ and $\mathbb{R}$ itself) with $\subseteq$ as $\leq$. The upper bound $\oplus$ of any (finite or infinite) collection of intervals is the smallest interval which contains the whole collection, that is, it is the convex hull of the union of all the intervals in the collection. The lower bound $\wedge$ is simply $\cap$. Then, consider

$a = [-3, -2], b = [2, 3]$ and $c = [-1, 1]$. We have that $(a \oplus b) \wedge c = c$, whereas $(a \wedge c) \oplus (b \oplus c) = \varnothing$. ■

The following theorem can be found in [57, p. 207].

Theorem 4.38 *A necessary and sufficient condition for a lattice to be distributive is that*

$$\forall a, b \ , \qquad \left\{ \exists c : \begin{array}{ccc} a \wedge c & = & b \wedge c \\ a \oplus c & = & b \oplus c \end{array} \right\} \Rightarrow \{a = b\} \ .$$

In [57] it is also shown that if $\mathcal{G}$ is a multiplicative 'lattice-ordered group', which means that, in addition to being a group and a lattice, the multiplication is isotone, then

- the multiplication is necessarily distributive with respect to both the upper and the lower bounds ($\mathcal{G}$ is called a 'reticulated group'),
- moreover, the lattice is distributive (that is, upper and lower bounds are distributive with respect to one another).

Also, one has the remarkable formulæ:

$$(a \wedge b)^{-1} = a^{-1} \oplus b^{-1} \ , \tag{4.9}$$

$$(a \oplus b)^{-1} = a^{-1} \wedge b^{-1} \ , \tag{4.10}$$

$$a \wedge b = a(a \oplus b)^{-1} b \ , \tag{4.11}$$

which should remind us of the De Morgan laws in Boolean algebra, and also the simple formula $\min(a, b) = -\max(-a, -b)$. However, this situation is far from being representative for the general case as shown by Examples 4.15 (total order) and 4.17 (partial order). Nevertheless, these examples correspond to 'distributive dioids' in the following sense.

Definition 4.39 (Distributive dioid) *A dioid $\mathcal{D}$ is distributive if it is complete and, for all subsets $\mathcal{C}$ of $\mathcal{D}$,*

$$\forall a \in \mathcal{D} \ , \qquad \begin{array}{l} \left(\bigwedge_{c \in \mathcal{C}} c \right) \oplus a = \bigwedge_{c \in \mathcal{C}} (c \oplus a) \ , \\ \left(\bigoplus_{c \in \mathcal{C}} c \right) \wedge a = \bigoplus_{c \in \mathcal{C}} (c \wedge a) \ . \end{array}$$

Notice that here distributivity is required to extend to infinite subsets. Both properties should be required now since one does not imply the other in the infinite case [57, p. 189]. Using the terminology of §4.4.1, we may state the preceding definition in other words by saying that a dioid is distributive if and only if the mappings $x \mapsto a \wedge x$ and $x \mapsto a \oplus x$ are both continuous for every a. All complete dioids considered so far are distributive. Example 4.37 can be extended to provide a nondistributive dioid. It suffices to define $\otimes$ as the 'sum' of intervals (conventional arithmetic sum). Of course, $\oplus$ is the upper bound as defined in that example (the reader may check the distributivity of $\otimes$ with respect to $\oplus$).

Remark 4.40 A distributive dioid may also be considered as a dioid with the two operations $\widehat{\oplus} \stackrel{\text{def}}{=} \oplus$ and $\widehat{\otimes} \stackrel{\text{def}}{=} \wedge$. But one can also choose $\widetilde{\oplus} \stackrel{\text{def}}{=} \wedge$ and $\widetilde{\otimes} \stackrel{\text{def}}{=} \oplus$. Special features of these dioid structures are that $\forall x, \widehat{\varepsilon} \leq x \leq \widehat{e}$ and that multiplication is commutative and idempotent. Examples 4.15 and 4.16 are instances of such dioids. ■

4.4 ISOTONE MAPPINGS AND RESIDUATION

Most of the material in this section is classical in Lattice Theory. The structure added by $\otimes$ in dioids plays virtually no role, except of course when the mappings considered themselves involve multiplication (for example when considering the residual of the mapping $x \mapsto ax$). A basic reference is the book by Blyth and Janowitz [24].

4.4.1 Isotony and Continuity of Mappings

We are going to characterize isotone mappings in terms of 'lower' and 'upper sets'.

Definition 4.41 (Lower, upper set) *A lower set is a nonempty subset L of $\mathcal{D}$ such that*

$$(x \in L \text{ and } y \leq x) \Rightarrow y \in L \ .$$

A closed lower set (generated by x) is a lower set denoted $[\leftarrow, x]$ of the form $\{y \mid y \leq x\}$. An upper set is a subset U such that

$$(x \in U \text{ and } y \geq x) \Rightarrow y \in U \ .$$

A closed upper set (generated by x) is an upper set denoted $[x, \rightarrow]$ of the form $\{y \mid y \geq x\}$.

The names '(principal) ideal' and '(principal) filter' are used for '(closed) lower set' and '(closed) upper set', respectively, in [24].

A closed lower set is a lower set which contains the upper bound of its elements. Similarly, a closed upper set is an upper set containing the lower bound of its elements. For a chain, say $\mathbb{R}_{\max}$, closed lower sets correspond to closed half-lines $(-\infty, x]$, lower sets are open or closed half-lines, whereas closed upper sets are of the type $[x, +\infty)$. Figure 4.3 gives examples of such sets in a partially-ordered lattice.

Obviously, if Π is a $\oplus$- or a $\wedge$-morphism, it is isotone (see §4.3.1). For example, for every $a \in \mathcal{D}$, the mapping $x \mapsto ax$ from $\mathcal{D}$ into itself is a $\oplus$-morphism, hence it is isotone. But, conversely, if Π is isotone, it is neither necessarily a $\oplus$- nor necessarily a $\wedge$-morphism.

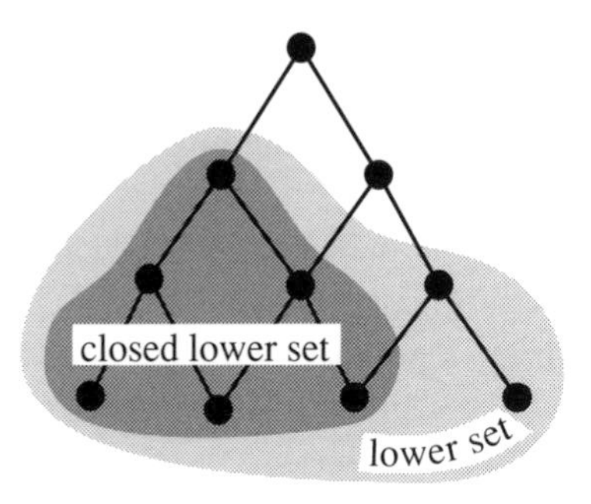

Figure 4.3: Lower set and closed lower set

Lemma 4.42 *Let Π be a mapping from a dioid $\mathcal{D}$ into another dioid $\mathcal{C}$. The following statements are equivalent:*

1. *the mapping Π is isotone;*
2. *the 'pre-image' $\Pi^{-1}([\leftarrow, x])$ of every closed lower set is a lower set or it is empty;*
3. *the pre-image $\Pi^{-1}([x, \rightarrow])$ of every closed upper set is an upper set or it is empty;*
4. *the mapping Π is a $\oplus$-supermorphism, that is,*
$$\forall a, b \in \mathcal{D} \ , \quad \Pi(a \oplus b) \geq \Pi(a) \oplus \Pi(b) \ ; \tag{4.12}$$
5. *if lower bounds exist in $\mathcal{D}$ and $\mathcal{C}$, Π is a $\wedge$-submorphism, that is,*
$$\forall a, b \in \mathcal{D} \ , \quad \Pi(a \wedge b) \leq \Pi(a) \wedge \Pi(b) \ . \tag{4.13}$$

Proof Suppose that Π is isotone. Let $a \in \Pi^{-1}([\leftarrow, x])$ if this subset is nonempty. Then $\Pi(a) \leq x$. Let $b \leq a$. Then $\Pi(b) \leq \Pi(a)$, hence $\Pi(b) \leq x$ and $b \in \Pi^{-1}([\leftarrow, x])$. Therefore $\Pi^{-1}([\leftarrow, x])$ is a lower set. Conversely, let $b \leq a$. Since obviously $a \in \Pi^{-1}([\leftarrow, \Pi(a)])$ and since this latter is a lower set by assumption, then b belongs to this subset. Hence $\Pi(b) \leq \Pi(a)$ and Π is isotone. A similar proof involving upper sets is left to the reader.

Suppose that Π is isotone. Since a and b are less than $a \oplus b$, then $\Pi(a)$ and $\Pi(b)$, and thus their upper bound $\Pi(a) \oplus \Pi(b)$, are less than $\Pi(a \oplus b)$ proving (4.12). Conversely, let $b \leq a$, or equivalently, $a = a \oplus b$. Then, under the assumption that (4.12) holds true, $\Pi(a) = \Pi(a \oplus b) \geq \Pi(a) \oplus \Pi(b)$, proving that $\Pi(b) \leq \Pi(a)$. Thus Π is isotone. A similar proof involving $\wedge$ instead of $\oplus$ can be given. ∎

If $\mathcal{D}$ is a chain, $a \oplus b$ is equal to either a or b, hence $\Pi(a \oplus b)$ is equal to either $\Pi(a)$ or $\Pi(b)$, hence to $\Pi(a) \oplus \Pi(b)$. That is, Π is a $\oplus$-morphism in this case. Similarly, it is a $\wedge$-morphism too.

If $\mathcal{D}$ and $\mathcal{C}$ are complete dioids, it is easy to see that (4.12) (respectively (4.13)) extends to $\oplus$ (respectively $\wedge$) operating over infinite subsets of $\mathcal{D}$.

Definition 4.43 (Continuity) *A mapping Π from a complete dioid $\mathcal{D}$ into a complete dioid $\mathcal{C}$ is lower-semicontinuous, abbreviated as l.s.c. respectively, upper-semicontinuous, abbreviated as u.s.c. if, for every (finite or infinite) subset X of $\mathcal{D}$,*

$$\Pi\left(\bigoplus_{x\in X} x\right) = \bigoplus_{x\in X} \Pi(x) \ ,$$

respectively,

$$\Pi\left(\bigwedge_{x\in X} x\right) = \bigwedge_{x\in X} \Pi(x) \ .$$

The mapping Π is continuous if it is both l.s.c. and u.s.c.

Of course, a l.s.c. (respectively an u.s.c.) mapping is a $\oplus$- (respectively a $\wedge$-) morphism. If Π is a $\oplus$-morphism, it is isotone and thus it is a $\wedge$-submorphism, but not necessarily a $\wedge$-morphism. This has already been illustrated by Example 4.31.

To justify the terminology 'l.s.c.', one should consider the example of a *nondecreasing* mapping Π from $\mathbb{R}$ to itself (because here we are interested in *isotone* mappings between ordered sets), and check that, in this case, the lower-semicontinuity in the previous sense coincides with the conventional notion of lower-semicontinuity (which requires that $\liminf_{x_i\to x}\Pi(x_i)\geq \Pi(x)$). The same observation holds true for upper-semicontinuity.

Lemma 4.44 *One has the following equivalences:*

- *Π is l.s.c.;*
- *the pre-image $\Pi^{-1}([\leftarrow, x])$ of every closed lower set is a closed lower set or it is empty.*

Similarly, the following two statements are equivalent:

- *Π is u.s.c.;*
- *the pre-image $\Pi^{-1}([x, \rightarrow])$ of every closed upper set is a closed upper set or it is empty.*

Proof We prove the former equivalence only. Suppose that Π is l.s.c. In particular it is isotone and the pre-image $X = \Pi^{-1}([\leftarrow, x])$ of every closed lower set, if nonempty, is a lower set. If $a \in X$, then $\Pi(a) \leq x$. Thus

$$\Pi\left(\bigoplus_{a\in X} a\right) = \bigoplus_{a\in X} \Pi(a) \leq x \ .$$

Hence the upper bound of X belongs to X: X is closed.

Conversely, suppose that the pre-image of every closed lower set is a closed lower set. In particular, according to Lemma 4.42, Π is isotone and for every nonempty subset $X \subseteq \mathcal{D}$, one has that

$$\Pi\left(\bigoplus_{x\in X} x\right) \geq \bigoplus_{x\in X} \Pi(x) \ . \tag{4.14}$$

On the other hand, it is obvious that

$$X \subseteq \Pi^{-1}\left([\leftarrow, \bigoplus_{x\in X} \Pi(x)]\right) .$$

But the latter subset is a closed lower set by assumption, hence it contains the upper bound of its elements and a fortiori it contains the upper bound of X. This implies the reverse inequality in (4.14). Therefore equality holds true and Π is l.s.c. ∎

Example 4.45 Let $\mathcal{D} = \overline{\mathbb{R}}_{\max}$ and $\mathcal{C} = \overline{\mathbb{N}}_{\max}$, let $\Pi : \mathcal{D} \to \mathcal{C}$ defined by

$$\Pi : x \mapsto y = \bigoplus_{y\in\mathcal{C}, y\leq x} y \ ,$$

where the symbol $\leq$ has its usual meaning. Indeed, Π is the residual of the mapping $x \mapsto x$ from $\mathcal{C}$ into $\mathcal{D}$ (not from $\mathcal{D}$ to $\mathcal{C}$!)—see §4.4.2 below. More simply, $\Pi(x)$ is just the integer part of the real number x. Then, Π is a $\oplus$- *and* a $\wedge$-morphism, it is u.s.c. (this is a consequence of being a residual) but not l.s.c. ∎

Lemma 4.46 *The set of l.s.c. mappings from a complete dioid $\mathcal{D}$ into itself is a complete dioid when endowed with the following addition $\widehat{\oplus}$ and multiplication $\widehat{\otimes}$:*

$$\begin{array}{rcll} \Pi\widehat{\oplus}\Phi : & x & \mapsto & \Pi(x)\oplus\Phi(x) \ ; \\ \Pi\widehat{\otimes}\Phi : & x & \mapsto & \Pi(\Phi(x)) \ . \end{array} \tag{4.15}$$

Similarly, the set of u.s.c. mappings from $\mathcal{D}$ into $\mathcal{D}$ is a complete dioid when endowed with the following addition $\widetilde{\oplus}$ and $\widetilde{\otimes}$:

$$\begin{array}{rcll} \Pi\widetilde{\oplus}\Phi : & x & \mapsto & \Pi(x)\wedge\Phi(x) \ ; \\ \Pi\widetilde{\otimes}\Phi : & x & \mapsto & \Pi(\Phi(x)) \ . \end{array} \tag{4.16}$$

Proof We only prove the former statement. It is easy to check that l.s.-continuity is preserved by addition and composition of mappings. The other axioms of dioids are also easily checked. In particular, $\widehat{\varepsilon}$ is the mapping identically equal to ε and $\widehat{e} = I_{\mathcal{D}}$ (identity of $\mathcal{D}$). Distributivity of right multiplication with respect to addition is straightforward from the very definitions of $\widehat{\oplus}$ and $\widehat{\otimes}$, whereas left distributivity involves the assumption of l.s.-continuity. ∎

Remark 4.47 Observe that since Lemma 4.46 defines a complete dioid structure for the set of l.s.c. mappings from a complete dioid $\mathcal{D}$ into itself, this dioid of mappings has a lower bound operation (Theorem 4.27) also denoted $\wedge$. However, in general, for two mappings Π and Φ, $(\Pi \wedge \Phi)(x) \neq \Pi(x) \wedge \Phi(x)$ since the right-hand side is in general not a l.s.c. function of x. ∎

Example 4.48 Consider $\mathcal{D} = \left(\overline{\mathbb{R}}_{\max}\right)^2$ (operations of $\mathcal{D}$ are those of $\overline{\mathbb{R}}_{\max}$ operating componentwise). Let $\mathcal{C} = \left(\overline{\mathbb{R}}^2, \widehat{\oplus}, \otimes\right)$ for which the underlying set is the same as that of $\mathcal{D}$ and $\otimes$ is also the same. But $\widehat{\oplus}$ is defined as follows:

$$\forall x, y \in \overline{\mathbb{R}}^2 \;, \quad x \widehat{\oplus} y = \begin{cases} x & \text{if } (x_1 > y_1) \text{ or } (x_1 = y_1 \text{ and } x_2 \geq y_2) \;; \\ y & \text{if } (x_1 < y_1) \text{ or } (x_1 = y_1 \text{ and } x_2 \leq y_2) \;. \end{cases}$$

The order in $\mathcal{D}$ is the usual partial order in $\overline{\mathbb{R}}^2$ whereas $\mathcal{C}$ is totally ordered by the lexicographic order. Let $\Pi : \mathcal{D} \to \mathcal{C}$ be simply the canonical bijection $x \mapsto x$. This is an isotone mapping since $x \leq y$ in $\mathcal{D}$ implies $x \leq y$ in $\mathcal{C}$. However, this mapping is neither l.s.c. nor u.s.c. Figure 4.4 depicts the shape of a closed lower set generated by the point $x = (2, 3)$ in $\mathcal{C}$ (shaded area). It is a half-plane including

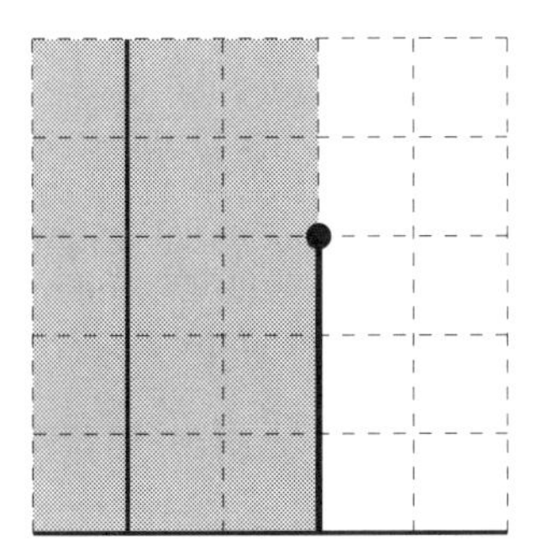

Figure 4.4: A closed lower set for the lexicographic order

the border for $x_2 \leq 3$ but not for $x_2 > 3$. Since Π^{-1} is also the canonical bijection in $\overline{\mathbb{R}}^2$, the pre-image of this closed lower set is itself and it is not a closed lower set in $\mathcal{D}$: closed lower sets in $\mathcal{D}$ consist of *closed* south-west orthants. Hence Π is not l.s.c.

This example is also interesting because it shows that an isotone bijection Π may have an inverse mapping Π^{-1} which is not isotone. As a matter of fact, $x \leq y$ in $\mathcal{C}$ does not imply the same inequality in $\mathcal{D}$ in general. However, we leave to the reader to prove that an isotone bijection from a *totally ordered* dioid $\mathcal{D}$ onto another dioid $\mathcal{C}$ has an isotone inverse and, moreover, if both dioids are complete, the mapping and its inverse are continuous. ∎

Lemma 4.49 *Let $\mathcal{D}$ and $\mathcal{C}$ be complete dioids and Π be a homomorphism from $\mathcal{D}$ into $\mathcal{C}$. Consider the congruence defined in Example 4.25. If Π is l.s.c. (respectively u.s.c.), then every equivalence class has a maximum (respectively minimum) element, which is therefore a canonical representative of the equivalence class.*

Proof We consider the case of a l.s.c. mapping. Let $[x]$ denote the equivalence class of any $x \in \mathcal{D}$, then $\widehat{x} \stackrel{\text{def}}{=} \bigoplus_{y\in[x]} y$ is the upper bound of $[x]$, and it belongs to $[x]$ since, by lower-semicontinuity,

$$\Pi(\widehat{x}) = \Pi\left(\bigoplus_{y\in[x]} y\right) = \bigoplus_{y\in[x]} \Pi(y) = \Pi(x) \ .$$

■

4.4.2 Elements of Residuation Theory

Residuation has to do with 'inverting' isotone mappings and with solving equations. Let Π be an isotone mapping from a dioid $\mathcal{D}$ into another dioid $\mathcal{C}$. To guarantee the existence of upper and lower bounds, we assume throughout this subsection that $\mathcal{D}$ and $\mathcal{C}$ are complete. If Π is not surjective, the equation in x: $\Pi(x) = b$ will have no solution for some values of b, and if Π is not injective, the same equation may have nonunique solutions. One way to always give a unique answer to this problem of equation solving is to consider the subset of so-called 'subsolutions', that is, values of x satisfying $\Pi(x) \leq b$, if this subset is nonempty, and then to take the upper bound of the subset, if it exists: it remains to be checked whether the upper bound itself is a subsolution, namely, that it is the maximum element of the subset of subsolutions, which has to do with l.s.-continuity of Π. In this case, this maximum element will be denoted $\Pi^\sharp(b)$ and we have

$$\Pi^\sharp(b) = \bigoplus_{\{x|\Pi(x)\leq b\}} x \qquad \text{and} \qquad \Pi\left(\Pi^\sharp(b)\right) \leq b \ . \tag{4.17}$$

Dually, one may consider 'supersolutions' satisfying $\Pi(x) \geq b$, if again this subset is nonempty, and then take the lower bound assuming it exists: again it remains to check whether the lower bound is itself a supersolution, namely, that it is the minimum element of the subset of supersolutions, which has to do with u.s.-continuity of Π. In this case this minimum element will be denoted $\Pi^\flat(b)$ and we have

$$\Pi^\flat(b) = \bigwedge_{\{x|\Pi(x)\geq b\}} x \qquad \text{and} \qquad \Pi\left(\Pi^\flat(b)\right) \geq b \ . \tag{4.18}$$

Theorem 4.50 *Let Π be an isotone mapping from the complete dioid $\mathcal{D}$ into the complete dioid $\mathcal{C}$. The following three statements are equivalent:*

1. *For all $b \in \mathcal{C}$, there exists a greatest subsolution to the equation $\Pi(x) = b$ (by this we mean that the subset of subsolutions is nonempty and that it has a maximum element).*

2. *$\Pi(\varepsilon) = \varepsilon$ and Π is l.s.c. (or equivalently, the pre-image of every closed lower set is nonempty and it is a closed lower set).*

3. *There exists a mapping $\Pi^\sharp$ from $\mathcal{C}$ into $\mathcal{D}$ which is isotone and u.s.c. such that*

$$\Pi \circ \Pi^\sharp \le I_{\mathcal{C}} \quad \text{(identity of } \mathcal{C}\text{)} ; \tag{4.19}$$

$$\Pi^\sharp \circ \Pi \ge I_{\mathcal{D}} \quad \text{(identity of } \mathcal{D}\text{)} . \tag{4.20}$$

Consequently, $\Pi^\sharp$ is unique. When Π satisfies these properties, it is said to be residuated and $\Pi^\sharp$ is called its residual.

Proof First of all, it should be clear that the two statements in '2.' above are equivalent. In the rest of the proof, we always refer to the former of these two statements.

1 ⇒ 3: As a matter of fact, $\forall b \in \mathcal{C}$, there exists a greatest subsolution that we denote $\Pi^\sharp(b)$. It is obvious that the mapping $\Pi^\sharp$ thus defined is isotone. Inequality (4.19) is immediate from the definition of a subsolution. Now, $\forall x \in \mathcal{D}$, let $b = \Pi(x)$. Since x is a subsolution corresponding to that b, from the definition of $\Pi^\sharp(b)$, $x \le \Pi^\sharp(b) = \Pi^\sharp \circ \Pi(x)$, from which (4.20) follows.

We now prove that $\Pi^\sharp$ is u.s.c. Since $\Pi^\sharp$ is isotone, using (4.13), for a subset $B \subseteq \mathcal{C}$, one has that

$$\Pi^\sharp\left(\bigwedge_{b \in B} b\right) \le \bigwedge_{b \in B} \Pi^\sharp(b) . \tag{4.21}$$

Using (4.13) again, we obtain

$$\Pi\left(\bigwedge_{b \in B} \Pi^\sharp(b)\right) \le \bigwedge_{b \in B} \Pi \circ \Pi^\sharp(b) \le \bigwedge_{b \in B} b ,$$

in which the latter inequality follows from (4.19). Hence $\bigwedge_{b \in B} \Pi^\sharp(b)$ is a subsolution corresponding to the right-hand side $\bigwedge_{b \in B} b$. Thus, the reverse inequality also holds true in (4.21), and equality is obtained, proving that $\Pi^\sharp$ is u.s.c.

3 ⇒ 2: From (4.19), $\Pi \circ \Pi^\sharp(\varepsilon) \le \varepsilon \Rightarrow \Pi \circ \Pi^\sharp(\varepsilon) = \varepsilon$. But $\Pi^\sharp(\varepsilon) \ge \varepsilon \Rightarrow \Pi \circ \Pi^\sharp(\varepsilon) \ge \Pi(\varepsilon)$. If we combine the two facts, it follows that $\varepsilon \ge \Pi(\varepsilon)$, proving the equality of the two sides.

Let $X \subseteq \mathcal{D}$. Since Π is isotone, it follows from (4.12) that

$$\bigoplus_{x \in X} \Pi(x) \le \Pi\left(\bigoplus_{x \in X} x\right) . \tag{4.22}$$

For all $x \in X$, let $b_x = \Pi(x)$. Because of (4.20), $\Pi^\sharp(b_x) \ge x$, hence

$$\Pi\left(\bigoplus_{x \in X} x\right) \le \Pi\left(\bigoplus_{x \in X} \Pi^\sharp(b_x)\right) \le \Pi \circ \Pi^\sharp\left(\bigoplus_{x \in X} b_x\right) \le \bigoplus_{x \in X} b_x = \bigoplus_{x \in X} \Pi(x) ,$$

where we used (4.12) for $\Pi^\sharp$ and then (4.19). This is the reverse inequality of (4.22), hence equality holds true and the l.s.-continuity of Π is proved.

2 $\Rightarrow$ 1: Since $\Pi(\varepsilon) = \varepsilon$, the subset of subsolutions $X_b \subseteq \mathcal{D}$ is nonempty $\forall b \in \mathcal{C}$. Then, by l.s.-continuity of Π, and since every $x \in X_b$ is a subsolution,

$$\Pi\left(\bigoplus_{x \in X_b} x\right) = \bigoplus_{x \in X_b} \Pi(x) \leq b \ .$$

This proves that $\bigoplus_{x \in X_b} x$ is a subsolution too.

Finally, since the greatest subsolution is unique by definition, the equivalences above imply that $\Pi^\sharp$ is unique as well. ■

Remark 4.51 It is clear that $\Pi^{-1}([\leftarrow, x]) = [\leftarrow, \Pi^\sharp(x)]$. Moreover, since $\Pi(\top) \leq \top$, then $\Pi^\sharp(\top) \geq \top$, hence $\Pi^\sharp(\top) = \top$. ■

Now, instead of being interested in the greatest subsolution, we may search for the least supersolution. This is dual residuation. The dual of Theorem 4.50 can be stated.

Theorem 4.52 *Let Π be an isotone mapping from the complete dioid $\mathcal{D}$ into the complete dioid $\mathcal{C}$. The following three statements are equivalent:*

1. *For all $b \in \mathcal{C}$, there exists a least supersolution to the equation $\Pi(x) = b$ (by this we mean that the subset of supersolutions is nonempty and that it has a minimum element).*
2. *$\Pi(\top) = \top$ and Π is u.s.c. (or equivalently, the pre-image of every closed upper set is nonempty and it is a closed upper set).*
3. *There exists a mapping $\Pi^\flat$ from $\mathcal{C}$ into $\mathcal{D}$ which is isotone and l.s.c. such that*

$$\Pi \circ \Pi^\flat \geq I_{\mathcal{C}} \quad \textit{(identity of } \mathcal{C}\textit{)} \ ; \tag{4.23}$$

$$\Pi^\flat \circ \Pi \leq I_{\mathcal{D}} \quad \textit{(identity of } \mathcal{D}\textit{)} \ . \tag{4.24}$$

Consequently, $\Pi^\flat$ is unique. When Π satisfies these properties, it is said to be dually residuated and $\Pi^\flat$ is called its dual residual.

Remark 4.53 One has that

$$\Pi^{-1}([x, \rightarrow]) = [\Pi^\flat(x), \rightarrow] \ ,$$

and

$$\Pi^\flat(\varepsilon) = \varepsilon \ .$$

It should also be clear that if Π is residuated, its residual is dually residuated and

$$(\Pi^\sharp)^\flat = \Pi \ .$$

■

Example 4.54 An example of a residuated mapping was encountered in Example 4.45. Indeed, if we let $\mathcal{D} = \overline{\mathbb{N}}_{\max}$ and $\mathcal{C} = \overline{\mathbb{R}}_{\max}$, the canonical injection from $\overline{\mathbb{N}}$ into $\overline{\mathbb{R}}$ is residuated and its residual is the mapping described in that example, that is, the 'integer part' of a real number 'from below'. The same injection is also dually residuated and its dual residual is the 'integer part from above'. ■

Example 4.55 Another interesting example is provided by the mapping $\Pi : x \mapsto (x, x)$ from a complete dioid $\mathcal{D}$ into $\mathcal{D}^2$. This mapping again is residuated and dually residuated and it is easy to check that

$$\Pi^\sharp(x, y) = x \wedge y \qquad \text{and} \qquad \Pi^\flat(x, y) = x \oplus y \ .$$

■

Subsection 4.4.4 provides other examples on residuation. The following theorem lists additional properties of residuated mappings and residuals, and dual properties when appropriate.

Theorem 4.56

- *If Π is a residuated mapping from $\mathcal{D}$ into $\mathcal{C}$, then*

$$\Pi \circ \Pi^\sharp \circ \Pi = \Pi \ ; \tag{4.25}$$

$$\Pi^\sharp \circ \Pi \circ \Pi^\sharp = \Pi^\sharp \ . \tag{4.26}$$

  One has the following equivalences:

$$\Pi^\sharp \circ \Pi = I_{\mathcal{D}} \Leftrightarrow \Pi \textit{ injective} \Leftrightarrow \Pi^\sharp \textit{ surjective} \ ; \tag{4.27}$$

$$\Pi \circ \Pi^\sharp = I_{\mathcal{C}} \Leftrightarrow \Pi^\sharp \textit{ injective} \Leftrightarrow \Pi \textit{ surjective} \ . \tag{4.28}$$

  The same statements hold true for dually residuated mappings by changing $\sharp$ into $\flat$.

- *If $\Pi : \mathcal{D} \to \mathcal{C}$ and $\Phi : \mathcal{C} \to \mathcal{B}$ are residuated mappings, then $\Phi \circ \Pi$ is also residuated and*

$$(\Phi \circ \Pi)^\sharp = \Pi^\sharp \circ \Phi^\sharp \ . \tag{4.29}$$

  Again, the same statement holds true with $\flat$ instead of $\sharp$.

- *If Π, Φ, Ψ and Θ are mappings from $\mathcal{D}$ into itself, and if Π and Θ are residuated, then*

$$\Pi \circ \Phi \le \Psi \circ \Theta \Leftrightarrow \Phi \circ \Theta^\sharp \le \Pi^\sharp \circ \Psi \ . \tag{4.30}$$

  As corollaries, one has that

$$\Pi \le \Theta \Leftrightarrow \Theta^\sharp \le \Pi^\sharp \ , \tag{4.31}$$

and

$$\Pi \le I_{\mathcal{D}} \quad \Leftrightarrow \quad \Pi^\sharp \ge I_{\mathcal{D}} \ , \tag{4.32}$$

$$\Pi \ge I_{\mathcal{D}} \quad \Leftrightarrow \quad \Pi^\sharp \le I_{\mathcal{D}} \ . \tag{4.33}$$

Similar statements hold true for dual residuals with appropriate assumptions; in particular, the analogue of (4.30) is

$$\Phi \circ \Pi \le \Theta \circ \Psi \Leftrightarrow \Theta^\flat \circ \Phi \le \Psi \circ \Pi^\flat \ . \tag{4.34}$$

- *If* Π *and* Φ *are two residuated mappings from a dioid* $\mathcal{D}$ *(in which* $\wedge$ *exists) into itself, then* $\Pi \oplus \Phi$ *is residuated and*

$$(\Pi \oplus \Phi)^\sharp = \Pi^\sharp \wedge \Phi^\sharp \ . \tag{4.35}$$

If Π *and* Φ *are dually residuated, then* $\Pi \wedge \Phi$ *is dually residuated and*

$$(\Pi \wedge \Phi)^\flat = \Pi^\flat \oplus \Phi^\flat \ . \tag{4.36}$$

- *If* Π *and* Φ *are two residuated mappings from a dioid* $\mathcal{D}$ *(in which* $\wedge$ *exists) into itself and if* $\Pi \wedge \Phi$ *is residuated, then*

$$(\Pi \wedge \Phi)^\sharp \ge \Pi^\sharp \oplus \Phi^\sharp \ . \tag{4.37}$$

With dual assumptions,

$$(\Pi \oplus \Phi)^\flat \le \Pi^\flat \wedge \Phi^\flat \ . \tag{4.38}$$

Proof

About (4.25)–(4.26): One has that

$$\Pi \circ \Pi^\sharp \circ \Pi = \Pi \circ \left(\Pi^\sharp \circ \Pi\right) \ge \Pi \ ,$$

which follows from (4.20). But one also has that

$$\Pi \circ \Pi^\sharp \circ \Pi = \left(\Pi \circ \Pi^\sharp\right) \circ \Pi \le \Pi \ ,$$

by making use of (4.19), hence (4.25) follows. Equation (4.26) is similarly proved by remembering that $\Pi^\sharp$ is isotone.

About (4.27)–(4.28): Assume that $\Pi^\sharp \circ \Pi = I_{\mathcal{D}}$ and suppose that $\Pi(x) = \Pi(y)$. Applying $\Pi^\sharp$, we conclude that $x = y$, hence Π is injective. Also, since $\Pi^\sharp \circ \Pi(x) = x$, it means that every x belongs to $\mathrm{Im}\Pi^\sharp$, hence $\Pi^\sharp$ is surjective. Conversely, if $\Pi^\sharp \circ \Pi \ne I_{\mathcal{D}}$, there exists y such that $x = \Pi^\sharp \circ \Pi(y) \ne y$. However, because of (4.25), $\Pi(x) = \Pi(y)$. Hence Π cannot be injective. On the other hand, if $\Pi^\sharp$ is surjective, $\forall x \in \mathcal{D}, \exists b \in \mathcal{C} : \Pi^\sharp(b) = x$. Since x is a subsolution corresponding to the right-hand side b, $\Pi(x) \le b$, hence $\Pi^\sharp \circ \Pi(x) \le \Pi^\sharp(b) = x$. We conclude that $\Pi^\sharp \circ \Pi \le I_{\mathcal{D}}$, but equality must hold true because of (4.20). This completes the proof of (4.27). The proof of (4.28) is similar.

About (4.29): As already noticed l.s.- or u.s.-continuity is preserved by composition of two similarly semicontinuous mappings and the same holds true for the property of being residuated (consider the conditions stated in item 2 of Theorem 4.50). Also $\Pi^\sharp \circ \Phi^\sharp$ is an isotone and u.s.c. mapping. Finally,

$$\Phi \circ \Pi \circ \Pi^\sharp \circ \Phi^\sharp = \Phi \circ \left(\Pi \circ \Pi^\sharp\right) \circ \Phi^\sharp \leq \Phi \circ \Phi^\sharp \leq I_C \ ,$$

by repeated applications of (4.19), showing that $\Pi^\sharp \circ \Phi^\sharp$ satisfies (4.19) together with $\Phi \circ \Pi$. Likewise, it can be proved that (4.20) is met by the two composed functions. From the uniqueness of the residual, we conclude that (4.29) holds true.

About (4.30)–(4.34): If $\Pi \circ \Phi \leq \Psi \circ \Theta$, then $\left(\Pi^\sharp \circ \Pi\right) \circ \Phi \circ \Theta^\sharp \leq \Pi^\sharp \circ \Psi \circ \left(\Theta \circ \Theta^\sharp\right)$ which implies that $\Phi \circ \Theta^\sharp \leq \Pi^\sharp \circ \Psi$ using (4.19)–(4.20). The converse proof is left to the reader (use a dual trick). Then (4.31) is obvious whereas to prove (4.32)–(4.33), we use the straightforward fact that $I_{\mathcal{D}}$ is residuated and is its own residual. The proof of (4.34) is similar to that of (4.30).

About (4.35)–(4.36): We give a proof of (4.35) only. First it is clear that the sum of two residuated mappings is residuated (l.s.-continuity is preserved by $\oplus$). Now consider the composition $\Gamma = \Psi_3 \circ \Psi_2 \circ \Psi_1$ of the following three mappings:

$$\begin{array}{llllccc}
\Psi_1 : & \mathcal{D} & \rightarrow & \mathcal{D}^2 \ , & x & \mapsto & \begin{pmatrix} x & x \end{pmatrix} \ , \\
\Psi_2 : & \mathcal{D}^2 & \rightarrow & \mathcal{D}^2 \ , & \begin{pmatrix} x & y \end{pmatrix} & \mapsto & \begin{pmatrix} \Pi(x) & \Phi(y) \end{pmatrix} \ , \\
\Psi_3 : & \mathcal{D}^2 & \rightarrow & \mathcal{D} \ , & \begin{pmatrix} x & y \end{pmatrix} & \mapsto & x \oplus y \ .
\end{array}$$

Thus

$$\Gamma : \mathcal{D} \rightarrow \mathcal{D} \ , \quad x \mapsto (\Pi \oplus \Phi)(x) \ .$$

Then,

$$\begin{array}{lll}
\Psi_1^\sharp : \mathcal{D}^2 \rightarrow \mathcal{D} \ , & \begin{pmatrix} x & y \end{pmatrix} \mapsto x \wedge y & \text{(see Example 4.55),} \\
\Psi_2^\sharp : \mathcal{D}^2 \rightarrow \mathcal{D}^2 \ , & \begin{pmatrix} x & y \end{pmatrix} \mapsto \begin{pmatrix} \Pi^\sharp(x) & \Phi^\sharp(y) \end{pmatrix} & \text{(trivial),} \\
\Psi_3^\sharp : \mathcal{D} \rightarrow \mathcal{D}^2 \ , & x \mapsto \begin{pmatrix} x & x \end{pmatrix} \ , &
\end{array}$$

the last statement following also from Example 4.55 since it was explained there that $(\Psi_3^\sharp)^\flat$ is indeed Ψ_3. Then, it suffices to calculate $\Gamma^\sharp$ by using (4.29) repeatedly to prove (4.35).

About (4.37)–(4.38): We prove (4.37) only. Observe first that it is necessary to assume that $\Pi \wedge \Phi$ is residuated since this is not automatically true. Then

$$(\Pi \wedge \Phi) \circ \left(\Pi^\sharp \oplus \Phi^\sharp\right) = (\Pi \wedge \Phi) \circ \Pi^\sharp \oplus (\Pi \wedge \Phi) \circ \Phi^\sharp \ ,$$

since $\Pi \wedge \Phi$ is assumed residuated and hence l.s.c. The former term at the right-hand side is less than $\Pi \circ \Pi^\sharp$ which is less than $I_{\mathcal{D}}$; the latter term is less than $\Phi \circ \Phi^\sharp$ which again is less than $I_{\mathcal{D}}$, and so is the left-hand side. This suffices to prove (4.37). ■

Remark 4.57 Returning to Lemma 4.49, if Π is residuated, it should be clear that $\widehat{x}$ considered in the proof of this lemma is nothing but $\Pi^\sharp \circ \Pi(x)$. ■

4.4.3 Closure Mappings

Here we study a special class of mappings of a dioid into itself which will be of interest later on.

Definition 4.58 (Closure mapping) *Let $\mathcal{D}$ be an ordered set and $\Pi : \mathcal{D} \to \mathcal{D}$ be an isotone mapping such that $\Pi = \Pi \circ \Pi \geq I_{\mathcal{D}}$, then Π is a called a closure mapping. If $\Pi = \Pi \circ \Pi \leq I_{\mathcal{D}}$, then Π is called a dual closure mapping.*

Theorem 4.59 *If $\Pi : \mathcal{D} \to \mathcal{D}$ is a residuated mapping, then the following four statements are equivalent:*

$$\Pi \circ \Pi = \Pi \geq I_{\mathcal{D}} \quad (\textit{i.e. } \Pi \textit{ is a closure mapping}), \tag{4.39}$$

$$\Pi^\sharp \circ \Pi^\sharp = \Pi^\sharp \leq I_{\mathcal{D}} \quad (\textit{i.e. } \Pi^\sharp \textit{ is a dual closure mapping}), \tag{4.40}$$

$$\Pi^\sharp = \Pi \circ \Pi^\sharp \ , \tag{4.41}$$

$$\Pi = \Pi^\sharp \circ \Pi \ . \tag{4.42}$$

Proof

(4.39) ⇒ (4.40): This follows from (4.29) and (4.33).

(4.40) ⇒ (4.41): From (4.40) it follows that $\Pi \circ \Pi^\sharp \circ \Pi^\sharp = \Pi \circ \Pi^\sharp$. The left-hand side is less than or equal to $\Pi^\sharp$ because of (4.19). The right-hand side is greater than or equal to $\Pi^\sharp$ because $\Pi^\sharp \leq I_{\mathcal{D}} \Rightarrow \Pi \geq I_{\mathcal{D}}$ (see (4.33)). Hence (4.41) is proved.

(4.41) ⇒ (4.42): From (4.41), it follows that $\Pi \circ \Pi^\sharp \circ \Pi = \Pi^\sharp \circ \Pi$. But the left-hand side is equal to Π (see (4.25)). Hence (4.42) results.

(4.42) ⇒ (4.39): Since $\Pi = \Pi^\sharp \circ \Pi$, then $\Pi \geq I_{\mathcal{D}}$ because of (4.20). On the other hand, (4.42) $\Rightarrow \Pi \circ \Pi^\sharp \circ \Pi = \Pi \circ \Pi$ but the left-hand side is equal to Π (see (4.25)). ■

Theorem 4.59 states that *all residuated* closure mappings can be expressed as in (4.42). Indeed, *all* closure mappings Π on $\mathcal{D}$ can be factored as $\Psi^\sharp \circ \Psi$ for some $\Psi : \mathcal{D} \to \mathcal{C}$, where $\mathcal{C}$ is another ordered set [24, Theorem 2.7]. Another characterization of l.s.c. closure mappings will be given in Corollary 4.69.

Theorem 4.60 *If $\Pi : \mathcal{D} \to \mathcal{D}$ is a dually residuated mapping, then the following four statements are equivalent:*

$$\Pi \circ \Pi = \Pi \leq I_{\mathcal{D}} \quad (\textit{i.e. } \Pi \textit{ is a dual closure mapping}), \tag{4.43}$$

$$\Pi^\flat \circ \Pi^\flat = \Pi^\flat \geq I_{\mathcal{D}} \quad (\textit{i.e. } \Pi^\flat \textit{ is a closure mapping}), \tag{4.44}$$

$$\Pi = \Pi \circ \Pi^\flat \ , \tag{4.45}$$

$$\Pi^\flat = \Pi^\flat \circ \Pi \ . \tag{4.46}$$

Lemma 4.61 *If Π and Φ are closure mappings on $\mathcal{D}$ and if they are $\wedge$-morphisms, then $\Pi \wedge \Phi$ also is a closure mapping. Likewise, if Π and Φ are dual closure mappings and if they are $\oplus$-morphisms, then $\Pi \oplus \Phi$ is a dual closure mapping. These statements extend to infinite numbers of mappings if the mappings are u.s.c., respectively l.s.c.*

Proof Let us prove the former statement. Clearly, $\Pi \wedge \Phi \geq I_{\mathcal{D}}$. Moreover,

$$(\Pi \wedge \Phi)\circ(\Pi \wedge \Phi) = \Pi \wedge \Pi\circ\Phi \wedge \Phi\circ\Pi \wedge \Phi = \Pi \wedge \Phi \ ,$$

since Π and Φ are greater than $I_{\mathcal{D}}$. ■

4.4.4 Residuation of Addition and Multiplication

In this subsection we consider the following mappings from a dioid $\mathcal{D}$ into itself:

$$\begin{aligned} T_a &: x \mapsto a \oplus x && \text{(translation by } a\text{);} \\ L_a &: x \mapsto a \otimes x && \text{(left multiplication by } a\text{);} \\ R_a &: x \mapsto x \otimes a && \text{(right multiplication by } a\text{).} \end{aligned}$$

Observe that

$$T_a\circ T_b = T_b\circ T_a = T_{a\oplus b} = T_a \oplus T_b \ . \tag{4.47}$$

Moreover, if $\mathcal{D}$ is a distributive dioid,

$$T_a \wedge T_b = T_{a\wedge b} \ . \tag{4.48}$$

As for multiplication, the associativity of $\otimes$ implies that

$$L_a\circ L_b = L_{ab} \ , \tag{4.49}$$

and also that

$$L_a\circ R_b = R_b\circ L_a \ . \tag{4.50}$$

The distributivity of $\otimes$ with respect to $\oplus$ implies that

$$L_a \oplus L_b = L_{a\oplus b} \ , \tag{4.51}$$

and also that

$$L_a\circ T_b = T_{ab}\circ L_a \ . \tag{4.52}$$

Observe that L_a is l.s.c. if and only if (left) multiplication is distributive with respect to addition of infinitely many elements, which we assume here, and, since moreover $L_a(\varepsilon) = \varepsilon$, L_a is residuated. The same considerations apply to right multiplication R_a.

Notation 4.62 We use the one-dimensional display notation $L_a^\sharp(x) = a \backslash x$ ('left division' by a—reads 'a (left) divides x'), respectively, $R_a^\sharp(x) = x / a$ ('right division' by a—reads 'x (right) divided by a'), and the two-dimensional display notation

$$L_a^\sharp(x) = a \backslash x \ , \qquad R_a^\sharp(x) = x / a \ .$$

■

As for T_a, since $T_a(\varepsilon) \neq \varepsilon$ unless $a = \varepsilon$, this mapping is not residuated. Actually, by restraining the range of T_a to $a \oplus \mathcal{D}$, that is, to the subset of elements greater than or equal to a (call this new mapping $\widehat{A}_a : \mathcal{D} \to a \oplus \mathcal{D}$), we could define a residual $\widehat{A}_a^\sharp$ with domain equal to $a \oplus \mathcal{D}$. However, this is not very interesting since $\widehat{A}_a^\sharp$ is simply the identity of $a \oplus \mathcal{D}$. Indeed, since $\widehat{A}_a$ is surjective (by definition), $\widehat{A}_a \circ \widehat{A}_a^\sharp$ is the identity according to (4.28). On the other hand, since $\widehat{A}_a$ is obviously a closure mapping, $\widehat{A}_a \circ \widehat{A}_a^\sharp = \widehat{A}_a^\sharp$ (see (4.41)). This is why we assume that $\mathcal{D}$ is a *distributive* dioid—see Definition 4.39—and, as a consequence of that, T_a is u.s.c. Since moreover $T_a(\top) = \top$, T_a is dually residuated.

Notation 4.63 We use the notation $T_a^\flat(x) = x \ominus a$. ■

It should be clear that:

$$x \ominus a = \varepsilon \Leftrightarrow a \geq x \ .$$

We are going to list a collection of formulæ and properties for these two new operations, 'division' and 'subtraction', which are direct consequences of the general properties enumerated in §4.4.2. For the sake of easy reference, the main formulæ have been gathered in Tables 4.1 and 4.2. In Table 4.1, left and right multiplication and division are both considered.

> Remember that, when we consider properties involving $\ominus$, the dioid $\mathcal{D}$ is tacitly assumed to be complete (hence multiplication is infinitely distributive) and also distributive.

For Table 4.1, we only prove the left-hand side versions of the formulæ. Formulæ (f.1) and (f.2) are consequences of the fact that $L_a^\sharp$ is u.s.c. and isotone. Dually, Formulæ (f.14) and (f.15) result from $T_a^\flat$ being l.s.c. and isotone. Inequalities (4.19)–(4.20) imply (f.5)–(f.6). However, if multiplication is cancellative, then L_a is injective and (see (4.27)) $a \backslash (ax) = x$ and $x \mapsto a \backslash x$ is surjective. If a is invertible, then $a \backslash x = a^{-1}x$. Dually, (4.23)–(4.24) yields

$$(x \ominus a) \oplus a \leq x \quad \text{and} \quad (x \oplus a) \ominus a \leq x \ .$$

But this is weaker than what results from Theorem 4.60. Clearly T_a is a closure mapping, hence from (4.44),

$$(x \ominus a) \ominus a = x \ominus a \ ,$$

and one also gets (f.16)–(f.17) from (4.45)–(4.46). It follows that

$$x \geq a \Rightarrow (x \ominus a) \oplus a = x \quad \text{and} \quad (x \geq a, y \geq a, x \ominus a = y \ominus a) \Rightarrow (x = y) \ ,$$

Table 4.1: Formulæ involving division

$\frac{x \wedge y}{a} = \frac{x}{a} \wedge \frac{y}{a}$	$\frac{x \wedge y}{a} = \frac{x}{a} \wedge \frac{y}{a}$	(f.1)
$\frac{x \oplus y}{a} \geq \frac{x}{a} \oplus \frac{y}{a}$	$\frac{x \oplus y}{a} \geq \frac{x}{a} \oplus \frac{y}{a}$	(f.2)
$\frac{x}{a \oplus b} = \frac{x}{a} \wedge \frac{x}{b}$	$\frac{x}{a \oplus b} = \frac{x}{a} \wedge \frac{x}{b}$	(f.3)
$\frac{x}{a \wedge b} \geq \frac{x}{a} \oplus \frac{x}{b}$	$\frac{x}{a \wedge b} \geq \frac{x}{a} \oplus \frac{x}{b}$	(f.4)
$a\frac{x}{a} \leq x$	$\frac{x}{a}a \leq x$	(f.5)
$\frac{ax}{a} \geq x$	$\frac{xa}{a} \geq x$	(f.6)
$a\frac{ax}{a} = ax$	$\frac{xa}{a}a = xa$	(f.7)
$\frac{a(a \backslash\!\!\circ\, x)}{a} = \frac{x}{a}$	$\frac{(x \circ\!\!\!/\, a)a}{a} = \frac{x}{a}$	(f.8)
$\frac{x}{ab} = \frac{a \backslash\!\!\circ\, x}{b}$	$\frac{x}{ba} = \frac{x \circ\!\!\!/\, a}{b}$	(f.9)
$\frac{a \backslash\!\!\circ\, x}{b} = \frac{x \circ\!\!\!/\, b}{a}$	$\frac{x \circ\!\!\!/\, a}{b} = \frac{b \backslash\!\!\circ\, x}{a}$	(f.10)
$b\frac{x}{a} \leq \frac{x}{a \circ\!\!\!/\, b}$	$\frac{x}{a}b \leq \frac{x}{b \backslash\!\!\circ\, a}$	(f.11)
$\frac{x}{a}b \leq \frac{xb}{a}$	$b\frac{x}{a} \leq \frac{bx}{a}$	(f.12)
$\frac{x}{a} \oplus b \leq \frac{x \oplus ab}{a}$	$\frac{x}{a} \oplus b \leq \frac{x \oplus ba}{a}$	(f.13)

Table 4.2: Formulæ involving subtraction

$(x \oplus y) \ominus a = (x \ominus a) \oplus (y \ominus a)$	(f.14)
$(x \wedge y) \ominus a \leq (x \ominus a) \wedge (y \ominus a)$	(f.15)
$(x \ominus a) \oplus a = x \oplus a$	(f.16)
$(x \oplus a) \ominus a = x \ominus a$	(f.17)
$x \ominus (a \oplus b) = (x \ominus a) \ominus b = (x \ominus b) \ominus a$	(f.18)
$x \ominus (a \wedge b) = (x \ominus a) \oplus (x \ominus b)$	(f.19)
$ax \ominus ab \leq a(x \ominus b)$	(f.20)
$x = (x \wedge y) \oplus (x \ominus y)$	(f.21)

which may also be viewed as consequences of the dual of (4.28) by observing that T_a is surjective if its range is restrained to $a \oplus \mathcal{D}$.

Formulæ (f.7)–(f.8) are consequences of (4.25)–(4.26). The dual result stated for $\oplus$ and $\ominus$ is weaker than (f.16)–(f.17).

As a consequence of (4.31) or its dual,

$$a \leq b \Leftrightarrow \left\{ \frac{x}{a} \geq \frac{x}{b} \ , \quad \forall x \right\} \Leftrightarrow \{ x \ominus a \geq x \ominus b \ , \quad \forall x \} \ .$$

In particular, $a \geq e \Leftrightarrow a \backslash x \leq x, \forall x$ and $x \ominus a \leq x, \forall a, \forall x$ since a is always greater than or equal to ε and $x \ominus \varepsilon = x$.

Using (4.29) and (4.49), one gets (f.9). Dually, using (4.47), Formula (f.18) is derived. Formula (f.10) is a consequence of (4.50) and (4.29). To obtain (f.12), one makes use of (4.30) with $\Pi = \Theta = L_a$ and $\Phi = \Psi = R_b$, and also of (4.50). The proof of (f.11) essentially uses (f.5) twice: $x \geq a \left(a \backslash x\right) \geq \left(a / b\right) b \left(a \backslash x\right)$ (the latter inequality arising from the version of (f.5) written for *right* multiplication and division applied to the pair (a, b) instead of (x, a)); by associativity of the product, and from the very definition of $L^\sharp_{a/b}(x)$, one obtains (f.11).

Equations (4.35) and (4.51) yield Formula (f.3) (which should be compared with (4.10)), whereas (4.36) and (4.48) yield (f.19). Because of (4.8), $L_{a \wedge b} \leq L_a \wedge L_b$. If $L_a \wedge L_b$ were residuated, we could use Inequality (4.37) to get that $L^\sharp_{a \wedge b} \geq (L_a \wedge L_b)^\sharp \geq L^\sharp_a \oplus L^\sharp_b$, which would prove (f.4). Unfortunately, $L_a \wedge L_b$ is not residuated in general, unless multiplication is distributive with respect to $\wedge$. A direct proof of (f.4) is as follows: $L_{a \wedge b} \leq L_a$, hence $L^\sharp_{a \wedge b} \geq L^\sharp_a$; similarly $L^\sharp_{a \wedge b} \geq L^\sharp_b$; hence $L^\sharp_{a \wedge b} \geq L^\sharp_a \oplus L^\sharp_b$.

As for (4.38) applied to $T_a \oplus T_b$, it would yield a weaker result than Equality (f.18). Finally, consider (4.52) and use (4.30) with $\Pi = \Theta = L_a, \Phi = T_b, \Psi =$

T_{ab}; this yields (f.13). Considering (4.52) again, but now in connection with (4.34), and setting $\Phi = \Psi = L_a, \Pi = T_b, \Theta = T_{ab}$ yield (f.20).

An interesting consequence of some of these formulæ is the decomposition of any x with respect to any y as given by (f.21). Indeed,

$$\begin{aligned}(x \wedge y) \oplus (x \ominus y) &= (x \oplus (x \ominus y)) \wedge (y \oplus (x \ominus y)) \\ &= x \wedge (x \oplus y) \\ &= x \ ,\end{aligned}$$

the first equality following from the assumption of distributivity, the second based on the fact that $x \ominus y \le x$ on the one hand, and (f.16) on the other hand, the last equality being obvious. As a corollary,

$$x \oplus y = (x \ominus y) \oplus (x \wedge y) \oplus (y \ominus x) \ ,$$

which is straightforward using the decompositions of x with respect to y and of y with respect to x.

Remark 4.64 Formula (f.3) can be written $L^\sharp_{a \oplus b}(x) = L_a(x)^\sharp \wedge L^\sharp_b(x)$, whereas (f.9) can be written $L^\sharp_{ab}(x) = L^\sharp_b \circ L^\sharp_a(x)$. Then considering the dioid structure of u.s.c. mappings from $\mathcal{D}$ into $\mathcal{D}$ described in Lemma 4.46 (see (4.16)), it is realized that the mapping $a \mapsto L^\sharp_a$ is a homomorphism from $\mathcal{D}$ into that dioid of u.s.c. mappings.

Likewise, (f.19) can be written $T^\flat_{a \wedge b}(x) = T^\flat_a(x) \oplus T^\flat_b(x)$, whereas (f.18) can be written $T^\flat_{a \oplus b}(x) = T^\flat_b(x) \circ T^\flat_a(x)$. Remember that now $\mathcal{D}$ is supposed to be a distributive dioid. Consider the dioid of l.s.c. mappings with the operations defined by (4.15). Observe that $\widehat{\otimes}$ is commutative and idempotent when restricted to elements of the form $T^\flat_a$. For the mapping $a \mapsto T^\flat_a$ to be a homomorphism, we must supply $\mathcal{D}$ with the addition $\overline{\oplus} \stackrel{\text{def}}{=} \wedge$ and the multiplication $\overline{\otimes} \stackrel{\text{def}}{=} \oplus$ (see Remark 4.40). ■

Example 4.65 Let us consider the complete dioid $\overline{\mathbb{R}}_{\max}$. From the very definition of $\ominus$, we have that

$$\forall a, b \in \overline{\mathbb{R}} \ , \quad a \ominus b = \begin{cases} a & \text{if } b < a \ ; \\ \varepsilon & \text{otherwise.} \end{cases}$$

As for $a \not/ b$ (or $b \backslash\!\!\!\!\not\ \ a$, which is the same since the multiplication is commutative), it is equal to $a - b$ (conventional subtraction) whenever there is no ambiguity in this expression, that is, in all cases except when $a = b = \varepsilon = -\infty$ and when $a = b = \top = +\infty$. Returning to the definition of $\not/$, it should be clear that $\varepsilon \not/ \varepsilon = \top \not/ \top = \top$, which yields the rule $\infty - \infty = +\infty$ in conventional notation.

Note that the conventional notation should be avoided because it may be misleading. As a matter of fact, we also have that $\top \otimes \varepsilon = \varepsilon$ (according to Axiom 4.7), which yields the rule $\infty - \infty = -\infty$: this seems to contradict the previous rule, at least when using conventional notation. ■

Example 4.66 In order to illustrate the operations $\ominus$ and $\not/$ in the case of the *commutative* dioid of Example 4.17, consider first Figure 4.5 in which A and B

are two disks (respectively, transparent and grey). The subset $C = B \ominus A$ is the smallest subset such that $C \cup A \supseteq B$: it is depicted in the figure which also illustrates Formula (f.16) in this particular case. Consider now Figure 4.6 in which A is a disk centered at the origin, whereas B is a square. Then, $C = B \not\phi A$ is the largest subset such that $C + A \subseteq B$: it is the dark small square in the middle of the figure. The right-hand side of this figure illustrates Formula (f.5) (written here for *right* division and multiplication). ■

Figure 4.5: The operation $\ominus$ in $\left(2^{\mathbb{R}^2}, \cup, +\right)$

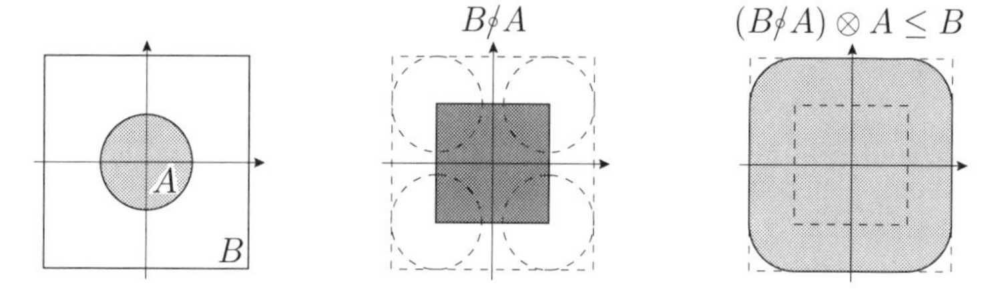

Figure 4.6: The operation $\not\phi$ in $\left(2^{\mathbb{R}^2}, \cup, +\right)$

We conclude this subsection by considering the problem of 'solving' equations of the form

$$ax \oplus b = c \tag{4.53}$$

in the sense of the greatest subsolution. This amounts to computing $\Pi^\sharp(c)$ with $\Pi = \widehat{A}_b \circ L_a$. Notice that, as discussed at the beginning of this subsection, we have to restrain the image of T_b to $b \oplus \mathfrak{D}$ for the mapping $\widehat{A}_b$ to be residuated. More directly, it is obvious that the subset of subsolutions of (4.53) is nonempty if and only if $c \geq b$. Then $\Pi^\sharp(c) = L_a^\sharp$ because of (4.29) and since $\widehat{A}_b^\sharp = I_{b \oplus \mathfrak{D}}$ as already discussed. We summarize this discussion in the following lemma.

Lemma 4.67 *There exists a greatest subsolution $\widehat{x}$ to (4.53) if and only if $b \leq c$. Then $\widehat{x} = a \backslash\!\!\!\phi\, c$.*

Of course, other similar equations may be considered as well, for example those involving the residuated mappings $T_b^\flat \circ L_a$ and $L_a \circ T_b^\flat$, or the dually residuated mappings $T_b \circ L_a^\sharp$ and $L_a^\sharp \circ T_b$.

4.5 FIXED-POINT EQUATIONS, CLOSURE OF MAPPINGS AND BEST APPROXIMATION

In this section we first study general fixed-point equations in complete dioids. The general results are then applied to special cases of interest. These special cases are motivated by some problems of 'best approximation' of elements of a dioid by other elements subject to 'linear constraints'. This kind of questions will arise frequently in Chapter 5 and therefore they are discussed here in some detail.

4.5.1 General Fixed-Point Equations

Let $\mathcal{D}$ be a complete dioid and consider the following 'fixed-point' equation and inequalities:

$$\Pi(x) = x , \tag{4.54}$$
$$\Pi(x) \leq x , \tag{4.55}$$
$$\Pi(x) \geq x , \tag{4.56}$$

where Π is an isotone mapping from $\mathcal{D}$ into $\mathcal{D}$. Observe that the notion of 'inequality' is somewhat redundant in complete dioids: indeed, Inequalities (4.55) and (4.56) can be written as equations, namely $\Pi(x) \oplus x = x$ and $\Pi(x) \wedge x = x$, respectively.

Let us again consider the operations defined by (4.15) or (4.16). We will use $\oplus$ instead of $\widehat{\oplus}$ and $\wedge$ instead of $\widetilde{\oplus}$. Accordingly, $\Pi \geq \Phi$ will mean $\Pi = \Pi \oplus \Phi$, or else $\Phi = \Pi \wedge \Phi$ (which would not be the case if $\wedge$ were considered as the 'sum'); Π^k will denote $\underbrace{\Pi \circ \ldots \circ \Pi}_{k \text{ times}}$ and $\Pi^0 = I_{\mathcal{D}}$. We introduce the following additional notation:

$$\Pi^* = \bigoplus_{k=0}^{+\infty} \Pi^k \; ; \qquad \Pi_* = \bigwedge_{k=0}^{+\infty} \Pi^k \; . \tag{4.57}$$

Although (4.15) (respectively (4.16)) has been defined only for l.s.c. (respectively u.s.c.) mappings, there is no difficulty in defining Π^* or Π_* for mappings Π which are neither l.s.c. nor u.s.c. We will also use the following notation:

$$\Pi^+ = \bigoplus_{k=1}^{+\infty} \Pi^k \; ; \qquad \Pi_+ = \bigwedge_{k=1}^{+\infty} \Pi^k \; .$$

Obviously

$$\Pi^* = I_{\mathcal{D}} \oplus \Pi^+ \; , \qquad \text{hence} \qquad \Pi^* \geq \Pi^+ \; , \tag{4.58}$$

but equality holds true if $\Pi \geq I_{\mathcal{D}}$. Also, because of (4.12),

$$\Pi \circ \Pi^* \geq \Pi^* \circ \Pi = \Pi^+ \; , \tag{4.59}$$

but equality holds true if Π is l.s.c. Similarly,

$$\Pi_* = I_{\mathcal{D}} \wedge \Pi_+ \; , \qquad \text{hence} \qquad \Pi_* \leq \Pi_+ \; , \tag{4.60}$$

but equality holds true if $\Pi \leq I_{\mathcal{D}}$. Also, because of (4.13),

$$\Pi \circ \Pi_* \leq \Pi_* \circ \Pi = \Pi_+ \ , \tag{4.61}$$

but equality holds true if Π is u.s.c.

If Π is a closure mapping, then $\Pi^* = \Pi$, and if Π is a dual closure mapping, then $\Pi_* = \Pi$. Indeed, when Π is not a closure (respectively dual closure) mapping, Π^* (respectively Π_*) is its 'closure' (respectively 'dual closure') under a semicontinuity assumption.

Lemma 4.68 *Let Π be a mapping from a complete dioid $\mathcal{D}$ into itself. If Π is l.s.c., then Π^* is the least closure mapping which is greater than Π. Likewise, if Π is u.s.c., then Π_* is the greatest dual closure mapping which is less than Π.*

Proof We give the proof of the first statement only. By direct calculations using the assumption of lower-semicontinuity, it is first checked that

$$\left(\Pi^*\right)^2 = \Pi^* \ . \tag{4.62}$$

Since moreover $\Pi^* \geq I_{\mathcal{D}}$, Π^* meets properties (4.39) and it is a closure mapping greater than Π.

Then, assume that Ψ is another closure mapping greater than Π. We have $\Psi \geq I_{\mathcal{D}}$, and successively, $\Psi \geq \Pi$, $\Psi = \Psi^2 \geq \Psi \circ \Pi \geq \Pi^2$ and $\Psi^k \geq \Pi^k, \forall k \in \mathbb{N}$. Therefore, by summing up these inequalities,

$$\Psi \geq \bigoplus_{k=0}^{+\infty} \Pi^k = \Pi^* \ ,$$

which completes the proof. ∎

Corollary 4.69 *Let Ψ be a mapping from a complete dioid $\mathcal{D}$ into itself. If Ψ is l.s.c., then*

$$\Psi \text{ is a closure mapping} \Leftrightarrow \Psi = \Psi^* \ .$$

If Ψ is u.s.c., then

$$\Psi \text{ is a dual closure mapping} \Leftrightarrow \Psi = \Psi_* \ .$$

This corollary provides an additional equivalent statement to Theorem 4.59, respectively Theorem 4.60.

We now return to (4.54), (4.55) and (4.56) and we set

$$\mathcal{D}_{\Pi}^{\natural} = \{x \mid \Pi(x) = x\}, \ \ \mathcal{D}_{\Pi}^{\flat} = \{x \mid \Pi(x) \leq x\}, \ \ \mathcal{D}_{\Pi}^{\sharp} = \{x \mid \Pi(x) \geq x\}. \tag{4.63}$$

Obviously, $\mathcal{D}_{\Pi}^{\natural} = \mathcal{D}_{\Pi}^{\flat} \cap \mathcal{D}_{\Pi}^{\sharp}$. With the only assumption that Π is isotone, Tarski's fixed-point theorem (see [22]) states that $\mathcal{D}_{\Pi}^{\natural}$ is nonempty (thus, $\mathcal{D}_{\Pi}^{\flat}$ and $\mathcal{D}_{\Pi}^{\sharp}$ are also nonempty).

Theorem 4.70

1. *Given two mappings* Π *and* Ψ, *if* $\Pi \geq \Psi$, *then* $\mathcal{D}^\flat_\Pi \subseteq \mathcal{D}^\flat_\Psi$.

2. *If* $\mathcal{C} \subseteq \mathcal{D}^\flat_\Pi$, *then* $\bigwedge_{x \in \mathcal{C}} x \in \mathcal{D}^\flat_\Pi$; *otherwise stated, the set* $\mathcal{D}^\flat_\Pi$ *with the order induced by that of* $\mathcal{D}$ *is a complete inf-semilattice having the same lower bound operation* $\wedge$ *as* $\mathcal{D}$. *Moreover,* $\top \in \mathcal{D}^\flat_\Pi$. *Hence (by Theorem 4.27, or rather by its dual),* $\mathcal{D}^\flat_\Pi$ *is also a complete lattice, but the upper bound operation does not need to be the same as that of* $\mathcal{D}$, *the latter being denoted* $\oplus$.

3. *If* Π *is l.s.c., then* $\mathcal{C} \subseteq \mathcal{D}^\flat_\Pi$ *implies* $\bigoplus_{x \in \mathcal{C}} x \in \mathcal{D}^\flat_\Pi$; *otherwise stated, the set* $\mathcal{D}^\flat_\Pi$ *with the order induced by that of* $\mathcal{D}$ *is a complete sup-semilattice having the same upper bound operation* $\oplus$ *as* $\mathcal{D}$.

4. *Statement 3 holds true also for* $\mathcal{D}^\natural_\Pi$.

5. *In general,* $\mathcal{D}^\flat_\Pi = \mathcal{D}^\flat_{\Pi^*} = \mathcal{D}^\natural_{\Pi^*}$. *Otherwise stated,*

$$\Pi(x) \leq x \Leftrightarrow \Pi^*(x) \leq x \Leftrightarrow \Pi^*(x) = x \ . \tag{4.64}$$

6. *If* Π *is l.s.c., then* $\mathcal{D}^\flat_\Pi = \Pi^*(\mathcal{D})$. *The minimum element is* $\Pi^*(\varepsilon)$ *which also belongs to* $\mathcal{D}^\natural_\Pi$, *and thus is the minimum element of this subset too.*

Proof

1. Straightforward.

2. Since Π is isotone, if $x, y \in \mathcal{D}^\flat_\Pi$, then $x \wedge y \geq \Pi(x) \wedge \Pi(y) \geq \Pi(x \wedge y)$ (see (4.13)). Hence $x \wedge y \in \mathcal{D}^\flat_\Pi$. This result obviously extends to any finite or infinite subset of $\mathcal{D}^\flat_\Pi$. Also $\top \geq \Pi(\top)$, hence $\top \in \mathcal{D}^\flat_\Pi$. The dual of Theorem 4.27 shows that any (finite or infinite) number of elements of $\mathcal{D}^\flat_\Pi$ admit an upper bound in $\mathcal{D}^\flat_\Pi$. But this does not mean that, if $x, y \in \mathcal{D}^\flat_\Pi$, then $x \oplus y \in \mathcal{D}^\flat_\Pi$, where of course $\oplus$ denotes the addition of $\mathcal{D}$.

3. Since Π is assumed to be l.s.c., for $x, y \in \mathcal{D}^\flat_\Pi$, $x \oplus y \geq \Pi(x) \oplus \Pi(y) = \Pi(x \oplus y)$. Hence $x \oplus y \in \mathcal{D}^\flat_\Pi$. This result obviously extends to any finite or infinite subset of $\mathcal{D}^\flat_\Pi$.

4. Same argument.

5. If $\Psi \geq I_\mathcal{D}$ and if $x \in \mathcal{D}^\flat_\Psi$, then $x \geq \Psi(x) \geq x$, hence $x = \Psi(x)$ and $x \in \mathcal{D}^\natural_\Psi$, thus $\mathcal{D}^\natural_\Psi = \mathcal{D}^\flat_\Psi$. In particular, this is true for $\Psi = \Pi^*$. Now, if $x \in \mathcal{D}^\flat_\Pi$, $x \geq \Pi(x) \geq \Pi^2(x) \geq \cdots$, and by summing up, $x \geq \Pi^*(x)$, hence $x \in \mathcal{D}^\flat_{\Pi^*}$ and $\mathcal{D}^\flat_\Pi \subseteq \mathcal{D}^\flat_{\Pi^*}$. But since $\Pi^* \geq \Pi$, the reverse inclusion also holds true and thus $\mathcal{D}^\flat_\Pi = \mathcal{D}^\flat_{\Pi^*} = \mathcal{D}^\natural_{\Pi^*}$.

6. From its very definition, $\mathcal{D}^{\natural}_{\Pi^*} \subseteq \Pi^*(\mathcal{D})$. On the other hand, let $x \in \Pi^*(\mathcal{D})$, hence $\exists y \in \mathcal{D} : x = \Pi^*(y)$. Then, $\Pi(x) = \Pi \circ \Pi^*(y) = \Pi^+(y)$, the latter equality being true because Π is assumed to be l.s.c. From (4.58), it follows that $\Pi(x) = \Pi^+(y) \leq \Pi^*(y) = x$, hence $x \in \mathcal{D}^{\flat}_{\Pi}$ and $\Pi^*(\mathcal{D}) \subseteq \mathcal{D}^{\flat}_{\Pi}$. But since it has been proved that $\mathcal{D}^{\flat}_{\Pi} = \mathcal{D}^{\natural}_{\Pi^*}$, finally $\mathcal{D}^{\flat}_{\Pi} = \mathcal{D}^{\natural}_{\Pi^*} = \mathcal{D}^{\flat}_{\Pi^*} = \Pi^*(\mathcal{D})$.

   Since $\xi = \Pi^*(\varepsilon)$ is clearly the minimum element of $\Pi^*(\mathcal{D})$, it is also that of $\mathcal{D}^{\flat}_{\Pi}$, but since $\mathcal{D}^{\natural}_{\Pi} \subseteq \mathcal{D}^{\flat}_{\Pi}$, ξ is a minorant of $\mathcal{D}^{\natural}_{\Pi}$. It remains to be proved that ξ indeed belongs to $\mathcal{D}^{\natural}_{\Pi}$. This follows from the fact that $\Pi^*(\varepsilon) = \Pi^+(\varepsilon) = \Pi \circ \Pi^*(\varepsilon)$, hence $\xi = \Pi(\xi)$. ∎

Lemma 4.71 *If Π is residuated, then Π^* is also residuated, and*

$$\left[x \geq \Pi(x) \Leftrightarrow \Pi^{\sharp}(x) \geq x\right], \quad \textit{that is,} \quad \mathcal{D}^{\flat}_{\Pi} = \mathcal{D}^{\sharp}_{\Pi^{\sharp}} \ . \tag{4.65}$$

Moreover,

$$\left(\Pi^*\right)^{\sharp} = \left(\Pi^{\sharp}\right)_* \ . \tag{4.66}$$

Proof If Π is residuated, the fact that Π^* is also residuated is immediate from its definition. Then (4.65) is a direct consequence of (4.19)–(4.20). Equality (4.66) can be proved using (4.29) and (4.35) (or rather its extension to infinitely many operations $\oplus$ and $\wedge$). ∎

The following dual of Theorem 4.70 is stated without proof.

Theorem 4.72

1. *If $\Pi \geq \Psi$, then $\mathcal{D}^{\sharp}_{\Pi} \supseteq \mathcal{D}^{\sharp}_{\Psi}$.*

2. *If $\mathcal{C} \subseteq \mathcal{D}^{\sharp}_{\Pi}$, then $\bigoplus_{x \in \mathcal{C}} x \in \mathcal{D}^{\sharp}_{\Pi}$; otherwise stated, the set $\mathcal{D}^{\sharp}_{\Pi}$ with the order induced by that of $\mathcal{D}$ is a complete sup-semilattice having the same upper bound operation $\oplus$ as $\mathcal{D}$. Moreover, $\varepsilon \in \mathcal{D}^{\sharp}_{\Pi}$. Hence (by Theorem 4.27), $\mathcal{D}^{\sharp}_{\Pi}$ also is a complete lattice, but the lower bound operation does not need to be the same as that of $\mathcal{D}$, the latter being denoted $\wedge$.*

3. *If Π is u.s.c., then $\mathcal{C} \subseteq \mathcal{D}^{\sharp}_{\Pi}$ implies $\bigwedge_{x \in \mathcal{C}} x \in \mathcal{D}^{\sharp}_{\Pi}$; otherwise stated, the set $\mathcal{D}^{\sharp}_{\Pi}$ with the order induced by that of $\mathcal{D}$ is a complete inf-semilattice having the same lower bound operation $\wedge$ as $\mathcal{D}$.*

4. *The same statement holds true for $\mathcal{D}^{\natural}_{\Pi}$.*

5. *In general, $\mathcal{D}^{\sharp}_{\Pi} = \mathcal{D}^{\sharp}_{\Pi_*} = \mathcal{D}^{\natural}_{\Pi_*}$.*

6. *If Π is u.s.c., then $\mathcal{D}^{\sharp}_{\Pi} = \Pi_*(\mathcal{D})$. The maximum element is $\Pi_*(\top)$ which also belongs to $\mathcal{D}^{\natural}_{\Pi}$, and thus is the maximum element of this subset too.*

4.5.2 The Case $\Pi(x) = a \mathbin{\backslash\!\!\!\circ} x \wedge b$

Given a and b in a complete dioid $\mathcal{D}$, we consider the equation

$$x = a \mathbin{\backslash\!\!\!\circ} x \wedge b \ , \tag{4.67}$$

and the inequality

$$x \le a \mathbin{\backslash\!\!\!\circ} x \wedge b \ . \tag{4.68}$$

We may use all the conclusions of Theorem 4.72 since the corresponding Π is u.s.c. (Π is the composition of two u.s.c. mappings, namely $L_a^\sharp$ introduced in §4.4.4, and $x \mapsto x \wedge b$). Let us evaluate $\Pi_*(x)$. By using (f.1) and (f.9) of Table 4.1, it follows that

$$\begin{aligned}
\Pi^2(x) &= \frac{a \mathbin{\backslash\!\!\!\circ} x \wedge b}{a} \wedge b = \frac{x}{a^2} \wedge b \wedge \frac{b}{a} \ , \\
&\vdots \\
\Pi^k(x) &= \frac{x}{a^k} \wedge b \wedge \frac{b}{a} \wedge \cdots \wedge \frac{b}{a^{k-1}} \ .
\end{aligned}$$

Taking the lower bound on both sides of these equalities for $k = 0, 1, 2, \ldots$, and using (f.3) (more properly, using its extension to infinitely many operations), it follows that

$$\begin{aligned}
\Pi_*(x) &= x \wedge \frac{x}{a} \wedge \frac{x}{a^2} \wedge \cdots \wedge b \wedge \frac{b}{a} \wedge \frac{b}{a^2} \wedge \cdots \\
&= \frac{x}{e \oplus a \oplus a^2 \oplus \cdots} \wedge \frac{b}{e \oplus a \oplus a^2 \oplus \cdots} \\
&= \frac{x \wedge b}{e \oplus a \oplus a^2 \oplus \cdots} \\
&= \frac{x \wedge b}{a^*} \ ,
\end{aligned} \tag{4.69}$$

with

$$a^* = e \oplus a \oplus a^2 \oplus \cdots \ . \tag{4.70}$$

Returning to Theorem 4.72, we know that $\Pi_*(\top) = a^* \mathbin{\backslash\!\!\!\circ} b$ is the maximum element of both subsets of solutions to (4.67) and (4.68). On the other hand, ε solves (4.68), but it also solves (4.67), unless $a = \varepsilon$ and $b \neq \varepsilon$ (note that $\varepsilon \mathbin{\backslash\!\!\!\circ} \varepsilon = \top$). Because of statement 5 of Theorem 4.72, if x solves (4.68), and a fortiori if it solves (4.67), that is, if $x \in \mathcal{D}_\Pi^\sharp$, then $x \in \mathcal{D}_{\Pi_*}^\natural$, that is, $x = \Pi_*(x) = a^* \mathbin{\backslash\!\!\!\circ} (x \wedge b)$. This implies that $x = a^* \mathbin{\backslash\!\!\!\circ} x$ since $x \le b$ as a solution of (4.68). We summarize these results in the following theorem.

Theorem 4.73 *Consider Equation (4.67) and Inequality (4.68) with a and b given in a complete dioid $\mathcal{D}$. Then,*

1. *$a^* \mathbin{\backslash\!\!\!\circ} b$ is the greatest solution of (4.67) and (4.68);*
2. *every solution x of (4.67) and (4.68) satisfies $x = a^* \mathbin{\backslash\!\!\!\circ} x$;*

3. *ε is the least solution of (4.68), and it is also the least solution of (4.67) provided that $a \neq \varepsilon$ or $b = \varepsilon$.*

Remark 4.74 (Some identities involving a^*) Observe that the notation a^* of (4.70) may be justified by the fact that

$$L_a^* = L_{a^*} \ , \tag{4.71}$$

which is a consequence of (4.57), (4.49) and (4.51). Since L_a^* is a closure mapping, $\left(L_a^*\right)^2 = L_a^*$ and with (4.49) and (4.71), it follows that

$$(a^*)^2 = a^*, \quad \text{hence} \quad (a^*)^* = a^* \ . \tag{4.72}$$

Consider again $\Pi(x) = a \mathbin{\circ\!\!\!\!\backslash} x$ (derived from the previous Π by letting $b = \top$). This mapping Π is nothing but $L_a^\sharp$. By using (4.69) in this case, we see that

$$(L_a^\sharp)_* = (L_{a^*})^\sharp = \left(L_a^*\right)^\sharp \ , \tag{4.73}$$

the latter equality following from (4.71). Indeed, this is a particular instance of Equation (4.66). Since $\left(L_a^*\right)^\sharp$ is a dual closure mapping, it is equal to its square, hence, with (4.73),

$$\forall x \ , \quad \frac{x}{a^*} = \frac{a^* \mathbin{\circ\!\!\!\!\backslash} x}{a^*} \ . \tag{4.74}$$

Since L_{a^*} is a closure mapping, and since its residual is $(L_{a^*})^\sharp$, from (4.42), it follows that

$$\forall x \ , \quad a^* x = \frac{(a^* x)}{a^*} \quad \left(\text{in particular, } a^* = \frac{a^*}{a^*}\right) \ . \tag{4.75}$$

From (4.41), we obtain

$$\forall x \ , \quad \frac{x}{a^*} = a^* \left(\frac{x}{a^*}\right) \ . \tag{4.76}$$

■

4.5.3 The Case $\Pi(x) = ax \oplus b$

Given a and b in a complete dioid $\mathcal{D}$, we consider the equation

$$x = ax \oplus b \ , \tag{4.77}$$

and the inequality

$$x \geq ax \oplus b \ . \tag{4.78}$$

We may use all the conclusions of Theorem 4.70 since the corresponding Π is l.s.c. A direct calculation shows that

$$\Pi^*(x) = a^*(x \oplus b) \ .$$

Then, $\Pi^*(\varepsilon) = a^* b$ is the minimum element of both subsets of solutions to (4.77) and (4.78). On the other hand, $\top$ solves (4.78), but it also solves (4.77) if $\mathcal{D}$ is Archimedian, unless $a = \varepsilon$ and $b \neq \top$. Because of (4.64), if x solves (4.78), and a fortiori if it solves (4.77), then $x = a^*(x \oplus b)$. This implies that $x = a^* x$ since $x \geq b$ as a solution of (4.78). We summarize these results in the following theorem.

Theorem 4.75 *Consider Equation (4.77) and Inequality (4.78) with a and b given in a complete dioid $\mathcal{D}$. Then,*

1. *a^*b is the least solution of (4.77) and (4.78);*

2. *every solution x of (4.77) and (4.78) satisfies $x = a^*x$;*

3. *$\top$ is the greatest solution of (4.78), and, if $\mathcal{D}$ is Archimedian, it is also the greatest solution of (4.77) provided that $a \neq \varepsilon$ or $b = \top$.*

We conclude this subsection by showing a result which is analogous to a classical result in conventional linear algebra. Namely, in conventional algebra, let A be an $n \times n$ matrix and b be an n-dimensional column vector, it is known that all the solutions of $Ax = b$ can be obtained by summing up a particular solution of this equation with all solutions of the 'homogeneous' equation $Ax = 0$. More precisely, if $Ax = b$ and if $Ay = 0$, then, by summing up the two equations, one obtains $A(x+y) = b$. This statement and proof also hold true for equation (4.77) in a dioid, where

$$x = ax \tag{4.79}$$

plays the part of the homogeneous equation.

Conversely, in conventional algebra, if $Ax = b$ and $Ax' = b$, by subtraction, $y = x - x'$ satisfies $Ay = 0$. This latter argument cannot be translated straightforwardly to the dioid situation. Indeed, one should first observe that, since 'adding' also means 'increasing' in a dioid, one cannot recover *all* solutions of (4.77) by adding something to a particular solution, unless this is the least solution. Moreover, the proof by subtraction has to be replaced by another argument. We are going to see that the 'minus' operation $\ominus$ indeed plays a part in proving essentially the expected result, although, admittedly, things are somewhat more tricky. Since we are playing with $\ominus$, we recall that $\mathcal{D}$ has to be assumed distributive.

Theorem 4.76 *Let $\mathcal{D}$ be a distributive dioid (which may as well be a matrix dioid—see §4.6). A necessary and sufficient condition for x to be a solution of (4.77) is that x can be written $y \oplus a^*b$, where y is a solution of (4.79).*

Proof Let x be a solution of (4.77). Consider the decomposition of x with respect to a^*b (see (f.21) of Table 4.2), that is,

$$\begin{aligned} x &= (x \wedge a^*b) \oplus (x \ominus a^*b) \\ &= a^*b \oplus (x \ominus a^*b) \end{aligned}$$

since $x \geq a^*b$ by Theorem 4.75. Let $r = x \ominus a^*b$. One has that

$$\begin{aligned} r &= (ax \oplus b) \ominus a^*b && \text{owing to (4.77),} \\ &= (ax \ominus a^*b) \oplus (b \ominus a^*b) && \text{using (f.14),} \\ &= ax \ominus a^*b && \text{since } b \ominus a^*b = \varepsilon \text{ because } e \leq a^*, \\ &\leq ax \ominus a^+b && \text{since } a^* \geq a^+, \\ &\leq a(x \ominus a^*b) = ar && \text{using (f.20).} \end{aligned}$$

Since $x = a^*b \oplus r$, one also has $a^*x = a^*a^*b \oplus a^*r = a^*b \oplus a^*r$ from (4.62). But $x = a^*x$ (by Theorem 4.75) and thus $x = a^*b \oplus y$ with $y \stackrel{\text{def}}{=} a^*r$. Observe that $y = a^*y$ (from (4.62) again). Since $r \leq ar$, then $y \leq ay$, and hence, multiplying by a^*, we obtain $y = a^*y \leq a^+y \leq a^*y = y$. Finally, we have proved that $y = ay$ and that $x = a^* \oplus y$. ■

4.5.4 Some Problems of Best Approximation

Let us give the practical rationale behind solving inequalities such as (4.68) or (4.78) in the sense of finding an 'extremal' solution (respectively the maximum or the minimum). This motivation will be encountered several times in Chapter 5. In a complete dioid $\mathcal{D}$, for some given a, $\mathcal{D}^\flat_{L_a}$ is the subset $\{x \in \mathcal{D} \mid x \geq ax\}$. Such subsets enjoy nice properties that will be described later on. Let I be the 'canonical injection' from $\mathcal{D}^\flat_{L_a}$ into $\mathcal{D}$, namely $\mathrm{I} : x \mapsto x$. Given any b, if $b \notin \mathcal{D}^\flat_{L_a}$, there is no solution to $\mathrm{I}(x) = b, x \in \mathcal{D}^\flat_{L_a}$. However, residuation theory provides an answer by looking for the maximum element in $\mathcal{D}^\flat_{L_a}$ which is less than b, or the minimum element in $\mathcal{D}^\flat_{L_a}$ which is greater than b, as long as I is both residuated and dually residuated (which we will check later on). In some sense, these solutions can be viewed as 'best approximations from above or from below' of b by elements of $\mathcal{D}^\flat_{L_a}$. It will be shown that these two residuation problems are directly related to the problems of §4.5.2 and §4.5.3.

We first study several equivalent characterizations of $\mathcal{D}^\flat_{L_a}$ and the structure of this subset.

Lemma 4.77

1. We have the following equivalences:

$$\underbrace{x \geq ax}_{\text{(i)}} \Leftrightarrow \underbrace{x = a^*x}_{\text{(ii)}} \Leftrightarrow \underbrace{x \leq a \backslash\!\!\!\circ\, x}_{\text{(iii)}} \Leftrightarrow \underbrace{x = a^* \backslash\!\!\!\circ\, x}_{\text{(iv)}} \ . \tag{4.80}$$

2. The subset $\mathcal{D}^\flat_{L_a}$ contains ε and $\top$; it is closed for addition; it is a left multiplicative ideal, that is,

$$\forall x \in \mathcal{D}^\flat_{L_a} \ , \quad \forall y \in \mathcal{D} \ , \quad xy \in \mathcal{D}^\flat_{L_a} \ ;$$

a fortiori, it is closed for multiplication.

3. The subset $\mathcal{D}^\flat_{L_a}$ is the image of $\mathcal{D}$ by L_{a^*} and also by $(L_{a^*})^\sharp$, that is, $\forall y \in \mathcal{D}$, $a^*y \in \mathcal{D}^\flat_{L_a}$ and $a^* \backslash\!\!\!\circ\, y \in \mathcal{D}^\flat_{L_a}$; the subset $\mathcal{D}^\flat_{L_a}$ is a complete dioid with a^* as its identity element (it is a subdioid of $\mathcal{D}$ only if $a \leq e$ and $\mathcal{D}^\flat_{L_a} = \mathcal{D}$).

Proof

1. The equivalences (i) $\Leftrightarrow$ (ii) and (iii) $\Leftrightarrow$ (iv) are direct consequences of Theorems 4.70 and 4.72 (statement 5), respectively, applied to $\Pi = L_a$ and $\Pi = L_a^\sharp$. The equivalence (i) $\Leftrightarrow$ (iii) comes from (4.65).

2. We may use any of the equivalent characterizations of $\mathcal{D}^\flat_{L_a}$ to prove the rest of the statements of this lemma. For each statement, we choose the most adequate characterization, but, as an exercise, we invite the reader to use the other ones to prove the same statements. Of course $\varepsilon \geq a\varepsilon$ and $\top \geq a\top$. We have $[x \geq ax, y \geq ay] \Rightarrow x \oplus y \geq a(x \oplus y)$. Also $x \geq ax \Rightarrow \forall y \in \mathcal{D}, xy \geq a(xy)$.

3. The first part of the statement is a direct consequence of statements 6 of Theorems 4.70 and 4.72 (applied to $\Pi = L_a$ and to $\Pi = L_a^\sharp$, respectively), and of (4.73). For all $x \in \mathcal{D}^\flat_{L_a}$, $a^*x = x$, and hence a^* behaves as the identity element in $\mathcal{D}^\flat_{L_a}$. Therefore, $\mathcal{D}^\flat_{L_a}$ satisfies all the axioms of a dioid; it is even a complete dioid since it is also closed for infinite sums. It is a subdioid of $\mathcal{D}$ if a^* coincides with e. Since $a \leq a^*$, this implies that $a \leq e$. In this case, $\mathcal{D}^\flat_{L_a}$ coincides with $\mathcal{D}$, which is a rather trivial situation. ■

Since $\mathcal{D}^\flat_{L_a} = L_{a^*}(\mathcal{D})$, from now on we will prefer the more suggestive notation $a^*\mathcal{D}$ instead of $\mathcal{D}^\flat_{L_a}$.

Let us now return to the problem of the best approximation of b by the 'closest' element of $a^*\mathcal{D}$ among those which are either 'below' or 'above' b. More precisely, we look for the greatest $x \in a^*\mathcal{D}$ such that $\mathrm{I}(x) \leq b$ or for the least $x \in a^*\mathcal{D}$ such that $\mathrm{I}(x) \geq b$. Such problems are well-posed if I is residuated or dually residuated respectively. This is indeed the case thanks to the fact that $a^*\mathcal{D}$ is a complete dioid containing ε and $\top$ which are mapped to the same elements of $\mathcal{D}$ (and all continuity assumptions needed are satisfied by I).

Consider the former problem of approximation from below, the solution of which is $\mathrm{I}^\sharp(b)$ by definition. We show that this problem is the same as that of finding the greatest element of $\mathcal{D}^\sharp_\Pi$ with $\Pi(x) = a \lndiv x \wedge b$. Indeed, x must be less than b; it must also belong to $a^*\mathcal{D}$, hence $x \leq a \lndiv x$, thus $x \leq a \lndiv x \wedge b$. Conversely, this inequality implies that x is less than b and less than $a \lndiv x$, hence it belongs to $a^*\mathcal{D}$. Therefore, from the results of §4.5.2, we conclude that $\mathrm{I}^\sharp(b) = a^* \lndiv b$.

Similarly, it can be shown that finding $\mathrm{I}^\flat(b)$ is the same problem as finding the least element of $\mathcal{D}^\flat_\Psi$ with $\Psi(x) = ax \oplus b$. The solution has been given in §4.5.3 and therefore $\mathrm{I}^\flat(b) = a^*b$.

We consider the mapping which associates with any $b \in \mathcal{D}$ its best approximation from below (or from above) in $a^*\mathcal{D}$. This mapping is of course surjective (any element of $a^*\mathcal{D}$ is its own approximation) but not injective: several b having the same best approximation are said to be 'equivalent'. We can partition $\mathcal{D}$ into equivalence classes. The following theorem summarizes and completes this discussion.

Theorem 4.78

1. Let $\mathrm{I} : a^*\mathcal{D} \to \mathcal{D}$ be such that $\mathrm{I}(x) = x$. The canonical injection I is both residuated and dually residuated and

$$\mathrm{I}^\sharp(b) = \frac{b}{a^*} \; , \qquad \mathrm{I}^\flat(b) = a^*b \; .$$

2. *The mapping* $\mathrm{I}^\flat : \mathcal{D} \to a^*\mathcal{D}$ *is a surjective l.s.c. dioid homomorphism. Considering the equivalence relation* $\overset{\mathrm{I}^\flat}{\equiv}$ *in* $\mathcal{D}$ *(see (4.3)), then for any* $b \in \mathcal{D}$*, its equivalence class* $[b]$ *contains one and only one element which can also be viewed as an element of* $a^*\mathcal{D}$ *and which, moreover, is the maximum element in* $[b]$*. This element is precisely given by* $\mathrm{I}\left(\mathrm{I}^\flat(b)\right) = a^*b$*.*

3. *The mapping* $\mathrm{I}^\sharp : \mathcal{D} \to a^*\mathcal{D}$ *is surjective and u.s.c. (it is not a homomorphism). Considering the equivalence relation* $\overset{\mathrm{I}^\sharp}{\equiv}$ *in* $\mathcal{D}$*, then for any* $b \in \mathcal{D}$*, its equivalence class* $[b]$ *contains one and only one element which can also be viewed as an element of* $a^*\mathcal{D}$ *and which, moreover, is the minimum element in* $[b]$*. This element is precisely given by* $\mathrm{I}\left(\mathrm{I}^\sharp(b)\right) = a^* \backslash\!\!\backslash b$*.*

Proof

1. Already done.

2. The fact that $\mathrm{I}^\flat$ is a homomorphism is obvious from its explicit expression; it is l.s.c. (as a dual residual) and surjective as already discussed. Each equivalence class by $\overset{\mathrm{I}^\flat}{\equiv}$ has a maximum element $\widehat{b}$ by Lemma 4.49, and an explicit expression for $\widehat{b}$ has been given in Remark 4.57: here $\Pi = \mathrm{I}^\flat$ and hence $\Pi^\sharp = \mathrm{I}$, thus $\widehat{b} = \mathrm{I}(a^*b)$. Clearly, $\widehat{b}$ may be considered as an element of $a^*\mathcal{D}$. If another b' belongs at the same time to $[b]$ (hence $a^*b' = a^*b$) and to $a^*\mathcal{D}$ (hence $b' = a^*b'$), then $b' = a^*b = a^*\widehat{b} = \widehat{b}$ and b' coincides with $\widehat{b}$.

3. Dual arguments can be used here. The main difference is that $\overset{\mathrm{I}^\sharp}{\equiv}$ is *not* a congruence because $\mathrm{I}^\sharp$ is only a $\wedge$-morphism (indeed it is u.s.c.), but it does not behave well with respect to $\oplus$ and $\otimes$. ■

Concrete applications of these results will be given in Chapter 6 and 5.

Remark 4.79 Most of the results of this subsection can be generalized to the situation when the subset $a^*\mathcal{D}$ characterized by (4.80) is replaced by $\mathcal{D}^\flat_\Pi$ (see (4.63)) with Π residuated. Theorems 4.70 and 4.72, and Lemma 4.71 show that other characterizations of $\mathcal{D}^\flat_\Pi$ are

$$x = \Pi^*(x) \Leftrightarrow x \leq \Pi^\sharp(x) \Leftrightarrow x = \left(\Pi^\sharp\right)_*(x) = \left(\Pi^*\right)^\sharp(x) \ ;$$

that this subset contains ε and $\top$; that it is closed for addition (but it is no longer a left multiplicative ideal, unless Π satisfies $\Pi(x)y \geq \Pi(xy), \forall x, y$, which is equivalent to $\Pi^\sharp(x)y \leq \Pi^\sharp(xy)$); that it is the image of the whole $\mathcal{D}$ by Π^* and also by $(\Pi^*)^\sharp$. The best approximation of some b from above in $\mathcal{D}^\flat_\Pi$ is given by $\Pi^*(b)$. It is the maximum representative of the equivalence class $[b]$ of b for the equivalence relation $\overset{\Pi^*}{\equiv}$ and the only element in $[b] \cap \mathcal{D}^\flat_\Pi$. Dual statements hold true for the best approximation from below given by $(\Pi^*)^\sharp(b)$. ■

4.6 MATRIX DIOIDS

4.6.1 From 'Scalars' to Matrices

Starting from a 'scalar' dioid $\mathcal{D}$, consider square $n \times n$ matrices with entries in $\mathcal{D}$. The sum and product of matrices are defined conventionally after the sum and product of scalars in $\mathcal{D}$.

The set of $n \times n$ matrices endowed with these two operations is also a dioid which is denoted $\mathcal{D}^{n\times n}$. The only point that deserves some attention is the existence of an identity element. Thanks to Axiom 4.7, the usual identity matrix with entries equal to e on the diagonal and to ε elsewhere is the identity element of $\mathcal{D}^{n\times n}$. This identity matrix will also be denoted e and the zero matrix will simply be denoted ε.

Remark 4.80 We prefer to move from 'scalars' directly to square matrices. In this way the product of two matrices is a matrix of the same type and $\mathcal{D}^{n\times n}$ can be given a dioid structure too (multiplication remains an 'internal' operation). In fact, from a practical point of view and for most issues that will be considered later on, in particular linear equations, we can deal with nonsquare matrices, and especially with row or column vectors, as well. This is just a matter of completing the nonsquare matrices by rows or columns with entries equal to ε in order to convert them into square matrices, and to check that, for the problem considered, this artificial part does not interfere with the real part of the problem and that it only adds a trivial part to that problem. ■

Notice that if $\mathcal{D}$ is a commutative dioid, this is not the case for $\mathcal{D}^{n\times n}$ in general. Even if $\mathcal{D}$ is entire, $\mathcal{D}^{n\times n}$ is not so.

Example 4.81 Let $n = 2$ and

$$A = \begin{pmatrix} \varepsilon & a \\ \varepsilon & \varepsilon \end{pmatrix} .$$

Then $A^2 = A \otimes A = \varepsilon$ although $A \neq \varepsilon$. ■

Of course

$$A \geq B \text{ in } \mathcal{D}^{n\times n} \Leftrightarrow \{A_{ij} \geq B_{ij} \text{ in } \mathcal{D} \ , \quad i = 1, \ldots, n \ , \quad j = 1, \ldots, n\} \ .$$

Even if $\mathcal{D}$ is a chain, $\mathcal{D}^{n\times n}$ is only partially ordered. If $\mathcal{D}$ is complete, $\mathcal{D}^{n\times n}$ is complete too. Moreover

$$(A \wedge B)_{ij} = A_{ij} \wedge B_{ij} \ .$$

If $\mathcal{D}$ is distributive, $\mathcal{D}^{n\times n}$ is also distributive. Even if $\mathcal{D}$ is Archimedian, $\mathcal{D}^{n\times n}$ is not Archimedian. Here is a counterexample.

Example 4.82 Let $n = 2$ and consider the matrices

$$A = \begin{pmatrix} a & \varepsilon \\ \varepsilon & \varepsilon \end{pmatrix} \quad \text{and} \quad B = \begin{pmatrix} \varepsilon & \varepsilon \\ \varepsilon & b \end{pmatrix} .$$

Then there is obviously no matrix C such that $AC \geq B$. ■

In §2.3 it was shown how weighted graphs can be associated with matrices, and moreover, in the case when the entries lie in sets endowed with two operations $\oplus$ and $\otimes$ satisfying certain axioms, how the sum and the product of two matrices can be interpreted in terms of those graphs (see §2.3.1). These considerations are valid for matrices with entries belonging to a general dioid. The only point that deserves some attention is the notion of 'circuit of maximum weight' in the case when the underlying dioid is not a chain. We will discuss this issue in the case of polynomial matrices in §4.7.3.

4.6.2 Residuation of Matrices and Invertibility

We consider the mapping L_A from $\mathcal{D}^n$ into $\mathcal{D}^n$ defined by $x \mapsto Ax$, where $A \in \mathcal{D}^{n\times n}$ and $\mathcal{D}$ is a dioid in which $\wedge$ exists. Returning to Remark 4.80, we could rather define a mapping from $\mathcal{D}^{n\times n}$ (which is a dioid unlike $\mathcal{D}^n$) into $\mathcal{D}^{n\times n}$, namely $X \mapsto AX$ and then use it for $X \in \mathcal{D}^{n\times n}$ having its first column equal to $x \in \mathcal{D}^n$ and its $n-1$ last columns identically equal to ε. The purpose here is to establish a formula for $L_A^\sharp$ and then to study conditions of exact invertibility to the left of matrix A.

Indeed, it is not more difficult to consider a 'matrix of operators' in the following way. To keep notation simple, we take $n = 3$ but the generalization is straightforward. Then, consider six dioids $\{\mathcal{D}_i\}_{i=1,2,3}$ and $\{\mathcal{C}_j\}_{j=1,2,3}$ and nine residuated mappings Π_{ij} from $\mathcal{D}_j$ to $\mathcal{C}_i$. The mapping Π maps $\mathcal{D}_1 \times \mathcal{D}_2 \times \mathcal{D}_3$ into $\mathcal{C}_1 \times \mathcal{C}_2 \times \mathcal{C}_3$ and is defined as follows:

$$\Pi : x = \begin{pmatrix} x_1 \\ x_2 \\ x_3 \end{pmatrix} \mapsto y = \begin{pmatrix} y_1 \\ y_2 \\ y_3 \end{pmatrix} = \begin{pmatrix} \Pi_{11}(x_1) \oplus \Pi_{12}(x_2) \oplus \Pi_{13}(x_3) \\ \Pi_{21}(x_1) \oplus \Pi_{22}(x_2) \oplus \Pi_{23}(x_3) \\ \Pi_{31}(x_1) \oplus \Pi_{32}(x_2) \oplus \Pi_{33}(x_3) \end{pmatrix} .$$

It is interesting to consider Π as the sum of the following three mappings:

$$\Pi_1(x) = \begin{pmatrix} \Pi_{11}(x_1) \\ \Pi_{22}(x_2) \\ \Pi_{33}(x_3) \end{pmatrix}, \quad \Pi_2(x) = \begin{pmatrix} \Pi_{12}(x_2) \\ \Pi_{23}(x_3) \\ \Pi_{31}(x_1) \end{pmatrix}, \quad \Pi_3(x) = \begin{pmatrix} \Pi_{13}(x_3) \\ \Pi_{21}(x_1) \\ \Pi_{32}(x_2) \end{pmatrix} .$$

The reason for considering these mappings is that their residuals should be obvious since each y_i depends upon a single x_j (or otherwise stated, they are 'diagonal' up to a permutation of 'rows'). For instance,

$$x = \Pi_3^\sharp(y) = \begin{pmatrix} \Pi_{21}^\sharp(y_2) \\ \Pi_{32}^\sharp(y_3) \\ \Pi_{13}^\sharp(y_1) \end{pmatrix} .$$

Then, since $\Pi = \Pi_1 \oplus \Pi_2 \oplus \Pi_3$, by application of (4.35), one obtains

$$\Pi^\sharp(y) = \begin{pmatrix} \Pi_{11}^\sharp(y_1) \wedge \Pi_{21}^\sharp(y_2) \wedge \Pi_{31}^\sharp(y_3) \\ \Pi_{12}^\sharp(y_1) \wedge \Pi_{22}^\sharp(y_2) \wedge \Pi_{32}^\sharp(y_3) \\ \Pi_{13}^\sharp(y_1) \wedge \Pi_{23}^\sharp(y_2) \wedge \Pi_{33}^\sharp(y_3) \end{pmatrix} .$$

Returning to the mapping $L_A : x \mapsto Ax$, we will use the natural notation $A\backslash\!\!\backslash y$ for $L_A^\sharp(y)$. It should be kept in mind that $L_A^\sharp$ is *not* a 'linear' operator in general, that is, it is not expressible as the left product by some matrix. The following lemma is indeed just a corollary of the considerations just made.

Lemma 4.83 *If* $A = (A_{ij}) \in \mathcal{D}^{n\times n}$ *where* $\mathcal{D}$ *is a dioid in which* $\wedge$ *exists, and* $y \in \mathcal{D}^{n\times 1}$, *then*

$$\left(A\backslash\!\!\backslash y\right)_i = \bigwedge_{j=1}^{n} \left(A_{ji}\backslash\!\!\backslash y_j\right) \ . \tag{4.81}$$

Therefore, calculating $A\backslash\!\!\backslash y$ amounts to performing a kind of (left) matrix product of the vector y by the *transpose* of matrix A where multiplication is replaced by (left) division and addition is replaced by lower bound. Recall that $\backslash\!\!\backslash$ is distributive with respect to $\wedge$ as shown by Formula (f.1).

With $A, D \in \mathcal{D}^{m\times n}, B \in \mathcal{D}^{m\times p}, C \in \mathcal{D}^{n\times p}$, it is straightforward to obtain the following more general formulæ for $C = A\backslash\!\!\backslash B$ and $D = B /\!\!/ C$:

$$C_{ij} = \bigwedge_{k=1}^{m} \left(A_{ki}\backslash\!\!\backslash B_{kj}\right) \ , \quad D_{ij} = \bigwedge_{k=1}^{p} \left(B_{ik} /\!\!/ C_{jk}\right) \ . \tag{4.82}$$

We now consider conditions under which there exists a left inverse to $A \in \mathcal{D}^{n\times n}$, that is, an operator B from $\mathcal{D}^n$ to $\mathcal{D}^n$ such that $B \circ A = I$ (here we use I instead of $I_{\mathcal{D}^n}$ to denote identity). If $\mathcal{D}$ is a commutative dioid, and $B \in \mathcal{D}^{n\times n}$ (as A does), Reutenauer and Straubing [118] proved that $BA = I \Leftrightarrow AB = I$. In what follows we do not assume that the operator B can be expressed as the left product by a matrix (see Remark 4.85 below) nor that there exists a right inverse to A.

Lemma 4.84 *Let* $\mathcal{D}$ *be a complete Archimedian dioid and let* A *be an* $n \times n$ *matrix with entries in* $\mathcal{D}$. *A necessary and sufficient condition for the existence of a left inverse operator to* A *is that there is one and only one entry in each row and column of* A *which is different from* ε *and each such an entry has a left inverse.*

Proof Notice first that if $B \circ A = I$, it can be proved that A is injective by using a similar argument as that used in the proof of (4.27). Then (4.27) again shows that $A\backslash\!\!\backslash A = I$. Hence $x = A\backslash\!\!\backslash(Ax), \forall x$. Fix any $i \in \{1, \ldots, n\}$ and set $x_i = \varepsilon$ and $x_j = \top, \forall j \neq i$. Using (4.81) and the conventional matrix product formula, one obtains

$$x_i = \varepsilon = \bigwedge_{k=1}^{n} \left(\frac{\bigoplus_{j\neq i} A_{kj} x_j}{A_{ki}} \right) \ . \tag{4.83}$$

For the lower bound, we can limit ourselves to the indices $k \in K(i)$, where $K(i) = \{k \mid A_{ki} \neq \varepsilon\}$, since $A_{ki} = \varepsilon \Rightarrow \left(A_{ki}\backslash\!\!\backslash y = \top, \forall y\right)$. This subset is nonempty for all i since otherwise the right-hand side of (4.83) would be equal to $\top$ for the reason just given, yielding a contradiction (or equivalently, because A is injective and there is no column of A which is identically equal to ε). For all $k \in K(i)$, we consider

$J(i,k) = \{j \mid j \neq i, A_{kj} \neq \varepsilon\}$. If none of these $J(i,k)$ was empty, we would again reach a contradiction since the right-hand side of (4.83) would again be equal to $\top$.

Therefore we have proved that for every column i, there exists at least one row k such that $A_{ki} \neq \varepsilon$ and all other entries in the same row are equal to ε. Since such a row can obviously be associated with only one index i, there are exactly n rows in A with a single nonzero entry. Hence A contains exactly n nonzero entries, but since it has no column identically zero, each column must also contain exactly one nonzero entry. Therefore, up to a permutation of rows and columns, A is a diagonal matrix. Then, using vectors x which are columns of the identity matrix, it is easy to prove that each diagonal term has a left inverse. ■

This result generalizes similar results by Wedderburn [128] and Rutherford [120] for Boolean matrices (observe that the Boole algebra is a complete Archimedian dioid). Other extensions using different assumptions on the dioid $\mathcal{D}$ are discussed in the notes section.

Remark 4.85 Of course, the mapping $y \mapsto A \backslash\!\!\!\circ\, y$ (denoted $A \backslash\!\!\!\circ\, \cdot$) is a $\wedge$-morphism, but when A has a left inverse, $A \backslash\!\!\!\circ\, \cdot$ is also a $\oplus$-morphism when restricted to the image of A. As a matter of fact,

$$x \oplus y = A \backslash\!\!\!\circ\, (Ax \oplus Ay) \geq A \backslash\!\!\!\circ\, (Ax) \oplus A \backslash\!\!\!\circ\, (Ay) \ ,$$

since $A \backslash\!\!\!\circ\, \cdot$ is isotone. But the last term is also equal to $x \oplus y$, hence equality holds throughout. However, when $\mathcal{D}$ is not commutative, this is not sufficient to associate a matrix with this operator. ■

4.7 DIOIDS OF POLYNOMIALS AND POWER SERIES

4.7.1 Definitions and Properties of Formal Polynomials and Power Series

Starting from a 'scalar' dioid $\mathcal{D}$, we can consider the set of formal polynomials and power series in one or several variables with coefficients in $\mathcal{D}$. If several variables are involved (e.g. z_1 and z_2), we only consider the situation of *commutative* variables (e.g. $z_1 z_2$ and $z_2 z_1$ are considered to be the same object). Exponents k_i of z_i can be taken in $\mathbb{N}$ or in $\mathbb{Z}$: in the latter case, one usually speaks of 'Laurent series'.

Definition 4.86 (Formal power series) *A formal power series in p (commutative) variables with coefficients in $\mathcal{D}$ is a mapping f from $\mathbb{N}^p$ or $\mathbb{Z}^p$ into $\mathcal{D}$: $\forall k = (k_1, \ldots, k_p) \in \mathbb{N}^p$ or $\mathbb{Z}^p$, $f(k)$ represents the coefficient of $z_1^{k_1} \ldots z_p^{k_p}$. Another equivalent representation is*

$$f = \bigoplus_{k \in \mathbb{N}^p \text{ or } \mathbb{Z}^p} f(k_1, \ldots, k_p) z_1^{k_1} \ldots z_p^{k_p} \ . \tag{4.84}$$

> Remember that e.g. $f(3)$ denotes the coefficient of z^3, not the 'numerical' value of the series for $z = 3$. First, this has no meaning if $\mathcal{D}$ is not a dioid of numbers but just an abstract dioid. Second, even if $\mathcal{D}$ is a set of numbers, we are not dealing here with *numerical functions* defined by either polynomials or series, we only deal with formal objects. The relationship between a formal polynomial and its related numerical function was discussed in Chapter 3.

Definition 4.87 (Support, degree, valuation) *The support* $\mathrm{supp}(f)$ *of a series* f *in* p *variables is defined as*

$$\mathrm{supp}(f) = \{k \in \mathbb{Z}^p \mid f(k) \neq \varepsilon\} \ .$$

The degree $\deg(f)$ *(respectively valuation* $\mathrm{val}(f)$*) is the upper bound (respectively lower bound) of* $\mathrm{supp}(f)$ *in the completed lattice* $\overline{\mathbb{Z}}^p$*, where* $\overline{\mathbb{Z}}$ *denotes* $\mathbb{Z} \cup \{-\infty\} \cup \{+\infty\}$.

Example 4.88 For $p = 2$ and $f = z_1 z_2^4 \oplus z_1^2 z_2^3$, $\deg(f) = (2,4)$ and $\mathrm{val}(f) = (1,3)$. ■

Definition 4.89 (Polynomial, monomial) *A polynomial (respectively a monomial) is a series with a finite support (respectively with a support reduced to a singleton).*

The set of formal series is endowed with the following two operations:

$$\left.\begin{array}{lrcl} f \oplus g: & (f \oplus g)(k) & = & f(k) \oplus g(k) \ , \\ f \otimes g: & (f \otimes g)(k) & = & \displaystyle\bigoplus_{i+j=k} f(i) \otimes g(j) \ . \end{array}\right\} \qquad (4.85)$$

These are the conventional definitions of sum and product of power series. The product is nothing other than a 'convolution'. As usual, there is no ambiguity in using the same $\oplus$ symbol in (4.84) and for the sum of series. It is easy to see that the set of series endowed with these two operations is a dioid denoted $\mathcal{D}[\![z_1, \ldots, z_p]\!]$. In particular, its zero element, still denoted ε, is defined by $f(k) = \varepsilon, \forall k$, and its identity element e corresponds to $f(0, \ldots, 0) = e$ and $f(k) = \varepsilon$ otherwise. Most of the time, we will consider exponents $k_i \in \mathbb{Z}$; we will not use a different notation when $k_i \in \mathbb{N}$ but we will state it explicitly when necessary. Notice that when k lies in $\mathbb{Z}^p$, the definition of $f \otimes g$ involves infinite sums: for this definition to make sense, it is then necessary to assume that $\mathcal{D}$ is complete. This is not required for polynomials. The subset of polynomials is a subdioid of $\mathcal{D}[\![z_1, \ldots, z_p]\!]$ denoted $\mathcal{D}[z_1, \ldots, z_p]$.

One has that

$$f \geq g \Leftrightarrow \{f(k) \geq g(k) \ , \quad \forall k\} \ .$$

Of course, $\mathcal{D}[\![z_1, \ldots, z_p]\!]$ is only partially ordered even if $\mathcal{D}$ is a chain. The dioid $\mathcal{D}[\![z_1, \ldots, z_p]\!]$ is commutative if $\mathcal{D}$ is commutative (this holds true because we consider commutative variables only). If $\mathcal{D}$ is complete, $\mathcal{D}[\![z_1, \ldots, z_p]\!]$ is complete, but $\mathcal{D}[z_1, \ldots, z_p]$ is not. Here is a counterexample.

Example 4.90 For $p = 1$, consider the infinite subset of polynomials $\{z^k\}_{k \in \mathbb{N}}$. Their sum is not a polynomial. ■

However, if lower bounds can be defined in $\mathcal{D}$, in particular when $\mathcal{D}$ is complete, these lower bounds extend to $\mathcal{D}[z_1, ..., z_p]$ and $\mathcal{D}[\![z_1, ..., z_p]\!]$ 'coefficientwise'. Distributivity of $\mathcal{D}$ implies distributivity of $\mathcal{D}[\![z_1, ..., z_p]\!]$. But even if $\mathcal{D}$ is Archimedian, $\mathcal{D}[\![z_1, ..., z_p]\!]$ and $\mathcal{D}[z_1, ..., z_p]$ are not necessarily so when exponents are in $\mathbb{N}^p$. Here is a counterexample.

Example 4.91 Let $p = 1$, $f = z$ and $g = e$. Obviously, there is no h such that $fh \geq g$, since z is always a factor of fh, that is, $(fh)(0) = \varepsilon$, which cannot dominate $g(0) = e$. ■

Lemma 4.92 *If $\mathcal{D}$ is Archimedian, $\mathcal{D}[z_1, ..., z_p]$ and $\mathcal{D}[\![z_1, ..., z_p]\!]$ are Archimedian too provided the exponents lie in $\mathbb{Z}^p$.*

Proof Given $f \neq \varepsilon$ and g (Laurent series or polynomials), we must find h such that $fh \geq g$. Since $f \neq \varepsilon$, there exists at least one ℓ such that $f(\ell) \neq \varepsilon$. Let f' denote the corresponding monomial, that is, $f'(\ell) = f(\ell)$ and $f'(k) = \varepsilon$ when $k \neq \ell$. Of course, $f \geq f'$, hence it suffices to find h such that $f'h \geq g$. One has that $(f'h)(k) = f'(\ell)h(k - \ell)$. Since $\mathcal{D}$ is Archimedian, for all k, there exists an a_k such that $f'(\ell)a_k \geq g(k)$. It suffices to set $h(k) = a_{k+\ell}$. Of course, if g is a polynomial, h can be a polynomial too. ■

Lemma 4.93 *We consider* supp(.) *as a mapping from the dioid $\mathcal{D}[\![z_1, ..., z_p]\!]$ into the dioid $\left(2^{\mathbb{Z}^p}, \cup, +\right)$ in which $\wedge$ is $\cap$, and* deg(.) *and* val(.) *as mappings from the dioid $\mathcal{D}[\![z_1, ..., z_p]\!]$ into the dioid $\left(\overline{\mathbb{Z}}, \max, +\right)^p$ in which all operations are componentwise, in particular $\wedge$ is* min *componentwise. Then*

$$\operatorname{supp}(f \oplus g) = \operatorname{supp}(f) \oplus \operatorname{supp}(g) , \tag{4.86}$$
$$\operatorname{supp}(f \wedge g) = \operatorname{supp}(f) \wedge \operatorname{supp}(g) , \tag{4.87}$$
$$\operatorname{supp}(f \otimes g) \leq \operatorname{supp}(f) \otimes \operatorname{supp}(g) , \tag{4.88}$$
$$\deg(f \oplus g) = \deg(f) \oplus \deg(g) , \tag{4.89}$$
$$\deg(f \wedge g) = \deg(f) \wedge \deg(g) , \tag{4.90}$$
$$\deg(f \otimes g) \leq \deg(f) \otimes \deg(g) , \tag{4.91}$$
$$\operatorname{val}(f \oplus g) = \operatorname{val}(f) \wedge \operatorname{val}(g) , \tag{4.92}$$
$$\operatorname{val}(f \wedge g) = \operatorname{val}(f) \oplus \operatorname{val}(g) , \tag{4.93}$$
$$\operatorname{val}(f \otimes g) \geq \operatorname{val}(f) \otimes \operatorname{val}(g) . \tag{4.94}$$

Of course, equalities and inequalities involving the lower bound in $\mathcal{D}[\![z_1, ..., z_p]\!]$ are meaningful only if this lower bound exists. Moreover, all inequalities become equalities if $\mathcal{D}$ is entire, and then supp *and* deg *are homomorphisms, whereas* val *would be a homomorphism if considered as a mapping from $\mathcal{D}[\![z_1, ..., z_p]\!]$ into $\left(\overline{\mathbb{Z}}, \min, +\right)^p$.*

Proof Equation (4.86)—respectively, (4.87)—results from the fact that

$$f(k) \oplus g(k) \neq \varepsilon \Leftrightarrow \{f(k) \neq \varepsilon \text{ or } g(k) \neq \varepsilon\}$$

—respectively,

$$f(k) \wedge g(k) \neq \varepsilon \Leftrightarrow \{f(k) \neq \varepsilon \text{ and } g(k) \neq \varepsilon\} \ .$$

Inequality (4.88) results from the fact that

$$(f \otimes g)(k) \neq \varepsilon \Rightarrow \{\exists i, j : i + j = k \ , \quad f(i) \neq \varepsilon \ , \quad g(j) \neq \varepsilon\} \ .$$

But the converse statement is also true if $\mathcal{D}$ is entire, proving equality in (4.88).

Now, to prove the corresponding statements for deg (respectively, val), it suffices to take the upper bound (respectively, the lower bound) at both sides of (4.86)–(4.88) and to observe that, in the particular case of $\left(\overline{\mathbb{Z}}, \max, +\right)^p$, $\oplus, \otimes, \wedge$ are distributive with respect to one another. ■

Remark 4.94 Since $\oplus$, and therefore $\leq$, operate componentwise for power series, it is clear that $\wedge$ operates also componentwise, as was claimed in Lemma 4.93. However, there is another interesting way of viewing this question. Consider a family $\{f_j\}_{j \in J} \subseteq \mathcal{D}[\![z]\!]$ (we limit ourselves to a single variable z simply to alleviate the notation) and the expression

$$\bigwedge_{j \in J} f_j = \bigwedge_{j \in J} \bigoplus_{k \in \mathbb{Z}} f_j(k) z^k \ .$$

Note that the general formula of distributivity of any abstract operation $\prod$ with respect to some other operation $\coprod$ is

$$\prod_{j \in J} \coprod_{k \in K} a_{jk} = \coprod_{\varphi \in K^J} \prod_{j \in J} a_{j\varphi(j)} \ , \tag{4.95}$$

where K^J is the set of mappings from J into K. Applying this formula to our situation, we obtain

$$\bigwedge_{j \in J} f_j = \bigoplus_{\varphi \in \mathbb{Z}^J} \bigwedge_{j \in J} f_j(\varphi(j)) z^{\varphi(j)} \ .$$

Then, since for any $a, b \in \mathcal{D}$, $az^k \wedge bz^\ell = \varepsilon$ whenever $k \neq \ell$, we can limit ourselves to *constant* mappings φ in the above formula. Therefore, we finally obtain

$$\bigwedge_{j \in J} f_j = \bigoplus_{k \in \mathbb{Z}} \bigwedge_{j \in J} f_j(k) z^k \ , \tag{4.96}$$

which is the expected result. ■

4.7.2 Subtraction and Division of Power Series

Since $\oplus$ operates componentwise, so does $\ominus$ for power series. Let us consider $\diamond\!\!\setminus$ which is more involved since $\otimes$ is a 'convolution'. We again limit ourselves to a single variable z without loss of generality. We also assume that the exponent k ranges in $\mathbb{Z}$ rather than in $\mathbb{N}$. A power series f with exponents in $\mathbb{N}$ is nothing but a series with exponents in $\mathbb{Z}$ for which $f(k) = \varepsilon$ for $k < 0$. However, if one considers $f = z$, for example, it should be clear, from the very definition of $\diamond\!\!\setminus$, that $z \diamond\!\!\setminus e = z^{-1}$ if exponents are allowed to range in $\mathbb{Z}$ and $z \diamond\!\!\setminus e = \varepsilon$ if exponents are restricted to belong to $\mathbb{N}$.

Since we consider $k \in \mathbb{Z}$, recall that $\mathcal{D}$ should be complete.

Lemma 4.95 *Under the foregoing assumptions, for any given f and h in $\mathcal{D}[\![z]\!]$, one has*

$$g \stackrel{\text{def}}{=} \frac{h}{f} = \bigoplus_{k\in\mathbb{Z}} \bigwedge_{\ell\in\mathbb{Z}} \frac{h(\ell)}{f(\ell - k)} z^k \; . \tag{4.97}$$

Proof This is another consequence of the considerations preceding Formula (4.81). If $h = f \otimes g$, then $h(\ell) = \bigoplus_k \Pi_{\ell k}(g(k))$, where $\Pi_{\ell k}(x) = f(\ell - k)x$. Therefore, $g(k) = \bigwedge_\ell \Pi^\sharp_{\ell k}(h(\ell))$, which yields (4.97). ■

Remark 4.96 There is another way to derive (4.97), which makes use of Formula (f.3) of Table 4.1, plus a remark concerning the division by monomial, namely:

$$\frac{\bigoplus_\ell h(\ell) z^\ell}{f(m) z^m} = \bigoplus_\ell \frac{h(\ell)}{f(m)} z^{\ell - m} \; . \tag{4.98}$$

This formula should be obvious, but note that it is stronger than the *inequality* derived from (f.2). Now, to derive (4.97), we have

$$\begin{aligned}
\bigoplus_k g(k) z^k &= \frac{\bigoplus_n h(n) z^n}{\bigoplus_m f(m) z^m} && \text{by definition,}\\
&= \bigwedge_m \frac{\bigoplus_n h(n) z^n}{f(m) z^m} && \text{by (f.3),}\\
&= \bigwedge_m \bigoplus_n \frac{h(n)}{f(m)} z^{n-m} && \text{by (4.98),}\\
&= \bigwedge_m \bigoplus_k \frac{h(m+k)}{f(m)} z^k && \text{by setting } n = m + k,\\
&= \bigoplus_k \bigwedge_m \frac{h(m+k)}{f(m)} z^k && \text{by (4.96),}\\
&= \bigoplus_k \bigwedge_\ell \frac{h(\ell)}{f(\ell - k)} z^k && \text{by setting } m = \ell - k.
\end{aligned}$$

■

4.7.3 Polynomial Matrices

Since $\mathcal{D}[z_1, \ldots, z_p]$ is a dioid, we may consider square $n \times n$ matrices with entries in this dioid: this is the dioid $\left(\mathcal{D}[z_1, \ldots, z_p]\right)^{n\times n}$. Here, we just want to return to the interpretation of such matrices in terms of precedence graphs, and discuss the issue of 'path or circuit of maximum weight' through an example.

Example 4.97 Suppose $\mathcal{D}$ is the dioid of Example 4.12 and let $p = 1$ and $n = 2$. Consider the matrix

$$A = \begin{pmatrix} \varepsilon & e \oplus z \\ 3 \oplus z & e \oplus 2z \end{pmatrix},$$

Figure 4.7 features the weighted graph $\mathcal{G}(A)$. We have

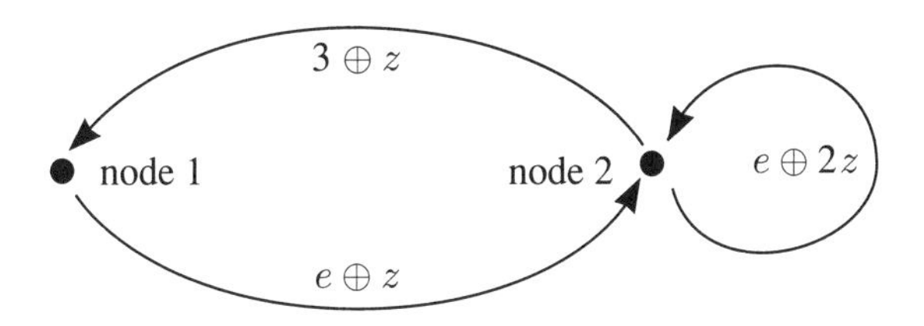

Figure 4.7: A graph representation of a polynomial matrix

$$A^2 = \begin{pmatrix} 3 \oplus 3z \oplus z^2 & e \oplus 2z \oplus 2z^2 \\ 3 \oplus 5z \oplus 2z^2 & 3 \oplus 3z \oplus 4z^2 \end{pmatrix}.$$

The term $(A^2)_{22} = 3 \oplus 3z \oplus 4z^2$ gives the upper bound of weights of circuits of length 2 passing through node 2. But *no* circuit of length 2 corresponds to this weight in Figure 4.7. This is due to the fact that $\mathcal{D}[z_1, ..., z_p]$ is only partially ordered. To figure out what happens, one may adopt the alternative representation shown in Figure 4.8, which amounts to viewing A as being equal to the sum $B \oplus C$

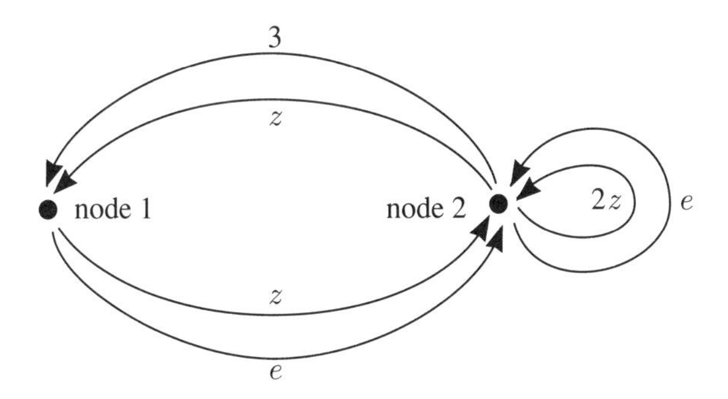

Figure 4.8: Another graph representation of the same matrix

of two matrices with *monomial* entries (according to the rule of parallel composition of graphs explained in §2.3.1—the pair (B, C) is of course not uniquely defined). The advantage is that monomials of the same degree can always be compared. It is then seen that the monomial 3 of $(A^2)_{22}$ is obtained by going from node 2 to node 1 using the arc weighted 3 and coming back using the arc weighted e; the monomial $3z$ of $(A^2)_{22}$ is obtained by going from node 2 to node 1 using the arc weighted 3 and coming back using the arc weighted z; finally, the monomial $4z^2$ is obtained by using the loop weighted $2z$ twice.

Therefore, each entry of A^2 can always be interpreted as the weight of a path or a circuit made up of arcs belonging to either $\mathcal{G}(B)$ or $\mathcal{G}(C)$. ∎

4.8 RATIONAL CLOSURE AND RATIONAL REPRESENTATIONS

The main motivation of this section arises from system theory over dioids, and in particular from realization theory. Therefore, this material forms a bridge to Chapter 5. However, since the results hold true in general dioids, this theory of rational calculus has its natural place in the present chapter.

4.8.1 Rational Closure and Rational Calculus

We consider a complete dioid $\mathcal{D}$ and a subset $\mathcal{T} \subseteq \mathcal{D}$ which contains ε and e. For example, think of $\mathcal{D}$ as the set of formal power series in one or several variables with coefficients in a complete dioid, and of $\mathcal{T}$ as the corresponding subset of polynomials. In general, $\mathcal{T}$ does not need to be a subdioid.

Definition 4.98 (Dioid closure) *The dioid closure of a subset $\mathcal{T}$ of a dioid $\mathcal{D}$, denoted $\mathcal{T}^\diamond$, is the least subdioid of $\mathcal{D}$ containing $\mathcal{T}$.*

This definition is well-posed since the set of subdioids containing $\mathcal{T}$ is nonempty (it contains $\mathcal{D}$ itself) and this set has a minimum element (for the order relation $\subseteq$) since the intersection (lower bound) of a collection of subdioids is a subdioid. The terminology 'closure' is justified because $\mathcal{T}^\diamond \supseteq \mathcal{T}$ and $\mathcal{T}^{\diamond\diamond} = \mathcal{T}^\diamond$. Notice that we do not require $\mathcal{T}^\diamond$ to be complete. It should be clear that $\mathcal{T}^\diamond$ contains, and is indeed reduced to, all elements of $\mathcal{D}$ which can be obtained by *finite* sets of operations $\oplus$ and $\otimes$ involving elements of $\mathcal{T}$ only.

The idea is now to consider 'scalar' equations like (4.77), subsequently called 'affine equations', with data a and b in $\mathcal{T}$ (or, equivalently, in $\mathcal{T}^\diamond$). The least solution a^*b exists in $\mathcal{D}$ since $\mathcal{D}$ is complete, but it does not necessarily belong to $\mathcal{T}^\diamond$ since the star operation involves an infinite sum. Thus, one may produce elements out of $\mathcal{T}^\diamond$ from data in $\mathcal{T}$ or $\mathcal{T}^\diamond$. One can then use these new elements as data of other affine equations, and so on and so forth. The 'rational closure' of $\mathcal{T}$, hereafter defined, is essentially the stable structure that contains all elements one can produce by repeating these operations a *finite* number of times. We shall see that if we consider *matrix*, instead of *scalar*, affine equations (with data in $\mathcal{T}$), but of arbitrary large, albeit finite, dimensions, it is not necessary to repeat the process of using solutions as data for further equations. In the case when $\mathcal{D}$ is a commutative dioid, it is even enough to limit ourselves to weighted sums of solutions to sets of decoupled scalar equations (the weights belonging to $\mathcal{T}$).

Definition 4.99 (Rational closure) *The rational closure of a subset $\mathcal{T}$ of a complete dioid $\mathcal{D}$, denoted $\mathcal{T}^\star$, is the least subdioid of $\mathcal{D}$ containing $\mathcal{T}$ and all finite sums,* *products and star operations over its elements. A subset $\mathcal{T}$ is rationally closed if $\mathcal{T}^\star = \mathcal{T}$.*

This definition is well-posed for the same reason as previously. Moreover, it is clear that $(\mathcal{T}^\diamond)^\star = \mathcal{T}^\star$ and that $\mathcal{T}^{\star\star} = \mathcal{T}^\star$ (hence the terminology 'closure').

If we go from scalars to matrices, we may first consider the subset $\mathcal{T}^{n\times n} \subseteq \mathcal{D}^{n\times n}$ of $n \times n$ matrices with entries in $\mathcal{T}$ and its rational closure $\left(\mathcal{T}^{n\times n}\right)^\star$. This is a subdioid of $\mathcal{D}^{n\times n}$. On the other hand, we may consider the subdioid $(\mathcal{T}^\star)^{n\times n} \subseteq \mathcal{D}^{n\times n}$. We state a first result which will be needed soon in its present form, but which will be improved later on (Theorem 4.104).

Lemma 4.100 *The subdioid $\left(\mathcal{T}^{n\times n}\right)^\star$ is included in the subdioid $(\mathcal{T}^\star)^{n\times n}$.*

The proof is based on the following technical lemma.

Lemma 4.101 *For $a \in \mathcal{D}^{n\times n}$ partitioned into four blocks, namely*

$$a = \begin{pmatrix} a_{11} & a_{12} \\ a_{21} & a_{22} \end{pmatrix}, \tag{4.99}$$

a^* is equal to

$$\begin{pmatrix} a_{11}^* \oplus a_{11}^* a_{12}(a_{21}a_{11}^* a_{12} \oplus a_{22})^* a_{21} a_{11}^* & a_{11}^* a_{12}(a_{21}a_{11}^* a_{12} \oplus a_{22})^* \\ (a_{21}a_{11}^* a_{12} \oplus a_{22})^* a_{21} a_{11}^* & (a_{21}a_{11}^* a_{12} \oplus a_{22})^* \end{pmatrix}. \tag{4.100}$$

Proof We use the fact that a^* is the least solution of equation $x = ax \oplus e$, which yields the system

$$x_{11} = a_{11}x_{11} \oplus a_{12}x_{21} \oplus e \ , \tag{4.101}$$

$$x_{12} = a_{11}x_{12} \oplus a_{12}x_{22} \ , \tag{4.102}$$

$$x_{21} = a_{21}x_{11} \oplus a_{22}x_{21} \ , \tag{4.103}$$

$$x_{22} = a_{21}x_{12} \oplus a_{22}x_{22} \oplus e \ . \tag{4.104}$$

We can solve this system in a progressive manner, using Gaussian elimination. From (4.101) and (4.102), we first calculate

$$x_{11} = a_{11}^*(a_{12}x_{21} \oplus e) \quad \text{and} \quad x_{12} = a_{11}^* a_{12} x_{22} \ ,$$

which we substitute into (4.103)–(4.104). These equations are then solved for x_{21}, x_{22}, and the solutions are placed back in the equations above, yielding the claimed formulæ. Note that placing partial least solutions in other equations preserves the objective of getting overall least solutions since all operations involved are isotone. ■

Another path to solve the system is to first get x_{21} and x_{22} from (4.103)–(4.104), and then to calculate x_{11} and x_{12}. This amounts to interchanging the roles of a_{11} and a_{12} with those of a_{22} and a_{21}, respectively. Identifying the expression of the solution which one gets in this way with the previous expression, the following identity is obtained:

$$(a_{21}a_{11}^* a_{12} \oplus a_{22})^* = a_{22}^* \oplus a_{22}^* a_{21}(a_{12}a_{22}^* a_{21} \oplus a_{11})^* a_{12}a_{22}^* \ . \tag{4.105}$$

Proof of Lemma 4.100 The sums and products of matrices in $\mathcal{T}^{n\times n}$ belong to $(\mathcal{T}^\star)^{n\times n}$. To prove that $\left(\mathcal{T}^{n\times n}\right)^\star$ is included in $(\mathcal{T}^\star)^{n\times n}$, it remains to be proved that one stays in $(\mathcal{T}^\star)^{n\times n}$ when performing star operations over elements of $\mathcal{T}^{n\times n}$. This is done by induction over the dimension n. The statement holds true for $n = 1$. Assuming it holds true up to some $n-1$, let us prove it is also true for n. It suffices to consider a partitioning of an element of $\mathcal{T}^{n\times n}$ into blocks as in (4.99) such that a_{11} is $(n-1)\times(n-1)$-dimensional. By inspection of (4.100), and by using the induction assumption, the proof is easily completed. ∎

4.8.2 Rational Representations

We are going to establish some results on representations of rational elements. Here the connection with realization theory of rational transfer functions should be clear. For reasons that will become more apparent in Chapter 5, we distinguish two particular subsets of $\mathcal{T}$, namely $\mathcal{B}$ and $\mathcal{C}$. There is no special requirement about these subsets except that they both must contain ε and e. Hence, we allow $\mathcal{B}$ and $\mathcal{C}$ to be overlapping and even identical. The extreme cases are $\mathcal{B} = \mathcal{C} = \{\varepsilon, e\}$ and $\mathcal{B} = \mathcal{C} = \mathcal{T}$.

Theorem 4.102 *The rational closure $\mathcal{T}^\star$ coincides with the set of elements x which can be written as*

$$x = c_x A_x^* b_x \ , \tag{4.106}$$

where $A_x \in \mathcal{T}^{n_x\times n_x}$, $b_x \in \mathcal{B}^{n_x\times 1}$ (column vector), and $c_x \in \mathcal{C}^{1\times n_x}$ (row vector). The dimension n_x is finite but may depend on x. For short, a representation of x like in (4.106) will be called a $(\mathcal{B}, \mathcal{C})$-representation.

Proof Let $\mathcal{F}$ be the subset of all elements of $\mathcal{D}$ having a $(\mathcal{B}, \mathcal{C})$-representation. This subset includes $\mathcal{T}$ because of the following identity:

$$x = \begin{pmatrix} e & \varepsilon \end{pmatrix} \begin{pmatrix} \varepsilon & x \\ \varepsilon & \varepsilon \end{pmatrix}^* \begin{pmatrix} \varepsilon \\ e \end{pmatrix} .$$

Suppose that we have already proved that $\mathcal{F}$ is stable by addition, multiplication and star operation, which we postpone to the end of this proof, then $\mathcal{F}$ is of course equal to its rational closure $\mathcal{F}^\star$. Since $\mathcal{F}$ includes $\mathcal{T}$, $\mathcal{F}^\star = \mathcal{F}$ includes $\mathcal{T}^\star$. On the other hand, from Lemma 4.100, A_x^* has its entries in $\mathcal{T}^\star$. From (4.106), it is thus clear that $\mathcal{F}$ is included in $\mathcal{T}^\star$. Finally, we conclude that $\mathcal{F} = \mathcal{F}^\star = \mathcal{T}^\star$.

For the proof to be complete, we have to show that, considering two elements of $\mathcal{F}$, say x and y, which, by definition, have $(\mathcal{B}, \mathcal{C})$-representations, $x \oplus y$, $x \otimes y$

and x^* also have $(\mathcal{B}, \mathcal{C})$-representations. This is a consequence of the following formulæ:

$$c_x A_x^* b_x \oplus c_y A_y^* b_y = \begin{pmatrix} c_x & c_y \end{pmatrix} \begin{pmatrix} A_x & \varepsilon \\ \varepsilon & A_y \end{pmatrix}^* \begin{pmatrix} b_x \\ b_y \end{pmatrix},$$

$$c_x A_x^* b_x \otimes c_y A_y^* b_y = \begin{pmatrix} c_x & \varepsilon & \varepsilon \end{pmatrix} \begin{pmatrix} A_x & b_x & \varepsilon \\ \varepsilon & \varepsilon & c_y \\ \varepsilon & \varepsilon & A_y \end{pmatrix}^* \begin{pmatrix} \varepsilon \\ \varepsilon \\ b_y \end{pmatrix},$$

$$(c_x A_x^* b_x)^* = \begin{pmatrix} \varepsilon & e \end{pmatrix} \begin{pmatrix} A_x & b_x \\ c_x & \varepsilon \end{pmatrix}^* \begin{pmatrix} \varepsilon \\ e \end{pmatrix}.$$

These formulæ can be proved by making repeated use of (4.100). However, the reader already familiar with system theory will have recognized the arithmetics of transfer functions in parallel, series and feedback. ■

Remark 4.103 As already mentioned, $\mathcal{B}$ and $\mathcal{C}$ can be any subsets of $\mathcal{T}$ ranging from $\{\varepsilon, e\}$ to $\mathcal{T}$ itself. Let $\mathbb{B} = \{\varepsilon, e\}$. For a fixed $x \in \mathcal{T}^\star$, and for two pairs $(\mathcal{B}', \mathcal{C}')$ and $(\mathcal{B}, \mathcal{C})$ such that $\mathcal{B}' \subseteq \mathcal{B}$ and $\mathcal{C}' \subseteq \mathcal{C}$, a $(\mathcal{B}', \mathcal{C}')$-representation can also be considered as a $(\mathcal{B}, \mathcal{C})$-representation. Conversely, every $(\mathcal{B}, \mathcal{C})$-representation can yield a $(\mathbb{B}, \mathbb{B})$-representation thanks to the formula (which is again a consequence of (4.100) used repeatedly)

$$cA^*b = \begin{pmatrix} \varepsilon & \varepsilon & e \end{pmatrix} \begin{pmatrix} A & b & \varepsilon \\ \varepsilon & \varepsilon & \varepsilon \\ c & \varepsilon & \varepsilon \end{pmatrix}^* \begin{pmatrix} \varepsilon \\ e \\ \varepsilon \end{pmatrix}.$$

However, we note that the corresponding inner dimension n increases when passing from the $(\mathcal{B}, \mathcal{C})$- to the $(\mathbb{B}, \mathbb{B})$-representation (which is also a $(\mathcal{B}', \mathcal{C}')$-representation).

In fact, this discussion cannot be pursued satisfactorily until one is able to clarify the issue of 'minimal representation', that is, for a given pair $(\mathcal{B}, \mathcal{C})$, and for a given $x \in \mathcal{T}^\star$, a representation yielding the minimal (canonical) value of n_x. This problem is yet unsolved. ■

Theorem 4.104 *The subdioids $\left(\mathcal{T}^{n \times n}\right)^\star$ and $(\mathcal{T}^\star)^{n \times n}$ are identical. Consequently, $(\mathcal{T}^\star)^{n \times n}$ is rationally closed.*

Proof The inclusion in one direction has been stated in Lemma 4.100. Therefore, we need only to prove the reverse inclusion. Let $X \in (\mathcal{T}^\star)^{n \times n}$ and assume that $n = 2$ for the sake of simplicity and without loss of generality. Then X can be written as

$$X = \begin{pmatrix} x_1 & x_2 \\ x_3 & x_4 \end{pmatrix}$$

with entries $x_i \in \mathcal{T}^\star$. Every x_i has a $(\mathcal{B}, \mathcal{C})$-representation consisting of a triple $(A_{x_i}, b_{x_i}, c_{x_i})$, with $A_{x_i} \in \mathcal{T}^{n_i \times n_i}$. Then

$$\begin{aligned} X &= \begin{pmatrix} c_{x_1} A_{x_1}^* b_{x_1} & c_{x_2} A_{x_2}^* b_{x_2} \\ c_{x_3} A_{x_3}^* b_{x_3} & c_{x_4} A_{x_4}^* b_{x_4} \end{pmatrix} \\ &= \begin{pmatrix} c_{x_1} & c_{x_2} & \varepsilon & \varepsilon \\ \varepsilon & \varepsilon & c_{x_3} & c_{x_4} \end{pmatrix} \begin{pmatrix} A_{x_1} & \varepsilon & \varepsilon & \varepsilon \\ \varepsilon & A_{x_2} & \varepsilon & \varepsilon \\ \varepsilon & \varepsilon & A_{x_3} & \varepsilon \\ \varepsilon & \varepsilon & \varepsilon & A_{x_4} \end{pmatrix}^* \begin{pmatrix} b_{x_1} & \varepsilon \\ \varepsilon & b_{x_2} \\ b_{x_3} & \varepsilon \\ \varepsilon & b_{x_4} \end{pmatrix} . \end{aligned}$$

The inner dimension is $\sum_{i=1}^{4} n_i$, but it can be artificially augmented to the next multiple of 2 (and more generally of n) by adding enough rows and columns with entries equal to ε in the matrices. Then, since the outer dimension is 2 and the inner dimension is now a multiple of 2, by appropriately partitioning these matrices in 2×2 blocks, one may consider this representation as a $(\mathcal{B}^{2\times 2}, \mathcal{C}^{2\times 2})$-representation. Application of Theorem 4.102 in the dioid $\mathcal{D}^{2\times 2}$ proves that X belongs to $(\mathcal{T}^{2\times 2})^\star$. ■

4.8.3 Yet Other Rational Representations

So far, we have considered representations of elements of $\mathcal{T}^\star$ by triples (A, b, c), such that the entries of A are taken in $\mathcal{T}$, whereas those of b and c are allowed to lie in subsets $\mathcal{B}$ and $\mathcal{C}$ of $\mathcal{T}$ which are arbitrary, up to the fact that they must contain $\mathbb{B} = \{\varepsilon, e\}$. Recall that $\mathcal{B}$ and $\mathcal{C}$ need to be neither distinct nor disjoint.

As an example to be encountered in Chapter 5, consider again $\mathcal{T}$ as the subset of polynomials of $\mathcal{D}$ which is the dioid of formal power series in one or several variables. Then $\mathcal{B}$ and $\mathcal{C}$ may be subsets of particular polynomials, or they may be reduced to $\mathbb{B}$. Since formal variables are going to be interpreted as 'shift' or 'delay' operators in the system theory setting, it means that no 'dynamics' is allowed in b and c in the latter case, whereas 'some' dynamics is allowed in the former case. In Chapter 5, we are going to consider a two-dimensional domain description involving two shift operators γ and δ in the event, respectively the time, domain. To describe the connection between this two-dimensional description and the more classical one-dimensional description (either in the event or in the time domain), it is necessary to study other rational representations. They correspond to other choices for the subsets in which the entries of A, b, c assume their values.

Let us introduce the following notation. For two subsets $\mathcal{U}$ and $\mathcal{V}$ of $\mathcal{D}$, let

$$\mathcal{U}^\star \otimes \mathcal{V} \stackrel{\text{def}}{=} \left\{ x \;\middle|\; \exists k \in \mathbb{N} : x = \bigoplus_{i=1}^{k} c_i b_i, c_i \in \mathcal{U}^\star, b_i \in \mathcal{V} \right\} .$$

The notation $\mathcal{V} \otimes \mathcal{U}^\star$ is similarly defined. Notice that ε belongs to the subsets so defined.

We now consider a 'covering' $(\mathcal{U}, \mathcal{V})$ of $\mathcal{T}$ (that is, $\mathcal{T} = \mathcal{U} \cup \mathcal{V}$ but $\mathcal{U} \cap \mathcal{V}$ does not need to be empty). We always assume that $\mathbb{B} \subseteq \mathcal{U}$ when considering $\mathcal{U}^\star$.

Theorem 4.105 *The rational closure $\mathcal{T}^\star$ coincides with the set of elements x which can be written as in (4.106), but with entries of A_x lying in $\mathcal{U}^\star \otimes \mathcal{V}$, those of b_x in $\mathcal{U}^\star \otimes \mathcal{B}$ and those of c_x in $\mathcal{C}$ (we call this an observer representation).*

Alternatively, there exist other representations such that the entries of A_x are in $\mathcal{V} \otimes \mathcal{U}^\star$, those of b_x are in $\mathcal{B}$, and those of c_x are in $\mathcal{C} \otimes \mathcal{U}^\star$ (we call these controller representations).

Proof Only the former statement will be proved. The latter can be proved similarly. We first prove that if $x \in \mathcal{T}^\star$, then x does have an observer representation. From Theorem 4.102, we know that x has a $(\mathcal{B}, \mathcal{C})$-representation, say (A, b, c). The matrix A can be written $A_\mathcal{V} \oplus A_\mathcal{U}$ in such a way that $A_\mathcal{V}$ contains only entries which are elements of $\mathcal{V}$, and $A_\mathcal{U}$ only elements of $\mathcal{U}$. If $\mathcal{V} \cap \mathcal{U}$ is nonempty, entries of A which lie in the intersection of those sets may be arbitrarily put either in $A_\mathcal{V}$ or in $A_\mathcal{U}$, or even in both matrices thanks to Axiom 4.9. Therefore, we have $x = c(A_\mathcal{V} \oplus A_\mathcal{U})^* b$. Consider (4.105) with $a_{11} = \varepsilon, a_{12} = e, a_{21} = a$ and $a_{22} = b$. We obtain

$$\begin{aligned}(a \oplus b)^* &= b^* \oplus b^* a (b^* a)^* b^* \ , \\ &= (e \oplus (b^* a)^+) b^* \ ,\end{aligned}$$

hence the identity

$$(a \oplus b)^* = (b^* a)^* b^* \ . \tag{4.107}$$

If we use this with $a = A_\mathcal{V}$ and $b = A_\mathcal{U}$, we obtain

$$x = c\left(A_\mathcal{U}^* A_\mathcal{V}\right)^* A_\mathcal{U}^* b \ ,$$

which is an observer representation.

Conversely, if x has an observer representation (A_x, b_x, c_x), then $x \in \mathcal{T}^\star$. As a matter of fact, it is easy to realize that the entries of A_x, b_x, c_x lie in subsets of $\mathcal{T}^\star$ (in particular, remember that $\mathcal{T}^{\star\star} = \mathcal{T}^\star$). The conclusion follows from Theorem 4.102. ■

Remark 4.106 Another form of (4.107) is obtained by letting $a_{11} = \varepsilon, a_{21} = e, a_{12} = a, a_{22} = b$ in (4.105), which yields

$$(a \oplus b)^* = b^* (a b^*)^* \ . \tag{4.108}$$

■

If we return to our example of $\mathcal{D}$ being the dioid of power series in two variables γ and δ (say, with exponents in $\mathbb{N}$), we may for example assume that $\mathcal{T} = \{\varepsilon, e, \gamma, \delta\}$—the dioid closure of which is the dioid of polynomials in γ, δ—and we may choose $\mathcal{B} = \mathcal{C} = \{\varepsilon, e\}$, $\mathcal{U} = \{\varepsilon, e, \gamma\}$ and $\mathcal{V} = \{\delta\}$. A more explicit interpretation of this situation will be discussed in Chapter 5.

4.8.4 Rational Representations in Commutative Dioids

We have defined 'rational elements' (i.e. elements of $\mathcal{T}^*$) as those elements which can be obtained by a finite number of operations such as sums, products and stars, starting from elements of $\mathcal{T}$. This can also be viewed as the process of obtaining (least) solutions from equations like (4.77), which in turn serve as coefficients of further equations of the same type, this process being repeated a finite number of times, starting with coefficients in $\mathcal{T}$. The results of the previous subsections showed that, indeed, all rational elements can also be obtained by solving equations with coefficients in $\mathcal{T}$ *only once*, but these should be *matrix* equations—or systems of equations—of arbitrary, albeit finite, dimensions.

What we are going to discuss here is the possibility, in the context of commutative dioids (Definition 4.10), of limiting ourselves to linear combinations of solutions of *scalar* equations with coefficients in $\mathcal{T}$, or otherwise stated, of solving only 'decoupled' systems of equations with coefficients in $\mathcal{T}$.

Lemma 4.107 *Let $\mathcal{D}$ be a complete commutative dioid, then*

$$\forall a, b \in \mathcal{D} \ , \quad (a \oplus b)^* = a^* b^* \ . \tag{4.109}$$

Proof One way to prove this is by direct calculations, starting from the very definition of the left-hand side above, and reducing it to the right-hand side using commutativity. Alternatively, one may start from (4.107) (for scalars) and remark that $(a \oplus b)^* = (b^* a)^* b^* = (b^* a^*) b^* = a^* (b^*)^2 = a^* b^*$ when commutativity holds true.

A third, maybe more involved, but interesting, argument is based on considering an equation like (4.54) with $\Pi(x) = ax \oplus xb \oplus c$. With or without commutativity, the least solution $\Pi^*(\varepsilon)$ is easily proved to be equal to $a^* c b^*$. But, with commutativity, the same equation can be written $x = (a \oplus b)x \oplus c$, the least solution of which is $(a \oplus b)^* c$. Setting $c = e$, we obtain the identity (4.109). ∎

With this formula at hand, (4.100) can be given a new useful form, at least when a_{22} is a scalar (i.e. a 1×1 block).

Lemma 4.108 *In a commutative dioid, for a matrix a partitioned into four blocks as in (4.99), where a_{22} is 1×1, and a_{12} and a_{21} are respectively column and row vectors, then a^* is equal to*

$$\begin{pmatrix} a_{11}^* (e \oplus a_{22}^* a_{12} a_{21} (a_{11} \oplus a_{12} a_{21})^*) & a_{22}^* (a_{11} \oplus a_{12} a_{21})^* a_{12} \\ a_{22}^* a_{21} (a_{11} \oplus a_{12} a_{21})^* & a_{22}^* (e \oplus a_{21} (a_{11} \oplus a_{12} a_{21})^* a_{12}) \end{pmatrix} . \tag{4.110}$$

Proof Since a_{22} and $a_{21} a_{11}^* a_{12}$ are scalars, using (4.109), one obtains

$$(a_{21} a_{11}^* a_{12} \oplus a_{22})^* = (a_{21} a_{11}^* a_{12})^* a_{22}^* \ .$$

Moreover, from (4.105) with $a_{22} = \varepsilon$, we find that

$$(a_{21} a_{11}^* a_{12})^* = e \oplus a_{21} (a_{11} \oplus a_{12} a_{21})^* a_{12} \ .$$

Therefore

$$(a_{21}a_{11}^*a_{12} \oplus a_{22})^* = a_{22}^* \oplus a_{22}^*a_{21}(a_{11} \oplus a_{12}a_{21})^*a_{12} \ .$$

These are the lower right-hand blocks of (4.100) and (4.110), respectively.

Consider now the upper right-hand block of (4.100) which is equal (see (4.100)) to the lower right-hand block premultiplied by $a_{11}^*a_{12}$. Using (4.108),

$$\begin{aligned} a_{11}^*a_{12}(a_{21}a_{11}^*a_{12} \oplus u)^* &= a_{22}^*a_{11}^*a_{12}\left(e \oplus a_{21}(a_{11} \oplus a_{12}a_{21})^*a_{12}\right) \\ &= a_{22}^*a_{11}^*a_{12}\left(e \oplus a_{21}a_{11}^*(a_{12}a_{21}a_{11}^*)^*a_{12}\right) \\ &= a_{22}^*a_{11}^*\left(e \oplus (a_{12}a_{21}a_{11}^*)^+\right)a_{12} \\ &= a_{22}^*a_{11}^*(a_{12}a_{21}a_{11}^*)^*a_{12} \\ &= a_{22}^*(a_{11}^* \oplus a_{12}a_{21})^*a_{12} \ . \end{aligned}$$

Similar calculations yield the left-hand blocks of (4.110). ■

Theorem 4.109 *Let $a \in \mathcal{D}^{n\times n}$ where $\mathcal{D}$ is a complete commutative dioid. Then all entries of a^* are finite sums of the form $\bigoplus_i c_i(b_i)^*$, where each c_i is a finite product of entries of a and each b_i is a finite sum of weights of circuits of the precedence graph $\mathcal{G}(a)$.*

Proof The proof is by induction. The statement is true for $n = 1$. Suppose that it also holds true up to dimension $n-1$. Consider the partitioning (4.99) of A with a_{22} scalar. In the graph associated with a, matrix $a_{12}a_{21}$ describes the weights of paths of length 2 which start from one of the first $n-1$ nodes, then go to the n-th node, and finally come back to one of the first $n-1$ nodes. The paths coming back to their initial nodes are circuits of length 2, among other circuits of the graph associated with a. Matrix $a_{12}a_{21}$ can be considered as describing a graph with $n-1$ nodes in which the previous paths or circuits of length 2 can be considered as arcs (i.e. paths of length 1) or loops. As for a_{11}, it describes the subgraph associated with the first $n-1$ nodes. Matrix $a_{11} \oplus a_{12}a_{21}$ corresponds to a graph with the same $n-1$ nodes but with weights calculated as upper bounds of the weights of the two previous graphs. The weights of paths of this graph are among the weights of paths of the graph of a. The induction assumption applies to $(a_{11} \oplus a_{12}a_{21})^*$. The conclusion follows easily by considering the expressions of the four blocks in (4.110) and by remembering that products of stars of scalar elements can be converted to stars of sums of these elements using (4.109). ■

Theorem 4.110 *Let $\mathcal{T}$ be a subset of the complete commutative dioid $\mathcal{D}$. Then, $\mathcal{T}^\star$ coincides with the set of elements x which can be written as*

$$x = \bigoplus_{i=1}^{k_x} c_i(b_i)^* \ , \tag{4.111}$$

where k_x is an arbitrary finite integer and $c_i, b_i \in \mathcal{T}^\diamond$ (the dioid closure of $\mathcal{T}$).

This is a straightforward consequence of Theorems 4.60 and 4.109.

4.9 NOTES

4.9.1 Dioids and Related Structures

Dioids, as defined and studied in this chapter, are members of a larger family of algebraic structures that stem from various fields of mathematics and from several works motivated by a wide range of applications. We shall not attempt to be exhaustive in describing the origins of these theories. The interested may refer e.g. to [66] where some references are given. In all these works, the set of axioms and the terminology are subject to some variations. The notion of 'semiring' has already been defined in Chapter 3. 'Absorbing semiring' is sometimes used when the first operation is supposed to be idempotent (Axiom 4.9), but 'idempotent semiring' would be a more appropriate denomination in this case. As already discussed, this axiom prevents the addition from being cancellative. This is why Gondran and Minoux reject the name 'semiring' which may suggest that the structure can be embedded into that of a ring. Hence they propose the appellation 'dioid' which they attribute to Kuntzmann [80]. In French (or Latin), 'di' is a prefix for 'two' as 'mono' is a prefix for 'one'. A 'dioid' is thus 'twice a monoid'.

As discussed in §4.3.2, Axiom 4.9 is closely related to the introduction of a partial order relation and to a semilattice structure. However, weaker axioms may serve the same purpose. The following axiom is proposed in [66]:

$$\{a = b \oplus c \quad \text{and} \quad b = a \oplus d\} \Rightarrow a = b \ , \tag{4.112}$$

and this axiom is sufficient for stating Theorem 4.28. We retained Axiom 4.9 because all dioids of interest to us naturally satisfy it. An example of a dioid satisfying (4.112) but not Axiom 4.9 is $(\mathbb{R}^+, +, \times)$. However, this example corresponds to a cancellative addition and it is natural to embed this structure in $(\mathbb{R}, +, \times)$, that is, in the conventional algebra.

Helbig [69], who himself refers to Zimmermann [130], defines an 'extremal algebra' with axioms which are very close to but stronger than ours on two points:

- the multiplication is commutative;
- Axiom 4.9 is replaced by the stronger one:

$$x \oplus y = \text{either } x \text{ or } y \ .$$

As stated by Lemma 4.30, the latter axiom corresponds to a total order.

Cuninghame-Green [49] studies structures that we called $\mathbb{R}_{\max}$ and $\mathbb{R}_{\min}$ under the name of 'minimax algebra'. The term 'path algebra' may also be found, owing to the relevance of these particular dioids in graph theory. Reference [34] is about 'incline algebra' which is a structure close to our dioid algebra, but with the following additional axiom:

$$\forall a, b \ , \quad a \oplus ab = a \ , \tag{4.113}$$

which says that $ab \leq a$. This suggests that the multiplication is close to the lower bound (although these two operations may be different), and that every element is less than e (the identity element—although the existence of an identity element is not required a priori). Indeed, Proposition 1.1.1 of [34] states that an incline algebra is exactly a distributive lattice (that is, multiplication and lower bounds are the same) if $a^2 = a$ (that is, the multiplication itself is idempotent). The dioid of Example 4.15 is an incline algebra. The structure $([0, 1], \max, \times)$ is an example of an incline algebra for which multiplication and lower bound do not coincide. Observe that Axiom (4.113) prevents the corresponding dioid from being Archimedian, unless it is isomorphic to the Boole algebra (Example 4.16).

Finally, since an idempotent addition can indirectly be introduced through the introduction of a semilattice or a lattice structure, in the literature on ordered sets, owing to the properties of the second operation (multiplication), the name 'lattice-ordered semigroup' is frequently encountered.

4.9.2 Related Results

Results of §4.3 and §4.4, which are not very specific to dioid theory, are largely based on the corresponding quoted references, with a few variations with respect to terminology (these variations have been indicated) and to presentation.

The main topic of §4.5 is about solving implicit equations like $x = ax \oplus b$ for example. Unlike [67] or Chapter 3 of this book, we only considered the case of complete dioids (in which a^* always exists), which makes the problem of the existence of a solution easier, but at the price of losing uniqueness in general (for example, in an Archimedian dioid, $\top$ is a trivial solution of $x = ax \oplus b$). Theorem 4.76 is an original result, first published in [44] with a slightly different proof. In this same reference, a discussion of the form of the general solution of the homogeneous equation (4.79) can be found.

The problem of invertibility of matrices (§4.6.2) has been considered by several authors, first for Boolean matrices ([128], [120]), then for more general dioids ([23], [34]). Formula (4.81) appears in [49] in the special case of $\mathbb{R}_{\max}$. As for the condition of exact invertibility (see Lemma 4.84 which appears here for the first time), it is similar to that obtained in the above mentioned references, but under quite different assumptions: like [34], reference [23] is more or less in the context of an incline algebra—or at least of an algebra in which every element lies between ε and e—whereas our result deals with Archimedian dioids.

Finally, the rational theory of §4.8, which appeared first in [44], is largely inspired by the use of it we are going to make in Chapter 5 in a system theoretic context.

PART III

Deterministic System Theory

CHAPTER 5

Two-Dimensional Domain Description of Event Graphs

5.1 INTRODUCTION

In Chapter 2 a class of Petri nets called event graphs has been discussed. This class pictorially describes discrete event systems. The dynamics of such systems is essentially driven by synchronization phenomena. In §2.5, it was shown that linear equations can be obtained for event graphs by appealing to some descriptive variables and to some associated dioid algebras.

To be precise, we will call an 'event' any occurrence which is instantaneous, such as the beginning of a transition firing, the end of a transition firing (these two events are simultaneous if transition firings are themselves immediate), the arrival of a token at, or the departure of a token from, a place, etc. In fact, we distinguish 'events', which are unique since they occur only once, from 'types of events' which refer to families of events of the same nature. For example, 'a message pops up on the screen of my computer' is a type of event, whereas 'a message pops up on the screen of my computer at five o'clock' is a particular event of this type. In the context of event graphs, a type of event will very often correspond to the successive firings of a particular transition (we assume that firings have a zero duration).

In the 'dater' description, one essentially deals with variables $d(k)$ associated with types of events such that, for a given type:

- k is an index in $\mathbb{Z}$ which numbers successive events of this type (from an initial, possibly negative, value onwards);
- $d(k)$ is the epoch (or 'date') at which the event numbered k takes place.

The mapping $k \mapsto d(k)$ is called the *dater* associated with the type of event. Because of the meaning of the index k, one may call this an 'event-domain description'. For this description, the appropriate underlying dioid is $\mathbb{R}_{\max}$ in continuous time or $\mathbb{Z}_{\max}$ in discrete time. Using the γ-transform (which is analogous to the z-transform of conventional system theory—see Chapter 1), daters can be represented by formal power series with exponents in $\mathbb{Z}$ and with coefficients in $\mathbb{R}_{\max}$ or $\mathbb{Z}_{\max}$.

In conventional system theory, a 'time-domain' description is rather used. For event graphs, this description involves variables $c(t)$ such that:

- t has the usual meaning of time (either in a continuous or in a discrete domain);

- $c(t)$ is the number[1] of the last event of the considered type which happens before or at time t.

In fact, there is a discrepancy between the definitions of daters and counters. To each k, at least from a certain k_0 (the initial value of the numbering process) to a certain k_1 which can be infinite, corresponds a unique $d(k)$ which is well defined. On the contrary, for any t, it may be that no event takes place at t, a single event happens at t, or several events occur simultaneously at t. Consequently, the definition of $c(t)$ adopted above is just one among several possible definitions. A purpose of this chapter is to discuss two 'canonical' definitions and their relationship with daters. In any case, the mapping $t \mapsto c(t)$, defined over the whole time domain, will be called a *counter*. The appropriate dioid algebra of counters turns out to be $\mathbb{Z}_{\min}$ (see e.g. Example 2.48). In order to enhance the symmetry between counter and dater descriptions, from now on in this chapter, time will be discrete. Then, the δ-transform of $c(\cdot)$ is classically defined as the formal power series $\bigoplus_{t\in\mathbb{Z}} c(t)\delta^t$ with coefficients in $\mathbb{Z}_{\min}$.

In view of what happens in conventional system theory, this dual possibility of describing event graphs by models written down either in the event domain or in the time domain is not usual. This arises because of the fact that trajectories exhibit a monotonic behavior, due to the numbering of events in the order they take place. Roughly speaking, the mappings $k \mapsto d(k)$ and $t \mapsto c(t)$ are inverses of each other. Indeed, to give to this statement a precise meaning, it will be necessary to appeal to residuation theory (see §4.4). Anyway, this inversion is a nonlinear operation. Nevertheless, the dater and counter descriptions are both 'linear', but of course not in the same dioid.

We will discuss the fact that neither description has a definite superiority over the other one. Then, we will study another description, namely in a two-dimensional domain which is the cartesian product of the event and time domains. In this new domain, a description involving formal power series in (γ, δ) will be proposed. Unlike $\mathbb{Z}_{\max}$ and $\mathbb{Z}_{\min}$, the corresponding dioid is no longer totally ordered, and it is not the straightforward product of these two dioids.

Section 5.6 addresses the issue of obtaining equations for 'dual' systems. We assume that desired outputs of an event graph are given and we wish to find the 'best possible' inputs which meet this target, that is, to compute the latest input dates which cause output dates to be less than or equal to the given target. This problem of 'inverting a system' is solved via residuation and the equations so obtained are reminiscent of adjoint- or co-state equations in conventional optimal control.

Section 5.7 discusses the equivalence of three notions related to transfer functions, namely rationality, periodicity and realizability. Finally, §5.8 studies the response of rational systems to some periodic inputs which are shown to be eigenfunctions of rational transfer functions (in the same way as sine functions are eigenfunctions in conventional system theory). The notions of phase shift, amplification gain and Black plots can then be demonstrated for timed event graphs.

[1] In French, 'numéro' rather than 'nombre', the former being a numerical label assigned to each event.

5.2 A COMPARISON BETWEEN COUNTER AND DATER DESCRIPTIONS

We consider the simple example of Figure 5.1 and we compare the equations obtained

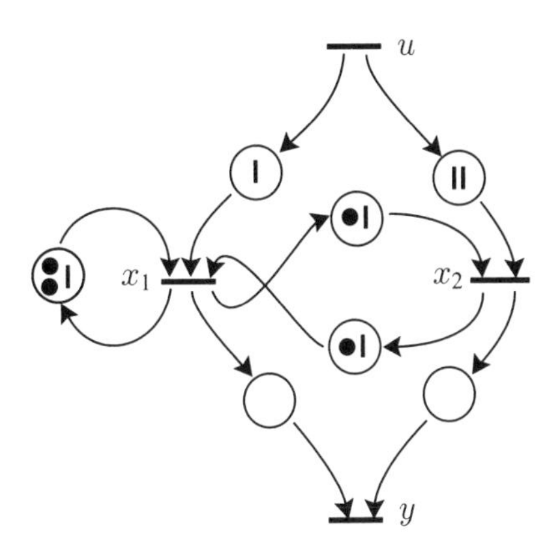

Figure 5.1: An event graph

for daters and counters.

> Bars in places indicate the holding times of these places (in time units). Each transition receives a name (indicated in the figure) and this name is also that of the descriptive variable attached to this transition, be it a dater or a counter. The name of the argument, either k or t, will indicate whether we are dealing with a dater or with a counter description. It should also be remembered that the symbol '$\oplus$' has a different meaning in each context: it stands for the max operation when used in conjunction with daters, and for the min operation in conjunction with counters.

According to §5.1, we consider that, e.g., $x(t)$ is the number of the last firing of transition x occurring before or at time t. The numbering of firings starts with 1, say, for all transitions. For the event graph of Figure 5.1, the following equations are then obtained (we do not discuss the issue of initial conditions at this moment—see §5.4.4.1, page 245).

Dater equations:

$$\begin{aligned} &x_1(k) = 1x_1(k-2) \oplus 1x_2(k-1) \oplus 1u(k) \ ; \\ &x_2(k) = 1x_1(k-1) \oplus 2u(k) \ ; \\ &y(k) = x_1(k) \oplus x_2(k) \ . \end{aligned} \tag{5.1}$$

Counter equations:

$$\begin{aligned} &x_1(t) = 2x_1(t-1) \oplus 1x_2(t-1) \oplus u(t-1) \ ; \\ &x_2(t) = 1x_1(t-1) \oplus u(t-2) \ ; \\ &y(t) = x_1(t) \oplus x_2(t) \ . \end{aligned} \tag{5.2}$$

Using the former representation, we derive

$$\begin{aligned}
y(k) &= x_1(k) \oplus x_2(k) \\
&= 1(x_1(k-1) \oplus x_1(k-2)) \oplus 1x_2(k-1) \oplus (1 \oplus 2)u(k) \\
&= 1x_1(k-1) \oplus 1x_2(k-1) \oplus 2u(k) \\
&= 1y(k-1) \oplus 2u(k) \ .
\end{aligned}$$

Thus a first order input-output relation has been obtained. It should be noticed that we have used two different rules in our simplifications. On the one hand, $2 \oplus 1 = 2$ because we are working with the dioid $\mathbb{Z}_{\max}$. On the other hand, we have used that $x_1(k-1) \oplus x_1(k-2) = x_1(k-1)$ because we are interested only in trajectories of x_1 which are nondecreasing functions of k.

Remark 5.1 The nondecreasingness is not an intrinsic property of solutions of (5.1). For example, if $u(k) = \varepsilon$ for $k < 0$ and $u(k) = e(=0)$ for $k \geq 0$ (such inputs will be interpreted as 'impulses' in §5.4.4.1), then one can check that

$$\forall k \in \mathbb{Z} \ , \quad \begin{pmatrix} x_1(k) & x_2(k) \end{pmatrix}' = \begin{cases} \begin{pmatrix} k+1 & k+3 \end{pmatrix}' & \text{if } k \text{ even;} \\ \begin{pmatrix} k+3 & k+1 \end{pmatrix}' & \text{if } k \text{ odd,} \end{cases}$$

is a *nonmonotonic* solution to (5.1). ■

In terms of γ-transforms, the preceding simplification rules can be summarized as follows:

$$t\gamma^\ell \oplus \tau\gamma^\ell = \max(t, \tau)\gamma^\ell \ ; \qquad t\gamma^\ell \oplus t\gamma^m = t\gamma^{\min(\ell,m)} \ . \tag{5.3}$$

In terms of event graphs, this corresponds to the graph reductions displayed in Figure 5.2.

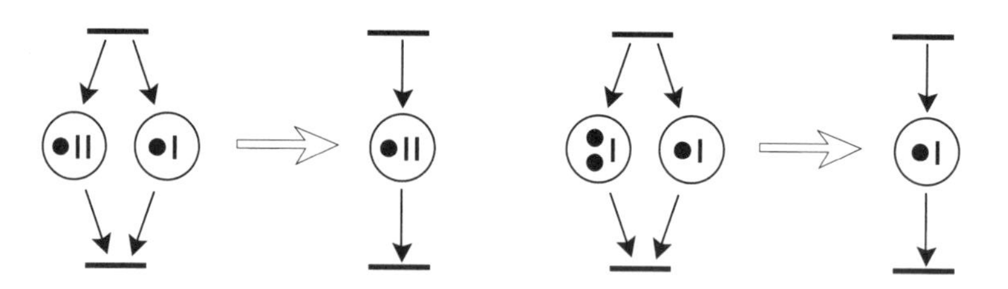

Figure 5.2: Two rules for graph reduction

Remark 5.2 Since we left apart the issue of initial conditions, one should be aware of the fact that the reduction shown on the right-hand side of Figure 5.2 is only valid for certain initial conditions (in particular, it holds true for canonical initial conditions discussed at §5.4.4.1, page 245). ■

Now, using the counter representation, we derive

$$\begin{aligned} y(t) &= x_1(t) \oplus x_2(t) \\ &= (2 \oplus 1)x_1(t-1) \oplus 1x_2(t-1) \oplus u(t-1) \oplus u(t-2) \\ &= 1(x_1(t-1) \oplus x_2(t-1)) \oplus u(t-2) \\ &= 1y(t-1) \oplus u(t-2) \ . \end{aligned}$$

We have used that $1 \oplus 2 = 1$ in $\mathbb{Z}_{\min}$, and that $u(t-1) \oplus u(t-2) = u(t-2)$ because u is a nondecreasing function of t. In terms of δ-transforms, these rules can be summarized by

$$k\delta^\tau \oplus \ell\delta^\tau = \min(k,\ell)\delta^\tau \ ; \qquad k\delta^\tau \oplus k\delta^\theta = k\delta^{\max(\tau,\theta)} \ . \tag{5.4}$$

These rules are similar to those of (5.3) but the roles of the exponents and coefficients are, roughly speaking, interchanged. In terms of event graphs, the rules (5.4) also express the graph reductions of Figure 5.2 (in reverse order).

The above example also shows that in both approaches we reach a kind of 'ARMA' (Auto-Regressive-Moving-Average) equation which, in this specific case, involves the same delay in the AR part in both representations, but different delays in the MA part. Consequently, we would need state vectors of different dimensions in both cases to convert this ARMA equation into standard state space equations (with only unit delays on the right-hand side). Otherwise stated, the same physical system appears to be of a different order in the dater and in the counter descriptions.

These discrepancies and dissymmetries are not very satisfactory and we could further accumulate remarks in the same vein. Let us just mention another intriguing fact. Figure 5.3 represents an event graph before and after the firing of the transition named x_1 or ξ_1. The following equations are obtained for the dater description before and after firing.

Before firing	**After firing**
$x_1(k) = 1x_1(k-1) \oplus x_2(k-1)$,	$\xi_1(k) = 1\xi_1(k-1) \oplus \xi_2(k)$,
$x_2(k) = x_1(k) \oplus u(k)$,	$\xi_2(k) = \xi_1(k-1) \oplus u(k)$,
$y(k) = x_2(k)$,	$y(k) = \xi_2(k)$.

Some substitutions yield the following equivalent descriptions:

Before firing

$$\begin{pmatrix} x_1(k) \\ x_2(k) \end{pmatrix} = \begin{pmatrix} 1 & e \\ 1 & e \end{pmatrix} \begin{pmatrix} x_1(k-1) \\ x_2(k-1) \end{pmatrix} \oplus \begin{pmatrix} \varepsilon \\ e \end{pmatrix} u(k) \ ,$$

$$y(k) = \begin{pmatrix} \varepsilon & e \end{pmatrix} \begin{pmatrix} x_1(k) \\ x_2(k) \end{pmatrix},$$

After firing

$$\begin{pmatrix} \xi_1(k) \\ \xi_2(k) \end{pmatrix} = \begin{pmatrix} 1 & \varepsilon \\ e & \varepsilon \end{pmatrix} \begin{pmatrix} \xi_1(k-1) \\ \xi_2(k-1) \end{pmatrix} \oplus \begin{pmatrix} e \\ e \end{pmatrix} u(k) \ ,$$

$$y(k) = \begin{pmatrix} \varepsilon & e \end{pmatrix} \begin{pmatrix} \xi_1(k) \\ \xi_2(k) \end{pmatrix}.$$

These are two state space realizations of the same γ-transfer function (which can be proved to be equal to $e \oplus \gamma(1\gamma)^*$ provided that all possible simplification rules be used). In matrix notation, we have

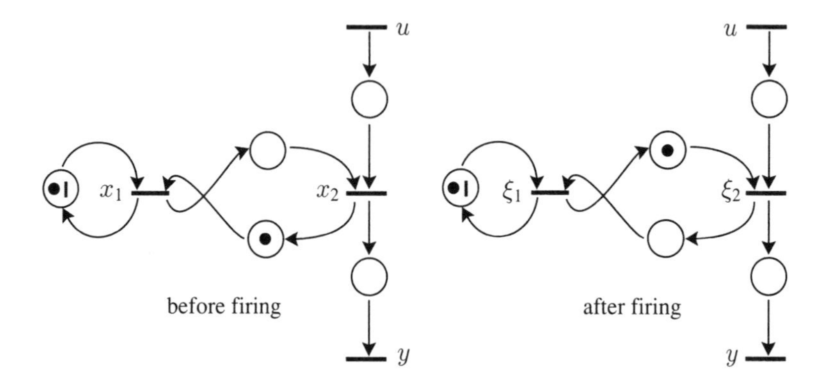

Figure 5.3: Firing a transition

$$
\begin{aligned}
x(k) &= Ax(k-1) \oplus Bu(k) \ , \qquad & y(k) &= Cx(k) \ , \\
\xi(k) &= \overline{A}\xi(k-1) \oplus \overline{B}u(k) \ , \qquad & y(k) &= \overline{C}\xi(k) \ .
\end{aligned}
$$

But one cannot find a linear coordinate transformation to pass from one realization to the other. As a matter of fact, this would require that an invertible 2×2 matrix T exists such that $x = T\xi$, implying for example that

$$
B = T\overline{B} \ , \quad \text{i.e.} \quad \begin{pmatrix} \varepsilon \\ e \end{pmatrix} = \begin{pmatrix} T_{11} & T_{12} \\ T_{21} & T_{22} \end{pmatrix} \begin{pmatrix} e \\ e \end{pmatrix} .
$$

The first row of this matrix relation implies that $T_{11} \oplus T_{12} = \varepsilon$, hence $T_{11} = T_{12} = \varepsilon$, which is not compatible with the fact that T is invertible.

Indeed, from the physical interpretation of this situation (remember that an internal transition fired once), or directly from the equations, it is apparent that the true relationship between ξ and x is $\xi_2(k) = x_2(k); \xi_1(k) = x_1(k-1)$. However, this cannot be captured by a (static) linear change of basis in the state space. Because, in the counter description, coefficients and delays are, roughly speaking, exchanged, this issue of finding a linear change of basis in the state space *can* be solved positively when moving to the counter description for the same example. In this description, entries of matrices correspond to numbers of tokens in the initial marking. Firing an internal transition removes one token from each upstream place: this subtracts 1—in conventional algebra—from each entry of the row of matrix A corresponding to that transition, and it does the same on the corresponding row of B. Similarly, the same transition firing adds one token to each downstream place: algebraically, this adds 1 to each entry of the corresponding column of A and C. These operations can be realized in $\mathbb{Z}_{\min}$ by pre-, respectively post-multiplication by appropriate matrices which are inverse of each other. For the above example, the pre-multiplication involves the matrix:

$$\begin{pmatrix} -1 & \varepsilon \\ \varepsilon & e \end{pmatrix} .$$

We let the reader works out this example completely in the counter description and check this claim.

Remark 5.3 From the above example, one should not conclude any superiority of the counter over the dater description. When it is possible, consider removing one bar from all places upstream of a given *internal* transition, and adding one bar to all downstream places: this leaves the input-output relation unchanged, and this is indeed the dual situation of firing a transition (which moves tokens instead of bars). Therefore, playing with bars instead of tokens will correspond to a change of basis in the dater description, but not in the counter description. ■

In the next sections we move smoothly to a two-dimensional description in which, roughly speaking, monomials such as $t\gamma^k$ in γ-transforms and $k\delta^t$ in δ-transforms will be represented by monomials of the form $\gamma^k\delta^t$; the basic objects will be power series in (γ, δ) with Boolean coefficients; and, in addition to the conventional sum and product of series, we will be allowed to use the rules

$$\gamma^k\delta^t \oplus \gamma^\ell\delta^t = \gamma^{\min(k,\ell)}\delta^t \; , \qquad \gamma^k\delta^t \oplus \gamma^k\delta^\tau = \gamma^k\delta^{\max(t,\tau)} \; , \tag{5.5}$$

which are just the synthesis of (5.3) and (5.4). However, this requires an algebraic foundation which appeals to isomorphisms, quotients of dioids by congruences and residuation theory. As an introduction to these algebraic manipulations, the next section discusses how the set of daters can be embedded into a more general set of nonmonotonic functions from $\mathbb{Z}$ to $\overline{\mathbb{Z}}_{\max}$.

5.3 DATERS AND THEIR EMBEDDING IN NONMONOTONIC FUNCTIONS

5.3.1 A Dioid of Nondecreasing Mappings

Recall that with each type of event is associated a numbering mechanism which assigns a number to each event of this type in the order of these events taking place, starting from an initial finite (but possibly negative) value $k_0 \in \mathbb{Z}$. We also consider a special type of event which corresponds to ticks of an absolute clock. These ticks are also numbered in increasing order, starting from an initial value $t_0 \in \mathbb{Z}$ (the origin of time). At each event of a given type, the current clock value is instantaneously read and the pair (event number, clock value) is saved. The dater associated with a type of event is just the mapping from $\mathbb{Z}$ into $\mathbb{Z}$ the graph of which is the set of all such pairs.

Obviously, daters are nondecreasing functions, but they may not be strictly increasing since several events of the same type may occur simultaneously. We use

d as a generic notation for a dater. Strictly speaking, the function d is defined over an interval of $\mathbb{Z}$ possibly extending to infinity to the right (if events of the same type occur infinitely often), and, wherever it is defined, d assumes finite, but a priori unbounded values in $\mathbb{Z}$. Indeed, in order to extend the definition of d to the whole domain $\mathbb{Z}$, it is convenient to assume that the range set is $\overline{\mathbb{Z}} \stackrel{\text{def}}{=} \mathbb{Z} \cup \{-\infty\} \cup \{+\infty\}$. The convention is that

$$d(k) = \begin{cases} -\infty & \text{if } k < k_0 \text{ (the initial value of the numbering);} \\ +\infty & \text{if the } k\text{-th event of the considered type never} \\ & \text{took place;} \\ \text{any finite value} & \text{otherwise.} \end{cases}$$

From a mathematical point of view, it may sometimes be useful to see daters as mappings from a complete dioid into a complete dioid. For this reason, we may extend the domain of d by setting

$$d(-\infty) = -\infty \quad \text{and} \quad d(+\infty) = +\infty \ . \tag{5.6}$$

Obviously, these end-point conditions are always compatible with the nondecreasingness property of d.

As already discussed, the natural algebra for the range space of d is $\overline{\mathbb{Z}}_{\max}$, that is, $(\overline{\mathbb{Z}}, \max, +)$. It should be remembered that, in $\overline{\mathbb{Z}}_{\max}$,

$$(-\infty) + (+\infty) = \varepsilon \otimes \top = \varepsilon = -\infty \tag{5.7}$$

according to Axiom 4.7. As for the domain space of d, the algebraic structure we need consists of the conventional order relation of $\overline{\mathbb{Z}}$ (this is necessary in order to speak of the nondecreasingness property of d), and the conventional addition (which will be needed for defining the product of daters). At this stage, it is immaterial to decide whether the domain will be called $\overline{\mathbb{Z}}_{\min}$ or $\overline{\mathbb{Z}}_{\max}$. Indeed, if we adopt the former option, the only consequence is that we should speak of 'nonincreasing', rather than 'nondecreasing' functions d with regard to the order relations implied by the dioid structures in the domain and in the range. There is however a more important criterion to decide which name is to be given to the domain of daters. In this dioid, do we wish that $+\infty - \infty = +\infty \otimes (-\infty) = -\infty$ or $+\infty$? This question involves $+$, i.e. $\otimes$, rather than $\oplus$ which is related to the order relation. We leave the answer open until Remark 5.4 below.

The next stage is to endow the set of daters with a dioid structure which already appeared to be appropriate for our purpose. Namely,

- addition is just the conventional pointwise maximum, or otherwise stated

$$\forall k \in \overline{\mathbb{Z}} \ , \quad (d_1 \oplus d_2)(k) = d_1(k) \oplus d_2(k) \ ,$$

  in which the symbol '$\oplus$' on the left-hand side denotes the addition of daters, whereas it denotes addition in the range dioid $\overline{\mathbb{Z}}_{\max}$ on the right-hand side; this definition is extended to infinite sums without difficulty since the range is a complete dioid;

- multiplication is the conventional 'sup-convolution', that is, for all $k \in \overline{\mathbb{Z}}$,

$$(d_1 \otimes d_2)(k) = \bigoplus_{\ell \in \overline{\mathbb{Z}}} \big(d_1(\ell) \otimes d_2(k-\ell)\big) = \sup_{\ell \in \overline{\mathbb{Z}}} \big(d_1(\ell) + d_2(k-\ell)\big) \ .$$

Remark 5.4 The above formula can be written

$$(d_1 \otimes d_2)(k) =$$
$$\sup\left(d_1(-\infty) + d_2(k+\infty), d_1(+\infty) + d_2(k-\infty), \sup_{\ell \in \mathbb{Z}} \big(d_1(\ell) + d_2(k-\ell)\big)\right) .$$

Using (5.6) and (5.7), it can be proved by inspection that

- for finite k,
$$(d_1 \otimes d_2)(k) = \sup_{\ell \in \mathbb{Z}} \big(d_1(\ell) + d_2(k-\ell)\big) \ ,$$
that is, the result is the same whether we consider that the domain is $\mathbb{Z}$ or $\overline{\mathbb{Z}}$;
- for $k = -\infty$, we obtain $(d_1 \otimes d_2)(-\infty) = -\infty$, whatever we decide upon the value to be given to $+\infty - \infty$ in the domain of ℓ (event domain);
- for $k = +\infty$, one has that
$$(d_1 \otimes d_2)(+\infty) = \sup\left(-\infty, +\infty + d_2(+\infty - \infty), \sup_{\ell \in \mathbb{Z}} \big(d_1(\ell) + \infty\big)\right) .$$
For the class of functions satisfying (5.6) to be closed by multiplication (it is obviously closed by addition), we want to ensure that $(d_1 \otimes d_2)(+\infty) = +\infty$, even if $d_1(\ell) = -\infty, \forall \ell < +\infty$. Then, we *must* decide that
$$+\infty - \infty = +\infty \text{ in the event domain.} \tag{5.8}$$

In conclusion,

- we should consider that the event domain is $\overline{\mathbb{Z}}_{\min}$ rather than $\overline{\mathbb{Z}}_{\max}$ (however, we will keep on speaking of 'nondecreasing' functions);
- we also observed that one may first consider that addition and multiplication operate on functions from $\mathbb{Z}$ (instead of $\overline{\mathbb{Z}}$) into $\overline{\mathbb{Z}}_{\max}$, and then complete the results of these operations by the end-point conditions (5.6). ∎

We summarize this subsection with the following definition.

Definition 5.5 (Daters) *Daters are nondecreasing mappings from $\overline{\mathbb{Z}}_{\min}$ into $\overline{\mathbb{Z}}_{\max}$ obeying the end-point conditions (5.6) ('nondecreasing' refers to the conventional order of $\overline{\mathbb{Z}}$ in both the domain and the range). The set of daters is endowed with the pointwise maximum of functions as the addition, and with the sup-convolution as the multiplication.*

One can check that the zero and identity elements of the dioid of daters are respectively:

$$\varepsilon(k) = \begin{cases} -\infty & \text{if } k < +\infty \ ; \\ +\infty & \text{otherwise;} \end{cases} \qquad e(k) = \begin{cases} -\infty & \text{if } k < 0 \ ; \\ 0 & \text{if } 0 \le k < +\infty \ ; \\ +\infty & \text{otherwise.} \end{cases} \tag{5.9}$$

5.3.2 γ-Transforms of Daters and Representation by Power Series in γ

5.3.2.1 *Power Series in γ and the Nondecreasingness Property*

A convenient way to manipulate daters is to encode them using their γ-transforms. This yields formal power series with coefficients in $\overline{\mathbb{Z}}_{\max}$. As for exponents, owing to the last observation in Remark 5.4, we may restrict them to belong to $\mathbb{Z}$. For a dater d, D will denote its γ-transform and we have

$$D = \bigoplus_{k \in \mathbb{Z}} d(k)\gamma^k \ .$$

As is usual, if some monomial γ^k is missing in the explicit expression of some D, this just means that the corresponding coefficient is 'zero', that is, it is equal to ε.

If the set of γ-transforms of daters is endowed with the addition and multiplication introduced in Chapter 4 (see (4.85)), then daters and their γ-transforms constitute two, isomorphic dioids. The latter will be denoted $\mathcal{D}[\![\gamma]\!]$. In $\mathcal{D}[\![\gamma]\!]$, the zero element can be denoted simply ε because, owing to (5.9), it is the zero series with all coefficients equal to $\varepsilon = -\infty$. As for the identity element, it is the γ-transform of $\acute{e}$ given in (5.9), and this is $\gamma^* = \gamma^0 \oplus \gamma \oplus \gamma^2 \oplus \cdots$.

Remark 5.6 Observe that the interpretation of γ is that of the 'backward shift operator in numbering' (or 'in the event domain') since the series γD corresponds to the γ-transform of the dater $k \mapsto d(k-1)$. The expression 'backward shift' is traditional in system theory as is the name 'forward shift' for the operator z (see [72]). However, this appellation is somewhat misleading since it should be realized that, if we plot the graphs of $k \mapsto d(k)$ and $k \mapsto d(k-1)$, then the latter is shifted *to the right* with respect to the former. ■

Note that γ itself, viewed as a formal power series which has all its coefficients equal to ε except that of γ^1 which is equal to e, may be considered as the γ-transform of the function

$$k \mapsto \gamma(k) \stackrel{\text{def}}{=} \begin{cases} e & \text{if } k = 1 \ ; \\ \varepsilon & \text{otherwise.} \end{cases} \tag{5.10}$$

Shifting a dater may be considered as achieving its sup-convolution with γ; with γ-transforms, this operation amounts to 'multiplying by γ'. Of course, the function γ itself is not a dater since it is not monotonic, hence $\gamma \notin \mathcal{D}[\![\gamma]\!]$. Therefore, to give a meaning to this 'multiplication by γ', we must embed elements of $\mathcal{D}[\![\gamma]\!]$ into a larger set, namely the set of (general) formal power series with coefficients in $\overline{\mathbb{Z}}_{\max}$ and exponents in $\mathbb{Z}$. According to §4.7.1, once endowed with the same operations as $\mathcal{D}[\![\gamma]\!]$ (see (4.85)), this set is a complete commutative distributive Archimedian dioid denoted $\overline{\mathbb{Z}}_{\max}[\![\gamma]\!]$.

The zero element ε of $\overline{\mathbb{Z}}_{\max}[\![\gamma]\!]$ is again the zero series (all coefficients equal to ε), but the identity element e of $\overline{\mathbb{Z}}_{\max}[\![\gamma]\!]$ is the series which has only one coefficient different from ε, namely that of γ^0 which is equal to $e = 0$ (in $\overline{\mathbb{Z}}_{\max}$). It is realized that this $e = \gamma^0$ of $\overline{\mathbb{Z}}_{\max}[\![\gamma]\!]$ is not formally equal to γ^* which is the identity element in the dioid $\mathcal{D}[\![\gamma]\!]$. Hence $\mathcal{D}[\![\gamma]\!]$ is *not* a subdioid of $\overline{\mathbb{Z}}_{\max}[\![\gamma]\!]$.

Actually, this situation is pretty much related to that considered in Theorem 4.78. To show this, let us first observe that the property of a function $f : \mathbb{Z} \to \overline{\mathbb{Z}}_{\max}$ to be nondecreasing can be characterized by

$$\forall k \in \mathbb{Z} \ , \quad f(k) \geq f(k-1) \ .$$

In terms of the γ-transform $F \in \overline{\mathbb{Z}}_{\max}[\![\gamma]\!]$, this translates into

$$f \text{ nondecreasing } \Leftrightarrow F \geq \gamma F \ . \tag{5.11}$$

This should be compared with (4.80) which provides other characterizations of nondecreasing functions.

Remark 5.7 If we let k range in $\overline{\mathbb{Z}}$ instead of $\mathbb{Z}$, without imposing the end-point conditions (5.6), then (5.11) is no longer a characterization of nondecreasing functions. For instance, consider the function f such that $f(-\infty) = 2$ and $f(k) = 1$ for $k > -\infty$: it satisfies $f(k) \geq f(k-1)$, and thus also (5.11), although it is *not* nondecreasing over $\overline{\mathbb{Z}}$. If (5.11) cannot be retained as a characterization of nondecreasing functions, then it is not clear how to solve in a simple way the best approximation problems addressed below. ■

It is thus realized that, as a subset of elements of $\overline{\mathbb{Z}}_{\max}[\![\gamma]\!]$ meeting condition (5.11), $\mathcal{D}[\![\gamma]\!]$ is nothing but what we have denoted $\gamma^*\overline{\mathbb{Z}}_{\max}[\![\gamma]\!]$ in §4.5.4. The following theorem is just a rephrasing of Theorem 4.78 in the present context.

Theorem 5.8 *Let* I *denote the canonical injection from* $\mathcal{D}[\![\gamma]\!]$ *into* $\overline{\mathbb{Z}}_{\max}[\![\gamma]\!]$, *and consider some* $F \in \overline{\mathbb{Z}}_{\max}[\![\gamma]\!]$.

1. *The greatest element* $\breve{F}$ *in* $\mathrm{I}\left(\mathcal{D}[\![\gamma]\!]\right)$ *which is less than or equal to* F *is given by*

$$\breve{F} = \gamma^* \backslash F = F \wedge \gamma^{-1} F \wedge \gamma^{-2} F \wedge \cdots \ . \tag{5.12}$$

   In the equivalence class of elements F *of* $\overline{\mathbb{Z}}_{\max}[\![\gamma]\!]$ *which have the same 'best approximation from below'* $\breve{F}$, *this* $\breve{F}$ *is the unique element which belongs to* $\mathrm{I}\left(\mathcal{D}[\![\gamma]\!]\right)$ *and it is also the minimum representative in the equivalence class.*

2. *The least element* $\widehat{F}$ *in* $\mathrm{I}\left(\mathcal{D}[\![\gamma]\!]\right)$ *which is greater than or equal to* F *is given by*

$$\widehat{F} = \gamma^* F \ . \tag{5.13}$$

   In the equivalence class of elements F *of* $\overline{\mathbb{Z}}_{\max}[\![\gamma]\!]$ *which have the same 'best approximation from above'* $\widehat{F}$, *this* $\widehat{F}$ *is the unique element which belongs to* $\mathrm{I}\left(\mathcal{D}[\![\gamma]\!]\right)$ *and it is also the maximum representative in the equivalence class.*

Corollary 5.9 *The greatest dater $\check{f}$ which is less than or equal to a given (not necessarily monotonic) mapping f from the event domain into $\overline{\mathbb{Z}}_{\max}$ is obtained by the formula*

$$\forall k \in \mathbb{Z} \ , \quad \check{f}(k) = \bigwedge_{\ell \geq k} f(\ell) = \inf_{\ell \geq k} f(\ell) \ . \tag{5.14}$$

The least dater $\widehat{f}$ which is greater than or equal to f is obtained by

$$\forall k \in \mathbb{Z} \ , \quad \widehat{f}(k) = \bigoplus_{\ell \leq k} f(\ell) = \sup_{\ell \leq k} f(\ell) \ . \tag{5.15}$$

Of course, these formulæ should be completed by the end-point conditions (5.6).

Proof The formulæ (5.14) and (5.15) are straightforward consequences of (5.12) and (5.13). ■

The mapping $\mathrm{I}^\sharp$ which associates with $F \in \overline{\mathbb{Z}}_{\max}[\![\gamma]\!]$ its best approximation from below in $\mathcal{D}[\![\gamma]\!]$ is u.s.c., but it is neither a $\oplus$- nor a $\otimes$-morphism. On the contrary, the mapping $\mathrm{I}^\flat$ which selects the best approximation from above is a l.s.c. surjective dioid homomorphism. This is why in what follows we concentrate on this type of approximation.

Remark 5.10 Because of (5.13) and of Lemma 4.77, statement 2, it should be clear that $\mathcal{D}[\![\gamma]\!]$ is a multiplicative ideal and that

$$\forall F, G \in \overline{\mathbb{Z}}_{\max}[\![\gamma]\!] \ , \quad \widehat{F \otimes G} = \widehat{F} \otimes \widehat{G} = F \otimes \widehat{G} = \widehat{F} \otimes G \ .$$

■

Figure 5.4 explains how to geometrically construct the graph of the mapping

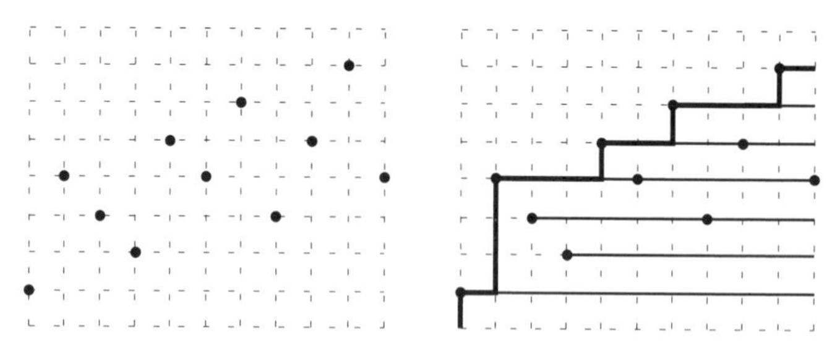

Figure 5.4: Featuring the construction of $\gamma^* F$

$k \mapsto \widehat{f}(k)$ associated with the power series $\widehat{F}$ for a given F (represented by the graph of $k \mapsto f(k)$): to each point of this discrete graph is attached a 'horizontal half line' extending to the right (corresponding to the multiplication by γ^*) and then, he graph of $\widehat{f}$ is obtained as the 'upper hull' of this set of half lines. Of course, when we speak of 'lines', only the trace of those lines over $\mathbb{Z}^2$ is significant.

The practical consequence of the preceding results is that an element of $\mathcal{D}[\![\gamma]\!]$ can be viewed as a particular representative (indeed the maximum one) of an equivalence class of elements of $\overline{\mathbb{Z}}_{\max}[\![\gamma]\!]$ for the equivalence relation

$$F \equiv G \Leftrightarrow \gamma^* F = \gamma^* G \ . \tag{5.16}$$

Calculations in $\mathcal{D}[\![\gamma]\!]$ can be performed using the general rules of power series with *any* representative of an equivalence class (i.e. not necessarily a series with nondecreasing coefficients); however, the second simplification rule (5.3) is now available (the first one is of course also valid since it arises from the fact that the coefficients lie in $\overline{\mathbb{Z}}_{\max}$). The symbol '$=$' in the second rule must be understood as the fact that both sides are in the same equivalence class.

Because of (5.16), there is no meaning in speaking of the degree of an element of $\mathcal{D}[\![\gamma]\!]$, because an element is an equivalence class, and two representatives of the same class may have different degrees. For example, $e = \gamma^0$ (which is of degree zero in $\overline{\mathbb{Z}}_{\max}[\![\gamma]\!]$) and γ^* (which is of infinite degree in $\overline{\mathbb{Z}}_{\max}[\![\gamma]\!]$) are the same element in $\mathcal{D}[\![\gamma]\!]$. The situation is better for the valuation which is invariant in an equivalence class. This is stated by the next lemma which also exhibits another invariant of equivalence classes.

Lemma 5.11 *Consider* $F = \bigoplus_{k\in\mathbb{Z}} f(k)\gamma^k$ *and* $G = \bigoplus_{k\in\mathbb{Z}} g(k)\gamma^k$ *in* $\overline{\mathbb{Z}}_{\max}[\![\gamma]\!]$. *If* F *and* G *are in the same equivalence class, then*

1. $\mathrm{val}(F) = \mathrm{val}(G)$, *i.e.* $\inf\{k \mid f(k) \neq \varepsilon\} = \inf\{k \mid g(k) \neq \varepsilon\}$;
2. $\bigoplus_{k\in\mathbb{Z}} f(k) = \bigoplus_{k\in\mathbb{Z}} g(k)$, *i.e.* $\sup_{k\in\mathbb{Z}} f(k) = \sup_{k\in\mathbb{Z}} g(k)$.

Proof

1. We have $\gamma^* F = \gamma^* G$. But $\mathrm{val}(\gamma^* F) = \mathrm{val}(\gamma^*) \otimes \mathrm{val}(F)$ from (4.94) (equality holds true since $\overline{\mathbb{Z}}_{\max}$ is entire), but $\mathrm{val}(\gamma^*) = e$ and hence $\mathrm{val}(F) = \mathrm{val}(\gamma^* F) = \mathrm{val}(\gamma^* G) = \mathrm{val}(G)$.
2. Since the two formal power series in (5.16) must be equal, the corresponding values in $\overline{\mathbb{Z}}_{\max}$, obtained by substituting a numerical value in $\overline{\mathbb{Z}}_{\max}$ for γ, must also be the same. Therefore, set $\gamma = e$ and the result follows. ■

5.3.2.2 *Minimum Representative*

In general, there is no minimum representative of an equivalence class because $\mathrm{I}^\flat$ is not a $\wedge$-morphism (check that $\mathrm{I}^\flat(F \wedge G) < \mathrm{I}^\flat(F) \wedge \mathrm{I}^\flat(G)$ with $F = e \oplus 1\gamma$ and $G = 1 \oplus \gamma$). However, it turns out that some equivalence classes do have a minimum representative. We address this question now. Let F define an equivalence class. If there were to exist a minimum member, say $\widetilde{F}$, of that equivalence class, then we would have that

$$\gamma^* F = \gamma^* \widetilde{F} = \widetilde{F} \oplus \gamma\gamma^* \widetilde{F} = \widetilde{F} \oplus \gamma\gamma^* F \ ,$$

where the following identities have been used:

$$\gamma^* = e \oplus \gamma^+ \quad \text{and} \quad \gamma^+ = \gamma\gamma^* \ . \tag{5.17}$$

Hence, if such an $\widetilde{F}$ were to exist, it should be the smallest one satisfying $\gamma^* F = \widetilde{F} \oplus \gamma^+ F$, and therefore it should be equal to $\gamma^* F \ominus \gamma^+ F$.

Theorem 5.12 (Minimum representative) *Let* $F = \bigoplus f(k)\gamma^k \in \overline{\mathbb{Z}}_{\max}[\![\gamma]\!]$ *and* $\widetilde{F} \stackrel{\text{def}}{=} \gamma^* F \ominus \gamma^+ F$. *Then, one also has that*

$$\widetilde{F} = F \ominus \gamma^+ F \tag{5.18}$$

(equality in $\overline{\mathbb{Z}}_{\max}[\![\gamma]\!]$*). Moreover, this* $\widetilde{F}$ *depends only on the equivalence class of* F *of which it is a minorant. Finally, the following three statements are equivalent:*

1. $\widetilde{F}$ *belongs to the equivalence class of* F *of which it is the minimum representative;*
2. $\text{val}(\widetilde{F}) = \text{val}(F)$;
3. $\lim_{k\to-\infty} f(k) = \varepsilon$.

Before giving a proof, let us consider Figure 5.5 which illustrates how $\widetilde{F}$ is obtained in practice using a geometric construction of $\gamma^* F \ominus \gamma^+ F$: the black graph represents

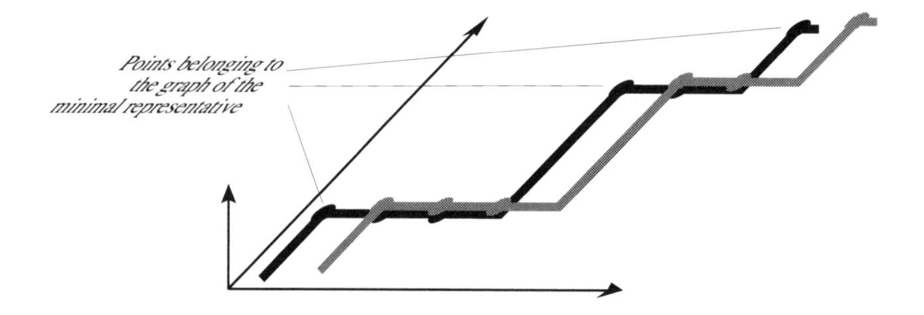

Figure 5.5: Featuring the construction of $\gamma^* F \ominus \gamma^+ F$

$\gamma^* F$ (a nondecreasing mapping from $\mathbb{Z}$ to $\overline{\mathbb{Z}}$); the grey graph represents $\gamma^+ F$ and is obtained from the previous one by a unit shift along the x-axis; finally, only the coefficients corresponding to points where the black graph differs from the grey graph are nonzero coefficients of $\widetilde{F}$.

Proof of Theorem 5.12 First, we have that

$$\widetilde{F} = \gamma^* F \ominus \gamma^+ F = (F \oplus \gamma^+ F) \ominus \gamma^+ F = F \ominus \gamma^+ F$$

according to Formula (f.17) of Table 4.2.

The former expression shows that $\widetilde{F}$ depends only on the equivalence class of F (since $\gamma^+ F = \gamma\gamma^* F$ and $\gamma^* F$ characterizes an equivalence class). The latter expression shows that $\widetilde{F} \leq F$, hence $\widetilde{F}$ is a minorant of the equivalence class of F since this inequality can be obtained for any F in this subset.

1 ⇒ 2 If $\widetilde{F}$ belongs to the equivalence class of F, then Lemma 5.11 shows that $\text{val}(\widetilde{F}) = \text{val}(F)$.

2 ⇒ 3 Suppose that $\text{val}(\widetilde{F}) = \text{val}(F)$. We also assume that $F \neq \varepsilon$ since otherwise $\widetilde{F} = \varepsilon$ and the theorem is trivial. Then, either $\text{val}(F) > -\infty$ —in this case the equality with $\text{val}(\widetilde{F})$ need not be assumed since it can indeed be proved, see Remark 5.13 below—or it is equal to $-\infty$. In the former case, clearly $f(k) = \varepsilon$ for all $k < \text{val}(F)$ and statement 3 is trivially true. In the latter case, we are going to prove that statement 3 also holds true. Indeed, since $\text{val}(\widetilde{F}) = -\infty$, for all k_0, there exists $k_0' \leq k_0$ such that $\widetilde{f}(k_0') > \varepsilon$. Since (5.18) says that $\widetilde{f}(k_0') = f(k_0') \ominus \sup_{\ell \leq k_0'-1} f(\ell)$, it is necessary that $\sup_{\ell \leq k_0'-1} f(\ell) \leq f(k_0') - 1$. Let $k_1 = k_0' - 1 < k_0$. By repeating the same argument, we can construct a strictly decreasing subsequence $\{k_i\}$ such that $\sup_{\ell \leq k_i} f(\ell) \leq f(k_i+1) - 1$. This clearly shows that $\lim_{k_i \to -\infty} \sup_{\ell \leq k_i} f(\ell) = \varepsilon$, and since the mapping $k \mapsto \sup_{\ell \leq k} f(l)$ is nondecreasing, then $\lim_{k \to -\infty} \sup_{\ell \leq k} f(\ell) = \varepsilon$. This property is equivalent to statement 3.

3 ⇒ 1 Statement 1 is equivalent to the fact that $A = B$ with $A \stackrel{\text{def}}{=} \gamma^* \widetilde{F}$ and $B \stackrel{\text{def}}{=} \gamma^* F$, which is also equivalent to the fact that $B \ominus A = \varepsilon$ because $\widetilde{F} \leq F$ and thus $A \leq B$. From (5.17), we have that $A = \gamma A \oplus \widetilde{F}$ and $B = \gamma B \oplus F$. With the help of Formula (f.16) of Table 4.2, we have that $B = \gamma B \oplus (F \ominus \gamma B) = \gamma B \oplus \widetilde{F}$. Moreover,

$$
\begin{aligned}
B \ominus A &= (\gamma B \oplus \widetilde{F}) \ominus (\gamma A \oplus \widetilde{F}) \\
&= (\gamma B \ominus (\gamma A \oplus \widetilde{F})) \oplus (\widetilde{F} \ominus (\gamma A \oplus \widetilde{F})) && \text{using (f.14),} \\
&= \gamma B \ominus (\gamma A \oplus \widetilde{F}) && \text{since } \widetilde{F} \ominus (\gamma A \oplus \widetilde{F}) = \varepsilon, \\
&= (\gamma B \ominus \gamma A) \ominus \widetilde{F} && \text{using (f.18),} \\
&\leq \gamma(B \ominus A) \ominus \widetilde{F} && \text{using (f.20),} \\
&\leq \gamma(B \ominus A) && \text{(obvious).}
\end{aligned}
$$

It follows that $X \stackrel{\text{def}}{=} B \ominus A$ satisfies $X \leq \gamma X$ which means that $x(\cdot) = b(\cdot) \ominus a(\cdot)$ is *nonincreasing*. In addition, for all k, $x(k) \leq b(k) = \sup_{\ell \leq k} f(\ell)$ and $\lim_{k \to -\infty} b(k) = \varepsilon$ from the assumption that statement 3 holds true. Therefore, x being nonincreasing and tending to ε at $-\infty$ is always equal to ε. ■

Remark 5.13 If $-\infty < \text{val}(F) < +\infty$, then $\text{val}(\gamma^+ F) = \text{val}(\gamma^+) \otimes \text{val}(F) = 1 + \text{val}(F) > \text{val}(F)$. From (5.18), we have that $F \leq \gamma^+ F \oplus \widetilde{F}$, hence, with (4.92), $\text{val}(F) \geq \min(\text{val}(\gamma^+ F), \text{val}(\widetilde{F}))$. But $\text{val}(\gamma^+ F) > \text{val}(F)$, hence $\text{val}(F) \geq \text{val}(\widetilde{F})$. On the other hand, since $\widetilde{F} \leq F$, $\text{val}(\widetilde{F}) \geq \text{val}(F)$ and finally $\text{val}(\widetilde{F}) =$

$\text{val}(F)$. If $\text{val}(F) = +\infty$, then $\widetilde{F} = F = \varepsilon$. Therefore, statement 2 of the theorem can be replaced by the statement $\{\text{val}(F) = -\infty \Rightarrow \text{val}(\widetilde{F}) = -\infty\}$. ■

5.4 MOVING TO THE TWO-DIMENSIONAL DESCRIPTION

In (5.3), the simplification rule on exponents is dual to the one which applies to coefficients. Therefore, it seems more natural to preserve the symmetry between exponents and coefficients. This is realized by a new coding of daters using *two* shift operators instead of one, the so-called two-dimensional domain description. Then a nice interpretation of this new coding will be given in terms of 'information' about events.

5.4.1 The $\overline{\mathbb{Z}}_{\max}$ Algebra through Another Shift Operator

In Examples 4.17 and 4.18 we observed that a dioid of vectors or scalars can be made isomorphic to a dioid of some subsets of this vector or scalar set in which $\cup$ plays the role of addition and $+$ ('vector sum') that of multiplication. For our present purpose, we consider $\overline{\mathbb{Z}}_{\max}$ on the one hand, and $(\mathcal{L}, \cup, +)$ on the other hand, where $\mathcal{L}$ is the subset of $2^{\mathbb{Z}}$ consisting of 'half lines of $\mathbb{Z}$' extending to the left, and including $\emptyset$ and $\mathbb{Z}$ itself. More precisely, we consider the mapping

$$\Lambda : \overline{\mathbb{Z}} \to 2^{\mathbb{Z}} \ , \quad t \longmapsto \begin{cases} \{s \in \mathbb{Z} \mid s \leq t\} & \text{if } t \in \mathbb{Z} \ ; \\ \emptyset & \text{if } t = \varepsilon = -\infty \ ; \\ \mathbb{Z} & \text{if } t = \top = +\infty \ . \end{cases} \tag{5.19}$$

Hence $\Lambda\left(\overline{\mathbb{Z}}\right) = \mathcal{L}$ and Λ is a dioid isomorphism between the two complete dioids $\overline{\mathbb{Z}}_{\max}$ and $(\mathcal{L}, \cup, +)$ (in the latter, $\varepsilon = \emptyset$ and $e = (-\infty, 0])$.

We now consider the set of power series in one variable δ, with 'Boolean coefficients' (denoted ε and e) belonging to the dioid of Example 4.16, and with exponents in $\mathbb{Z}$, this set of series being endowed with the conventional sum and product of series; this dioid is denoted $\mathbb{B}[\![\delta]\!]$. With any subset S of $\mathbb{Z}$, we associate a power series via the mapping

$$S = \{t\}_{t \in J_S} \mapsto \bigoplus_{t \in J_S} \delta^t \ . \tag{5.20}$$

This expression should be interpreted as a series in which only coefficients equal to e are explicitly mentioned, the missing monomials having a coefficient equal to ε. Clearly, $S \cup S'$ is represented by the series obtained by summing up the series related to S and S'. The empty subset is represented by the zero series (all coefficients equal to ε) and the subset $\mathbb{Z}$ is represented by the series $\top$ having all coefficients equal to e. Also, if

$$S \otimes S' \stackrel{\text{def}}{=} S + S' = \{t + t' \mid t \in J_S, t' \in J_{S'}\} \ ,$$

then the product of the series associated with S and S' is the series associated with $S \otimes S'$. The identity element consists of the subset $\{0\}$, and is represented by the series δ^0 also denoted e.

The mapping (5.20) is an isomorphism between the two complete dioids $(2^{\mathbb{Z}}, \cup, +)$ and $\mathbb{B}[\![\delta]\!]$. The subset $\mathcal{L}$ of $2^{\mathbb{Z}}$ is mapped to some subset of $\mathbb{B}[\![\delta]\!]$ which we are going to characterize. Note first that δ is the series representing the subset $\{1\}$, and 'multiplying by δ' amounts to shifting a subset to the right by one (later on, δ will be called a 'backward shift operator in timing' or 'in the time domain' for reasons akin to those put forward in Remark 5.6). Then, a half line $L \in \mathcal{L}$ is a subset characterized by the fact that it is included in its own image obtained by translation to the right: in terms of associated series, and keeping the same letter to denote the half line and its coding series, this means that $L \leq \delta L$ or

$$L \geq \delta^{-1} L \ . \tag{5.21}$$

Given any subset S, we may look for the smallest half line L larger than (i.e. containing) S: in the algebraic setting, this amounts to solving the algebraic problem of the 'best approximation from above' of a series S by a series L satisfying (5.21). By direct application of Theorem 4.78, the solution of this problem is obtained by using the formula $L = (\delta^{-1})^* S$. The dioid $\mathcal{L}[\![\delta]\!]$ of series representing half lines L is isomorphic to the quotient of $\mathbb{B}[\![\delta]\!]$ by the congruence

$$\forall S, S' \in \mathbb{B}[\![\delta]\!] \ , \quad S \equiv S' \Leftrightarrow (\delta^{-1})^* S = (\delta^{-1})^* S' \ , \tag{5.22}$$

and it is also isomorphic to the multiplicative ideal $(\delta^{-1})^* \mathbb{B}[\![\delta]\!]$. Calculations in $\mathcal{L}[\![\delta]\!]$, which amount to manipulations of half lines—and hence also of numbers in $\overline{\mathbb{Z}}_{\max}$ according to the mapping (5.19)—can be done with any representative of an equivalence class in $\mathbb{B}[\![\delta]\!]$, provided that the following simplification rule be remembered (which should remind us of the second rule in (5.4)):

$$\delta^t \oplus \delta^\tau = \delta^{\max(t,\tau)} \ .$$

This indeed expresses the equivalence of both sides of the equation.

Remark 5.14 The composition of (5.19) and (5.20) (direct correspondence from $\overline{\mathbb{Z}}_{\max}$ to $(\delta^{-1})^* \mathbb{B}[\![\delta]\!]$) is given by

$$t \mapsto \begin{cases} \delta^t (\delta^{-1})^* & \text{if } t \in \mathbb{Z} \ ; \\ \varepsilon \text{ (zero series)} & \text{if } t = \varepsilon = -\infty \ ; \\ (\delta^{-1})^* \delta^* = (\delta^{-1} \oplus \delta)^* = (\delta^{-1})^* \oplus \delta^* & \text{if } t = \top = +\infty \ . \end{cases} \tag{5.23}$$

In the first two cases there exist minimum representatives in the corresponding equivalence classes of $\mathbb{B}[\![\delta]\!]$ which are respectively δ^t and ε (the latter class contains only this element), but there is no minimum representative in the class of $\top$ (the last case). If we attempt to allow infinite exponents for power series in $\mathbb{B}[\![\delta]\!]$ in order to say that $\delta^{+\infty}$ is a minimum representative of $\top$, then expression (5.22) of the congruence is no longer valid since $\delta^{+\infty}$ and δ^*, which should both represent $\top$, do not appear to be algebraically equivalent through (5.22), that is, $\delta^{+\infty} \left(\delta^{-1}\right)^* \neq \delta^* \left(\delta^{-1}\right)^*$. The reason is that a subset of $\overline{\mathbb{Z}}$ which is a left half line can no longer be characterized by the fact that this subset is included in its image by a right unit shift: this fails for the subset $\{+\infty\}$. This observation is similar to that of Remark 5.7. ■

5.4.2 The $\mathcal{M}_{\rm in}^{\rm ax}[\![\gamma,\delta]\!]$ Algebra

We start from the set of formal power series in two variables (γ, δ) with Boolean coefficients and with exponents in $\mathbb{Z}$, this set being endowed with the conventional addition and multiplication of series: this dioid is called $\mathbb{B}[\![\gamma,\delta]\!]$. In two stages, that is, by two successive quotients by equivalence relations, we reach an algebraic structure, called $\mathcal{M}_{\rm in}^{\rm ax}[\![\gamma,\delta]\!]$ (pronounced 'min max $\gamma\delta$'), which is isomorphic to $\mathcal{D}[\![\gamma]\!]$ (the dioid of γ-transforms of *nondecreasing* functions from $\mathbb{Z}$ to $\overline{\mathbb{Z}}_{\max}$). At the first stage, we reach a dioid which is isomorphic to $\overline{\mathbb{Z}}_{\max}[\![\gamma]\!]$ (γ-transforms of *general* functions). We also show that the two steps can be combined into a single one. For each stage, we give algebraic and geometric points of view.

5.4.2.1 *From Sets of Points in the Plane to Hypographs of Functions*

The dioid $\mathbb{B}[\![\gamma,\delta]\!]$ is complete, commutative, distributive and Archimedian. It is isomorphic to the dioid $\left(2^{\mathbb{Z}^2}, \cup, +\right)$ via the one-to-one correspondence:

$$F \in 2^{\mathbb{Z}^2} \quad , F = \{(k,t)\}_{(k,t)\in J_F} \longmapsto \bigoplus_{(k,t)\in J_F} \gamma^k\delta^t \in \mathbb{B}[\![\gamma,\delta]\!] \ . \tag{5.24}$$

The lower bound operation $\wedge$ in $\mathbb{B}[\![\gamma,\delta]\!]$ corresponds to the intersection $\cap$ in $2^{\mathbb{Z}^2}$.

Instead of subsets of points in $\mathbb{Z}^2$, we can manipulate their indicator functions over $\mathbb{Z}^2$ which assume the value e at a point belonging to the corresponding subset and the value ε elsewhere. This set of Boolean functions is a complete dioid once endowed with the pointwise maximum as the addition and the two-dimensional max-convolution as the multiplication. Then, elements of $\mathbb{B}[\![\gamma,\delta]\!]$ appear as (γ,δ)-transforms of these functions in an obvious sense.

From an algebraic point of view, $\mathbb{B}[\![\gamma,\delta]\!]$ is also isomorphic to $\mathbb{B}[\![\delta]\!][\![\gamma]\!]$ which is the dioid of power series in one variable γ with coefficients in $\mathbb{B}[\![\delta]\!]$. The equivalence relation (5.22) can be extended to elements of $\mathbb{B}[\![\gamma,\delta]\!]$ by using the same definition (note that $(\delta^{-1})^*$ is another notation for $(\delta^{-1})^*\gamma^0$ in $\mathbb{B}[\![\gamma,\delta]\!]$). The quotient of $\mathbb{B}[\![\gamma,\delta]\!]$ by this equivalence relation, denoted $(\delta^{-1})^*\mathbb{B}[\![\gamma,\delta]\!]$ because it is isomorphic to this multiplicative ideal, is also isomorphic to $\left((\delta^{-1})^*\mathbb{B}[\![\delta]\!]\right)[\![\gamma]\!]$ which is the dioid of power series in γ with coefficients in $(\delta^{-1})^*\mathbb{B}[\![\delta]\!]$. Since this one is isomorphic to $\overline{\mathbb{Z}}_{\max}$ by the correspondence (5.23), we are back to the dioid $\overline{\mathbb{Z}}_{\max}[\![\gamma]\!]$. We summarize these considerations with the following lemma.

Lemma 5.15 *The dioids $(\delta^{-1})^*\mathbb{B}[\![\gamma,\delta]\!]$ and $\overline{\mathbb{Z}}_{\max}[\![\gamma]\!]$ are isomorphic.*

Geometrically, if one starts from a collection of points in $\mathbb{Z}^2$ (coded by an element of $\mathbb{B}[\![\gamma,\delta]\!]$ as indicated by (5.24)), the quotient by (5.22) corresponds to 'hanging a vertical half line' (extending downwards) at each point as shown in Figure 5.6. This operation is the counterpart of the isomorphism described in §5.4.1 which associates a half line extending to the left with each number of $\overline{\mathbb{Z}}_{\max}$ (but now $\overline{\mathbb{Z}}_{\max}$ is disposed vertically along the y-axis). All the subsets of $\mathbb{Z}^2$ yielding the same collection of points under this transformation are equivalent. We obtain the geometric representation of an element of $\left((\delta^{-1})^*\mathbb{B}[\![\delta]\!]\right)[\![\gamma]\!]$ (in fact, a maximum representative

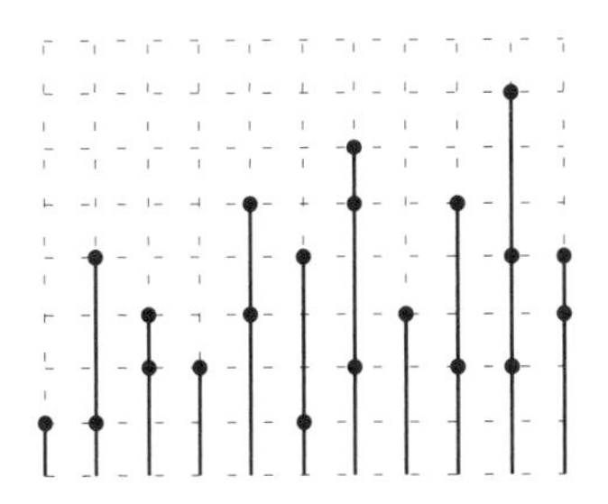

Figure 5.6: Hypograph

of an equivalence class). Since $((\delta^{-1})^*\mathbb{B}[\![\delta]\!])[\![\gamma]\!]$ is isomorphic to $\overline{\mathbb{Z}}_{\max}[\![\gamma]\!]$, this geometric figure is in turn in one-to-one correspondence with the γ-transform of some function $k \mapsto f(k)$ of which it is the 'hypograph' (see Definition 3.37). This function is determined by

$$f : k \mapsto f(k) = \sup_{(k,t)\in J_F} t \ . \tag{5.25}$$

Conversely, given a function $f : \mathbb{Z} \to \overline{\mathbb{Z}}_{\max}$, it follows from (5.23) that one representative of the corresponding element $F \in (\delta^{-1})^*\mathbb{B}[\![\gamma,\delta]\!]$ is obtained by

$$F = \left(\bigoplus_{\{k|-\infty<f(k)<+\infty\}} \gamma^k \delta^{f(k)} \right) \oplus \left(\bigoplus_{\{k|f(k)=+\infty\}} \gamma^k \delta^* \right) . \tag{5.26}$$

5.4.2.2 *From Hypographs of General Functions to Hypographs of Nondecreasing Functions*

The next step is to restrict ourselves to *nondecreasing* functions. This amounts to making the quotient of $(\delta^{-1})^*\mathbb{B}[\![\gamma,\delta]\!]$—isomorphic to $\overline{\mathbb{Z}}_{\max}[\![\gamma]\!]$—by the equivalence relation (5.16): the result is isomorphic to the multiplicative ideal $\gamma^*(\delta^{-1})^*\mathbb{B}[\![\gamma,\delta]\!]$, and it will be denoted $\mathcal{M}_{\text{in}}^{\text{ax}}[\![\gamma,\delta]\!]$. This dioid is isomorphic to $\mathcal{D}[\![\gamma]\!]$.

Lemma 5.16 *The dioids $\mathcal{D}[\![\gamma]\!]$ (dioid of γ-transforms of daters) and $\mathcal{M}_{\text{in}}^{\text{ax}}[\![\gamma,\delta]\!] \stackrel{\text{def}}{=} \gamma^*(\delta^{-1})^*\mathbb{B}[\![\gamma,\delta]\!]$ are isomorphic.*

Geometrically, this new quotient amounts to attaching a horizontal right half line to each point of the hypograph of a general function (as we did in Figure 5.4) to obtain the hypograph of a nondecreasing function d which is determined by

$$d : k \mapsto d(k) = \sup_{\substack{(\ell,t)\in J_F \\ \ell\le k}} t \ . \tag{5.27}$$

This formula is derived from (5.25) and (5.15). Conversely, given a nondecreasing function d, one representative in $\mathcal{M}_{\text{in}}^{\text{ax}}[\![\gamma,\delta]\!]$ of this dater is obtained by (5.26) with

d replacing f, that is,

$$F = \left(\bigoplus_{\{k \mid -\infty < d(k) < +\infty\}} \gamma^k \delta^{d(k)} \right) \oplus \left(\bigoplus_{\{k \mid d(k) = +\infty\}} \gamma^k \delta^* \right) . \qquad (5.28)$$

5.4.2.3 *Directly from $\mathbb{B}[\![\gamma,\delta]\!]$ to $\mathcal{M}_{\text{in}}^{\text{ax}}[\![\gamma,\delta]\!]$*

It is realized that the quotients associated with $(\delta^{-1})^*$ and with γ^* done sequentially can be condensed into a single one using the new equivalence relation in $\mathbb{B}[\![\gamma,\delta]\!]$:

$$\forall A, B \in \mathbb{B}[\![\gamma,\delta]\!] \ , \quad A \equiv B \Leftrightarrow \gamma^*(\delta^{-1})^* \ A = \gamma^*(\delta^{-1})^* B \ . \qquad (5.29)$$

Because of Formula (4.109), $\mathcal{M}_{\text{in}}^{\text{ax}}[\![\gamma,\delta]\!]$ is also equal to $(\gamma \oplus \delta^{-1})^* \mathbb{B}[\![\gamma,\delta]\!]$.

Geometrically, starting from a collection of points in $\mathbb{Z}^2$ (in one-to-one correspondence with an element of $\mathbb{B}[\![\gamma,\delta]\!]$), one first attaches vertical half lines down from each point, and then horizontal right half lines to all the points so obtained: this amounts to fixing a cone extending in south-east directions with vertical and horizontal borders to each original point. Note that the cone with its vertex at the origin is coded by $\gamma^*(\delta^{-1})^*$ in $\mathbb{B}[\![\gamma,\delta]\!]$; it corresponds to the identity element in the quotient dioid.

Notation 5.17 We introduce the following notation: $\forall (k,t), (\ell,\tau) \in \mathbb{Z}^2$,

$$\{(\ell,\tau) \preceq (k,t) \text{ or } (k,t) \succeq (\ell,\tau)\} \Leftrightarrow \{\ell \geq k \text{ and } \tau \leq t\} \ ,$$

$$\{(\ell,\tau) \prec (k,t) \text{ or } (k,t) \succ (\ell,\tau)\} \Leftrightarrow \{(\ell,\tau) \preceq (k,t) \text{ and } (\ell,\tau) \neq (k,t)\} \ .$$

■

Geometrically, the point (ℓ,τ) lies in a south, east or south-east direction with respect to (k,t). The elements of $\mathcal{M}_{\text{in}}^{\text{ax}}[\![\gamma,\delta]\!]$ could have been obtained by raising the geometric problem of finding the smallest set of points in $\mathbb{Z}^2$ containing a given set of points and closed by translations to the right and downwards (that is, a set containing its own images by these translations). The corresponding algebraic formulation in $\mathbb{B}[\![\gamma,\delta]\!]$ is the following: for a given $A \in \mathbb{B}[\![\gamma,\delta]\!]$, find the 'best approximation from above' by a B satisfying

$$B \geq \gamma B \ , \quad B \geq \delta^{-1} B \quad \text{or equivalently} \quad B \geq (\gamma \oplus \delta^{-1}) B \ .$$

By application of Theorem 4.78, this B is equal to $\gamma^*(\delta^{-1})^* A$ and it is the maximum representative of the equivalence class of A. The problem of minimum representatives is addressed later on.

Note that there is another path to obtain $\mathcal{M}_{\text{in}}^{\text{ax}}[\![\gamma,\delta]\!]$ from $\mathbb{B}[\![\gamma,\delta]\!]$: it consists in making the quotient by γ^* *first*, followed by the quotient by $(\delta^{-1})^*$. This procedure may be interpreted in terms of functions $t \mapsto g(t)$, which amounts to inverting the role of the x- and y-axes. This is what we will do when considering counter descriptions in §5.5. Finally, we obtain the commutative diagram of Figure 5.7.

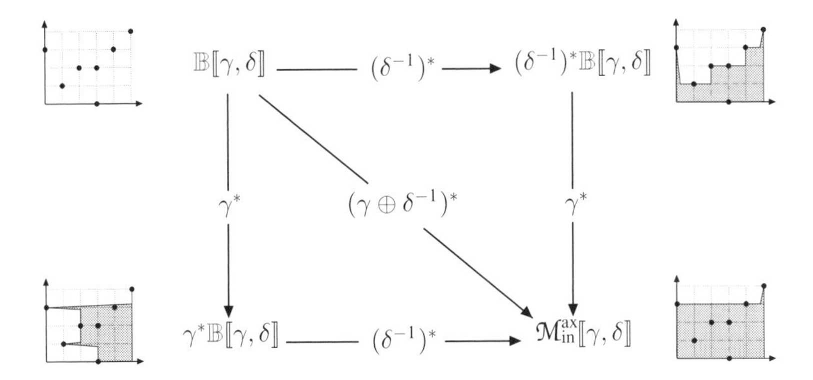

Figure 5.7: Commutative diagram

The practical rule for manipulating elements of $\mathcal{M}_{\text{in}}^{\text{ax}}[\![\gamma,\delta]\!]$ is to use any representative in each equivalence class and the usual rules of addition and multiplication of power series in two variables plus the rules (5.5) which should be understood as equivalences with respect to the congruence (5.29). The symbol $\mathcal{M}_{\text{in}}^{\text{ax}}[\![\gamma,\delta]\!]$ is supposed to suggest the rules (5.5) which involve min and max. These rules can be summarized by the following one:

$$(\ell,\tau) \preceq (k,t) \Rightarrow \gamma^k\delta^t \oplus \gamma^\ell\delta^\tau = \gamma^k\delta^t \ . \tag{5.30}$$

The comments preceding Lemma 5.11 can be repeated here with some adaptation. Because of (5.29), it is clearly meaningless to speak of the degree in γ, or of the valuation in δ, of an element of $\mathcal{M}_{\text{in}}^{\text{ax}}[\![\gamma,\delta]\!]$: two members of the same equivalence class may have different such characteristics. The next lemma is a rephrasing of Lemma 5.11 in the context of $\mathcal{M}_{\text{in}}^{\text{ax}}[\![\gamma,\delta]\!]$.

Lemma 5.18 *Consider F and G in $\mathbb{B}[\![\gamma,\delta]\!]$. If F and G represent the same element of $\mathcal{M}_{\text{in}}^{\text{ax}}[\![\gamma,\delta]\!]$, then the valuations of F and G in γ are equal, and so are the degrees of F and G in δ.*

Proof Essentially, the proof uses the same argument as at point 1 of the proof of Lemma 5.11. We start from the equality of $\gamma^*(\delta^{-1})^*F$ and $\gamma^*(\delta^{-1})^*G$ (equality in $\mathbb{B}[\![\gamma,\delta]\!]$). Then we apply (4.91), respectively (4.94), to the degree in δ, respectively the valuation in γ, of those series. These are equalities since $\{\varepsilon, e\}$ is an entire dioid. We finally observe that $\deg((\delta^{-1})^*) = \text{val}(\gamma^*) = e$. ■

Definition 5.19 *For any element F of $\mathcal{M}_{\text{in}}^{\text{ax}}[\![\gamma,\delta]\!]$, its* valuation *(still denoted* $\text{val}(F)$*), respectively its* degree *(still denoted* $\deg(F)$*), is the valuation in γ, respectively the degree in δ, of any representative (in $\mathbb{B}[\![\gamma,\delta]\!]$) of F. Such an F is a* polynomial *if it is equal to ε, or if its valuation and its degree are both finite. It is a* monomial *if it is a polynomial and, when it is not equal to ε, if it is equal (in $\mathcal{M}_{\text{in}}^{\text{ax}}[\![\gamma,\delta]\!]$) to $\gamma^{\text{val}(F)}\delta^{\deg(F)}$.*

Of course, we cannot claim here that $\mathrm{val}(F) \le \deg(F)$ as is the case for conventional nonzero polynomials or power series. However, it is straightforward to see that the relevant properties of Lemma 4.93 are still valid (with equality). Also, for a given polynomial F, any monomial which is greater than or equal to F must have a valuation not larger than $\mathrm{val}(F)$ and a degree not smaller than $\deg(F)$. Therefore, the smallest such monomial is $\gamma^{\mathrm{val}(F)}\delta^{\deg(F)}$.

5.4.2.4 *Minimum Representative*

Let us return to the problem of the minimum representative in each equivalence class.

Theorem 5.20 (Minimum representative in $\mathcal{M}_{\mathrm{in}}^{\mathrm{ax}}[\![\gamma,\delta]\!]$) *Let* $F = \bigoplus f(k,t)\gamma^k\delta^t \in \mathbb{B}[\![\gamma,\delta]\!]$ *(*$f(k,t) \in \mathbb{B}$*) and* $\widetilde{F} \stackrel{\mathrm{def}}{=} (\gamma \oplus \delta^{-1})^* F \ominus (\gamma \oplus \delta^{-1})^+ F$. *Then, one has that*

$$\widetilde{F} = F \ominus (\gamma \oplus \delta^{-1})^+ F \tag{5.31}$$

(equality in $\mathbb{B}[\![\gamma,\delta]\!]$*). Moreover,* $\widetilde{F}$ *depends only on the equivalence class of* F *of which it is a minorant (the equivalence relation is (5.29)). Finally, the following three statements are equivalent:*

1. $\widetilde{F}$ *belongs to the equivalence class of* F *of which it is the minimum representative;*

2. $\mathrm{val}(\widetilde{F}) = \mathrm{val}(F)$ *and* $\deg(\widetilde{F}) = \deg(F)$;

3. *the following two conditions are satisfied:*

$$\forall t \in \mathbb{Z}\ , \quad \exists k \in \mathbb{Z}: \ \forall (\ell,\tau) \succeq (k,t)\ , \quad f(\ell,\tau) = \varepsilon\ , \tag{5.32}$$
$$\forall k \in \mathbb{Z}\ , \quad \exists t \in \mathbb{Z}: \ \forall (\ell,\tau) \succeq (k,t)\ , \quad f(\ell,\tau) = \varepsilon\ . \tag{5.33}$$

Figure 5.8 illustrates the geometric construction of the minimum representative: the set of points of $\gamma^* F$ (in white) is shifted downwards (shift by δ^{-1} which yields the light grey set) and to the right (shift by γ which yields the dark grey set), and only the points of the white set which are not 'covered' by points of at least one of the grey sets are kept for the minimum representative.

Proof of Theorem 5.20 The fact that $\widetilde{F}$ only depends on the equivalence class of F and that it is a minorant of this equivalence class is proved in the same way as in Theorem 5.12, $\gamma \oplus \delta^{-1}$ now replacing γ.

1 ⇒ 2 This is an immediate consequence of Lemma 5.18.

2 ⇒ 3 We consider the equality of valuations and show that it implies (5.32). A similar proof, not given here, can be made for the equality of degrees implying (5.33). The case when $F = \varepsilon$ is trivial. Moreover, when $\mathrm{val}(F)$ is finite, it can be proved, as shown in Remark 5.13, that this implies that $\mathrm{val}(\widetilde{F})$ is equal to

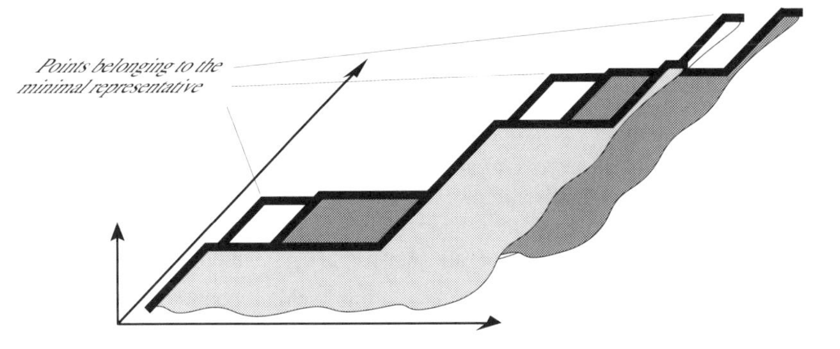

Figure 5.8: Featuring the construction of $(\gamma \oplus \delta^{-1})^* F \ominus (\gamma \oplus \delta^{-1})^+ F$

it, and (5.32) is again obvious. Finally, suppose that $\text{val}(\widetilde{F}) = \text{val}(F) = -\infty$ and $F \neq \varepsilon$. Pick some k_0; there must exist some $k'_0 \leq k_0$ and some t'_0 such that $\widetilde{f}(k'_0, t'_0) = e$, which implies that $f(k'_0, t'_0) = e$ and, for all $(\ell, \tau) \succ (k'_0, t'_0)$, $f(\ell, \tau) = \varepsilon$, since otherwise $\widetilde{f}(k'_0, t'_0)$ would be equal to ε by (5.31). Set $(k_1, t_1) = (k'_0 - 1, t'_0 - 1)$. We can repeat the same argument and find some (k'_1, t'_1) such that $k'_1 \leq k_1$, $f(k'_1, t'_1) = e$ and, for all $(\ell, \tau) \succ (k'_1, t'_1)$, $f(\ell, \tau) = \varepsilon$. Necessarily, $t'_1 \leq t'_0 - 1$ since otherwise, we would have found a $(k'_1, t'_1) \succ (k'_0, t'_0)$ such that $f(k'_1, t'_1) = e$, which is a contradiction. Hence we can construct a sequence $\{(k'_i, t'_i)\}$ with the mentioned property and such that $k'_{i+1} < k'_i$ and $t'_{i+1} < t'_i$. Hence $(k'_i, t'_i) \to (-\infty, -\infty)$ as $i \to +\infty$. Let any t be given. Pick the next (k'_i, t'_i) such that $t'_i \leq t$. Set $k = k'_i - 1$. This k fulfills the condition expressed by (5.32).

3 ⇒ 1 Let $X \stackrel{\text{def}}{=} B \ominus A$ with $A \stackrel{\text{def}}{=} (\gamma \oplus \delta^{-1})^* \widetilde{F}$ and $B \stackrel{\text{def}}{=} (\gamma \oplus \delta^{-1})^* F$. To prove that $\widetilde{F}$ belongs to the equivalence class determined by F, we need to prove that $A = B$. Since we know that $A \leq B$, it suffices to prove that $X = \varepsilon$. In a way similar to that of the proof of Theorem 5.12, it can be proved that

$$X \leq (\gamma \oplus \delta^{-1}) X \ . \tag{5.34}$$

Suppose that there exists (k_0, t_0) such that $x(k_0, t_0) = e$. Then, because of (5.34), either $x(k, t + 1)$ or $x(k - 1, t)$ is also equal to e. Call this new point (k_1, t_1), where x assumes the value e. This argument can be repeated at the point (k_1, t_1), providing the next point $(k_2, t_2) \succ (k_1, t_1)$ at which x is equal to e, etc. Therefore, we can construct an infinite sequence of points such that $(k_{i+1}, t_{i+1}) \succ (k_i, t_i)$ and at each point $x(k_i, t_i) = e$, which also implies that $b(k_i, t_i) = e$, since $x = b \ominus a$. One of the following three possibilities must then occur:

- *the sequence $\{t_i\}$ is bounded from above, hence it stays at some value $\bar{t}$ for i large enough*; then k_i must go to $-\infty$ when i increases; consequently, there exist arbitrarily small values of k such that

$b(k, \bar{t}) = e$; on the other hand, according to (5.32), there exists a $\bar{k}$ such that $f(\ell, \tau) = \varepsilon$ for all $(\ell, \tau) \succeq (\bar{k}, \bar{t})$ which also implies that $b(\ell, \bar{t}) \stackrel{\text{def}}{=} \bigoplus_{(\ell', \tau') \succeq (\ell, \bar{t})} f(\ell', \tau') = \varepsilon$; this yields a contradiction;

- *the sequence* $\{k_i\}$ *is bounded from below, hence it stays at some value* $\bar{k}$ *for* i *large enough*; then t_i must go to $+\infty$ when i increases; consequently, there exist arbitrarily large values of t such that $b(\bar{k}, t) = e$; on the other hand, according to (5.33), there exists a $\bar{t}$ such that $f(\ell, \tau) = \varepsilon$ for all $(\ell, \tau) \succeq (\bar{k}, \bar{t})$ which also implies that $b(\bar{k}, \tau) \stackrel{\text{def}}{=} \bigoplus_{(\ell', \tau') \succeq (\bar{k}, \tau)} f(\ell', \tau') = \varepsilon$; this yields a contradiction;
- *the sequences* $\{k_i\}$ *and* $\{t_i\}$ *are both unbounded and converge to* $-\infty$ *and* $+\infty$*, respectively*; this again yields a contradiction with both (5.32) and (5.33).

Finally, x cannot assume the value e anywhere. ■

Remark 5.21 With the aid of (5.27), it is seen that the condition (5.32) is equivalent to the fact that $\lim_{k \to -\infty} d(k) = \varepsilon$ which is statement 2 of Theorem 5.12. But now there is an extra condition, namely (5.33), which is equivalent to saying that $d(k)$ remains finite for all finite k. The reason for this extra condition is that, in $\overline{\mathbb{Z}}_{\max}[\![\gamma]\!]$, the point $+\infty$ does belong to the y-axis, whereas in $\mathbb{B}[\![\gamma, \delta]\!]$ it does not. About this issue, the reader should refer to Remark 5.14. ■

5.4.3 Algebra of Information about Events

We are going to provide an interpretation of the algebraic manipulation of power series of $\mathbb{B}[\![\gamma, \delta]\!]$ using the additional rule (5.30) in terms of the manipulation of information about events. Consequently, the relation order $\preceq$ introduced earlier will be interpreted as the domination of pieces of information over one another.

Given a power series F (see (5.24)), we may view each pair $(k, t) \in J_F$ as the coordinates of a 'pixel' in $\mathbb{Z}^2$ which is 'on' (whereas pairs of exponents of (γ, δ) corresponding to zero coefficients in F represent pixels which are 'off'). Each such pixel which is 'on' gives a piece of information about the associated dater d evaluated by (5.27): it says that $\forall \ell \geq k, d(\ell) \geq t$, or, in words, 'the event numbered k and the subsequent ones take place at the earliest at time t'. Geometrically, the graph of d cannot cross the region of $\mathbb{Z}^2$ delineated by the south-east cone $\{(ell, \tau) \mid (\ell, \tau) \preceq (k, t-1)\}$.

Now, given two pixels (k_1, t_1) and (k_2, t_2), the forbidden region is of course the union of the two corresponding cones. Obviously, if $(k_1, t_1) \preceq (k_2, t_2)$, the piece of information associated with the latter pixel is at least as informative as the two pieces of information together (for one cone is included in the other one) and hence the latter piece of information only may be kept. Indeed, we are just rephrasing the rule (5.30). In summary, power series in $\mathcal{M}_{\text{in}}^{\text{ax}}[\![\gamma, \delta]\!]$ can be interpreted as representations of collections of pieces of information about events, and summing up two power series consists in gathering all the pieces of information brought by the two series about the same type of event. At any stage, the reduction of the representation using the rule (5.30) amounts to canceling the pieces of information which are redundant. The relation order associated with this idempotent addition expresses the domination of collections of information over one another. The particular element ε, which corresponds to the power series with zero coefficients, has all its pixels 'off' and therefore it brings no information at all. It is the neutral element for the addition of information.

To complete our interpretation of the manipulations in $\mathcal{M}_{\mathrm{in}}^{\mathrm{ax}}[\![\gamma,\delta]\!]$, we discuss the product operation in the next subsection in which we return to event graphs.

5.4.4 $\mathcal{M}_{\mathrm{in}}^{\mathrm{ax}}[\![\gamma,\delta]\!]$ Equations for Event Graphs

5.4.4.1 Transfer Function

Let us refer back to Figure 5.1. With each transition is associated a power series in $\mathcal{M}_{\mathrm{in}}^{\mathrm{ax}}[\![\gamma,\delta]\!]$ (with the same name as the transition itself) which encodes the information available about the corresponding dater trajectory. For the sake of simplicity, in the same way as we have assumed that there is a global clock delivering the ticks numbered t for all the transitions, we assume that there is a *common initial value* of the numbering mechanisms at all transitions (assigning numbers k at successive transition firings). Then, each arc between two transitions, indeed the place on this arc, transmits information from upstream to downstream, but this information is 'shifted' by the number of 'bars' in terms of timing and by the number of 'dots' in terms of numbering. Algebraically, this shift is obtained by multiplication of the corresponding series by the appropriate monomial. For example, since the place between u and x_1 has one bar, and since e.g. u denotes the information available about the transition with the same name, the arc $u \to x_1$ carries the information δu. Hence $x_1 \geq \delta u$, that is, the information available at x_1 is *at least* δu. In the same way, $x_1 \geq \gamma^2\delta x_1$ and $x_1 \geq \gamma\delta x_2$. The transition x_1 gathers the information brought by all incoming arcs. Finally,

$$x_1 \geq \gamma^2\delta x_1 \oplus \gamma\delta x_2 \oplus \delta u \ .$$

In the same way, we can obtain inequalities for x_2 and y. In matrix form (remember that all elements belong to $\mathcal{M}_{\mathrm{in}}^{\mathrm{ax}}[\![\gamma,\delta]\!]$), we obtain

$$\begin{pmatrix} x_1 \\ x_2 \end{pmatrix} \geq \begin{pmatrix} \gamma^2\delta & \gamma\delta \\ \gamma\delta & \varepsilon \end{pmatrix} \begin{pmatrix} x_1 \\ x_2 \end{pmatrix} \oplus \begin{pmatrix} \delta \\ \delta^2 \end{pmatrix} u \ ,$$

$$y \geq \begin{pmatrix} e & e \end{pmatrix} \begin{pmatrix} x_1 \\ x_2 \end{pmatrix} ,$$

of the general form

$$x \geq Ax \oplus Bu \ , \qquad y \geq Cx \ . \tag{5.35}$$

These inequalities should be compared with the equations obtained in §5.2.

Remark 5.22 Without our assumption of a common initial value of all the numbering mechanisms, corrections should have been made as for the exponent of γ in the shift operator associated with each place in order to account for the difference in numbering initial values between the upstream and downstream transitions of this place. ■

We make the following assumption which is different from that made in §2.5.2.3, but which will find a justification later on in this section.

> The initial global clock value is $t = 0$ by convention, and the numbering mechanism at each transition assigns the value $k = 0$ to the first transition firing occurring at or after time 0.

This convention does not mean that tokens cannot be brought from the outside world before time 0: it suffices to include these tokens in the initial marking of the place connecting the input transition with other internal transitions.

Remark 5.23 (Interpretation of ε inputs) Because ε is the bottom element, an ε input is the least constraining input possible: it is less constraining than any input $\gamma^n\delta^{-t}$ for arbitrarily large n and t. Therefore, one may view ε inputs as those which correspond to bringing an infinity of tokens at time $-\infty$. ■

So far, only the inequalities (5.35) have been obtained. Without further information, the behavior of the system is not completely specified and nothing more can be said. In particular, lag times of tokens of the initial marking (see Definition 2.49) have not been stipulated. For the time being, we give a mathematical answer to this lack of information by selecting a 'canonical' solution to the system of inequalities. Later on, a more concrete interpretation of this uniquely defined solution in terms of arrival times of tokens of the initial marking in places will be given.

From Theorem 4.75, we know that the least solution of (5.35) is given by

$$x = A^*Bu \ , \qquad y = Cx = CA^*Bu \ , \tag{5.36}$$

and that it satisfies the equalities in (5.35). This solution corresponds to the earliest possible occurrences of all events. For the time being, let us assume that the least solution is indeed the one of interest. We will return to this issue later on. As an exercise, the reader may try to evaluate the expression CA^*B for the considered example. This can be done either by Gaussian elimination (hint: first express x_2 with respect to x_1 and u, then solve a fixed-point equation in x_1), or, equivalently, by using the formulæ (4.100), or preferably (4.110), and (4.109), plus the simplification rules (5.5). Finally, one obtains

$$y = \delta^2(\gamma\delta)^*u \ . \tag{5.37}$$

Under the assumption that the earliest possible behavior is the one which will occur, we reach the conclusion that the input-output relation of the event graph of Figure 5.1 is given by (5.37): in general CA^*B will be called the *transfer function* of the system (for single-input-single-output (SISO) systems).

Note that CA^*B encodes the output dater caused by the input dater e. One possible representation of e is $\delta^0(\gamma^0 \oplus \gamma^1 \oplus \gamma^2 \oplus \cdots)$. *Due to the convention adopted earlier* regarding the initial time and the initial numbering value of events, the input e may be interpreted as the action of firing the transition u an infinite number of times at time 0 (or putting an infinite amount of tokens at time 0 at the inlet of this transition)[2]. This is the analogue of an *impulse* in conventional system theory, and therefore the transfer function may be viewed as the coding of the *impulse response*.

We now return to the issue of selecting the 'earliest possible solution' of (5.35). This solution corresponds to the least constraining conditions. It does not only mean that transitions are fired immediately after being enabled, but also that 'the best initial condition' must be selected: this concerns the time at which tokens of the initial marking are available; these tokens must not determine the firing epochs of transitions they contribute to enable, whatever the input u is, and whatever the holding times are. For the relations (5.36) to be valid for all u, irrespective of the initial marking and holding times, we thus assume the following condition.

Tokens of the initial marking are available at time $-\infty$.

This convention corresponds to always choosing lag times equal to $-\infty$. These lag times may fail to fulfill item 2 of Definition 2.50 of weakly compatible lag times. If other lag times are desired, there is a way to introduce them without changing the above convention. This is the topic of the following discussion.

5.4.4.2 *Introduction of More General Lag Times*

Consider a place p of an event graph with, say, two tokens in the initial marking and a holding time equal to two time units (see Figure 5.9a). Suppose that for the

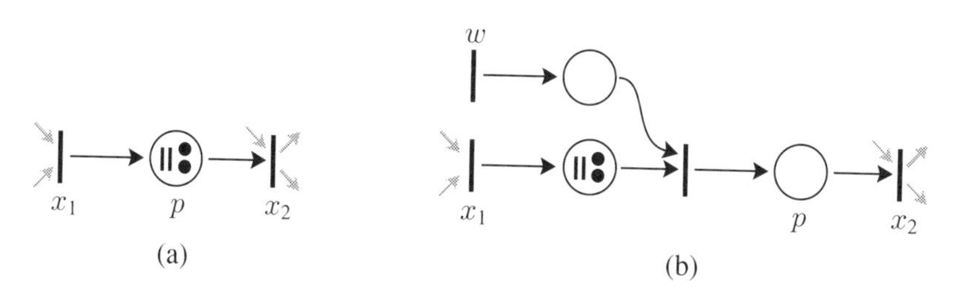

Figure 5.9: The introduction of lag times

two tokens of the initial marking, we wish to have the lag times $w(0)$ and $w(1)$, respectively. Then the event graph is modified locally in such a way as to introduce an additional place and two additional transitions, one of which is an input transition (labeled w) as shown in Figure 5.9b. The new additional place keeps the original

[2]With the convention of §2.5.2.3 regarding the numbering of events, the same sequence of events would be coded by $\delta^0(\gamma^1 \oplus \gamma^2 \oplus \cdots) = \gamma$.

initial marking and holding time. The original place p is now free of any initial marking and holding time. The lag times are forced by the additional input

$$w = \gamma^0 \delta^{w(0)} \oplus \gamma^1 \delta^{w(1)} \left(\oplus \gamma^2 \delta^{w(1)} \oplus \cdots \right) , \tag{5.38}$$

that is, the first token is introduced at time $w(0)$ and infinitely many tokens are introduced at time $w(1)$. Since the convention that tokens of the initial marking are available since $-\infty$ is still assumed, it is seen that indeed the first token of the initial marking starts enabling transition x_2 at time $w(0)$ and the second token does the same at time $w(1)$, which is consistent with the definition of lag times. After time $w(1)$, the input defined by (5.38) is no longer constraining for the rest of the life of the system.

Consider again Figure 5.9a, and assume now that this figure represents an isolated event graph, instead of a part of a larger event graph (that is, grey arrows are discarded). Rename x_1 and x_2 as u and y, respectively. The input-output relation of such a system is $y = \gamma^2 \delta^2 u$. If, say, $u = e$ (u is an impulse at time 0), then we obtain that the corresponding output is equal to

$$y = \delta^2 (\gamma^2 \oplus \gamma^3 \oplus \cdots) \ . \tag{5.39}$$

In terms of information, we thus learn that the *third* token (numbered 2) and the next ones get out at time 2. Nothing is said about the first two tokens (those of the initial marking). Alternatively, by completing the missing information in a canonical way (behavior 'at the earliest possible time'), it can be said that these two initial tokens went out at $-\infty$. This contradicts the fact that they are part of the initial marking, if this initial marking is meant to represent the exact position of tokens at time 0. This paradox occurs because the lag times, equal to $-\infty$ after our convention, are not weakly compatible in this case (transition y is enabled twice *before* the initial time). We now discuss two different ways of resolving this contradiction. The first one is described below. The second one is the topic of the next paragraph.

Along the lines of Chapter 2, we consider the modification shown in Figure 5.9b, and we only accept weakly compatible lag times (they serve to determine the additional input w as in (5.38)). In this specific case, these lag times must be nonnegative and less than or equal to 2, the holding time of p. Then, the input-output relation is given by

$$y_w = \gamma^2 \delta^2 u \oplus w \ . \tag{5.40}$$

Notice that this is now an *affine*, rather than a *linear*, function of u. For $u = e$ and for w given by (5.38), we obtain the information already provided by (5.39) for the output tokens numbered 2, 3,. . . , but we obtain additional information regarding the epochs at which tokens numbered 0 and 1 get out. We see that y_w, given by (5.40), is not less than y, given by (5.39), from both the $\mathcal{M}_{\text{in}}^{\text{ax}}[\![\gamma, \delta]\!]$ (algebraic) and the informational points of view (of course, these two points of view are consistent according to §5.4.3).

The input-output relation (5.40) can be considered to be linear, rather than affine, if we restrict ourselves to inputs u not less than $\gamma^{-2}\delta^{-2}w$ (w being given by (5.38)), which amounts to considering that the two tokens of the initial marking have also been produced by the input: this discussion will not be pursued further here but it obviously related to the notion of a *compatible initial condition* (see Definition 2.61).

5.4.4.3 System-Theoretic View of Event Graphs

From a different point of view, we may consider event graphs as playing the role of block-diagrams in conventional system theory. Recall that, for block-diagrams, the 'initial conditions' are not directly related to the operators shown in the blocks of the diagram, but they are either set to zero canonically (in which case the input-output relation is indeed linear), or they are forced by additional (Dirac-like) inputs which make the 'states' (i.e. the initial values of the integrators in continuous time models) jump to nonzero values at the initial time.

In an analogous way, we may view places of event graphs as serving the only purpose of representing elementary shift operators in the event domain (number of 'dots') and in the time domain (number of 'bars') in a pictorial way. In this more abstract (or more system-theoretic) point of view, there is no notion of 'circulation of tokens' involved, and therefore no applicability of any 'initial position of tokens'. The conventional rules of Petri nets which make tokens 'move' inside the net (and possibly get outside) are not viewed as describing any dynamic evolution, but they are rather considered as transformation rules affecting the internal representation but not the input-output relation (at least when tokens do not get outside the system; they add a shift in counting between input and output when some tokens get outside the system during these 'moves'). As discussed in Remark 5.3, there is a counterpart to transformations which move tokens, namely transformations which move bars, and both classes of transformations correspond to linear changes of basis in the internal representation. That is, the vectors x—see (5.35)—of two such equivalent representations can be obtained from each other by multiplication by an invertible matrix with entries in $\mathcal{M}_{\rm in}^{\rm ax}[\![\gamma, \delta]\!]$ (a shift of the output y may also be necessary when tokens or bars cross the output transitions during their moves).

To illustrate this point with an example which is even simpler than that of Figure 5.9, consider an output transition y connected to an input transition u by a place with one token and a zero holding time. This is the representation of the elementary shift operator γ. There is no way to represent this elementary input-output relation $y = \gamma u$ for *any* u by an event graph which at the same time preserves the elementary view of this object: for the token of the initial marking to be 'here' at time zero, we need an extra input $w = e$ (as in Figure 5.9b), but this modified graph represents the input-output relation $y = \gamma u \oplus e$ which coincides with $y = \gamma u$ only for $u \geq \gamma^{-1}$.

Remark 5.24 Note also that this is not the first time in this book that we meet a situation in which the naive interpretation of event graphs raises problems which do not appear in a purely algebraic conception: recall the discussion of Remark 2.86 on circuits with no tokens and no bars. ■

5.4.4.4 *Reduction of the Internal Representation*

It should be realized that (5.37) is also the input-output relation of the event graph shown in Figure 5.10 which is thus equivalent to the previous one from this 'external'

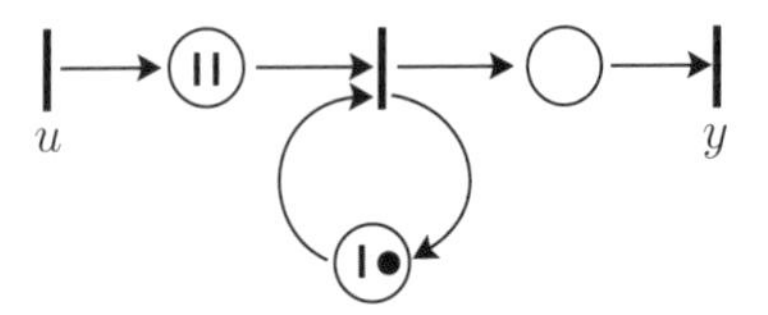

Figure 5.10: An equivalent simpler event graph

(i.e. input-output) point of view: this illustrates the dramatic simplifications provided by algebraic manipulations. As in conventional system theory, the internal structure is not uniquely determined by the transfer function, and several more or less complex realizations of the same transfer function can be given.

Of course, this equivalence of internal realizations assumes our convention that tokens of the initial marking are available since $-\infty$. To handle different lag times, one must first appeal to the transformation described above, then compute the transfer function, and finally find a reduced representation of this transfer function taking lag times (that is, additional inputs) into account. As an example, the left-hand side of Figure 5.11 displays the event graph of Figure 5.1 for which additional inputs allow

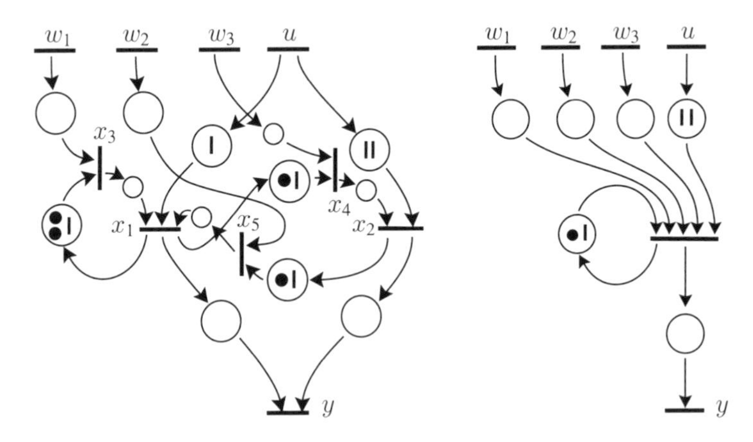

Figure 5.11: Event graph with lag times and its reduced representation

the choice of lag times for all tokens of the initial marking. For this new event graph, the input-output relation turns out to be $y = (\gamma\delta)^* \left(\delta^2 u \oplus w_1 \oplus w_2 \oplus w_3\right)$: another simpler event graph which has the same transfer function.

Remark 5.25 (Multiple-Input Multiple-Output—MIMO—systems) All the notions presented in this section extended without difficulty to the case when the input

u and the output y are (column) vectors of respective dimension m and p. In particular, CA^*B (see (5.36)) is then a $p \times m$ matrix (called the transfer matrix), the entries of which are polynomial elements of $\mathcal{M}_{\text{in}}^{\text{ax}}[\![\gamma, \delta]\!]$. ■

5.5 COUNTERS

5.5.1 A First Derivation of Counters

As already mentioned in §5.1, discrete event systems, and in particular event graphs, can be described in the event domain by daters, or in the time domain by counters. We also mentioned that for a counter $t \mapsto c(t)$ associated with some type of event, the precise meaning given to $c(t)$ deserves some attention: roughly speaking, it represents the value reached by the corresponding numbering mechanism of events at time t; however, several events may take place simultaneously at time t. On the contrary, there is no ambiguity in speaking of the epoch $d(k)$ of the event numbered k. From a mathematical point of view, this issue about counters will receive a natural solution in this section. Yet, from an intuitive point of view, the interpretation given to $c(t)$ might not appear as the most natural one. This interpretation will be given only after the relationship with the corresponding dater has been established.

At this stage, we can discuss the two possibilities of describing event graphs, by daters or by counters, from an abstract point of view. Consider two different ways of moving from the upper left-hand corner to the lower right-hand corner of the commutative diagram of Figure 5.7, namely via the eastern then southern directions on the one hand, and via the southern then eastern directions on the other hand. The former path has been described in some detail: the first move is developed in §5.4.2.1, and the second move in §5.4.2.2.

We can interpret the latter path in the same way. The first move goes from $\mathbb{B}[\![\gamma, \delta]\!]$ to $\gamma^*\mathbb{B}[\![\gamma, \delta]\!]$ which is isomorphic to $\left(\gamma^*\mathbb{B}[\![\gamma]\!]\right)[\![\delta]\!]$. For reasons dual to those given in §5.4.1, $\gamma^*\mathbb{B}[\![\gamma]\!]$ is isomorphic to $\overline{\mathbb{Z}}_{\text{min}}$ by the correspondence from $\overline{\mathbb{Z}}_{\text{min}}$ to $\gamma^*\mathbb{B}[\![\gamma]\!]$:

$$k \mapsto \begin{cases} \gamma^k\gamma^* & \text{if } k \in \mathbb{Z} \ ; \\ \varepsilon \text{ (zero series)} & \text{if } k = \varepsilon = +\infty \ ; \\ (\gamma^{-1})^*\gamma^* = (\gamma^{-1} \oplus \gamma)^* = (\gamma^{-1})^* \oplus \gamma^* & \text{if } k = \top = -\infty \ , \end{cases} \tag{5.41}$$

which is the counterpart of (5.23). Therefore, $\left(\gamma^*\mathbb{B}[\![\gamma]\!]\right)[\![\delta]\!]$ can be viewed as the subset of $\mathbb{B}[\![\gamma, \delta]\!]$ (the latter encodes collections of points in $\mathbb{Z}^2$) corresponding to epigraphs of mappings $g : t \mapsto g(t)$ from $\mathbb{Z}$ into $\overline{\mathbb{Z}}_{\text{min}}$, or alternatively, as the set of δ-transforms of those mappings g.

Remark 5.26 We still speak of 'epigraph', that is the part of the plane above the graph (see Definition 3.36), although, in $\overline{\mathbb{Z}}_{\text{min}}$, the dioid order is reversed with respect to the conventional order. ■

The second move in the diagram, namely that going from $\gamma^*\mathbb{B}[\![\gamma, \delta]\!]$ to $\mathcal{M}_{\text{in}}^{\text{ax}}[\![\gamma, \delta]\!]$ through the south, corresponds to selecting only the nondecreasing mappings from

$\mathbb{Z}$ to $\overline{\mathbb{Z}}_{\min}$ which are precisely the counters. The approximation of nonmonotonic functions by nondecreasing ones is 'from below'. Again, the words 'nondecreasing' and 'from below' should be understood with reference to the conventional order.

These two moves can be summarized by the following formulæ which are the counterparts of (5.25) and (5.27): the mappings g and c defined by

$$g : t \mapsto g(t) = \inf_{(k,t)\in J_F} k \ , \qquad c : t \mapsto \inf_{s\geq t} g(s) = \inf_{\substack{(k,s)\in J_F \\ s\geq t}} k \tag{5.42}$$

are successively associated with a power series $F \in \mathbb{B}[\![\gamma,\delta]\!]$, or with the corresponding collection of points in $\mathbb{Z}^2$ (see (5.24)). In terms of 'information', a pixel $(k,t) \in J_F$, tells that, *at time t, the counter reaches at most the value k* (since, from (5.42), $c(s) \leq k$ for all $s \leq t$). Conversely, to a counter $c : \mathbb{Z} \to \overline{\mathbb{Z}}_{\min}$ corresponds a power series in $\mathcal{M}_{\text{in}}^{\text{ax}}[\![\gamma,\delta]\!]$, namely

$$\left(\bigoplus_{\{t|-\infty<c(t)<+\infty\}} \gamma^{c(t)}\delta^t \right) \oplus \left(\bigoplus_{\{t|c(t)=-\infty\}} (\gamma^{-1})^*\delta^t \right) , \tag{5.43}$$

a formula which follows from (5.41). This formula is the counterpart of (5.28).

5.5.2 Counters Derived from Daters

Let us now discuss the relationship that exists between the dater and the counter associated with the same type of event. We are going to prove that under some mild condition, the counter is the dual residual of the dater, i.e. $c = d^\flat$.

However, for this to be possible, d must first be considered as an isotone function between two complete dioids. Indeed, as already discussed, d is a monotonic mapping from $\overline{\mathbb{Z}}_{\min}$ into $\overline{\mathbb{Z}}_{\max}$; as such, it is *antitone* rather than isotone (that is, with the dioid orders in the domain and the range, it is 'decreasing' rather than 'increasing'). However, we may as well consider the mapping $\widehat{d}$ from $\overline{\mathbb{Z}}$ (with the natural order) into itself having the same graph as d. This $\widehat{d}$ is an isotone mapping from a complete lattice into itself. Because of the end-point conditions (5.6) that we imposed on purpose, $\widehat{d}$ can be both residuated and dually residuated, provided that the required semicontinuity conditions be fulfilled (see Theorems 4.50 and 4.52, statements 2). The following theorem discusses the dual residuation. By abuse of expression, we speak of the residuation of d rather than of $\widehat{d}$.

Theorem 5.27 *A dater d is dually residuated if and only if*

$$\lim_{k\to-\infty} d(k) = -\infty \ . \tag{5.44}$$

Then, if a power series in $\mathcal{M}_{\text{in}}^{\text{ax}}[\![\gamma,\delta]\!]$ is associated with d by (5.28), and if c is derived from this series by (5.42), then $c = d^\flat$.

Proof Since d is a mapping between two totally ordered and discrete sets, the required upper-semicontinuity condition (see Definition 4.43) has to be checked only for subsets $\{k_i\}$ such that $\bigwedge_i k_i = -\infty$. Since we imposed $d(-\infty) = -\infty$, condition (5.44) follows.

Then, the set of pixels $\{(k,t)\}$ defined by

$$\left(\bigcup_{-\infty < d(k) < +\infty} (k, d(k)) \right) \cup \left(\bigcup_{\substack{d(k)=+\infty \\ \tau \geq 0}} (k, \tau) \right)$$

is associated with d. Indeed, this is *one* possible collection, since a south-east cone can be attached to any pixel of the collection. The above formula follows from (5.28). Finally, looking at (5.42), it should be clear that

$$c(t) = \inf_{d(k) \geq t} k \ , \tag{5.45}$$

which is nothing but a possible definition of $d^\flat$ (see statement 1 of Theorem 4.52). ■

Remark 5.28 If (5.44) is satisfied, then it follows from (4.23) and (4.24) that

$$\forall t \ , \quad d(c(t)) \geq t \qquad \text{and} \qquad \forall k \ , \quad c(d(k)) \leq k \ . \tag{5.46}$$

Moreover, since $c = d^\flat$, then $d = c^\sharp$, hence

$$d(k) = \sup_{c(t) \leq k} t \ . \tag{5.47}$$

The relation (5.45) always holds true, even if condition (5.44) is not fulfilled. But then, we cannot say that c is the dual residual of d since the other statements of Theorem 4.52 do not hold true, in particular (5.46) and (5.47) may be wrong. For example, consider the mapping defined by (5.6) and $d(k) = 0$ for all finite k. Then the corresponding c is such that $c(t) = +\infty$ for $t > 0$ and $c(t) = -\infty$ otherwise. Therefore, $d(c(0)) = d(-\infty) = -\infty < 0$, in contradiction with (5.46). Also, $\sup_{c(t) \leq -\infty} t = 0 > d(-\infty) = -\infty$, in contradiction with (5.47).

However, this discussion is of purely mathematical interest since, given our conventions, any realistic dater, even after some *finite* shift in numbering due to the initial marking, will be such that $d(k) = -\infty, \forall k < k_0$ for some finite k_0. That is, condition (5.44) will always be satisfied in practice. ■

Remark 5.29 (Interpretation of counters) In words, the relation (5.45) expresses that $c(t)$ is the smallest value the numbering mechanism will reach at or after time t; otherwise stated, $c(t)$ is the number of the *next event to come* at or after time t. This explains the inequalities (5.46) which may seem counterintuitive at first sight. An alternative definition of counters is considered in the next subsection. ■

5.5.3 Alternative Definition of Counters

Another possible definition of counters is $\widetilde{c} = d^\sharp$, provided that d be residuated. A necessary and sufficient condition is that d be l.s.c., which here is implied by the dual condition of (5.44), namely

$$\lim_{k \to +\infty} d(k) = +\infty \ . \tag{5.48}$$

This condition means that no infinite numbers of events occur in finite time. Then, the residuated mapping $\widetilde{c}$ of d is, by definition, equal to

$$\widetilde{c}(t) = \sup_{d(k) \leq t} k \ . \tag{5.49}$$

Even if (5.48) is not fulfilled, we can take (5.49) as the definition of $\widetilde{c}$ (but then, we cannot say that, conversely, $d(k) = \inf_{\widetilde{c}(t) \geq k} t$). The meaning of this new counter $\widetilde{c}(t)$ is *the number of the last event which occurs before or at* t. This might seem a more natural definition for counters than the one corresponding to $c(t)$. However, from a mathematical point of view, there is a drawback in manipulating this alternative notion of counters. To explain this point, we first establish the following result.

Lemma 5.30 *If* c *and* $\widetilde{c}$ *are derived from the same* d *by (5.45) and (5.49), respectively, then*

$$\forall t \ , \quad c(t) = \widetilde{c}(t-1) + 1 \ .$$

Proof Let us start from (5.45). A $c(t)$ so defined is characterized by the fact that $d(c(t)) \geq t$ but also $d(c(t) - 1) < t$, that is, $d(c(t) - 1) \leq t - 1$. Observe that if we set $k \stackrel{\text{def}}{=} c(t) - 1$, these inequalities also tell us that $k = \sup_{d(\ell) \leq t-1} \ell$, which is nothing but $\widetilde{c}(t-1)$. ■

Given c, a two-dimensional representative in $\mathcal{M}_{\text{in}}^{\text{ax}}[\![\gamma, \delta]\!]$ can be associated with c by (5.43). Therefore, owing to the above lemma, given $\widetilde{c}$, the corresponding two-dimensional representative is given by

$$\left(\bigoplus_{\{t \mid -\infty < \widetilde{c}(t) < +\infty\}} \gamma^{\widetilde{c}(t)+1} \delta^{t+1} \right) \oplus \left(\bigoplus_{\{t \mid c(t) = -\infty\}} (\gamma^{-1})^* \delta^{t+1} \right) ,$$

which is of course not as convenient as with c.

5.5.4 Dynamic Equations of Counters

We return to the definition (5.45) of counters. We introduce the following notation:

- Equation (5.45) defines a functional Z from the set of nondecreasing mappings $d : \overline{\mathbb{Z}} \to \overline{\mathbb{Z}}_{\max}$ (referred to as the 'dater set') to the set of nondecreasing mappings $c : \overline{\mathbb{Z}} \to \overline{\mathbb{Z}}_{\min}$ (referred to as the 'counter set'); if we restrict ourselves to mappings d which satisfy (5.6) and (5.44), then $Z(d)$ is simply $d^\flat$ (and then, $Z^{-1}(c) = c^\sharp$);

- in the dater set, the pointwise addition of mappings (denoted $\oplus$) is the operation of upper hull; in the counter set, the pointwise addition (denoted $\widehat{\oplus}$) is the lower hull;

- in the dater set, the 'shift' $\{k \mapsto d(k)\} \mapsto \{k \mapsto d(k-1)\}$ is denoted γ for obvious reasons (see Remark 5.6); similarly, the same kind of shift in the counter set is denoted δ;

- in the dater set, the 'gain' $\{k \mapsto d(k)\} \mapsto \{k \mapsto d(k)+1\}$ is denoted $\mathbf{1}_{\mathrm{d}}$; the analogous unit gain in the counter set is denoted $\mathbf{1}_{\mathrm{c}}$.

Lemma 5.31 *With the previous notation, we have, for all d or d_i in the dater set:*

$$Z(d_1 \oplus d_2) = Z(d_1)\widehat{\oplus}Z(d_2) \ , \tag{5.50}$$

$$Z(\gamma d) = \mathbf{1}_{\mathrm{c}} Z(d) \ , \tag{5.51}$$

$$Z(\mathbf{1}_{\mathrm{d}} d) = \delta Z(d) \ . \tag{5.52}$$

Proof A proof can be given by playing with Definition (5.45) of Z and with the meaning of the notation γ, δ, ... A smarter proof is obtained in the following way. For a given d, let D denote the element in $\mathcal{M}_{\mathrm{in}}^{\mathrm{ax}}[\![\gamma,\delta]\!]$ defined by the right-hand side of (5.28). Similarly, for a given c, let C denote the element in $\mathcal{M}_{\mathrm{in}}^{\mathrm{ax}}[\![\gamma,\delta]\!]$ defined by (5.43). If $c = Z(d)$, it should be clear that $C = D$ in $\mathcal{M}_{\mathrm{in}}^{\mathrm{ax}}[\![\gamma,\delta]\!]$ (that is, C and D are two representatives of the same equivalence class). The series γD corresponds to the dater γd, hence the series associated with the counter $Z(\gamma d)$ must be γC, but this series corresponds, through (5.43), to the series associated with the counter $\mathbf{1}_{\mathrm{c}} c = \mathbf{1}_{\mathrm{c}} Z(d)$. This proves (5.51). Formula (5.52) can be similarly proved (by noticing that the series associated with $\mathbf{1}_{\mathrm{d}} d$ is δD).

As for (5.50), let $D_i, i = 1, 2$, be the series associated with the daters $d_i, i = 1, 2$. Then, because of the second rule (5.5), which can be used in $\mathcal{M}_{\mathrm{in}}^{\mathrm{ax}}[\![\gamma,\delta]\!]$, $D_1 \oplus D_2$ (here $\oplus$ is the addition in $\mathcal{M}_{\mathrm{in}}^{\mathrm{ax}}[\![\gamma,\delta]\!]$) is associated with $d_1 \oplus d_2$ (pointwise maximum of the mappings d_1 and d_2). Similarly, for counters, $C_1 \oplus C_2$ is associated with $c_1 \widehat{\oplus} c_2$ because of the first rule (5.5). With these observations at hand, the proof of (5.50) is easily completed. ■

Remark 5.32 In the case when $Z(d) = d^\flat$, and with the necessary adaptation of notation, (5.50) can be viewed as a stronger version of (4.38). ■

As a consequence of Lemma 5.31, Equations (5.2) can be derived from (5.1). For example, $1x_1(k-2)$ is the value at k of the dater $\mathbf{1}_{\mathrm{d}}\gamma^2 x_1$ with which the counter $\delta(\mathbf{1}_{\mathrm{c}})^2 x_1$ is associated according to (5.51)–(5.52) (here, the dater and its associated counter are denoted with the same symbol, as we did in §5.2). Therefore, the term $2x_1(t-1)$ corresponds to the term $1x_1(k-2)$ in the counter equations. And $\oplus$ in dater equations (that is, max) is converted to $\widehat{\oplus}$ in counter equations (that is, min) according to (5.50). Afterwards, it is realized that $\mathbf{1}_{\mathrm{d}}$, respectively $\mathbf{1}_{\mathrm{c}}$, could have been denoted δ, respectively γ.

Using Lemma 5.30, once Equations (5.2) have been established using one notion of counters (given by (5.45)), it is clear that these equations are also valid with the alternative notion of counters (given by (5.49)).

5.6 BACKWARD EQUATIONS

So far we have been interested in computing outputs produced by given inputs. In the dater description, outputs are sequences of the *earliest* dates at which events (numbered sequentially) can occur. Sometimes, it may also be useful to derive inputs from outputs, which, roughly speaking, corresponds to 'inverse' the system. More precisely, and still in the dater setting, we may be given a sequence of dates at which one would like to see events occur *at the latest*, and we are asked to provide the *latest* input dates that would meet this objective. It is the topic of this section to discuss this problem.

From a mathematical point of view, as long as the transfer function (matrix, in the MIMO case) has to be inverted, it is no surprise that residuation plays an essential role. This inversion translates into the fact that recurrent equations in the event domain for daters, respectively in the time domain for counters, now proceed backwards in event numbering, respectively in time. Moreover, the 'algebra' $(\wedge, \between)$ is substituted for $(\oplus, \otimes)$ in these backward equations. These equations offer a strong analogy with the adjoint-state (or co-state) equations of optimal control theory.

5.6.1 $\mathcal{M}_{\mathrm{in}}^{\mathrm{ax}}[\![\gamma, \delta]\!]$ Backward Equations

Consider a system in $\mathcal{M}_{\mathrm{in}}^{\mathrm{ax}}[\![\gamma, \delta]\!]$ described by Equation (5.35) or (5.36) (recall that (5.36) yields the least solution y of (5.35) with either inequalities or equalities). Let y be given. The greatest u such that

$$z = Hu \stackrel{\text{def}}{=} CA^*Bu \leq y \tag{5.53}$$

is, by definition, obtained as

$$u = H^\sharp y = CA^*B \between y \ . \tag{5.54}$$

From the previous results, in practice (5.53) means that, in the dater description, the output events produced by u occur *not later* than those described by y; moreover, u being the 'greatest' input having property (5.53), the input events corresponding to u occur *not earlier* than with any other input having property (5.53).

Recall that $y = CA^*Bu$ can be also described as the least solution of

$$\left.\begin{array}{l} x = Ax \oplus Bu \ , \\ y = Cx \ . \end{array}\right\} \tag{5.55}$$

We are going to give a similar 'internal' representation for the mapping $H^\sharp$ defined by (5.54).

Lemma 5.33 *Let u be derived from y by Equation (5.54). Then u is the greatest solution of the system*

$$\xi = \frac{\xi}{A} \wedge \frac{y}{C} \,, \tag{5.56}$$

$$u = \frac{\xi}{B} \,. \tag{5.57}$$

This is equivalent to saying that ξ must be selected as the greatest solution of (5.56). Moreover, the symbol '$=$' can be replaced by '$\leq$' in (5.56)–(5.57).

Proof We have

$$\begin{aligned} u &= CA^*B \backslash\!\!\!\circ\, y && \text{owing to (5.54),} \\ &= A^*B \backslash\!\!\!\circ\, (C \backslash\!\!\!\circ\, y) && \text{thanks to (f.9),} \\ &= B \backslash\!\!\!\circ\, (A^* \backslash\!\!\!\circ\, (C \backslash\!\!\!\circ\, y)) && \text{(same reason).} \end{aligned}$$

Let $\xi \stackrel{\text{def}}{=} A^* \backslash\!\!\!\circ\, (C \backslash\!\!\!\circ\, y)$. By Theorem (4.73), we know that ξ is the greatest solution of (5.56) with equality or with the inequality $\leq$. ∎

If x (hence ξ) is n-dimensional, and if u, respectively y, is p-dimensional, respectively m-dimensional, then, by using Formula (4.81), (5.56)–(5.57) can be more explicitly written as

$$\left.\begin{aligned} &\forall i = 1, \ldots, n \,, && \xi_i = \left(\bigwedge_{j=1}^{n} \frac{\xi_j}{A_{ji}} \right) \wedge \left(\bigwedge_{r=1}^{p} \frac{y_r}{C_{ri}} \right) , \\ &\forall \ell = 1, \ldots, m \,, && u_\ell = \bigwedge_{s=1}^{n} \frac{\xi_s}{B_{s\ell}} \,. \end{aligned}\right\} \tag{5.58}$$

Observe again the transposition of matrices and the substitution of the operation $\wedge$, respectively $\backslash\!\!\!\circ$, to the operation $\oplus$, respectively $\otimes$. However, these equations are not 'linear'. Owing to (f.1) and (f.9), the mapping $H^\sharp$ rather obeys the dual properties:

$$H^\sharp(y \wedge z) = H^\sharp(y) \wedge H^\sharp(z) \,, \qquad H^\sharp(\alpha \backslash\!\!\!\circ\, y) = \alpha \backslash\!\!\!\circ\, H^\sharp(y) \,,$$

where α is a 'scalar' (i.e. $\alpha \in \mathcal{M}_{\text{in}}^{\text{ax}}[\![\gamma, \delta]\!]$).

5.6.2 Backward Equations for Daters

We are going to translate the preceding equations in the setting of the dater description. Consider a system described by equations of the form (2.36), but with matrices (that is, holding times) which do not depend on the event number k (hence we will write e.g. $A(\ell)$ for $A(k, k-\ell)$). More specifically, we consider a system described by $\overline{\mathbb{R}}_{\max}$ equations of the form:

$$\left.\begin{aligned} x(k) &= A(0)x(k) \oplus \cdots \oplus A(M)x(k-M) \\ &\quad \oplus B(0)u(k) \oplus \cdots \oplus B(M)u(k-M) \,, \\ y(k) &= C(0)x(k) \oplus \cdots \oplus C(M)x(k-M) \,. \end{aligned}\right\} \tag{5.59}$$

There is of course no loss of generality in assuming that there is the same delay M for x and u and in both equations (this possibly amounts to completing the expressions with terms having zero matrix coefficients). The γ-transforms of these equations yield (5.55) where x, u, y now denote the γ-transforms of signals, and similarly, e.g.

$$A = \bigoplus_{\ell=1}^{M} A(\ell)\gamma^{\ell} \ .$$

In the same way, (5.56)–(5.57) are still valid with the new interpretation of notation. Using (4.97), we can write these equations more explicitly in terms of power series in γ. Taking into account that A, B, C are in fact polynomials, i.e. power series for which coefficients are ε for powers of γ out of the set $\{0, \dots, M\}$, we finally obtain the relations, for all k:

$$\left.\begin{aligned} \xi(k) &= \frac{\xi(k)}{A(0)} \wedge \dots \wedge \frac{\xi(k+M)}{A(M)} \wedge \frac{y(k)}{C(0)} \wedge \dots \wedge \frac{y(k+M)}{C(M)} \ , \\ u(k) &= \frac{\xi(k)}{B(0)} \wedge \dots \wedge \frac{\xi(k+M)}{B(M)} \ . \end{aligned}\right\} \tag{5.60}$$

From these equations, the backward recursion is clear. To alleviate notation, let us now limit ourselves to $M = 1$, that is, consider the standard form (2.39), namely,

$$x(k+1) = Ax(k) \oplus Bu(k) \ ; \qquad y(k) = Cx(k) \ . \tag{5.61}$$

Then, combining (5.60) with (5.58), we obtain

$$\left.\begin{aligned} \forall i = 1, \dots, n \ , \quad & \xi_i(k) = \left(\bigwedge_{j=1}^{n} \frac{\xi_j(k+1)}{A_{ji}} \right) \wedge \left(\bigwedge_{r=1}^{p} \frac{y_r(k+1)}{C_{ri}} \right), \\ \forall \ell = 1, \dots, m \ , \quad & u_\ell(k) = \bigwedge_{s=1}^{n} \frac{\xi_s(k)}{B_{s\ell}} \ . \end{aligned}\right\} \tag{5.62}$$

Let us rewrite these equations with conventional notation. We refer the reader to Example 4.65, to the rule regarding the ambiguous expression $\infty - \infty$ that may show up if ratios such as $\varepsilon \backslash\!\!\!\! \circ \varepsilon$ are encountered, and finally to the warning about the ambiguity of conventional notation since the expression $\infty - \infty$ may also be obtained as the result of $\varepsilon \otimes \top$ which yields a different value. With this warning in mind, (5.62) can be written:

$$\left.\begin{aligned} \xi_i(k) &= \min\left[\min_{j=1}^{n} \left(\xi_j(k+1) - A_{ji} \right), \min_{r=1}^{p} \left(y_r(k+1) - C_{ri} \right) \right], \\ u_\ell(k) &= \min_{s=1}^{n} \left(\xi_s(k) - B_{s\ell} \right) \ . \end{aligned}\right\} \tag{5.63}$$

The reader is invited to establish these equations by a direct reasoning with an event graph for which the sequence $\{y(k)\}_{k\in\mathbb{Z}}$ of desired output dates is given. It is then realized that the recursion is not only backward in event numbering (index k) but also backward in the graph (from output transitions to input transitions through internal transitions).

A few words are in order regarding 'initial conditions' of the recursion (5.63). This problem arises when the backward recursion starts at some finite event number, say k_{f} ('f' for 'final'), because desired objectives $\{y(k)\}$ are only given up to this number k_{f}. In accordance with the idea of finding the 'latest input dates', that is the greatest subsolution of $CA^*Bu \leq y$ as supposed by residuation theory, the missing information must be set to the maximum possible value. This amounts to saying that $y(k)$ must be set to $\top = +\infty$ beyond k_{f}, and more importantly $\xi(k) = \top$ for $k > k_{\mathrm{f}}$. As for the tokens of the initial marking, they are still supposed to be available at time $-\infty$ (see page 245) since this is the assumption under which the 'direct' system obeys the input-output relation at the left-hand side of Inequality (5.53).

At the end of this section, let us consider the following situation. Suppose that some output trajectory $y(\cdot)$ has been produced by processing some input $v(\cdot)$ through a system obeying Equations (5.61). This output trajectory is taken as the desired latest output trajectory. Of course, it is feasible since it is an actual output of the system. Then, if we compute the latest possible input, say u, that meets the given objective by using Equations (5.62), this u will also produce the output y, and it will be greater than or equal to v. Therefore, the pairs (v, y) and (u, y) are both solutions of (5.61), but two different internal state trajectories, say χ and x, respectively, would be obtained with $\chi \leq x$. Moreover, the trajectory x is also different from ξ which is computed by (5.62) using y as the input. The differences $\xi_i(k) - x_i(k), i = 1, \ldots, n; k \in \mathbb{Z}$, are nonnegative since (5.62) corresponds to the backward operation at 'the latest time' whereas (5.61) describes the forward operation 'at the earliest time' for the same input-output pair (u, y). For the k-th firing of the i-th transition, the difference $\xi_i(k) - x_i(k)$ indicates the time margin which is available, that is, the maximum delay by which this firing may be postponed, with respect to its earliest possible occurrence, without affecting the output transition firing. This kind of information is of course very useful in practical situations.

Let us finally summarize the equations satisfied by the pair (x, ξ). The following is derived from (5.61)–(5.62) in which u and y are replaced by their expressions, namely $u = B \backslash\!\!\!\backslash \xi$ and $y = Cx$, respectively. Then we obtain the system

$$\left.\begin{aligned} x(k+1) &= Ax(k) \oplus B \frac{\xi(k)}{B} \ , \\ \xi(k) &= \frac{Cx(k+1)}{C} \wedge \frac{\xi(k+1)}{A} \ . \end{aligned}\right\} \tag{5.64}$$

This system is very reminiscent of state/co-state (or Hamiltonian) equations derived for example from Pontryagin's minimum principle for optimal control problems in conventional control theory. Moreover, the difference $\xi_i(k) - x_i(k)$ alluded to above is the i-th diagonal entry of the matrix $\xi(k) /\!\!\!/ x(k)$. Pursuing the analogy with conventional control problems, it is known that introducing the 'ratio' of the co-state vector by the state vector yields a matrix which satisfies a Riccati equation.

5.7 RATIONALITY, REALIZABILITY AND PERIODICITY

5.7.1 Preliminaries

At this point, we know that event graphs can be described by general equations of the form (5.55) in which the mathematical form of u, x, y, A, B, C depends on the description adopted. Essentially u, x, y may be:

- power series in γ with coefficients in $\overline{\mathbb{Z}}_{\max}$;
- power series in δ with coefficients in $\overline{\mathbb{Z}}_{\min}$;
- power series in (γ, δ) with coefficients in $\{\varepsilon, e\}$.

As for A, B, C, they are matrices with polynomial entries of the same nature as the power series u, x, y, but with only *nonnegative* exponents since tokens of the initial marking and holding times introduce nonnegative 'backward' shifts (see Remark 5.6) in event numbering, respectively in time. The input-output relation $u \mapsto y$ is given by $y = CA^*Bu$, hence the entries of this so-called transfer matrix belong to the rational closure (see Definition 4.99) of the corresponding class of polynomials with nonnegative exponents. One purpose of this section is to study the converse implication, namely that a system with a 'rational' transfer matrix does have a finite-dimensional 'realization' of the form (5.55) with polynomial matrices. In fact, this is an immediate consequence of Theorem 4.102. Moreover, playing with the various possibilities of realizations provided by Theorem 4.105, we will recover essentially the three possible descriptions of systems alluded to above by starting from a single framework, namely the one offered by the $\mathcal{M}_{\mathrm{in}}^{\mathrm{ax}}[\![\gamma, \delta]\!]$ algebra.

Remark 5.34 In an event graph, if there are direct arcs from input to output transitions, one obtains an output equation of the form $y = Cx \oplus Du$ and a transfer matrix of the form $CA^*B \oplus D$. However, by redefining the 'state' vector as $\widetilde{x} = \begin{pmatrix} x' & u' \end{pmatrix}'$, it is possible to come back to the form $\widetilde{C}\widetilde{A}^*\widetilde{B}$ with $\widetilde{C} = \begin{pmatrix} C & D \end{pmatrix}$, $\widetilde{B}' = \begin{pmatrix} B' & e \end{pmatrix}$ and $\widetilde{A} = \mathrm{diag}(A, \varepsilon)$, but at the price of increasing the 'state' dimensionality. Indeed, in what follows the issue of the 'minimal' realization will not be addressed since only partial results have been obtained so far. ■

The equivalence between rationality of the transfer matrix and its 'realizability' is a classical result in both conventional linear system theory and in automata and formal language theory. However, there is here a third ingredient coming in: rationality is also equivalent to some 'periodicity' property of the transfer function or the impulse response. This is analogous to the situation of rational numbers which have a periodic decimal expansion.

We will only address the SISO case. The MIMO case (e.g. 2 inputs, 2 outputs) can be dealt with in a trivial manner by considering all the individual scalar transfer functions $H_{ij} : u_j \mapsto y_i$, $j = 1, 2$, $i = 1, 2$, first. Suppose that 3-tuples (A_{ij}, B_{ij}, C_{ij}) have been found to realize H_{ij} in the form (5.55) (A_{ij} is in general a matrix, not a scalar). Then, it is easy to check that the 3-tuple

$$A = \begin{pmatrix} A_{11} & \varepsilon & \varepsilon & \varepsilon \\ \varepsilon & A_{12} & \varepsilon & \varepsilon \\ \varepsilon & \varepsilon & A_{21} & \varepsilon \\ \varepsilon & \varepsilon & \varepsilon & A_{22} \end{pmatrix}, \quad B = \begin{pmatrix} B_{11} & \varepsilon \\ \varepsilon & B_{12} \\ B_{21} & \varepsilon \\ \varepsilon & B_{22} \end{pmatrix},$$

$$C = \begin{pmatrix} C_{11} & C_{12} & \varepsilon & \varepsilon \\ \varepsilon & \varepsilon & C_{21} & C_{22} \end{pmatrix},$$

is a realization of the 2×2 transfer matrix. Of course, this way of handling MIMO systems does not consider the dimensionality of the realization explicitly. In the following, we will comment on the MIMO case when appropriate.

5.7.2 Definitions

We start with the following definitions.

Definition 5.35 (Causality) *An element h of $\mathcal{M}_{\text{in}}^{\text{ax}}[\![\gamma, \delta]\!]$ is causal either if $h = \varepsilon$ or if $\text{val}(h) \geq 0$ and $h \geq \gamma^{\text{val}(h)}$.*

This definition is somewhat technical, due to the fact that h, as an element of $\mathcal{M}_{\text{in}}^{\text{ax}}[\![\gamma, \delta]\!]$, has various formal representations, and among them e.g. the maximal one which involves the multiplication by $(\delta^{-1})^*$ (hence it may have monomials with negative exponents in δ). However, the definition, while using the language of $\mathcal{M}_{\text{in}}^{\text{ax}}[\![\gamma, \delta]\!]$, clearly says that the graph of the associated dater lies in the right-half plane and above the x-axis. It can then be formally checked that the set of causal elements of $\mathcal{M}_{\text{in}}^{\text{ax}}[\![\gamma, \delta]\!]$ is a subdioid of $\mathcal{M}_{\text{in}}^{\text{ax}}[\![\gamma, \delta]\!]$. For example, if p and q are causal, then $\text{val}(p \oplus q) = \min(\text{val}(p), \text{val}(q)) \geq 0$, and $\gamma^{\text{val}(p \oplus q)} = \gamma^{\min(\text{val}(p), \text{val}(q))} = \gamma^{\text{val}(p)} \oplus \gamma^{\text{val}(q)} \leq p \oplus q$, proving that $p \oplus q$ is also causal. A similar proof can be given for $p \otimes q$.

Definition 5.36 (Rationality) *An element of $\mathcal{M}_{\text{in}}^{\text{ax}}[\![\gamma, \delta]\!]$ is rational if it belongs to the rational closure of the subset $\mathcal{T} \stackrel{\text{def}}{=} \{\varepsilon, e, \gamma, \delta\}$. A vector or matrix is rational if its entries are all rational.*

Indeed, because of the choice of the basic set, the rational elements will also be causal.

Definition 5.37 (Realizability) *A matrix $H \in \left(\mathcal{M}_{\text{in}}^{\text{ax}}[\![\gamma, \delta]\!]\right)^{p \times m}$ is realizable if it can be written as*

$$H = C(\gamma A_1 \oplus \delta A_2)^* B \tag{5.65}$$

where A_1 and A_2 are $n \times n$ matrices, n being an arbitrary but finite integer (depending on H), C and B are $n \times m$ and $p \times n$ matrices respectively, and each entry of these matrices is equal to either ε or e.

Definition 5.38 (Periodicity) *An element h of $\mathcal{M}_{\mathrm{in}}^{\mathrm{ax}}[\![\gamma,\delta]\!]$ is periodic if there exist two polynomials p and q and a monomial m (all causal) such that*

$$h = p \oplus qm^* \ . \tag{5.66}$$

A matrix H is periodic if its entries are all periodic.

Here, we adopt a 'mild' definition of periodicity. It is however mathematically equivalent to seemingly other more sophisticated definitions which put further constraints on the polynomials p and q. This point will be discussed in §5.7.4. At this stage, it suffices to understand that, if one considers h as the (γ,δ)-transform of a trajectory, say an impulse response, the intuitive meaning of Formula (5.66) is that a certain pattern represented by q is reproduced indefinitely, since the multiplication by $m = \gamma^r\delta^s$ represents a shift by r units along the x-axis—event domain—and s units along the y-axis—time domain—and $qm^* = q \oplus qm \oplus qm^2 \oplus \cdots$ is the union of all these shifted versions of q. This periodic behavior occurs after a certain transient which is essentially (but not always exactly, as we shall see) represented by p. The ratio s/r represents the asymptotic slope of the graph of the dater associated with h, and thus the asymptotic output rate: on the average, r events occur every s time units. The extreme cases $s = 0$ and $r = 0$ will be discussed in §5.7.4.

5.7.3 Main Theorem

Theorem 5.39 *For $H \in \left(\mathcal{M}_{\mathrm{in}}^{\mathrm{ax}}[\![\gamma,\delta]\!]\right)^{p\times m}$, the following three statements are equivalent*

(i) H is realizable;

(ii) H is rational;

(iii) H is periodic.

Proof The implication (i) $\Rightarrow$ (ii) is straightforward. The converse (ii) $\Rightarrow$ (i) follows, at least for the SISO case, from Theorem 4.105 with $\mathcal{B} = \mathcal{C} = \mathcal{U} = \mathbb{B} = \{\varepsilon, e\}$ and $\mathcal{V} = \{\gamma, \delta\}$. We then observe that $\mathcal{U}^\star = \mathcal{U}$ hence $\mathcal{U}^\star \otimes \mathcal{V} = \{\varepsilon, \gamma, \delta, \gamma \oplus \delta\}$. It remains to split up matrix A_x, which appears in the $(\mathcal{B},\mathcal{C})$-representation (4.106) (with entries in $\mathcal{U}^\star \otimes \mathcal{V}$), into $\gamma A_1 \oplus \delta A_2$, which offers no difficulty. The MIMO case is handled as indicated previously.

We now outline the proof of the equivalence (ii) $\Leftrightarrow$ (iii). Since the definitions of periodicity and rationality refer to the entries of H individually, it suffices to deal with the SISO case. The implication (iii) $\Rightarrow$ (ii) is obvious: if h can be written as in (5.66), then clearly $h \in \mathcal{T}^\star$. Conversely, if h is rational, since $\mathcal{M}_{\mathrm{in}}^{\mathrm{ax}}[\![\gamma,\delta]\!]$ is a commutative dioid, we can use Theorem 4.110 (applied to the dioid closure of $\mathcal{T}$) to see that h can be written as

$$\begin{aligned} h &= \bigoplus_{i\in I} \gamma^{\alpha_i}\delta^{\beta_i}\left(\bigoplus_{j\in J_i}\gamma^{r_j}\delta^{s_j}\right)^* \\ &= \bigoplus_{i\in I} \gamma^{\alpha_i}\delta^{\beta_i}\bigotimes_{j\in J_i}(\gamma^{r_j}\delta^{s_j})^* \ , \end{aligned} \tag{5.67}$$

where I and the J_i are finite sets, $\alpha_i, \beta_i, r_j, s_j$ are nonnegative integers, and (5.67) follows from (4.109). The proof is then completed by showing that (5.67) is amenable to the form (5.66) where m is essentially the monomial $\gamma^{r_j}\delta^{s_j}$ with maximal 'slope' s_j/r_j. Indeed, the term $(\gamma^{r_j}\delta^{s_j})^*$ tends to asymptotically dominate all other similar terms in sums and products. When the monomial with maximal slope is nonunique, the precise rules for obtaining the monomial m used in (5.66) will be given in the proof of Theorem 6.32 in the next chapter. The precise derivation of this last part of the proof, which is rather technical, will be skipped here. The reader is referred to [44] for a more detailed outline and to [62] for a full treatment. ■

In the above proof, instead of attempting to prove the implication (ii) $\Rightarrow$ (iii), we might have proved the implication (i) $\Rightarrow$ (iii) using Theorem 4.109 and Formula (4.110). This would have provided some insight into how the monomial m_{ij} appearing in the representation (5.66) of each H_{ij} (in the form of m_{ij}^*) is related to the weights of circuits of $\mathcal{G}(A)$, where $A = \gamma A_1 \oplus \delta A_2$ appears in (5.65). The graph $\mathcal{G}(A)$ is drawn as explained in §4.7.3 (see Figure 4.8). Then, for an input-output pair (u_j, y_i), we consider all (oriented) paths connecting these transitions and all circuits which have at least one node in common with those paths. These are the circuits of interest to determine the maximal ratio s_{ij}/r_{ij}. If there is no such circuit, this means that the polynomial q of (5.66) is ε and thus m_{ij} is irrelevant. As a consequence, if matrix A is strongly connected (which in particular precludes any direct path from the input to the output, given that we have not included a term Du — see Remark 5.34), then the ratios s_{ij}/r_{ij} take a unique value for all the m_{ij}.

We could also have proved that (iii) $\Rightarrow$ (i) directly by providing an explicit realization of a periodic element. This essentially follows the scheme illustrated by Figure 6.4 in the next chapter.

5.7.4 On the Coding of Rational Elements

In this subsection, it is helpful to use the pictorial representation of (causal) elements of $\mathcal{M}_{\text{in}}^{\text{ax}}[\![\gamma, \delta]\!]$ by collection of points in the $\mathbb{N}^2$-plane: a monomial $\gamma^k\delta^t$ is represented by a point with coordinates (k, t) and we assume that polynomials are represented by their minimum representatives (which do exist—see Theorem 5.20).

Let the monomial m involved in (5.66) be equal to $\gamma^r\delta^s$.

- If $s = 0$, then $a = p \oplus qm^* = p \oplus q$ since $m^* = (\gamma^r)^* = e$, that is, the impulse response a is a polynomial : this is the behavior of an event graph having either no circuits or only circuits with zero holding time. Such a system is able to work at infinite speed. This 'finite' impulse response thus corresponds to the situation when an infinite number of tokens get out at time $t = \deg(a)$; the first tokens get out earlier as described by $p \oplus q$. From the point of view of the minimum coding of a, there no way but to retain the minimum representative of $p \oplus q$.

- If $s \neq 0$ but $r = 0$ (and $q \neq \varepsilon$), and since $m^* = (\delta^s)^*$ corresponds to an infinite slope, the impulse response is 'frozen' after a transient during which some tokens get out (those numbered from 0 to $\text{val}(q) - 1$): this is indeed the kind of behavior one would like to call a 'finite response'. It happens when there is a circuit without tokens but with a positive holding time: the system is 'deadlocked'. The nontrivial part of a is provided by $p \ominus \gamma^{\mathrm{val}(q)}\delta^*$ (the part of the plot of a which lies at the left hand of the vertical asymptote) which should again be coded by its minimum representative.

These two particular cases will no longer be considered in detail. Hence, from now on, we assume that $r > 0$ and $s > 0$ unless stated otherwise.

With (5.66), we have adopted a characterization of rational (or periodic) elements which is mathematically simple. The forthcoming lemma shows that this characterization is equivalent to another one which puts further constraints on the polynomials p and q. The latter definition will be used in Chapter 6 (see Theorem 6.32). It has the advantage of making the interpretation of the new p as the transient part and of the new q as the periodic pattern (see Figure 6.2) possible. Indeed, p can then be represented by points contained in a box of width $\nu - 1$ and height $\tau - 1$ with its lower left-hand corner located at the origin. The periodic pattern q can be encapsulated in a box of width $r - 1$ and height $s - 1$ with its lower left-hand corner located at the point (ν, τ). This box is translated indefinitely by the vector (r, s). These conditions are now expressed mathematically in the following lemma.

Lemma 5.40 *An element $a \in \mathcal{M}_{\mathrm{in}}^{\mathrm{ax}}[\![\gamma, \delta]\!]$ is rational if and only if it is a polynomial, or an element of the form $p \oplus \gamma^\nu \delta^*$ (p is a polynomial), or if there exist positive integers r and s, nonnegative integers ν and τ, and polynomials*

$$p = \bigoplus_{(k,t)\in J_p} \gamma^k \delta^t \quad \text{with} \quad k \leq \nu - 1, \quad t \leq \tau - 1, \quad \forall (k,t) \in J_p \ ,$$

$$q = \bigoplus_{(\kappa,\theta)\in J_q} \gamma^\kappa \delta^\theta \quad \text{with} \quad \nu \leq \kappa \leq \nu + r - 1, \ \tau \leq \theta \leq s + \tau - 1, \ \forall (\kappa, \theta) \in J_q \ ,$$

(J_p and J_q are necessarily finite) such that $a = p \oplus q\,(\gamma^r \delta^s)^*$.

Proof Only the case when $r > 0$, $s > 0$ and $q \neq \varepsilon$ is of interest here. Moreover, since the new characterization of rational elements is a particular case of (5.66), it suffices to prove that, conversely, (5.66) is amenable to the new characterization. Consider $a = p \oplus q(\gamma^r \delta^s)^*$ for *any* polynomials p and q. Let $\tau \stackrel{\text{def}}{=} \max(\deg(p) + 1, \deg(q) - s + 1)$. For any $(k, t) \in J_q$, there exists a unique $\ell_{(k,t)} \in \mathbb{Z}$ such that $\tau \leq t + \ell_{(k,t)} s \leq \tau + s - 1$. Because $\deg(q) \leq \tau + s - 1$ and $s > 0$, necessarily $\ell_{(k,t)} \geq 0$. We consider

$$\widetilde{q} \stackrel{\text{def}}{=} \bigoplus_{(k,t)\in J_q} \gamma^{k+\ell_{(k,t)} r} \delta^{t+\ell_{(k,t)} s} \ . \tag{5.68}$$

In addition, let $\alpha \stackrel{\text{def}}{=} \max_{(k,t)\in J_q} \ell_{(k,t)}$ and $\nu \stackrel{\text{def}}{=} \mathrm{val}\left(\widetilde{q}\right)$. Observe that all points representing monomials appearing at the right-hand side of (5.68) lie in a strip of height $s - 1$ delimited by the horizontal lines $y = \tau$ and $y = \tau + s - 1$, and at the right-hand closed half plane bordered by the vertical line at $x = \nu$.

Let a minimum representative of a be written $\bigoplus_{(k,t)\in J_a} \gamma^k\delta^t$ (J_a is countably infinite). This minimum representative does exist since we deal with causal elements and since we assume that $r > 0$ (see Theorem 5.20). Consider any $(k,t) \in J_a$. If $t \geq \tau$, then the corresponding monomial cannot belong to p since $\tau > \deg(p)$. Hence, it necessarily belongs to some qm^n. If $t < \tau$, then this monomial may belong to either p or to some qm^n but then $n < \alpha$: indeed, for $n \geq \alpha$, we have by construction that $t \geq \tau$. Hence, if we set $\widetilde{p} \stackrel{\text{def}}{=} p \oplus q \oplus qm \oplus \cdots \oplus qm^{\alpha-1}$, we can consider all pairs $(k,t) \in J_a$ with $t < \tau$ as coming from $\widetilde{p}$. We now prove that the other pairs can be explained by monomials of $\widetilde{q}m^*$. If $(k,t) \in J_a$ and $t \geq \tau$, then there exist $\left(\widetilde{k},\widetilde{t}\right) \in J_q$ and $\ell \geq 0$ such that $(k,t) = \left(\widetilde{k},\widetilde{t}\right) + \ell \times (r,s)$. Moreover, $\ell \geq \ell_{\left(\widetilde{k},\widetilde{t}\right)}$, hence

$$(k,t) = \left(\widehat{k},\widehat{t}\right) + \left(\ell - \ell_{\left(\widetilde{k},\widetilde{t}\right)}\right) \times (r,s) \enspace ,$$

where $\gamma^{\widehat{k}}\delta^{\widehat{t}}$ is one of the polynomials involved in (5.68).

At this point, we have proved that all monomials of $a = p \oplus qm^*$ are among the monomials of $\widetilde{p} \oplus \widetilde{q}m^*$, hence $a \leq \widetilde{p} \oplus \widetilde{q}m^*$. But the converse statement is also true since the monomials in $\widetilde{p}$ and $\widetilde{q}$ have been obtained from a. Hence $\widetilde{p} \oplus \widetilde{q}m^*$ is another expression for a.

To complete the proof, we must delete monomials which are useless from this expression (because they are dominated by other polynomials in the same expression in the sense of the order relation of $\mathcal{M}_{\text{in}}^{\text{ax}}[\![\gamma,\delta]\!]$). Firstly, concerning $\widetilde{p}$, if monomials of $\widetilde{p}$, thus also of a, have a degree greater than or equal to τ, they can also be obtained by other monomials of $\widetilde{q}m^*$ (proceed as previously) and thus they can be dropped from $\widetilde{p}$. Let $\widetilde{m} = \gamma^\nu\delta^{\widetilde{\tau}}$ be the monomial of $\widetilde{q}$ with valuation ν (recall that $\nu = \text{val}\left(\widetilde{q}\right)$). Observe that $\widetilde{\tau} \geq \tau$. This monomial dominates the monomials $\gamma^k\delta^t$ of $\widetilde{p}$ with $t < \tau$ but $k \geq \nu$ which can thus also be dropped from $\widetilde{p}$. Finally, the new $\widetilde{p}$ stays in the lower left-hand part of the plane delimited by the horizontal line $y = \tau - 1$ and the vertical line $x = \nu - 1$. Secondly, concerning $\widetilde{q}$, consider the monomials of $\widetilde{q}$ with valuation greater than $\nu + s - 1$: their degree being necessarily less than $\tau + s$, they are dominated by the monomial $\gamma^{\nu+r}\delta^{\widetilde{\tau}+s}$ contained in $\widetilde{q}m$. This observation is of course preserved by the successive translations (r,s). Therefore, a new $\widetilde{q}$ can be used which stays in the box given in the statement of the lemma. ■

We refer the reader to [62] in which an algorithm is provided to obtain the 'best' possible representation of the type described by the lemma. The following example shows that by redefining not only p and q but also m, more compact representations may be obtained (the reader is invited to make the drawing corresponding to each example, which is the best way to quickly grasp the situation).

Example 5.41 The expression $(e \oplus \gamma\delta)(\gamma^2\delta^2)^*$ is already in the form of Lemma 5.40 but it can be simplified to $(\gamma\delta)^*$ by redefining m as $\gamma\delta$ instead of $\gamma^2\delta^2$. ■

Consider now the following example.

Example 5.42 Let $a = p \oplus qm^*$ with $p = e \oplus \gamma^2\delta^2 \oplus \gamma^5\delta^3 \oplus \gamma^6\delta^4 \oplus \gamma^8\delta^6 \oplus \gamma^{11}\delta^7$, $q = \gamma^{12}\delta^8(e \oplus \gamma^2\delta)$ and $m = \gamma^3\delta^2$. Another representation of a as $\widetilde{p} \oplus \widetilde{q}m^*$ involves $\widetilde{p} = \gamma^2\delta^2 \oplus \gamma^8\delta^6$ and $\widetilde{q} = e \oplus \gamma^2\delta$. ■

In this example, what happens is that p can partly be explained by 'noncausal shifts' qm^{-l} of q. Algebraically, there exists some n (here $n = 4$) such that adding $b = \bigoplus_{l=1}^{n} qm^{-l}$ to a does not change a. Hence

$$a = a \oplus b = p \oplus qm^* \oplus b = (p \ominus b) \oplus b \oplus qm^* = \widetilde{p} \oplus \widetilde{q}m^* \ ,$$

where $\widetilde{p} = p \ominus b$ and $\widetilde{q} = qm^{-n}$. Now $\widetilde{p}$ does not any longer appear as the transient part, but rather as a transient perturbation of the periodic regime. Now the contributions of $\widetilde{p}$ and $\widetilde{q}$ to the transient part are interweaved.

5.7.5 Realizations by γ- and δ-Transforms

In the proof of Theorem 5.39, Theorem 4.105 has been used with the choice $\mathcal{B} = \mathcal{C} = \mathcal{U} = \mathbb{B}$ and $\mathcal{V} = \{\gamma, \delta\}$. Other possibilities have been suggested at the end of §4.8.3.

5.7.5.1 *Dater Realization*

If we consider the possibility $\mathcal{B} = \mathcal{C} = \mathbb{B}$, $\mathcal{U} = \mathbb{B} \cup \{\delta\}$ and $\mathcal{V} = \{\gamma\}$, then $\mathcal{U}^\star = \{\varepsilon, e, \delta, \delta^2, \ldots, \delta^*\}$ since all sums of such elements are reducible to one of them by the rule $\delta^t \oplus \delta^\tau = \delta^{\max(t,\tau)}$. Hence $\mathcal{U}^\star$, being obviously a complete dioid in this case, is isomorphic to (and will be identified with) $\overline{\mathbb{N}}_{\max}$ (with δ^* identified with $\top = +\infty$). Consequently, in an observer representation $(C, \widetilde{A}, B)$ of a rational matrix H as provided by Theorem 4.105, B has its entries in $\overline{\mathbb{N}}_{\max}$ and $\widetilde{A} = \gamma A$ where A is also a matrix with entries in $\overline{\mathbb{N}}_{\max}$; C is a Boolean matrix. This realization may be interpreted as the one directly derived, by the γ-transform, from the dater equations

$$x(k) = Ax(k-1) \oplus Bu(k) \ , \qquad y(k) = Cx(k) \ ,$$

where in addition y is a subvector of x (hence the name 'observer representation').

For a controller representation, B is a Boolean matrix whereas C makes any linear combination of the x_i with weights in $\overline{\mathbb{N}}_{\max}$.

Remark 5.43 It is thus always possible, starting from any timed event graph, to obtain an equivalent event graph with a structure corresponding to the observer (respectively, the controller) representation, that is, the initial marking consists of exactly one token in internal places, no tokens in the input and output places, and in addition, zero holding times for the output (respectively, the input) places. ■

It should be realized that there is a trick here to represent deadlocked systems in this way, i.e. event graphs circuits with no tokens and positive holding times. These systems will be realized in 'state space' form by matrices A, B or C having some entries equal to δ^*. Indeed, an arc with weight δ^* introduces an unbounded

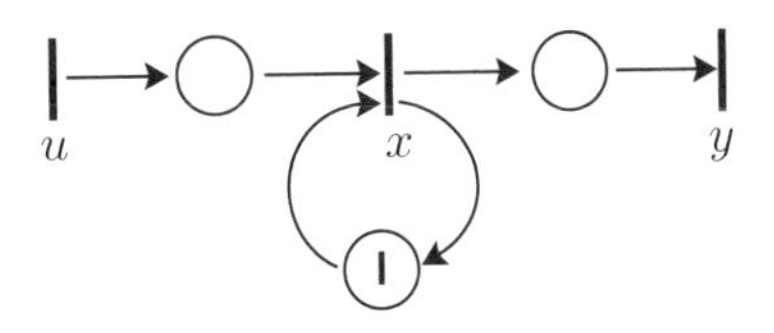

Figure 5.12: A deadlocked event graph

holding time, which is sufficient to block some parts of the system. For example, for the graph of Figure 5.12 (the transfer function of which is δ^*), an observer (dater) realization is $C = e$, $\widetilde{A} = \gamma$, $B = \delta^*$.

5.7.5.2 *Counter Realization*

The counter realization corresponds to the dual possibility offered by Theorem 4.105, namely to choose $\mathcal{U} = \mathbb{B} \cup \{\gamma\}$ and $\mathcal{V} = \{\delta\}$. Then $\mathcal{U}^\star = \{\varepsilon, e, \gamma, \gamma^2, \ldots\}$. This is a complete dioid. Due to the rule $\gamma^k \oplus \gamma^\kappa = \gamma^{\min(k,\kappa)}$, this $\mathcal{U}^\star$ is now identified with $\overline{\mathbb{N}}_{\min}$ (note also that $\gamma^* = e$).

Remark 5.44 In $\overline{\mathbb{N}}_{\max}$, all elements are greater than e except ε. In $\overline{\mathbb{N}}_{\min}$, all elements lie between $\varepsilon = +\infty$ and $\top = e = 0$. ■

In this new context, the realizations one obtains appear to be derived, by the δ-transform, from counter equations in 'state space' form with, in addition, Boolean matrices C or B.

Deadlocked systems can be represented directly since a weight of e means no tokens in the corresponding place (or arc). For the graph of Figure 5.12, an observer (counter) realization is $C = B = e$ and $\widetilde{A} = \delta$.

5.8 FREQUENCY RESPONSE OF EVENT GRAPHS

In conventional system theory, with a continuous time-domain, rational transfer functions may be viewed as rational expressions of a formal operator denoted s, which can be interpreted as the derivative with respect to time. However, transfer functions, say $H(s)$, are also used as *numerical* functions when an imaginary numerical value $j\omega$ is substituted for s. It is well known that when a signal of pure frequency ω is used as the input of a (stable) linear time-invariant system, the corresponding output is, after a possible transient, a signal of the same frequency ω which is phase-shifted by arg $H(j\omega)$ and amplified by the gain $|H(j\omega)|$ with respect to the input. The transient can be avoided by starting at time $-\infty$. Then sine functions of any frequency ω appear as eigenfunctions of any rational transfer function H with corresponding eigenvalue $H(j\omega)$.

The main purpose of this section is to discuss analogous results for event graphs. We will confine ourselves to the SISO case. Since we can consider transfer functions under three different forms as recalled at the beginning of §5.7 (referred to as dater, counter or two-dimensional representations), the following developments could be made from these three different points of view. We will favor the two-dimensional point of view, but the analogy of certain quantities to phase shift or amplification gain depends on which of the first two points of view is adopted. In §6.4, this topic will be revisited from the dater point of view and with a continuous domain.

5.8.1 Numerical Functions Associated with Elements of $\mathbb{B}[\![\gamma,\delta]\!]$

In this section, an element F of $\mathbb{B}[\![\gamma,\delta]\!]$ or of $\mathcal{M}_{\text{in}}^{\text{ax}}[\![\gamma,\delta]\!]$ will be written either as

$$\bigoplus_{(k,t)\in J_F} \gamma^k\delta^t \tag{5.69}$$

or as

$$\bigoplus_{(k,t)\in\mathbb{Z}^2} F(k,t)\gamma^k\delta^t \quad\text{with}\quad F(k,t)=\begin{cases} e & \text{if } (k,t)\in J_F\ ;\\ \varepsilon & \text{otherwise.}\end{cases} \tag{5.70}$$

Guided by the analogy with conventional system theory alluded to above, we are going to use such a formal expression as a numerical function. This means that numerical integer values will be substituted for the formal variables γ and δ, and the quantity thus obtained will be evaluated numerically. Therefore, the symbols $\oplus$ and $\otimes$ (the latter being implicit in the above expressions) must be given a meaning in order to operate on numbers. We choose the max-plus interpretation, that is, $\oplus$ will be interpreted as max (or sup) and $\otimes$ as $+$. Consistently, a coefficient e is interpreted as 0 and ε as $-\infty$. For the time being, let us consider F as an element of $\mathbb{B}[\![\gamma,\delta]\!]$ and define the associated numerical function, denoted $\mathcal{F}(F)$, as the mapping from $\mathbb{Z}^2$ into $\overline{\mathbb{Z}}$:

$$\begin{aligned}(g,d)\mapsto \bigoplus_{(k,t)\in\mathbb{Z}^2} F(k,t)g^kd^t &= \sup_{(k,t)\in\mathbb{Z}^2}\big(F(k,t)+gk+dt\big)\\ &= \sup_{(k,t)\in J_F}(gk+dt)\ .\end{aligned} \tag{5.71}$$

> In this section, according to our general convention, the multiplication, denoted by mere juxtaposition of elements, must be interpreted as the conventional $\times$ when the $+$, $-$, sup or inf symbols are involved in the same expression.

Lemma 5.45 *The set of mappings from $\mathbb{Z}^2$ into $\overline{\mathbb{Z}}$ endowed with the pointwise maximum as addition, and the pointwise (conventional) addition as multiplication, is a complete dioid. The mapping $\mathcal{F}$ introduced above is a l.s.c. dioid homomorphism from $\mathbb{B}[\![\gamma,\delta]\!]$ into this dioid of numerical functions.*

The proof is straightforward.

The mapping $\mathcal{F}$ will be referred to as the *evaluation homomorphism* in this chapter and the next one (see §6.4.1), although the context is somewhat different in the two chapters. Equations (5.71) show that not all numerical functions are in the range of $\mathcal{F}$ for at least two reasons.

- If the function $\mathcal{F}(F)$ was extended to the continuous domain $\mathbb{R}^2$ instead of $\mathbb{Z}^2$ (with range in $\overline{\mathbb{R}}$ instead of $\overline{\mathbb{Z}}$) in an obvious manner, then it would be a *convex* function as the supremum of a family of linear functions.

- The function $\mathcal{F}(F)$ is *positively homogeneous of degree* 1, that is, $[\mathcal{F}(F)](a \times g, a \times d) = a \times [\mathcal{F}(F)](g, d)$ for any nonnegative (integer) number a (in particular, $[\mathcal{F}(F)](0, 0) = 0$).

For the latter reason, it suffices practically to know the value of $\mathcal{F}(F)$ for all values of the ratio g/d which ranges in $\overline{\mathbb{Q}}$.

From the geometric point of view, since an element F of $\mathbb{B}[\![\gamma, \delta]\!]$ encodes a subset of points J_F in the $\mathbb{Z}^2$ plane, it is realized that $\mathcal{F}(F)$ is nothing but the so-called 'support function' of this subset [119]. It is well known that support functions characterize only the convex hulls of subsets: this amounts to saying that $\mathcal{F}$ is certainly not injective; its value at F depends only on the extreme points of the subset associated with F.

Being l.s.c. and such that $\mathcal{F}(\varepsilon) = \varepsilon$, the mapping $\mathcal{F}$ is residuated.

Lemma 5.46 *Let $\widehat{F}$ be a mapping from $\mathbb{Z}^2$ into $\overline{\mathbb{Z}}$. If $\widehat{F}$ is positively homogeneous of degree 1, then $F = \mathcal{F}^\sharp\left(\widehat{F}\right)$ is obtained by*

$$F(k, t) = \inf_{(g,d) \in \mathbb{Z}^2} \left(\widehat{F}(g, d) - gk - dt\right). \tag{5.72}$$

Proof Recall that F is the largest element in $\mathbb{B}[\![\gamma, \delta]\!]$ such that $\mathcal{F}(F) \le \widehat{F}$. Because of (5.71), for all $(k, t) \in \mathbb{Z}^2$, we must have

$$\forall (g, d) \in \mathbb{Z}^2 , \quad F(k, t) \le \widehat{F}(g, d) - (gk + dt) ,$$

hence,

$$F(k, t) \le \inf_{(g,d) \in \mathbb{Z}^2} \left(\widehat{F}(g, d) - (gk + dt)\right). \tag{5.73}$$

The largest such F is defined by the equality in (5.73). For this F, we must prove that:

1. $F(k, t)$ assumes only the values $-\infty$ and 0;

2. the inequality $\mathcal{F}(F) \le \widehat{F}$ is still verified for this F.

The first fact stems from the homogeneity of $\widehat{F}$. Indeed, if we set $g = d = 0$ at the right-hand side of (5.73), this shows that $F(k, t) \le 0$. Then, it suffices to realize that the inf cannot be equal to any *finite* and strictly negative value: as a matter of fact, for any positively homogeneous function φ, we have

$$\inf_{(g,d) \in \mathbb{Z}^2} \varphi(g, d) = \inf_{a \in \mathbb{N}} \left(a \times \left[\inf_{(g,d) \in \mathbb{Z}^2} \varphi(g, d)\right]\right),$$

and a contradiction would be obtained for any value of this inf different from 0 and $-\infty$.

As for item 2 above, according to (5.71) and the definition of F, we obtain

$$\begin{aligned}\left[\mathcal{F}(F)\right](g,d) &= \sup_{(k,t)\in\mathbb{Z}^2}\left(gk+dt+\inf_{(\overline{g},\overline{d})}\left(\widehat{F}(\overline{g},\overline{d})-\overline{g}k-\overline{d}t\right)\right)\\ &\le \inf_{(\overline{g},\overline{d})}\sup_{(k,t)\in\mathbb{Z}^2}\left((g-\overline{g})k+(d-\overline{d})t+\widehat{F}(\overline{g},\overline{d})\right)\\ &= \widehat{F}(g,d)\ .\end{aligned}$$

■

5.8.2 Specialization to $\mathcal{M}_{\text{in}}^{\text{ax}}[\![\gamma,\delta]\!]$

We are now interested in redefining a similar evaluation homomorphism, but for elements of $\mathcal{M}_{\text{in}}^{\text{ax}}[\![\gamma,\delta]\!]$. We have seen that elements of $\mathcal{M}_{\text{in}}^{\text{ax}}[\![\gamma,\delta]\!]$, which are indeed equivalence classes, may be represented by different formal expressions in $\mathbb{B}[\![\gamma,\delta]\!]$. All these expressions are characterized by the fact that they yield the same element of $\mathbb{B}[\![\gamma,\delta]\!]$ (the maximum representative) when they are multiplied by $\gamma^*(\delta^{-1})^*$. By mere application of (5.71), we have that

$$\left[\mathcal{F}\left(\gamma^*(\delta^{-1})^*\right)\right](g,d)=\begin{cases}0 & \text{if } g\le 0 \text{ and } d\ge 0\ ;\\ +\infty & \text{otherwise.}\end{cases}$$

Therefore, by application of the homomorphism property of $\mathcal{F}$, it is seen that, for all (g,d) such that $g\le 0$ (denoted $g\in(-\mathbb{N})$) and $d\ge 0$,

$$F\gamma^*(\delta^{-1})^* = G\gamma^*(\delta^{-1})^* \text{ (in } \mathbb{B}[\![\gamma,\delta]\!]) \Rightarrow [\mathcal{F}(F)](g,d)=[\mathcal{F}(G)](g,d)\ .$$

It is readily checked that we can repeat all the proofs and results of §5.8.1, using any representatives in $\mathbb{B}[\![\gamma,\delta]\!]$ of elements in $\mathcal{M}_{\text{in}}^{\text{ax}}[\![\gamma,\delta]\!]$, provided that we restrict the pairs (g,d) to belong to $(-\mathbb{N})\times\mathbb{N}$.

Remark 5.47 For $(g,d)\notin(-\mathbb{N})\times\mathbb{N}$, the value of $\mathcal{F}(F)$ can consistently be set to $+\infty$ whenever F is considered as an element of $\mathcal{M}_{\text{in}}^{\text{ax}}[\![\gamma,\delta]\!]$, except if $F=\varepsilon$, for this is the value obtained with the maximum representative of F. This may be explained by recalling the geometric interpretation of $\mathcal{F}(F)$ as a support function of a subset, and by observing that the 'cones of information' introduced in §5.4.3 extend indefinitely in the South and East directions (characterized by $g>0$ or $d<0$).

Observe that the subset of numerical functions equal to $+\infty$ outside $(-\mathbb{N})\times\mathbb{N}$, plus the function ε equal to $-\infty$ everywhere, is also a complete dioid for the operations defined at Lemma 5.45. ■

We will keep on using the notation $\mathcal{F}$ for this mapping defined over $\mathcal{M}_{\text{in}}^{\text{ax}}[\![\gamma,\delta]\!]$ (since the previous $\mathcal{F}$ defined over $\mathbb{B}[\![\gamma,\delta]\!]$ will no longer be in use). The following definition and lemma, which should now be clear, summarizes the situation.

Definition 5.48 (Evaluation homomorphism) *The mapping $\mathcal{F}$ from $\mathcal{M}_{\text{in}}^{\text{ax}}[\![\gamma,\delta]\!]$ into the dioid of numerical functions (introduced at Lemma 5.45) is defined as follows:*

- $\mathcal{F}(\varepsilon) = \varepsilon$;
- *if* $F \neq \varepsilon$,
  - *if* $(g,d) \in (-\mathbb{N}) \times \mathbb{N}$, *then* $[\mathcal{F}(F)](g,d)$ *is defined by Equations (5.71) using any representative of* F;
  - *if* $(g,d) \notin (-\mathbb{N}) \times \mathbb{N}$, *then* $[\mathcal{F}(F)](g,d) = +\infty$.

Lemma 5.49 *The mapping $\mathcal{F}$ just defined is a l.s.c. dioid homomorphism over $\mathcal{M}_{\text{in}}^{\text{ax}}[\![\gamma,\delta]\!]$ which is residuated, and $\mathcal{F}^\sharp$ can be defined by (5.72) in which the* inf *is restricted to $g \in (-\mathbb{N})$ and $d \in \mathbb{N}$.*

5.8.3 Eigenfunctions of Rational Transfer Functions

We now introduce particular elements of $\mathcal{M}_{\text{in}}^{\text{ax}}[\![\gamma,\delta]\!]$ which will be shown to play the role of sine functions in conventional system theory.

Definition 5.50 *For two positive integers k and t, we set*

$$L_{(k,t)} \stackrel{\text{def}}{=} \mathcal{F}^\sharp\left(\mathcal{F}\left(\left(\gamma^k\delta^t \oplus \gamma^{-k}\delta^{-t}\right)^*\right)\right).$$

It is easy to check that

$$M_{(k,t)} \stackrel{\text{def}}{=} \left(\gamma^k\delta^t \oplus \gamma^{-k}\delta^{-t}\right)^* = \left(\gamma^k\delta^t\right)^*\left(\gamma^{-k}\delta^{-t}\right)^* = \left(\gamma^k\delta^t\right)^* \oplus \left(\gamma^{-k}\delta^{-t}\right)^*.$$

Lemma 5.51 *The element $L_{(k,t)}$ depends only on the ratio $c = t/k > 0$ (therefore it will be denoted simply L_c with $c > 0$) and it is given explicitly by*

$$L_c = \bigoplus_{s \le c \times l} \gamma^l\delta^s .$$

If one draws the line of slope $c > 0$ in the $\mathbb{R}^2$-plane, L_c is the coding of the points of $\mathbb{Z}^2$ which lie below this line. In other words, L_c represents the best discrete approximation of this line from below. For example, with $k = 3$ and $t = 2$ hence $c = 2/3$ (see Figure 5.13), then

$$L_c = (e \oplus \gamma^2\delta)\left(\gamma^3\delta^2\right)^*\left(\gamma^{-3}\delta^{-2}\right)^*.$$

Proof of Lemma 5.51 We have $M_{(k,t)}(l,s) = e$ if $(l,s) = (n \times k, n \times t)$ with $n \in \mathbb{Z}$ and $M_{(k,t)}(l,s) = \varepsilon$ otherwise. Then, according to (5.71),

$$\left[\mathcal{F}\left(M_{(k,t)}\right)\right](g,d) = \sup_{n \in \mathbb{Z}} n(gk + dt) = \begin{cases} 0 & \text{if } gk + dt = 0\ ; \\ +\infty & \text{otherwise.} \end{cases}$$

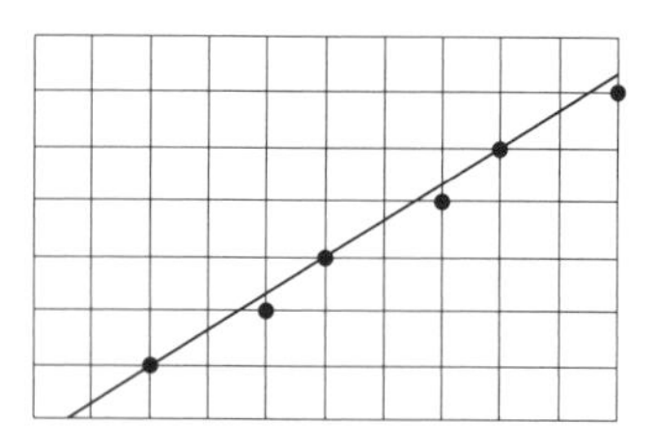

Figure 5.13: A 'linear' function

Obviously, this expression depends only on the ratio $c = t/k$. Finally, according to Lemma 5.49,

$$
\begin{aligned}
[\mathcal{F}^\sharp(\mathcal{F}(M_{(k,t)}))](l,s) &= \inf_{\substack{g\in(-\mathbb{N}),d\in\mathbb{N}\\ gk+dt=0}} (-(gl+ds)) \\
&= \inf_{d\in\mathbb{N}} d(lc-s) \\
&= \begin{cases} 0 & \text{if } s \le cl \ ; \\ -\infty & \text{otherwise.} \end{cases}
\end{aligned} \tag{5.74}
$$

■

The element L_c describes a sequence of events which occurs at the average rate of $1/c$ events per unit of time. Consider a SISO event graph with transfer function H. Since H is realizable, hence periodic (see Definition 5.38), it can be written as $P \oplus Q\left(\gamma^{\overline{k}}\delta^{\overline{t}}\right)^*$ (P and Q are polynomials). We will confine ourselves to the nontrivial cases when $Q \neq \varepsilon$ and $\overline{k} > 0$, $\overline{t} > 0$. The ratio $\overline{k}/\overline{t}$ (the inverse of the asymptotic 'slope' of the impulse response) characterizes the limit of the rate of events the system can process. If L_c is used as the input of the event graph, and if $1/c$ exceeds this limit, then there will be an indefinite accumulation of tokens inside the system and the output is indefinitely delayed with respect to the input. Otherwise, the following theorem states that, using L_c as the input produces an output $\gamma^{\kappa_c}\delta^{\theta_c}L_c$, that is, essentially the same trajectory as the input up to shifts by κ_c along the x-axis (event domain) and θ_c along the y-axis (time domain). The theorem also shows how (κ_c, θ_c) is related to $[\mathcal{F}(H)](g,d)$ for any (g,d) such that $g = -d \times c$.

Theorem 5.52 *Consider a SISO system with rational transfer function $H = P \oplus Q\left(\gamma^{\overline{k}}\delta^{\overline{t}}\right)^*$, where P and Q are polynomials in $\mathcal{M}_{\text{in}}^{\text{ax}}[\![\gamma,\delta]\!]$, Q is supposed to be different from ε, and $\overline{k}$ and $\overline{t}$ are supposed to be strictly positive. Then,*

1. *for all $g \le 0$ and $d \ge 0$, $[\mathcal{F}(H)](g,d)$ is different from $+\infty$ if and only if*

$$c = -g/d \ge \overline{t}/\overline{k} \ , \tag{5.75}$$

and then

$$[\mathcal{F}(H)](g,d) = [\mathcal{F}(P \oplus Q)](g,d) = \kappa_c g + \theta_c d \tag{5.76}$$

for some finite nonnegative integers κ_c *and* θ_c*;*

2. *those* κ_c *and* θ_c *are not necessarily unique, but any selection yields nonincreasing functions of* c*;*

3. *let* $c \stackrel{\text{def}}{=} t/k$ *and assume that* c *satisfies (5.75); then we have*

$$HL_c = \gamma^{\kappa_c}\delta^{\theta_c}L_c \ . \tag{5.77}$$

Proof

1. Let $R = \left(\gamma^{\bar{k}}\delta^{\bar{t}}\right)^*$. Since for the minimum representative of R, $R(l,s) = e$ when $(l,s) = n(\bar{k},\bar{t})$ with $n \in \mathbb{N}$, and $R(l,s) = \varepsilon$ otherwise, we have that

$$[\mathcal{F}(R)](g,d) = \sup_{n\in\mathbb{N}} n\left(g\bar{k} + d\bar{t}\right) = \begin{cases} 0 & \text{if } (g,d) \text{ satisfies (5.75);} \\ +\infty & \text{otherwise.} \end{cases}$$

   On the other hand, for a polynomial, say P, the minimum representative involves a finite number of points, and $\mathcal{F}(P)$ is the convex hull of a finite number of linear functions. Therefore, by the homomorphism property of $\mathcal{F}$, $\mathcal{F}(H) = \mathcal{F}(P) \oplus \mathcal{F}(Q) \otimes \mathcal{F}(R)$, and since $Q \neq \varepsilon$, $[\mathcal{F}(H)](g,d)$ is finite if and only if (g,d) satisfies (5.75), and then $\mathcal{F}(H) = \mathcal{F}(P) \oplus \mathcal{F}(Q)$. For any such pair $(g,d) \in (-\mathbb{N}) \times \mathbb{N}$, we thus have

$$[\mathcal{F}(H)](g,d) = \sup_{(l,s)\in J_P \cup J_Q} (gl + ds) \ ,$$

   and the supremum is reached at some (not necessarily unique) point (κ_c, θ_c) which clearly depends only on the ratio $c = -g/d > 0$.

2. By a well known result of convexity theory [119], (κ_c, θ_c) belongs to the subdifferential of the convex function $\mathcal{F}(H)$ at the point (g,d). Hence the mappings $(g,d) \mapsto (\kappa_c, \theta_c)$, for any choices of (κ_c, θ_c), are monotone, that is, for two pairs (g_i, d_i), $i = 1, 2$, and any associated subgradients (κ_i, θ_i), we have

$$(g_1 - g_2)(\kappa_1 - \kappa_2) + (d_1 - d_2)(\theta_1 - \theta_2) \geq 0 \ .$$

   Since we are only concerned with the ratios $c_i = -g_i/d_i$, we can either take $g_1 = g_2$ or $d_1 = d_2$ in the above inequality. This shows the monotonicity property claimed for κ and θ as functions of c.

3. Let $Y = HL_c$ with $c = t/k$ satisfying (5.75). Then, for all $(l,s) \in \mathbb{Z}^2$,

$$\begin{aligned} Y(l,s) &= \sup_{(m,r)\in\mathbb{Z}^2} \left(H(m,r) + L_c(l-m, s-r)\right) \\ &= \sup_{(m,r)\in\mathbb{Z}^2} \left(H(m,r) + \inf_{d\in\mathbb{N}} d((l-m)c - (s-r))\right) \qquad \text{(from (5.74))} \\ &= \sup_{(m,r)\in\mathbb{Z}^2} \inf_{d\in\mathbb{N}} \left(H(m,r) + d(-mc + r) + d(lc - s)\right) \ . \end{aligned} \tag{5.78}$$

On one hand, by inverting the sup and the inf, we obtain a new expression which is larger than (5.78), and which turns out to be equal to

$$\begin{aligned}\inf_{d\in\mathbb{N}}\left([\mathcal{F}(H)](-dc,d)+d(lc-s)\right) &= \inf_{d\in\mathbb{N}} d\left((l-\kappa_c)c+\theta_c-s\right)\\ &= L_c(l-\kappa_c, s-\theta_c)\ , \qquad (5.79)\end{aligned}$$

the latter equality being true because of (5.74). On the other hand, if we choose the particular value $(m,r)=(\kappa_c,\theta_c)$ instead of performing the supremum, we obtain an expression which is less than (5.78), and which turns out to be identical to (5.79) (the clue here is that (κ_c,θ_c) which realizes the maximum in the evaluation of $[\mathcal{F}(H)](-dc,d)$ does not depend on d indeed). Finally, we have proved that

$$\forall(l,s)\in\mathbb{Z}^2\ ,\quad Y(l,s)=L_c(l-\kappa_c,s-\theta_c)\ ,$$

which is equivalent to (5.77). ■

It is intuitively appealing that the 'shifts' κ_c and θ_c are nonincreasing with c. Indeed, recall that when c decreases, the average time between two successive events at the input decreases, hence the input is faster: the delays introduced by the system, in terms of counters or in terms of daters, are likely to increase. Moreover, there is a 'threshold' effect (or a 'low-pass' effect) in that, above a certain speed which is defined by the asymptotic slope of the impulse response, the system, driven by an input which is too fast, 'blows up', and the delays become infinite. This corresponds to an unstable situation (using the same calculations as in the above proof, it can be proved in this case that $Y=\top$). This is also similar to conventional system theory in which the sine functions are eigenfunctions only in the stable case. The difference is here that the stability property is not an intrinsic feature of the system (at least in the SISO case considered here), but it depends on the mutual speeds of the input and of the system itself.

Let us conclude this section by an example.

Example 5.53 Consider $H=\gamma\delta^2\oplus(\gamma^2\delta)^*$, the impulse response of which is represented at the left-hand side of Figure 5.14. This system cannot process events

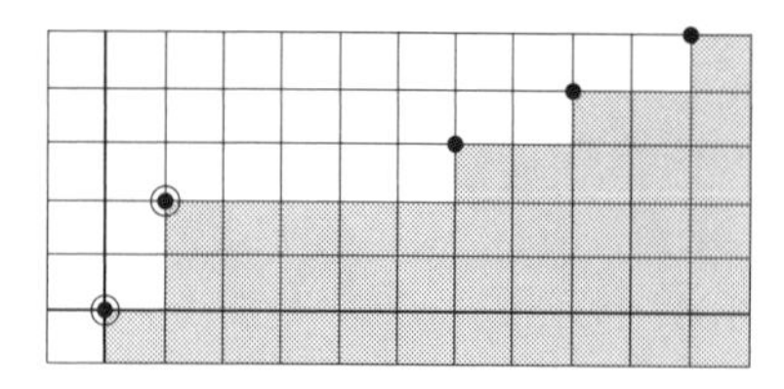

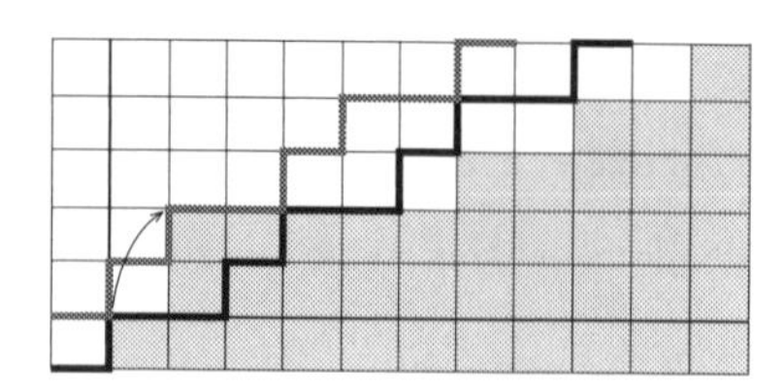

Figure 5.14: The impulse response of $H=\gamma\delta^2\oplus(\gamma^2\delta)^*$ and the response to $L_{2/3}$

faster than 2 events per time unit. Let us study the functions κ_c and θ_c with respect to c: the subset of points with coordinates (κ_c,θ_c) in the $\mathbb{N}^2$-plane, when c varies, may be considered as the Black plot by analogy with conventional system theory in which the Black plot is the curve generated by the points $\left(\arg H(j\omega), \log|H(j\omega)|\right)$ when ω varies. In this case, it is easy to see that

$$(\kappa_c, \theta_c) = \begin{cases} (0,0) & \text{if } 2 \leq c < +\infty \ ; \\ (1,2) & \text{if } 1/2 \leq c \leq 2 \ . \end{cases}$$

The points belonging to the Black plot are circled in the figure. At the right-hand side of this figure, the 'trajectory' of L_c is represented by a solid black line for $c = 2/3$ (see also Figure 5.13) and the response of the system to this input is indicated by a gray line. The shifts along the two axes is indicated by a curved arrow. In the dater point of view, one may say that the 'phase shift' is κ_c whereas the 'amplification gain' is θ_c. In the counter point of view (which is closer to conventional system theory since the time domain is usually represented as the x-axis), the role of κ_c and θ_c as phase shift and amplification gain are reversed. ■

5.9 NOTES

The first papers on a new linear theory of some discrete event systems have been published in early 1983 [37, 38]. They were based on the dater representation and, hence, on the max-plus algebra. The connection of these linear models with timed event graphs was established in 1984 [41]. In 1985, it has been realized that specific algebraic problems arise from the fact that daters, as time sequences, are nondecreasing [42]. At about the same time, the idea of the counter representation has been introduced by Caspi and Halbwachs [35] to whom this terminology is due. This finally resulted in the two-dimensional representation first published in [43]. A more detailed account of the so-called $\mathcal{M}_{\text{in}}^{\text{ax}}[\![\gamma, \delta]\!]$ algebra was given in [44] together with some of the formulæ about residuation of $\oplus$ and $\otimes$. A large part of the material of this (and the previous) chapter(s) is based on that paper—e.g. the sections on backward equations and on rationality, realizability and periodicity—although the presentation has been somewhat improved here. In particular, the role of residuation theory has been clarified in the process of establishing backward equations, and in the relationships between dater and counter representations. Only the results on 'minimum representatives' appear for the first time in this book. The idea of using the formal transfer function as a numerical function, the fact that the Fenchel transform plays a role similar to that played by the Laplace transform in conventional system theory, the parallel notion of eigenfunctions of linear transfer functions in discrete event and conventional system theories, ..., were all discovered and published in 1989 [40]. However, here, the presentation has been more tightly confined in the two-dimensional point of view.

CHAPTER 6

Max-Plus Linear System Theory

6.1 INTRODUCTION

In this chapter a linear system theory is developed for a class of max-plus linear systems with discrete or continuous domain and range. This class provides a generalization of the class of event graphs which have been considered in Chapter 2. We will start with an input-output point of view. We will view a max-plus linear system as a max-plus linear operator mapping an input, which is a function over some domain, to an output, which is another function over the same domain. For this max-plus linear operator, we will review the classical notions of conventional linear system theory. In particular, the notions of causality, shift-invariance, impulse response, convolution, transfer function, rationality, realization and stability will be considered.

The outline of the chapter is as follows. In §6.2 we give general definitions, we present the system algebra and we discuss some fundamental elementary systems. In §6.3 we define some subalgebras of the system algebra by progressively specializing the systems, starting with the most general ones, and finishing with causal shift-invariant systems with nondecreasing impulse responses. Most practical examples of discrete event systems fall into this last category. In the dater description, their output is the result of a sup-convolution between their input and their impulse response. In §6.4 we introduce the notion of transfer functions which are related to impulse responses by means of the Fenchel transform. In §6.5 we discuss rationality in the max-plus context, and characterize rational elements in terms of periodicity. We also discuss the problem of minimal realization of these max-plus systems. In §6.6 we give a definition of internally stable systems and characterize them in terms of equations which are the analogue of the conventional Lyapunov equation. In this chapter, mainly single-input single-output (SISO) linear max-plus systems are considered.

6.2 SYSTEM ALGEBRA

6.2.1 Definitions

Definition 6.1 (Signal) *A signal u is a mapping from $\mathbb{R}$ into $\overline{\mathbb{R}}_{\max}$. When a signal is a nondecreasing function it is called a dater.*

The signal set is $\overline{\mathbb{R}}_{\max}^{\mathbb{R}}$. This signal set is endowed with two operations, namely

- pointwise maximum of signals which plays the role of addition:

$$\forall k \in \mathbb{R} \ , \quad \forall u, v \in \overline{\mathbb{R}}_{\max}^{\mathbb{R}} \ , \quad (u \oplus v)(k) \stackrel{\text{def}}{=} u(k) \oplus v(k) = \max(u(k), v(k)) \ ;$$

- addition of a constant to a signal, which plays the role of the external product of a signal by a scalar:

$$\forall k \in \mathbb{R} \ , \quad \forall a \in \overline{\mathbb{R}}_{\max} \ , \quad \forall u \in \overline{\mathbb{R}}_{\max}^{\mathbb{R}} \ , \quad (au)(k) \stackrel{\text{def}}{=} a \otimes u(k) = a + u(k) \ .$$

Therefore the set of signals is endowed with a moduloid structure. This algebraic structure is called $\mathcal{U}$.

In previous chapters the domain of signals was $\mathbb{Z}$ (event domain) and trajectories were nondecreasing. In this chapter we develop the theory in the more general framework of Definition 6.1.

Definition 6.2 (Max-plus linear system) *A system is an operator* $S : \mathcal{U} \to \mathcal{U}, u \mapsto y$. *The signal* u *(respectively* y*) is called the* input *(respectively* output*) of the system. We say that the system is max-plus linear when the corresponding operator satisfies*

$$S\left(\bigoplus_{i \in I} u_i\right) = \bigoplus_{i \in I} S(u_i) \ , \tag{6.1}$$

for any finite or infinite set $\{u_i\}_{i \in I}$, *and*

$$S(au) = aS(u) \ , \quad \forall a \in \overline{\mathbb{R}}_{\max} \ , \quad \forall u \in \mathcal{U} \ .$$

Remark 6.3 Equation (6.1) is indeed the requirement of lower-semicontinuity of S and not only the requirement that S is an $\oplus$-morphism. Here is an example of a system which is an $\oplus$-morphism but which fails to be l.s.c.:

$$[S(u)](t) = \limsup_{s \to t} u(s) \ .$$

This system is clearly an $\oplus$-morphism, but to show that it is not l.s.c., consider

$$\forall n \geq 1, \quad u_n(k) = \begin{cases} 0 & \text{if } k \leq 0 \ ; \\ n \times k & \text{if } 0 < k < \frac{1}{n} \ ; \\ 1 & \text{if } \frac{1}{n} \leq k \ . \end{cases}$$

For all $n \geq 1$, we have $[S(u_n)](0) = 0$, and

$$\bigoplus_{n \geq 1} u_n(t) = \begin{cases} 0 & \text{if } t \leq 0 \ ; \\ 1 & \text{otherwise} \ . \end{cases}$$

This yields $\left[S\left(\bigoplus_n u_n\right)\right](0) = 1$, which is different from $\left[\bigoplus_n S(u_n)\right](0) = 0$. ■

The set of linear systems is endowed with two internal and one external operations, namely

parallel composition: $S = S_1 \oplus S_2$ is defined as follows:

$$[S(u)](k) = [S_1(u)](k) \oplus [S_2(u)](k) \ ; \tag{6.2}$$

series composition: $S = S_1 \otimes S_2$, [1] or more briefly, $S_1 S_2$ is defined as follows:

$$[S(u)](k) = [S_1(S_2(u))](k) \ ; \tag{6.3}$$

amplification: $T = a \otimes S$, $a \in \overline{\mathbb{R}}_{\max}$ is defined by:

$$T(k) = a \otimes S(k) \ .$$

In addition to these basic operations, we have another important one, the feedback:

feedback: S^* defined by the mapping from $\mathcal{U}$ into $\mathcal{U}$: $u \mapsto y$, where y (see Figure 6.1) is the least solution of

$$y = S(y) \oplus u \ . \tag{6.4}$$

The notation S^* is justified by the fact that the least solution of (6.4) does exist by Theorem 4.75 and it is given by $u \oplus S(u) \oplus S(S(u)) \oplus \cdots$.

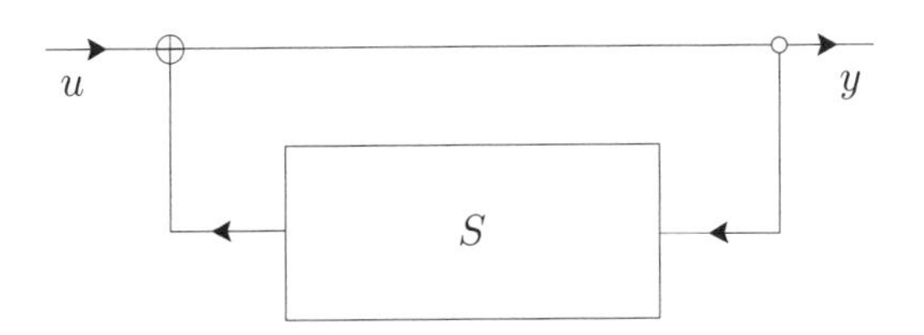

Figure 6.1: The feedback operation $y = S^* u$

The set of systems endowed with the first three operations defines an idempotent algebra called the algebra of systems. We do not lose anything by considering only the two internal operations because the amplification can be realized by a series composition where the downstream system is what we call a *gain* in the next section. Whenever we speak of the set of systems endowed with the two internal operations, we refer to it as the 'dioid of systems'. The second operation is not invertible, and therefore this set is not an idempotent semifield (see Chapter 3), it is only a dioid.

6.2.2 Some Elementary Systems

We have discussed how we can combine systems using compositions and feedbacks. Here, we describe some elementary though fundamental systems with which more complex systems can be built up.

[1] We make the usual abuse of notation which consists in using the same symbol for external multiplication by a scalar and for internal multiplication of systems. This will be justified later on.

We first introduce the following notation.

Notation 6.4 For $f : \mathbb{R} \to \overline{\mathbb{R}}_{\max}$, $\bigoplus_a^{b} f(s)$ denotes the supremum of $f(s)$ when s ranges in the interval $[a, b]$ (or e.g. $(a, b]$ if $a = -\infty$). We may also use the notation $\bigoplus_{\mathbb{R}} f(s)$ if $a = -\infty$ and $b = +\infty$. ■

The following elementary systems are now introduced.

Zero system ε**:** this system produces the constant output ε whatever the input is: $y(k) = \varepsilon$ for all k. It satisfies

$$\varepsilon \oplus \varepsilon = \varepsilon \otimes \varepsilon = \varepsilon^* = \varepsilon \ .$$

Identity e**:** this system produces an output equal to the input $y(k) = u(k)$ for all k. It satisfies

$$e \oplus e = e \otimes e = e^* = e \ .$$

Shift Γ^g**:** this system maps inputs to outputs according to the equation $y(k) = u(k - g)$ for all k. The notation Γ^g is justified by the following rule of series composition which should be obvious to the reader:

$$\Gamma^g \otimes \Gamma^{g'} = \Gamma^{g+g'} = \Gamma^{g \otimes g'} \ .$$

Therefore Γ^1 may be denoted Γ. If we restrict ourselves to signals that are nondecreasing signals (see the discussion just above Remark 5.1), we have the simplification rule

$$\Gamma^g \oplus \Gamma^{g'} = \Gamma^{\min(g,g')} \ . \tag{6.5}$$

In the context of event graphs, an initial stock of c tokens in a place introduces such a shift between inputs and outputs in the domain where we 'count' events. In the framework of the continuous system depicted in Figure 1.13, the same role is played by the initial amount of fluid in the reservoir at the outlet. Note however that, in that example, equations were written in a counter, rather than in a dater, representation, and consequently, this device operated as a gain rather than as a shift.

Gain Δ^d**:** this system maps inputs to outputs according to the equation $y(k) = d \otimes u(k) = d + u(k)$ for all k. Again, the notation Δ^d is justified by the following rule of series composition:

$$\Delta^d \otimes \Delta^{d'} = \Delta^{d+d'} = \Delta^{d \otimes d'} \ .$$

Therefore Δ^1 may be denoted Δ. We also have the simplification rule (which holds true for any input signal)

$$\Delta^d \oplus \Delta^{d'} = \Delta^{d \oplus d'} = \Delta^{\max(d,d')} \ .$$

In the context of timed event graphs, this is the general input-output relation induced by a place with holding time d. For the system of Figure 1.13, this input-output relation is that of along pipe at the inlet of a funnel.

Flow limiter Φ_a**:** this system maps inputs to outputs according to the relation

$$y(k) = \oint_{-\infty}^{k} u(s)a^{k-s} . \tag{6.6}$$

Unlike Γ^c and Δ^d, we use the notation Φ_a with a as a subscript, because a does not behave like an exponent. Indeed, the following parallel, series and feedback composition rules can be checked by direct calculation:

$$\Phi_a \oplus \Phi_{a'} = \Phi_a \otimes \Phi_{a'} = \Phi_{\max(a,a')} = \Phi_{a \oplus a'} . \tag{6.7}$$

Moreover, $\Phi_a > e$ and hence $\Phi_a = (\Phi_a)^*$.

Physically, this system corresponds to the input-output relation between the cumulated quantities traversing a pipe which limits the flow to $1/a$ (of course, here a is a positive number). This is the case of the aperture of the funnel in Figure 1.13 (recall that this example is worked out using counter rather than dater equations). This system plays the role of the SISO system governed by the differential equation

$$\dot{y} = ay + u$$

in conventional system theory, the solution of which is

$$y(t) = \int_{-\infty}^{t} u(s) \exp(a(t-s)) \, \mathrm{d}s ,$$

which is the analogue of (6.6).

Integrator $\acute{e}$**:** this is another notation for Φ_0. It maps inputs to outputs according to the equation $y(k) = \oint_{-\infty}^{k} u(s)$. The output of such a system is always nondecreasing. It plays the role of an identity element for shift-invariant systems with nondecreasing impulse responses as we shall see later on. This role justifies the notation $\acute{e}$. It satisfies

$$\acute{e} \oplus \acute{e} = \acute{e} \otimes \acute{e} = (\acute{e})^* = \acute{e} .$$

Local integrator Σ^w**:** this system maps inputs to outputs according to the relation $y(k) = \oint_{k-w}^{k} u(s)$. It is the analogue of a conventional system recursively averaging the input in a window of width w. The following series, parallel and feedback compositions of local integrators can be easily checked:

$$\Sigma^w \oplus \Sigma^{w'} = \Sigma^{w \oplus w'} ,$$

$$\Sigma^w \otimes \Sigma^{w'} = \Sigma^{w+w'} = \Sigma^{w \otimes w'} ,$$

$$(\Sigma^w)^* = \Sigma^\infty = \acute{e} , \quad \forall w > 0 .$$

6.3 IMPULSE RESPONSES OF LINEAR SYSTEMS

In this section we introduce the notion of impulse response for a max-plus linear system. The algebra of impulse responses is isomorphic to the algebra of systems. The former algebra is first specialized to the case of shift-invariant systems and subsequently to the case of systems with nondecreasing impulse responses.

6.3.1 The Algebra of Impulse Responses

We saw that the set of systems can be endowed with a moduloid structure. The next step is to introduce a kind of 'canonical basis' for this algebraic structure. Classically, for time functions, this basis is provided by the Dirac function at 0, and all its shifted versions at other time instants. Therefore, we now introduce

$$e(\cdot) : k \mapsto e(k) \stackrel{\text{def}}{=} \begin{cases} e & \text{if } k = 0 \ ; \\ \varepsilon & \text{otherwise,} \end{cases} \tag{6.8}$$

and

$$\gamma^s(\cdot) \stackrel{\text{def}}{=} \Gamma^s(e(\cdot)) \qquad \text{i.e.} \qquad \gamma^s(k) = e(k-s) \ , \quad \forall k \ . \tag{6.9}$$

The justification of the notation $e(\cdot)$ will come from the fact that this particular signal is the identity element for sup-convolution which will be the internal multiplication in the system set. Indeed, it can be checked by direct calculation that

$$\forall u \ , \forall k \ , \quad u(k) = \oint_{\mathbb{R}}^{s} u(s) e(k-s) \ . \tag{6.10}$$

In view of (6.9), this can be rewritten

$$u = \oint_{\mathbb{R}}^{s} u(s) \gamma^s \ , \tag{6.11}$$

which shows that u is obtained as a linear combination of the signals γ^s. This is the decomposition of signals with respect to the canonical basis. This decomposition is unique since, if there exists another function $v : \mathbb{R} \to \overline{\mathbb{R}}_{\max}$ such that $u = \oint_{\mathbb{R}}^{s} v(s)\gamma^s$, we conclude that $v(s) = u(s), \forall s$, because of Identity (6.10) applied to the function v.

Now we can state the following theorem which introduces the notion of *impulse response*.

Theorem 6.5 *Let S be a linear system, then there exists a unique function $h(k,s)$ (called the* impulse response*) such that $y = S(u)$ can be obtained by*

$$\forall k \ , \quad y(k) = \sup_{s \in \mathbb{R}} [h(k,s) + u(s)] = \oint_{\mathbb{R}}^{s} h(k,s) u(s) \ , \tag{6.12}$$

for all input-output pairs (u,y).

Proof We have

$$y(k) = [S(u)](k) = \left[S\left(\oint_{\mathbb{R}}^{s} u(s)\gamma^s\right)\right](k) \ ,$$

which, owing to the linearity assumption, implies

$$y(k) = \oint_{\mathbb{R}}^{s} \left(\left[S(\gamma^s)\right](k)\right) u(s) = \oint_{\mathbb{R}}^{s} h(k,s)u(s) \ ,$$

where we have set $h(k,s) \stackrel{\text{def}}{=} [S(\gamma^s)](k)$. To prove uniqueness, suppose that there exists another function $f(\cdot,\cdot)$ which satisfies (6.12). Then using inputs $u = \gamma^s$, we obtain

$$\begin{aligned} h(k,s) &\stackrel{\text{def}}{=} [S(\gamma^s)](k) \\ &= \oint_{\mathbb{R}}^{\tau} f(k,\tau)\gamma^s(\tau) \\ &= f(k,s) \ , \end{aligned}$$

for all $s, k \in \mathbb{R}$, where the last equality is (6.10) applied to the function $f(k,\cdot)$. ■

To the series, parallel, amplification and feedback compositions of systems correspond operations on the impulse responses.

Theorem 6.6 *Given $a \in \overline{\mathbb{R}}_{\max}$ and the systems S, S_1 and S_2 with respective impulse responses h, h_1 and h_2, then,*

- *the impulse response of $S_1 \oplus S_2$ is $[h_1 \oplus h_2](k,s) \stackrel{\text{def}}{=} h_1(k,s) \oplus h_2(k,s)$;*
- *the impulse response of $S_1 \otimes S_2$ is $[h_1 \otimes h_2](k,s) \stackrel{\text{def}}{=} \oint_{\mathbb{R}}^{r} h_1(k,r)h_2(r,s)$;*
- *the impulse response of aS is $[ah](k,s) \stackrel{\text{def}}{=} ah(k,s)$;*
- *the impulse response of S^* is $h^* \stackrel{\text{def}}{=} \bigoplus_{i\in\mathbb{N}} h^i$.*

The set of impulse responses endowed with the first three operations (respectively the first two operations) defines an idempotent algebra (respectively a dioid), called the algebra (respectively the dioid) of impulse responses which is denoted $\mathcal{H}$. Impulse responses are representations of systems written in a canonical basis, just like matrices are finite dimensional linear operators written in a particular basis.

Definition 6.7 *A linear system S is causal if, for all inputs u_1 and u_2 with corresponding outputs y_1 and y_2,*

$$\forall s \ , \quad \forall k \leq s \ , \quad u_1(k) = u_2(k) \Rightarrow y_1(s) = y_2(s) \ .$$

Theorem 6.8 *A system S is causal if its impulse response $h(k,s)$ equals ε for $k \leq s$.*

Proof If S is causal, $S(\gamma^s) = h(k, s)$ coincides with $S(\varepsilon) = \varepsilon$ for $k \leq s$. ■

Remark 6.9 The impulse response h of a series composition of two causal systems of impulse responses h_1 and h_2 has the simplified form

$$h(k, s) = \oint_{s}^{k} h_1(k, r) h_2(r, s) \ .$$

■

6.3.2 Shift-Invariant Systems

Let us specialize the algebra $\mathcal{H}$ to shift-invariant systems.

Definition 6.10 *A linear system S is called shift-invariant if it commutes with all shift operators, that is, if*

$$\forall u \ , \quad \forall c \ , \quad S(\Gamma^c(u)) = \Gamma^c(S(u)) \ .$$

Theorem 6.11 *A system S is shift-invariant if and only if its impulse response $h(k, s)$ depends only on the difference $k - s$. With the usual abuse of notation, the impulse response is denoted $h(k - s)$ in this case. It is equal to $h(\cdot) = [S(e)](\cdot)$.*

Proof We have

$$h(k, s) \stackrel{\text{def}}{=} [S(\gamma^s)](k) = [S(\Gamma^s(e))](k) = [\Gamma^s(S(e))](k) = [S(e)](k - s) \ .$$

■

Consequently, in the shift-invariant case, the kernel defining the impulse response is reduced to a function. The input-output relation can be expressed as follows:

$$y(k) = (h \otimes u)(k) \stackrel{\text{def}}{=} \oint_{\mathbb{R}}^{s} h(k - s) u(s) \ .$$

This new operation, also denoted $\otimes$, is nothing but the *sup-convolution* which plays the role of the convolution in conventional system theory. We also note that the series composition corresponds to the sup-convolution of the corresponding impulse responses.

Definition 6.12 *The algebra of shift-invariant impulse responses, denoted $\mathcal{S}$, is the set $\overline{\mathbb{R}}_{\max}^{\mathbb{R}}$ endowed with:*

- *the pointwise maximum of functions denoted $\oplus$;*
- *the sup-convolution denoted $\otimes$;*
- *the external operation which consists in adding a constant to the function.*

The zero element denoted $\varepsilon(\cdot)$ is defined by $\varepsilon(k) = \varepsilon, \forall k$. It is absorbing for multiplication. The identity element denoted $e(\cdot)$ is described by (6.8).

Remark 6.13

1. This idempotent algebra $\mathcal{S}$ can simply be considered as a dioid.
2. Because signals and SISO systems can be represented by functions, we do not have to distinguish them.
3. Impulse responses of shift-invariant causal systems satisfy $h(k) = \varepsilon$ for $k < 0$. ■

Example 6.14 The elementary systems introduced in §6.2.2 are shift-invariant linear systems. Their impulse responses are given later in Table 6.2. Notice that $\gamma^0 = \delta^0 = \phi_\varepsilon = e$ if ϕ_ε denotes the pointwise limit of ϕ_a when a goes to $-\infty$. ■

6.3.3 Systems with Nondecreasing Impulse Response

In the context of event graphs, input signals have the meaning of sequences of dates at which successive events occur, and therefore they are nondecreasing functions. In the case of the continuous system depicted in Figure 1.13, the input and the output are also nondecreasing.

A nondecreasing signal u can be characterized by the inequality $u \geq \Sigma^w u$ for any arbitrary positive w. From an algebraic point of view, this situation is identical to that described by Inequality (5.11) if Σ^w plays the role earlier played by γ, now that the domain is continuous. Hence from Theorem 5.8, we know that $\acute{v} \stackrel{\text{def}}{=} (\Sigma^w)^* v$ is the best approximation from above of a signal v in the subset of nondecreasing signals. Recall that $(\Sigma^w)^* = \acute{e}$ (see end of §6.2.2). In particular, a nondecreasing function u is characterized by $u = \acute{u} = \acute{e}u$.

Consider a system with impulse response h. Then, if only nondecreasing inputs u are considered, the outputs are also nondecreasing as shown by the following equalities:

$$y = h \otimes u = h \otimes (\acute{e} \otimes u) = \acute{e} \otimes (h \otimes u) = \acute{e} \otimes y \ .$$

We also notice that, for this class of nondecreasing inputs, the systems with impulse responses h and $\acute{h} = \acute{e} \otimes h$ yield the same outputs. This $\acute{h}$ is called the 'nondecreasing version' of the impulse response h. The subset of nondecreasing signals and impulse responses, denoted $\acute{\mathcal{S}}$, is a dioid with the same addition and multiplication as $\mathcal{S}$, but the identity element is $\acute{e}$.

The following are the nondecreasing versions of the impulse responses of some elementary systems encountered earlier, the nonmonotonic versions of which are given in Table 6.2 below:

$$\acute{e}(k) = \begin{cases} e & \text{if } k \geq 0 \ ; \\ \varepsilon & \text{otherwise;} \end{cases} \quad \acute{\gamma}^c(k) = \begin{cases} e & \text{if } k \geq c \ ; \\ \varepsilon & \text{otherwise;} \end{cases} \quad \acute{\delta}^d(k) = \begin{cases} d & \text{if } k \geq 0 \ ; \\ \varepsilon & \text{otherwise.} \end{cases}$$

6.4 TRANSFER FUNCTIONS

6.4.1 Evaluation Homomorphism

In this section we discuss the notion of transfer functions associated with shift-invariant max-plus linear systems. Transfer functions are related to impulse responses by a transformation which plays the role of the Fourier or Laplace transform in conventional system theory, and which, in our case, is similar to the Fenchel transform of convex analysis.

We saw that signals and impulse responses are functions belonging to the same idempotent algebra and that, in the canonical basis, they can be written

$$f = \oint_{\mathbb{R}}^{s} f(s)\gamma^s \ .$$

We associate a transfer function g, which will be a mapping from $\mathbb{R}_{\max}$ into $\overline{\mathbb{R}}_{\max}$, with such an impulse response viewed as a generalization of a formal polynomial introduced in Chapter 3. The value at a point of this latter function is obtained by substituting a numerical variable in $\mathbb{R}_{\max}$ for γ in the expression of f. The resulting expression is evaluated using the calculation rules of $\overline{\mathbb{R}}_{\max}$. This substitution of a numerical value for the generator should be compared with what one does in conventional system theory when substituting numerical values in $\mathbb{C}$ for the formal operator of the derivative (denoted s) in continuous time, or the shift operator (denoted z) in discrete time. To formalize this notion, we introduce the idempotent algebra of convex functions. Recall that a closed convex function is a function which is

1. l.s.c. in the conventional sense, that is, it satisfies $\lim_{x_n \to x} f(x_n) \geq f(x)$;

2. convex;

3. proper, that is, nowhere equal to $-\infty$;

or a function which is always equal to $-\infty$. It is exactly the set of the upper hulls of collections of affine functions [119, Theorem 12.1].

Definition 6.15 *The set of closed convex functions endowed with the pointwise maximum denoted $\oplus$, the pointwise addition denoted $\otimes$ and the addition of a scalar as external operation, is called the algebra of convex functions and is denoted C_{cx}.*

Once more, there is no loss of generality in considering the dioid of convex functions endowed with two internal operations only. Indeed, the product by a scalar or the pointwise product by a constant function gives the same result.

Definition 6.16 *For $f = \oint_{\mathbb{R}}^{s} f(s)\gamma^s \in \mathcal{S}$, let*

$$g : \mathbb{R} \to \overline{\mathbb{R}}_{\max} \ , \quad c \mapsto \oint_{\mathbb{R}}^{s} f(s) \otimes c^s \ . \tag{6.13}$$

Then g is called the numerical transfer function[2] *associated with f. The transform $\mathcal{F}$ which maps f to g is called the* evaluation homomorphism *(as will be justified by the forthcoming Theorem 6.17).*

Five different complete and commutative idempotent algebras and dioids have been considered, and consequently five different meanings of $\oplus$ and $\otimes$ have been used. As usual, the context should indicate which one is meant according to the nature of elements on which these binary operations operate. Table 6.1 recalls the meaning of these operations. The application of the evaluation homomorphism to the impulse

Table 6.1: Five dioids

Dioid	$\overline{\mathbb{R}}_{\max}$	$\mathcal{H}$	$\mathcal{S}$
$\oplus$	max	pointwise max	
$\otimes$	$+$	max-plus kernel product	sup-convolution
ε	$-\infty$	$\varepsilon(k,l) = -\infty \ , \quad \forall k,l$	$\varepsilon(k) = -\infty \ , \quad \forall k$
e	0	$e(k,l) = \begin{cases} 0 & \text{if } k = l \\ -\infty & \text{otherwise} \end{cases}$	$e(k) = \begin{cases} 0 & \text{if } k = 0 \\ -\infty & \text{otherwise} \end{cases}$

Dioid		$\acute{\mathcal{S}} = \acute{e} \otimes \mathcal{S}$	$\mathcal{C}_{\mathrm{cx}}$
$\oplus$		pointwise max	
$\otimes$		sup-convolution	pointwise addition
ε		$\varepsilon(k) = -\infty \ , \quad \forall k$	$\varepsilon(c) = -\infty \ , \quad \forall c$
e		$\acute{e}(k) = \begin{cases} 0 & \text{if } k \geq 0 \\ -\infty & \text{otherwise} \end{cases}$	$e(c) = 0 \ , \quad \forall c$

responses of the elementary systems is given in Table 6.2.

Theorem 6.17 *The evaluation homomorphism $\mathcal{F}$ is a l.s.c. (in the sense of Definition 4.43) epimorphism (surjective homomorphism) from $\mathcal{S}$ onto $\mathcal{C}_{\mathrm{cx}}$.*

Proof The homomorphism and l.s.c. properties are true by construction. Indeed, $\mathcal{F}(f \oplus f') = \mathcal{F}(f) \oplus \mathcal{F}(f')$, and its extension to infinite sums is true by commutativity of the sup operation. Finally $\mathcal{F}(f \otimes f') = \mathcal{F}(f) \otimes \mathcal{F}(f')$ is true by definition of the $\otimes$ operation in $\mathcal{S}$. Surjectivity arises from the fact that $\mathcal{F}(\mathcal{S})$ is the set of the upper hulls of families of affine functions which coincides with the set of closed convex functions. ∎

Remark 6.18

1. Clearly $\mathcal{F}$ is not injective, for example

$$c \oplus c^2 = [\mathcal{F}(\gamma \oplus \gamma^2)](c) = \left[\mathcal{F}\left(\oint_1^2 \gamma^s\right)\right](c) \ .$$

[2] In this chapter we will call it simply a transfer function because the notion of formal transfer is not used.

Table 6.2: Impulse responses and transfer functions of the elementary systems

System	Impulse response	Transfer function
ε	$\varepsilon(k) = \varepsilon \ , \quad \forall k$	$[\mathcal{F}(\varepsilon)](c) = \varepsilon \ , \quad \forall c$
e	$e(k) = \begin{cases} e & \text{if } k = e \\ \varepsilon & \text{otherwise} \end{cases}$	$[\mathcal{F}(e)](c) = e \ , \quad \forall c$
Γ^g	$\gamma^g(k) = \begin{cases} e & \text{if } k = g \\ \varepsilon & \text{otherwise} \end{cases}$	$[\mathcal{F}(\gamma^s)](c) = g^c \ , \quad \forall c$
Δ^d	$\delta^d(k) = \begin{cases} d & \text{if } k = 0 \\ \varepsilon & \text{otherwise} \end{cases}$	$[\mathcal{F}(\delta^d)](c) = d \ , \quad \forall c$
Φ_a	$\phi_a(k) = \begin{cases} a^k & \text{if } k \geq 0 \\ \varepsilon & \text{otherwise} \end{cases}$	$[\mathcal{F}(\phi_a)](c) = \begin{cases} e & \text{if } c \leq -a \\ \top & \text{otherwise} \end{cases}$
$\acute{e}$	$\acute{e}(k) = \begin{cases} e & \text{if } k \geq 0 \\ \varepsilon & \text{otherwise} \end{cases}$	$[\mathcal{F}(\acute{e})](c) = \begin{cases} e & \text{if } c \leq 0 \\ \top & \text{otherwise} \end{cases}$
Σ^w	$\varsigma^w(k) = \begin{cases} e & \text{if } w \geq k \geq 0 \\ \varepsilon & \text{otherwise} \end{cases}$	$[\mathcal{F}(\Sigma^w)](c) = \begin{cases} e & \text{if } c \leq 0 \\ w^c & \text{otherwise} \end{cases}$

2. The convex function
$$g(c) = \begin{cases} \varepsilon & \text{if } c = 0 \ , \\ \top & \text{otherwise,} \end{cases}$$
is not closed. Neither is it the upper hull of a set of affine functions. Indeed, each affine function would be below g and therefore would be equal to ε everywhere. Nevertheless this function is l.s.c. because the subsets $\{c \mid g(c) \leq a\}$, which are equal to $\{0\}$ for all $a \in \mathbb{R}$, are closed.

3. By returning to conventional notation, $\mathcal{F}$ can be interpreted in terms of the Fenchel transform. More precisely we have
$$[\mathcal{F}(f)](c) = \sup_k [kc + f(k)] = [\mathcal{F}_e(-f)](c) \ , \tag{6.14}$$
where $[\mathcal{F}_e(f)](c) \stackrel{\text{def}}{=} \sup_k (kc - f(k))$ denotes the classical Fenchel transform of convex analysis [58]. Recalling that the Fenchel transform converts inf-convolutions into pointwise (conventional) additions, we see that the choice of multiplication in $\mathcal{C}_{\text{cx}}$ is consistent with this property of the Fenchel transform.

■

6.4.2 Closed Concave Impulse Responses and Inputs

It is well known that the Fenchel transform only characterizes closed convex functions; or otherwise stated, all functions having the same convex hull have the same Fenchel transform. Rephrasing this result in terms of the evaluation homomorphism, we obtain that only the closed concave impulse responses are completely characterized by their transfer functions. For this subclass, the evaluation homomorphism is a tool as powerful as the Laplace transform in conventional system theory.

Theorem 6.19 *For $g \in \mathcal{C}_{\mathrm{cx}}$, the subset $\mathcal{F}^{-1}(g)$ admits a maximum element $\mathcal{F}^{\sharp}(g)$ defined by*

$$[\mathcal{F}^{\sharp}(g)](k) \stackrel{\text{def}}{=} \bigwedge_{c} g(x) \not{c}^{k} = \inf_{c}[g(c) - ck] \tag{6.15}$$

(where the latter expression in conventional notation requires the convention about $\infty - \infty$ discussed in Example 4.65). Moreover, $\mathcal{F}^{\sharp}(g)$ is the concave upper hull of any other element of $\mathcal{F}^{-1}(g)$.

Proof From the preceding considerations, we see that all the assumptions required in Theorem 4.50 are fulfilled. Then (6.15) is a straightforward extension of (3.11) to the continuous domain case. ∎

Definition 6.20 *The subset $\mathcal{S}_{\mathrm{cv}}$ of $\mathcal{S}$ consists of closed concave functions, that is, the functions which are concave, upper-semicontinuous (u.s.c. in the conventional sense) and either nowhere equal to $\top$ or always equal to $\top$.*

Remark 6.21 The set $\mathcal{S}_{\mathrm{cv}}$ is closed for multiplication (sup-convolutions of concave u.s.c. functions yield concave u.s.c. functions), but not for addition (the upper hull of concave functions is not in general a concave function). It is closed for pointwise infimum. Therefore, this subset is not a subdioid of $\mathcal{S}$. ∎

The next theorem tells us that the computation of the sup-convolution of two concave functions is equivalent to a pointwise addition and three Fenchel transforms. Knowing that there exists a fast Fenchel transform which is the analogue of the fast Fourier transform [31], this formula gives an efficient algorithm to compute sup-convolutions.

Theorem 6.22 *We have the formula*

$$\forall f, g \in \mathcal{S}_{\mathrm{cv}}\ , \quad h = f \otimes g = \mathcal{F}^{\sharp}\left(\mathcal{F}(f) \otimes \mathcal{F}(g)\right)\ ,$$

which, in conventional notation, means

$$h(k) = \sup_{x+y=k}[f(x) + g(y)] = \mathcal{F}^{\sharp}\left(\mathcal{F}(f) + \mathcal{F}(g)\right)\ .$$

Proof Equation (6.15) shows that $\mathcal{S}_{\mathrm{cv}}$ equals $\mathcal{F}^{\sharp}(\mathcal{C}_{\mathrm{cx}})$, since lower hulls of families of affine functions are closed concave functions. Therefore,

$$\forall f \in \mathcal{S}_{\mathrm{cv}}\ , \quad \mathcal{F}^{\sharp} \circ \mathcal{F}(f) = f\ .$$

Then, using the closedness of $\mathcal{S}_{\mathrm{cv}}$ and the homomorphism property of $\mathcal{F}$, we have

$$f \otimes g = \mathcal{F}^{\sharp} \circ \mathcal{F}(f \otimes g) = \mathcal{F}^{\sharp}\left(\mathcal{F}(f) \otimes \mathcal{F}(g)\right)\ .$$

∎

6.4.3 Closed Convex Inputs

In conventional system theory any L^2 function can be decomposed with respect to the basis of sine functions. In the present situation, any closed convex function can be decomposed with respect to conventional linear functions:

$$y = l_c(x) \stackrel{\text{def}}{=} c \times x = x^c \ ,$$

which may be considered as the max-plus exponentials (the last expression is in max-plus notation). This decomposition can be used to compute the outputs of a shift-invariant max-plus system driven by convex inputs. Indeed, the max-plus exponentials are eigenvectors for any shift-invariant max-plus linear system in the same way as the sine functions are eigenvectors for any shift-invariant linear system in conventional linear system theory.

Definition 6.23 *The subset $\mathcal{S}_{\text{cx}}$ of $\mathcal{S}$ consists of closed convex functions. The canonical injection of $\mathcal{S}_{\text{cx}}$ into $\mathcal{C}_{\text{cx}}$ is denoted I.*

Remark 6.24

1. The difference between $\mathcal{S}_{\text{cx}}$ and $\mathcal{C}_{\text{cx}}$ is the $\otimes$ operation (see Table 6.1).

2. Unlike $\mathcal{S}_{\text{cv}}$ which is closed for multiplication but not for addition, $\mathcal{S}_{\text{cx}}$ is closed for addition and multiplication. But in general the multiplication of two convex functions is equal to $\top$. The only exception is the product of two affine functions with the same slope.

3. The identity element $e(\cdot)$ is not convex. Therefore, $\mathcal{S}_{\text{cx}}$ is not a subdioid of $\mathcal{S}$ either.

4. The intersection of $\mathcal{S}_{\text{cv}}$ and $\mathcal{S}_{\text{cx}}$ is the subset of weighted exponentials in the max-plus sense ($b \otimes l_c(\cdot)$) or affine functions in the conventional sense. ■

In the max-plus framework, the decomposition of closed convex functions tells us that these functions are integrals of weighted exponentials. Moreover, the corresponding weights are explicitly given by the Fenchel transform.

Theorem 6.25 *For all $f \in \mathcal{S}_{\text{cx}}$, we have*

$$f = \mathrm{I}^{-1} \circ \mathcal{F} \circ \mathcal{F}^{\sharp} \circ \mathrm{I}(f) \ ,$$

which can be written

$$\forall k \ , \quad f(k) = \oint_{\mathbb{R}}^{c} c^k \left[\mathcal{F}^{\sharp} \circ \mathrm{I}(f)\right](c) = \oint_{\mathbb{R}}^{c} k^c \left[\mathcal{F}^{\sharp} \circ \mathrm{I}(f)\right](c) \ , \tag{6.16}$$

to emphasize the exponential decomposition.

Proof A function $f \in \mathcal{S}_{cx}$ may be viewed as a transfer function in $\mathcal{C}_{cx}$ because it is a closed convex function. Therefore, $\mathrm{I}(f)$ equals f but considered as an element of $\mathcal{C}_{cx}$. Because $\mathrm{I}(f) \in \mathcal{C}_{cx}$, we can solve $\mathcal{F}(g) = \mathrm{I}(f)$ for g. But we have an explicit formula for g, namely $g = \mathcal{F}^{\sharp}\left(\mathrm{I}(f)\right)$. Then using the fact that $\mathcal{F} \circ \mathcal{F}^{\sharp} = \mathrm{I}_{\mathcal{C}_{cx}}$, we have proved the result. ■

Let us show now that the max-plus exponentials (conventional linear functions) l_c are eigenvectors for the operator defined as the sup-convolution with a given impulse response.

Theorem 6.26 *For all impulse responses $h \in \mathcal{S}$ and all scalars c, we have*

$$h \otimes l_c = \left[\mathcal{F}(h)\right](-c)\, l_c \ . \tag{6.17}$$

Therefore $\left[\mathcal{F}(h)\right](-c)$ is the eigenvalue (called the gain *of h for the exponential l_c) associated with the eigenvector l_c of the operator $g \mapsto h \otimes g$.*

Proof The proof is the same as in conventional algebra:

$$[h \otimes l_c](k) = \oint_{\mathbb{R}}^{s} c^{k-s} h(s) = c^k \oint_{\mathbb{R}}^{s} c^{-s} h(s) = c^k \left[\mathcal{F}(h)\right](-c) \ .$$

■

We may use this property of the exponentials to compute the output of a shift-invariant system driven by a convex input.

Theorem 6.27 *We have*

$$\forall f \in \mathcal{S}_{cx} \ , \quad \forall h \in \mathcal{S} \ , \quad h \otimes f = \oint_{\mathbb{R}}^{c} \left[\mathcal{F}^{\sharp} \circ \mathrm{I}(f)\right](c) \ \left[\mathcal{F}(h)\right](-c)\, l_c \ ,$$

where $\left[\mathcal{F}^{\sharp} \circ \mathrm{I}(f)\right](c)$ is the weight of the exponential l_c in the spectral decomposition of f and $\left[\mathcal{F}(h)\right](-c)$ is the gain of h for the same exponential.

Proof Using the distributivity of $\otimes$ (sup-convolution in $\mathcal{S}$) with respect to $\oint$, we have

$$\begin{aligned}
h \otimes f &= h \otimes \oint_{\mathbb{R}}^{c} \left[\mathcal{F}^{\sharp} \circ \mathrm{I}(f)\right](c)\, l_c && \text{by (6.16),} \\
&= \oint_{\mathbb{R}}^{c} \left[\mathcal{F}^{\sharp} \circ \mathrm{I}(f)\right](c)\, (h \otimes l_c) && \text{by linearity,} \\
&= \oint_{\mathbb{R}}^{c} \left[\mathcal{F}^{\sharp} \circ \mathrm{I}(f)\right](c)\, [\mathcal{F}(h)](-c)\, l_c && \text{by (6.17),}
\end{aligned}$$

and the function $h \otimes f$ also belongs to $\mathcal{S}_{cx}$. ■

In conclusion, we have encountered two situations of special interest to compute the response of a system to an input

- if the input and the impulse response are concave, we can use the evaluation homomorphism to transform this inf-convolution into a pointwise conventional sum;
- if the input is convex, we can first decompose it as a sum of exponential functions (in the dioid sense), and then, using the linearity of the system, we can sum up the responses to these exponentials inputs.

6.5 RATIONAL SYSTEMS

The set $\mathcal{S}$ is a nice algebraic structure but its elements are functions and therefore cannot be coded by finite sets of numbers in general. It is useful to consider subsets of these functions which can be coded in a finite way. The algebraic functions would constitute such a set. Those functions are described as the solutions of a polynomial systems of equations in $\mathcal{S}$. But even in classical system theory the study of these systems is in its infancy. Therefore we restrict ourselves to a simpler situation. We only consider systems which can be described by a finite set of *special linear* equations $y = hy \oplus u$. These equations describe the input-output relation of systems obtained by series, parallel and feedback compositions of elementary systems for which the impulse responses are explicitly known. Such systems are called rational. Clearly this notion of rationality depends on the elementary systems considered. Rational systems can be described in terms of the star operation ($y = h^*u$). This story is not specific to max-plus algebra, but the rationals of these max-plus algebras have simple characterizations in terms of their periodic asymptotic behavior which is similar to the periodicity property of the decimal expansion of rational numbers. The aim of this section is to characterize max-plus rational systems by their asymptotic behavior.

6.5.1 Polynomial, Rational and Algebraic Systems

Let us consider

1. a subset K of $\mathcal{S}$ which also has a structure of idempotent algebra but not necessarily the same identity element as $\mathcal{S}$ (for example, nondecreasing functions define an algebra with $\acute{e}$ as the identity element—see §6.3.3);
2. a finite set $\alpha = \{\alpha_1, \ldots, \alpha_\ell\}$ of elements of $\mathcal{S}$.

Let us define five subsets of $\mathcal{S}$ which may be considered as extensions of K and which have a structure of idempotent algebra:

polynomial or dioid closure $K[\alpha]$ of $K \cup \alpha$: its elements are obtained by combining the elements of $K \cup \alpha$ using a finite number of $\oplus$ and $\otimes$ operations;

rational closure $K(\alpha)$ of $K \cup \alpha$: its elements are obtained by combining the elements of $K \cup \alpha$ using a finite number of $\oplus$, $\otimes$ and * operations;

algebraic closure $K\{\alpha\}$ of $K \cup \alpha$: its elements are obtained by combining the elements of $K \cup \alpha$ using a finite number of $\oplus$ and $\otimes$ operations and by solving polynomial equations with coefficients in $K \cup \alpha$;

series closure $K[\![\alpha]\!]$ of $K \cup \alpha$: its elements are obtained by combining the elements of $K \cup \alpha$ using a countable number of $\oplus$ and $\otimes$ operations: this is the completion of the polynomial closure;

topological closure $K\{\!\{\alpha\}\!\}$ of $K \cup \alpha$: its elements are obtained by combining the elements of $K \cup \alpha$ using an infinite number of $\oplus$ and $\otimes$ operations and by solving polynomial equations: this is the completion of the algebraic closure.

Note that $\mathcal{S}$ is a subset of $\overline{\mathbb{R}}_{\max}\{\!\{\gamma\}\!\}$. In general we have

$$K \subset K[\alpha] \subset K(\alpha) \subset \left\{ \begin{matrix} K[\![\alpha]\!] \\ K\{\alpha\} \end{matrix} \right\} \subset K\{\!\{\alpha\}\!\} \ .$$

For example, consider $K = \mathbb{B} = \{\varepsilon, e\}$ and $\alpha = \{\delta\}$, where δ is the impulse response mentioned in Table 6.2. Recall that in $\mathcal{S}$, $\delta^{d_1} \oplus \delta^{d_2} = \delta^{\max(d_1,d_2)}$. In this particular case we have

$$\mathbb{B}[\delta] \simeq \mathbb{N}_{\max} \subset \mathbb{B}(\delta) = \mathbb{B}[\![\delta]\!] \simeq \overline{\mathbb{N}}_{\max} \subset \mathbb{B}\{\delta\} \simeq \overline{\mathbb{Q}}_{\max} \subset \mathbb{B}\{\!\{\delta\}\!\} \simeq \overline{\mathbb{R}}_{\max} \ ,$$

where $A_{\max}$ for a set A means $A \cup \{\varepsilon\}$ endowed with the max and the $+$ operations, $\overline{A}$ is $A_{\max} \cup \{+\infty\}$, and the isomorphisms above identify scalars d with impulse responses δ^d.

Remark 6.28 Observe that in $\mathbb{B}[\delta]$ it is difficult to speak of the notion of valuation. For example, the valuation of $\delta^{d_1} \oplus \delta^{d_2}$ would formally be equal to $\min(d_1, d_2)$; but at the same time $\delta^{d_1} \oplus \delta^{d_2}$ is equal to $\delta^{\max(d_1,d_2)}$; the latter is a monomial the valuation of which is thus the same as the degree, namely $\max(d_1, d_2)$.

Similarly, consider $\acute{\mathbb{B}}[\acute{\gamma}]$ which is equal to $\acute{e} \otimes \mathbb{B}[\gamma]$, and observe that the notion of degree is equally difficult to define. Indeed, owing to (6.5) which holds true for nondecreasing signals, $\acute{\gamma}^{g_1} \oplus \acute{\gamma}^{g_2}$ is equal to $\acute{\gamma}^{\min(g_1,g_2)}$, whereas it would formally have a degree equal to $\max(g_1, g_2)$.

The same difficulties arise in other polynomial (or dioid) closures such as $\acute{\mathbb{B}}[\acute{\gamma}, \acute{\delta}] = \acute{e}\mathbb{B}[\gamma, \delta]$. This dioid is isomorphic to the polynomial subdioid of $\mathcal{M}_{\mathrm{in}}^{\mathrm{ax}}[\![\gamma, \delta]\!]$ ('polynomial' in the sense of Definition 5.19). Any element in $\acute{\mathbb{B}}[\acute{\gamma}, \acute{\delta}]$ can be represented, in a nonunique way, as the product of $\acute{e}$ by an element of $\mathbb{B}[\gamma, \delta]$: the latter may be called a 'representative' of the former. It is thus possible to speak of the valuations and degrees in γ and δ of such a representative in $\mathbb{B}[\gamma, \delta]$ of an element of $\acute{\mathbb{B}}[\acute{\gamma}, \acute{\delta}]$. However, these notions can be given an intrinsic meaning only if we restrict ourselves to the 'minimum representative' which exists for polynomials (see Theorem 5.20). ■

6.5.2 Examples of Polynomial Systems

Table 6.3 gives the main examples of polynomial closures of $\overline{\mathbb{R}}_{\max}$ or of $\mathbb{B}$ used in this book. They are obtained from the set of scalars (identified with impulse responses δ^d as explained earlier) augmented with the impulse responses of some of the elementary systems encountered previously. We set $\phi = \{\phi_{c_1}, \ldots, \phi_{c_\ell}\}$, where ϕ_{c_i} is defined in Table 6.2. In the following, the c_i are assumed to be positive and therefore ϕ_{c_i} is nondecreasing.

6.5.3 Characterization of Rational Systems

A characterization of elements of $K(\alpha)$ is given under the assumption that K is rationally closed (see Definition 4.99); this is the representation problem. In the present context, we defined a rational element as an element h obtained by a finite number of $\oplus$, $\otimes$ and * operations applied to elements of $K \cup \alpha$. The following result shows that only one * operation is needed with respect to the elements in α. Consequently, it is easy to obtain a linear system which admits h as its impulse response.

Theorem 6.29 (Representation of rational impulse responses) *We assume that K is rationally closed (for example, K is a complete dioid). Then, for all $h \in K(\alpha)$, there exist $n \in \mathbb{N}$, $B, C \in K^n$ and $A_i \in K^{n\times n}$, $i = 1, \ldots, \ell$, such that*

$$h = C' \left(\bigoplus_{i=1}^{\ell} \alpha_i A_i \right)^* B \ . \tag{6.18}$$

Proof Let us refer to Theorem 4.105 and set $\mathcal{B} = K$, $\mathcal{C} = K$, $\mathcal{U} = K$ and $\mathcal{V} = \alpha$. Since K is supposed to be rationally closed, then $\mathcal{U}^\star = K$, $\mathcal{U}^\star \otimes \mathcal{B} = K$ and $\mathcal{U}^\star \otimes \mathcal{V}$ consists of linear combinations of elements of α with coefficients in K. These observations lead to (6.18). ■

Example 6.30 Let us consider the element h of $\overline{\mathbb{R}}_{\max}(\gamma)$ defined by $h = ((1\gamma^3)^* \oplus (\gamma^2)^*)^*$. Observe that $\overline{\mathbb{R}}_{\max}$ is a complete dioid. Using (4.109) and the fact that $(a^*)^* = a^*$, we have $h = ((1\gamma^3)^*)^*((\gamma^2)^*)^* = (1\gamma^3)^*(\gamma^2)^* = (1\gamma^3 \oplus \gamma^2)^*$ for which we obtain the realization

$$x_2 = \gamma x_1 \ , \quad x_3 = \gamma x_2 \ , \quad x_1 = 1\gamma x_3 \oplus \gamma x_2 \oplus u \ , \quad y = x_1 \ .$$

■

In the case of nondecreasing impulse responses, the form of the rational functions may be explicited by specializing Theorem 6.29.

Corollary 6.31 *Every $h \in \acute{\overline{\mathbb{R}}}_{\max}(\phi)$ can be written*

$$h = \bigoplus_{i=1}^{\ell} h_i \phi_{c_i} \ , \quad h_i \in \overline{\mathbb{R}}_{\max} \ , \quad i = 1, \ldots, \ell \ .$$

Table 6.3: Polynomial extensions.

K	α	$h \in K[\alpha] \subset \overline{\mathbb{R}}_{\max}^{\mathbb{R}}$
$\overline{\mathbb{R}}_{\max}$ $\simeq$ $\mathbb{B}\{\!\{\delta\}\!\}$	$\{\gamma\}$	$h(k) \neq \varepsilon$ only for $k \in \mathbb{N}$. 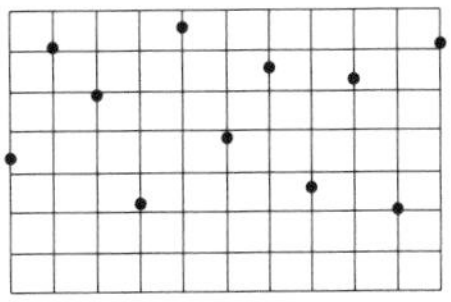
$\acute{\overline{\mathbb{R}}}_{\max}$ $\stackrel{\text{def}}{=}$ $\acute{e}\overline{\mathbb{R}}_{\max}$	$\{\acute{\gamma}\}$	$h(k) = \varepsilon$ for $k \in \mathbb{R}^-$; over $\mathbb{R}^+$, h is nondecreasing and piecewise constant with a finite number of discontinuities at integer abscissæ. 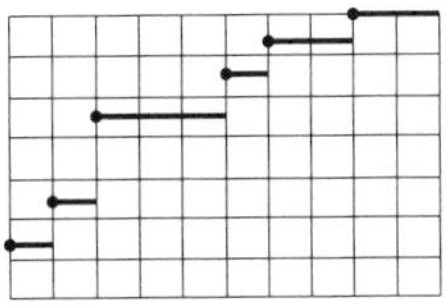
$\acute{\overline{\mathbb{R}}}_{\max}$	ϕ $=$ $\{\phi_{c_1}, \ldots, \phi_{c_\ell}\}$	$h(k) = \varepsilon$ for $k \in \mathbb{R}^-$; over $\mathbb{R}^+$, h is convex, nondecreasing and piecewise linear with slopes in $\{c_1, \ldots, c_\ell\}$. 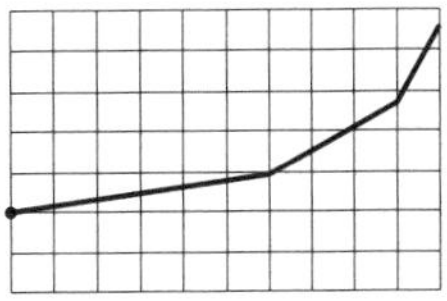
$\acute{\mathbb{B}}$ $\stackrel{\text{def}}{=}$ $\acute{e}\{\varepsilon, e\}$	$\{\acute{\gamma}, \acute{\delta}\}$	$h(k) = \varepsilon$ for $k \in \mathbb{R}^-$; over $\mathbb{R}^+$, h is nondecreasing, piecewise constant and integer-valued with a finite number of discontinuities at integer abscissæ. 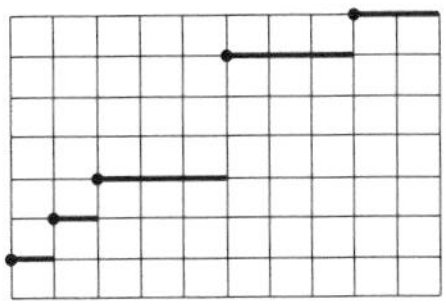
$\acute{\mathbb{B}}$	$\{\acute{\gamma}, \acute{\delta}\} \cup \phi$	$h(k) = \varepsilon$ for $k \in \mathbb{R}^-$; over $\mathbb{R}^+$, h is convex, nondecreasing, piecewise linear with slopes in $\{c_1, \ldots, c_\ell\}$ and with a finite number of discontinuities at integer abscissæ. 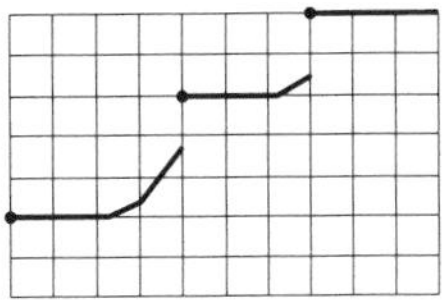

Proof Using Theorem 6.29 with $K = \acute{\overline{\mathbb{R}}}_{\max}$ (this is a complete, hence rationally closed, dioid) and $\alpha = \phi$, we can write

$$h = c' \left(\bigoplus_{i=1}^{\ell} a_i \phi_{c_i} \right)^* b \ ,$$

where the entries of b, c and a_i belong to $\acute{\overline{\mathbb{R}}}_{\max}$. By expanding the * expression and by using the simplification rules given in (6.7), we obtain the form claimed in the statement of the corollary, but with coefficients $\acute{h}_i$ of the ϕ_{c_i} belonging to $\acute{\overline{\mathbb{R}}}_{\max}$. As such, they can be written $\acute{h}_i = \acute{e} h_i$ for some $h_i \in \overline{\mathbb{R}}_{\max}$. On the other hand, recall that we assumed $c_i > 0$ for all i, which implies that the ϕ_{c_i} are nondecreasing. Hence $\acute{e}\phi_{c_i} = \phi_{c_i}$. This observation allows us to adopt the h_i as the coefficients of the ϕ_{c_i}. ∎

Theorem 6.32 *Every $h \in \acute{\mathbb{B}}(\acute{\gamma}, \acute{\delta})$ can be written $h = \acute{e}\left(p \oplus \gamma^\nu \delta^\tau (\gamma^r \delta^s)^* q\right)$, where*

- *$p \in \mathbb{B}[\gamma, \delta]$ is a polynomial of degree at most $\nu - 1$ in γ and $\tau - 1$ in δ;*
- *$q \in \mathbb{B}[\gamma, \delta]$ is a polynomial of degree at most $r - 1$ in γ and $s - 1$ in δ.*

For a given h, the ratio s/r is independent of the particular representation of this type.

The above form expresses a periodic behavior of the impulse response h. The polynomial p represents the transient part having a 'width' of ν and a 'height' of τ. The polynomial q represents a pattern having a 'width' of r and a 'height' of s. This pattern is reproduced indefinitely after the transient part (see Figure 6.2). The ratio s/r represents the 'asymptotic slope' (see Definition 6.46 below). For the extreme cases $r = 0$ or $s = 0$, the reader may return to the discussion in §5.7.4.

Proof Because we are in the commutative case, we can refer to Theorem 4.110 with $\mathcal{T} = \{\varepsilon, \acute{e}, \acute{\gamma}, \acute{\delta}\}$ and $\mathcal{D} = \acute{\mathbb{B}}[\![\acute{\gamma}, \acute{\delta}]\!]$. In fact, in the following, we will also need to use elements of $\acute{\mathbb{B}}\{\acute{\gamma}, \acute{\delta}\}$, hence we may embed all these structures into a larger one, namely $\acute{\mathbb{B}}\{\!\{\acute{\gamma}, \acute{\delta}\}\!\}$. From Theorem 4.110 we have that $h = \bigoplus_{i=1}^{l} \acute{a}_i \left(\acute{b}_i\right)^*$, for some $\acute{a}_i$ and $\acute{b}_i$ which are elements of $\mathcal{T}^\diamond = \acute{\mathbb{B}}[\acute{\gamma}, \acute{\delta}] = \acute{e}\mathbb{B}[\gamma, \delta]$ (see §6.3.3). Since $\acute{a}_i$ and $\acute{b}_i$ are polynomials, we may consider their minimum representatives in $\mathbb{B}[\gamma, \delta]$ (see Remark 6.28), denoted a_i and b_i, respectively, and thus obtain the new form $h = \acute{e} \bigotimes_{i=1}^{l} a_i (b_i)^*$. It remains to show that this form can be reduced to the form given in the theorem statement, which essentially uses the star of a single monomial in (γ, δ). This proof is outlined below.

Considering monomials $m = \gamma^r \delta^s$, we first introduce the rational number $\mathrm{sl}(m) \stackrel{\text{def}}{=} s/r$, called the 'slope' (with the convention that $\mathrm{sl}(e) \stackrel{\text{def}}{=} 0$). This notion is extended to polynomials (or power series) as follows. If m_1 and m_2 are two monomials, then

$$\mathrm{sl}(m_1 \oplus m_2) = \mathrm{sl}(m_1) \oplus \mathrm{sl}(m_2) \ .$$

The expression of $\mathrm{sl}(m_1 \otimes m_2)$ is a direct consequence of the definition since the product of two monomials is also a monomial. Using these rules, we notice that, if p is a polynomial, then $\mathrm{sl}(p^*) = \mathrm{sl}(p)$ and this is the maximum slope among the monomials which form the polynomial.

We now propose the following inequalities:

$$x \stackrel{\text{def}}{=} \acute{e}\delta^s(\gamma\delta^{s/r})^* \geq y \stackrel{\text{def}}{=} \acute{e}(\gamma^r\delta^s)^* \geq z \stackrel{\text{def}}{=} \acute{e}\gamma^r(\gamma\delta^{s/r})^*$$

(note that $\delta^{s/r}$ is an element of the algebraic closure of $\mathcal{T}$). Only the inequality $y \geq z$ will be proved. The other inequality can be proved using similar calculations. With $n = \alpha r + \beta$, $\alpha, \beta \in \mathbb{N}$ and $\beta < r$, all the monomials of z namely $\acute{e}\gamma^r(\gamma\delta^{s/r})^n$, $n \in \mathbb{N}$, can be written $\acute{e}(\gamma^r\delta^s)^{\alpha+1}\gamma^\beta\delta^{(\beta/r-1)s}$. The monomial $\acute{e}(\gamma^r\delta^s)^{\alpha+1}$ appear in y, whereas the multiplicative monomial $\acute{e}\gamma^\beta\delta^{(\beta/r-1)s}$ is less than $\acute{e}$ (it has a nonnegative exponent in γ and a negative exponent in δ), owing to the simplification rules for 'shifts' and 'gains' given in §6.2.2. Thus each monomial of z is dominated by a monomial of y.

From these inequalities and from Lemma 3.107, we can derive the following four rules.

Rule 1: $\mathrm{sl}(m_2) < \mathrm{sl}(m_1) \Rightarrow p_1(m_1)^* \oplus p_2(m_2)^* = p \oplus p_1(m_1)^*$, where p is a polynomial depending on the given polynomials p_i and monomials m_i.

Rule 2: $\mathrm{sl}(m_2) = \mathrm{sl}(m_1) \Rightarrow (m_1)^* \oplus (m_2)^* = p \otimes (\mathrm{lcm}(m_1, m_2))^*$, where $m_i = \acute{e}\gamma^{r_i}\delta^{s_i}$, $\mathrm{lcm}(m_1, m_2) \stackrel{\text{def}}{=} \acute{e}\gamma^{\mathrm{lcm}(r_1,r_2)}\delta^{\mathrm{lcm}(s_1,s_2)}$, and p is a polynomial depending on the m_i.

Rule 3: $\mathrm{sl}(m_2) < \mathrm{sl}(m_1) \Rightarrow (m_1)^* \otimes (m_2)^* = (m_1 \oplus m_2)^* = p \oplus q(m_1)^*$, where p and q are polynomials depending on the given monomials m_i. This rule can be derived from $(m_1)^* \otimes (m_2)^* = \bigoplus_r m_1^r m_2^*$.

Rule 4: $\mathrm{sl}(m_2) = \mathrm{sl}(m_1) \Rightarrow (m_1)^* \otimes (m_2)^* = (m_1 \oplus m_2)^* = p \oplus m \ \mathrm{gcd}(m_1, m_2)^*$, where the gcd of two monomials is defined in a similar way as the lcm previously, m is a monomial and p a polynomial, both depending on the m_i.

The possibility of reducing h to the claimed form comes from the recursive utilization of these four rules.

Finally, it should be clear that h cannot have two representations with different values of the ratio s/r. ■

Remark 6.33 The representation of rationals in $\overline{\acute{\mathbb{R}}}_{\max}(\acute{\gamma})$ is a simple extension of Theorem 6.32. Indeed $\acute{\mathbb{B}}(\acute{\gamma}, \acute{\delta}) \simeq \left[\mathbb{B}\left(\acute{\delta}\right)\right](\acute{\gamma}) \simeq \overline{\mathbb{N}}_{\max}(\acute{\gamma})$. Therefore we have to generalize the situation to the case when the coefficients of power series in γ are real instead of integer. This extension is straightforward. The result becomes: for each $h \in \overline{\acute{\mathbb{R}}}_{\max}(\acute{\gamma})$, there exist $p, q \in \overline{\mathbb{R}}_{\max}[\gamma]$ of degrees $\nu - 1$ and $r - 1$, respectively, and $a \in \mathbb{R}$ such that $h = \acute{e}\left(p \oplus q\gamma^\nu(a\gamma^r)^*\right)$. For a given h (recall this is a nondecreasing

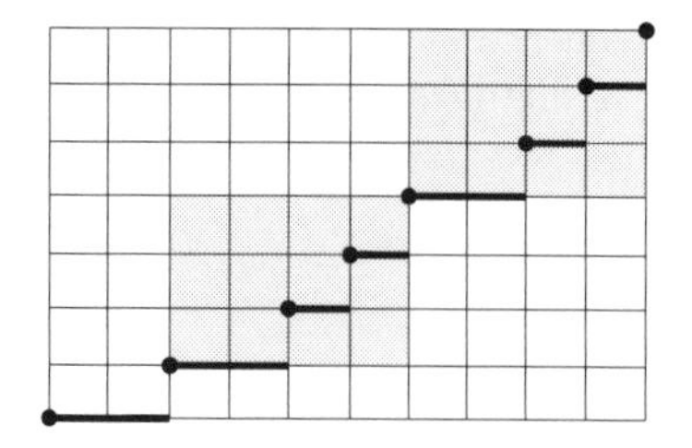

Figure 6.2: An element of $\acute{\mathbb{B}}(\acute{\gamma}, \acute{\delta})$

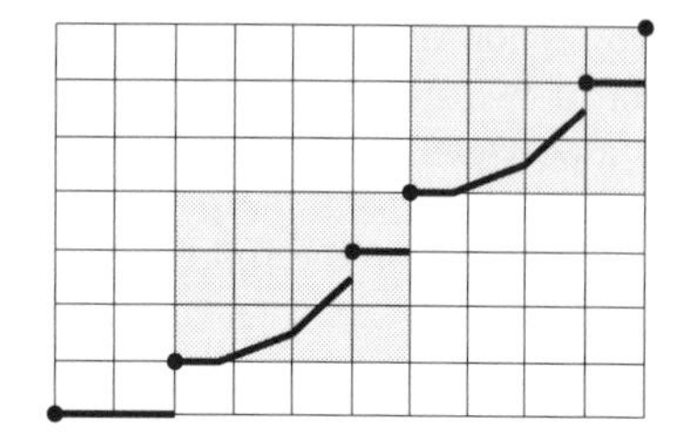

Figure 6.3: An element of $\acute{\mathbb{B}}(\acute{\gamma}, \acute{\delta}, \phi)$

impulse response), a can be restrained to be nonnegative, and then the nonnegative slope a/r is independent of the particular representation chosen for h. ■

Finally, the following corollary is just the synthesis of the previous results.

Corollary 6.34 *Every $h \in \acute{\mathbb{B}}(\acute{\gamma}, \acute{\delta}, \phi)$ can be written as $h = \acute{e}(p \oplus \gamma^\nu \delta^\tau (\gamma^r \delta^s)^* q)$, where*

- *$p \in \mathbb{B}[\gamma, \delta, \phi]$ is a polynomial of degree at most $\nu - 1$ in γ and $\tau - 1$ in δ, and it is linear in ϕ_{c_i};*
- *$q \in \mathbb{B}[\gamma, \delta, \phi]$ is a polynomial of degree at most $r - 1$ in γ and $s - 1$ in δ, and it is linear in ϕ_{c_i}.*

This theorem describes the asymptotically periodic behavior of the impulse response, the periodic pattern being a piecewise nondecreasing convex function (see Figure 6.3).

6.5.4 Minimal Representation and Realization

The minimal representation problem can be stated in different terms depending on the elementary subsystems that we consider. Let us discuss the two examples $\acute{\overline{\mathbb{R}}}_{\max}[\phi]$ and $\acute{\overline{\mathbb{R}}}_{\max}(\acute{\gamma})$.

Definition 6.35 (Minimal representation in $\acute{\overline{\mathbb{R}}}_{\max}[\phi]$) *Given $h \in \acute{\overline{\mathbb{R}}}_{\max}[\phi]$, where $\phi = \{\phi_{c_1}, \ldots, \phi_{c_\ell}\}$ (the c_i are nonnegative), the minimal representation problem consists in finding a subset of ϕ with minimal cardinality $\ell_{\min}$ such that $h = \bigoplus_{i=1}^{\ell_{\min}} h_i \phi_{c_i}$, with $h_i \in \overline{\mathbb{R}}_{\max}$, $i = 1, \ldots, \ell_{\min}$.*

In conventional system theory, this problem corresponds to finding the minimal number of exponentials of which the impulse response is a linear combination.

Observe that this representation directly corresponds to a realization (A, B, C) with a diagonal matrix A (see the theorem below). Indeed, in conventional system theory, the impulse response of a continuous-time shift-invariant system may contain functions of the form $t^n \exp(kt)$. The max-plus case is simpler because $a^t = t^a$ and therefore the impulse response is only composed of max-plus exponentials.

Theorem 6.36 *Given $h = \bigoplus_{i=1}^{\ell} h_i \phi_{c_i}$, the realization*

$$x_1 = \phi_{c_1} u \ , \ldots , \quad x_\ell = \phi_{c_\ell} u \ , \quad y = Cx \ ,$$

with $C = \begin{pmatrix} h_1 & \ldots & h_\ell \end{pmatrix}$, is minimal if and only if the points $(c_i, h_i) \in \mathbb{R}^+ \times \overline{\mathbb{R}}$ are the corners of the graph of a decreasing and concave piecewise linear function.

Proof Let $c_{i_{min}} = \min_i c_i$ and $c_{i_{\max}} = \max_i c_i$. Over $\mathbb{R}^+$, h is the upper hull of ℓ affine functions $x \mapsto c_i x + h_i$, whereas $h(x) = -\infty$ for $x < 0$. Since we are interested in determining whether the ℓ affine functions are all needed to represent h, it does not matter if we replace h by a new function H such that $H(x) = h(x)$ for $x \geq 0$ and $H(x) = +\infty$ for $x > 0$. This H is convex and is fully characterized by its Fenchel transform (see Remark 3.36). This latter function also is convex and piecewise linear, and it admits some of the points $(c_i, -h_i)$ as the corners of its graph. Moreover, owing to our assumption that $H(x) = +\infty$ for $x < 0$, this function is constant at the value $-h_{i_{\min}}$ on the left of $c_{i_{\min}}$ and is equal to $+\infty$ beyond $c_{i_{\max}}$. Because of the horizontal branch of the graph at the left-hand side, the first slope is zero and the next slopes are all positive since the slope of a convex function is nondecreasing in general, and moreover it strictly increases when a corner is traversed. Any pair $(c_i, -h_i)$ which is not a corner of the graph of the Fenchel transform can be discarded without changing this function. The corresponding affine function of x can also be discarded without changing h. The statement of the theorem expresses these conditions, which are obviously necessary and sufficient, up to the change of $-h_i$ into h_i. ∎

Definition 6.37 (Minimal realization in $\overline{\mathbb{R}}_{\max}^{\prime}(\acute{\gamma})$) *Given $h \in \overline{\mathbb{R}}_{\max}^{\prime}(\acute{\gamma})$, the minimal realization problem consists in finding a triple $(A, B, C) \in \overline{\mathbb{R}}_{\max}^{n \times n} \times \overline{\mathbb{R}}_{\max}^{n} \times \overline{\mathbb{R}}_{\max}^{n}$ with minimal n such that $h = C(\acute{\gamma} A)^* B$. Equivalently, if $y = hu$, then there exists $x \in \left(\overline{\mathbb{R}}_{\max}^{\prime}(\acute{\gamma})\right)^n$ such that (u, x, y) satisfy:*

$$\begin{pmatrix} x \\ y \end{pmatrix} = S \begin{pmatrix} x \\ u \end{pmatrix} = \acute{e} \begin{pmatrix} A & B \\ C & \varepsilon \end{pmatrix} \begin{pmatrix} x \\ u \end{pmatrix} .$$

Matrix S is called the system matrix.

The problem of finding a minimal realization is still open. The previous minimal representation theorem in the case of continuous impulse responses cannot be extended to the present discrete situation in a straightforward manner, essentially because it is difficult to precisely identify the underlying 'exponentials' (that is, the expressions of the form $(a\gamma^{r_i})^*$ which may contribute to the transient part of the impulse response). Many attempts to solve this problem have not been successful yet. In Chapter 9 partial results are given. Nevertheless the following theorem gives a realization which is not necessarily minimal but which contains only one star operation.

Theorem 6.38 (Realization in $\overline{\mathbb{R}}_{\max}(\acute{\gamma})$) *Every element of $\overline{\mathbb{R}}_{\max}(\acute{\gamma})$ can be realized with a system matrix S given by*

$$S = \acute{e} \begin{pmatrix} \varepsilon & \gamma & \varepsilon & \cdot & \cdot & \cdot & \cdot & \varepsilon \\ \varepsilon & \varepsilon & \gamma & \cdot & \cdot & \cdot & \cdot & \varepsilon \\ \cdot & \cdot & \cdot & \cdot & \cdot & \cdot & \cdot & \cdot \\ a\gamma & \varepsilon & \cdot & \cdot & \gamma & \varepsilon & \cdot & \cdot \\ \varepsilon & \cdot & \cdot & \cdot & \cdot & \gamma & \cdot & \varepsilon \\ \cdot & \cdot & \cdot & \cdot & \cdot & \cdot & \gamma & \varepsilon \\ \cdot & \cdot & \cdot & \cdot & \cdot & \cdot & \cdot & e \\ q(r-1) & \cdot & \cdot & q(0) & p(\nu-1) & \cdot & p(0) & \varepsilon \end{pmatrix}$$

corresponding to the event graph given in Figure 6.4.

Proof Remark 6.33 showed that if $h \in \overline{\mathbb{R}}_{\max}(\acute{\gamma})$, it can be represented as $h = \acute{e}\left(p \oplus q\gamma^{\nu}(a\gamma^{r})^{*}\right)$, with $p = \bigoplus_{i=0}^{\nu-1} p(i)\gamma^{i}$ and $q = \bigoplus_{j=0}^{r-1} q(j)\gamma^{j}$. By direct calculation, which consists in eliminating x in

$$\begin{pmatrix} x \\ y \end{pmatrix} = S \begin{pmatrix} x \\ u \end{pmatrix} ,$$

with the above expression of S, one can check that $y = hu$ with the given expression of h. ∎

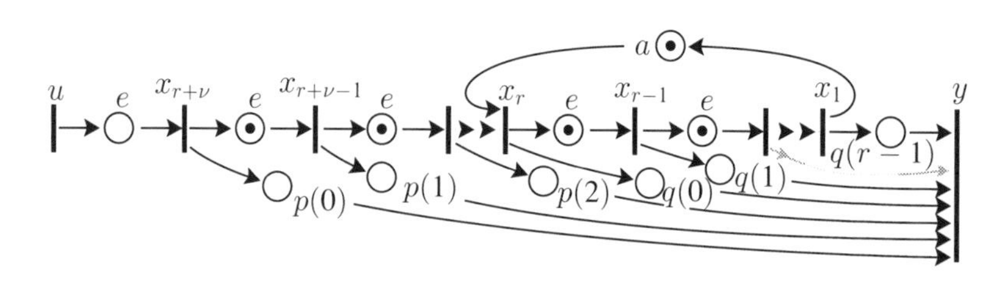

Figure 6.4: One-star realization of rational systems in $\overline{\mathbb{R}}_{\max}(\acute{\gamma})$

6.6 CORRELATIONS AND FEEDBACK STABILIZATION

In this section we develop a second-order max-plus system theory. This theory offers algebraic similarities with conventional second-order system theory. In the context of event graphs, its main application is the evaluation of sojourn times of tokens in places or in broader portions of the graph. A notion of internal stability is introduced by saying that a system is stable if all its sojourn times are bounded. Finally it is shown that a structurally observable and controllable system can be stabilized by a dynamic feedback while preserving the asymptotic open-loop performance.

6.6.1 Sojourn Time and Correlations

We consider the problem of computing the sojourn times in timed event graphs. Let v and u be two daters associated with a pair of transitions (also named v and u) surrounding a place p containing μ tokens initially (v corresponds to the upstream transition, u to the downstream). The token involved in the firing of transition u numbered k (this firing occurs at time $u(k)$) corresponds to the token which was produced by the firing of transition v numbered $k - \mu$ (occurring at $v(k - \mu)$). This is because we deal with deterministic event graphs with constant holding times and the FIFO rule may be assumed for places. Therefore, we define the sojourn time $T_{uv}(k, \mu)$ of this token in place p (along arc (v, u), marked with μ tokens initially) by

$$T_{uv}(k, \mu) = u(k) - v(k - \mu) \ .$$

More generally, for two transitions v and u connected by a path ρ containing μ_ρ tokens initially (i.e. $\mu_\rho = |\rho|_{\mathrm{t}}$), $T_{uv}(k, \mu_\rho) = u(k) - v(k - \mu_\rho)$ represents the time spent along the path ρ by the token numbered k at u. These notions can be generalized to continuous systems, like the one presented in §1.2.7, by considering that tokens are 'molecules' of fluid in pipes. More formally, we introduce the following notions.

Definition 6.39 *Let u, respectively v, be an n-dimensional, respectively p-dimensional, vector with entries in $\mathcal{S}$.*

Sojourn-time matrix *The sojourn time $(T_{uv})_{ij}(k, \mu)$ of the token participating in the k-th firing of transition u_i, and using a path from v_j to u_i which contains μ tokens initially, is defined as*

$$(T_{uv})_{ij}(k, \mu) = u_i(k) \not\phi v_j(k - \mu) = \big(u(k) \not\phi v(k - \mu)\big)_{ij} \ .$$

Here $\not\phi$ denotes the residuation of the instantaneous matrix product. The $n \times p$ matrix function $T_{uv}(\cdot, \cdot)$, which gives the sojourn times between paths going from the set v of p transitions to the set u of n transitions, is called the sojourn-time matrix.

Correlation matrix *Let R_{uv} be the matrix with entries in $\mathcal{S}$ defined by $R_{uv} = u \not\phi v$. Here $\not\phi$ is considered as the residuation of the (matrix) 'convolution' product in force in $\mathcal{S}$ (which is similar to power series product). Therefore (see (4.97)),*

$$\forall m \in \mathbb{R}, R_{uv}(\mu) = [u \not\phi v](\mu) = \bigwedge_k (u(k) \not\phi v(k - \mu)) = \bigwedge_k T_{uv}(k, \mu), \quad (6.19)$$

This R_{uv} is called the correlation matrix of u with v. If $u = v$, it is called the autocorrelation matrix of u.

There might be parallel paths with different initial markings.

Lemma 6.40 *If $v \in \acute{\mathcal{S}}$, the mappings $\mu \mapsto T_{uv}(k, \mu)$ and $\mu \mapsto R_{uv}(\mu)$ are nondecreasing.*

Proof Since v is nondecreasing,

$$\forall k \ , \quad \mu' \leq \mu \Rightarrow v(k-\mu') \geq v(k-\mu) \Rightarrow u(k) \phi v(k-\mu') \leq u(k) \phi v(k-\mu) \ .$$

The results follow immediately. ■

Remark 6.41 We refer the reader to Example 4.65 for the manipulation of ϕ in $\overline{\mathbb{R}}_{\max}$ and to §4.6.2 for the matrix formulæ involving ϕ. It may be useful to recall the point of view adopted in [49]. With the choice of primitives used therein, the residuation operator can be evaluated as follows in the case of vectors over $\mathbb{R}_{\max}$. Let us first introduce the following notation: an overlined (square or nonsquare) matrix or vector will denote the transposed matrix or vector in which, moreover, the (conventional) sign of all entries has been changed to the opposite. For example, if $a = \begin{pmatrix} 2 & 3 \end{pmatrix}$, then $\overline{a} = \begin{pmatrix} -2 & -3 \end{pmatrix}'$; if $h \in \mathcal{S}$, then $\overline{h}(t) = -h(-t)$.[3] Then we have $u \phi v = \overline{v \otimes \overline{u}}$ and $v \backslash u = \overline{\overline{u} \otimes v}$, where $\otimes$ still denotes the matrix product in $\mathbb{R}_{\max}$. These formulæ hold also true in $\overline{\mathbb{R}}_{\max}$: we have

$$\top = \varepsilon \phi \varepsilon = \overline{\varepsilon \otimes \overline{\varepsilon}} = \overline{\varepsilon \otimes \top} \ , \qquad \top = \top \phi \top = \overline{\top \otimes \overline{\top}} = \overline{\top \otimes \varepsilon} \ .$$

It is also useful to recall the De Morgan formulæ for the sup and the inf (see (4.9)–(4.11)) and in particular the following formula:

$$\overline{a \otimes b} = \overline{b} \odot \overline{a} \ ,$$

where $\odot$ denotes the matrix product based on min and $+$ (the absorbing element for scalar multiplication being now $\top = +\infty$: we have $\varepsilon \otimes \top = \varepsilon$ but $\varepsilon \odot \top = \top$). For example,

$$a \phi b = \overline{b \otimes \overline{a}} = a \odot \overline{b} \ .$$

■

Remark 6.42

1. Another interesting quantity is $T^{+}_{uv}(k,\mu) = u(k) - v((k-\mu)_{-})$ where $v(k_{-}) = \lim_{s \uparrow k} v(s)$. This T^{+} is different from T when v is not left-continuous.
2. Let us consider the case when u and v are scalar functions. The analogy between the conventional correlation and the max-plus correlation should be clear:

$$S_{uv}(\mu) = \lim_{T \to \infty} \frac{1}{2T} \int_{-T}^{T} u(s) v(s-\mu) \, \mathrm{d}s \ , \quad R_{uv}(\mu) = \bigwedge_{s \in \mathbb{R}} \left(u(s) \phi v(s-\mu) \right) \ .$$

3. For *finite* functions u_i and v_j, the classical distance $\sup_k |u_i(k) - v_j(k)|$ can be expressed as $-\inf\big((R_{uv})_{ij}(0), (R_{vu})_{ji}(0)\big)$, which shows some connection between the notions of distance and correlation.

[3] Indeed, if we view the convolution as an extension of the matrix product with infinite-dimensional elements (for special matrices in which entry (i,j) depends only on the difference $t = i - j$), then $\overline{h}(t) = -h(-t)$ is the composition of transposition and of change of sign.

4. From (6.19), it is clear that $(T_{uv})_{ij}(k,\mu) = u_i(k) \phi v_j(k-\mu)$ is bounded from below by $(R_{uv})_{ij}(\mu)$ for all i, j, k, m. On the other hand, $(R_{vu})_{ji}(-\mu) = \bigwedge_l v_j(l) \phi u_i(l+\mu) = \bigwedge_k v_j(k-\mu) \phi u_i(k)$, hence $e \phi \left(v_j(k-\mu) \phi u_i(k)\right)$ is bounded from above by $e \phi \left((R_{vu})_{ji}(-\mu)\right)$. This would provide an upper bound for $(T_{uv})_{ij}(k,\mu)$ if it were true that $e \phi (x \phi y) = y \phi x$. In $\overline{\mathbb{R}}_{\max}$ this equality obviously holds true whenever x and y are scalars assuming finite values. Otherwise it may not hold, as shown by the following example: let $x = y = \varepsilon$, then $x \phi y = \top$, $e \phi \top = \varepsilon$ but $y \phi x = \top$. ■

Let us now give the evolution equation of the sojourn time for a shift-invariant autonomous linear system.

Theorem 6.43 *For the system $x(k+1) = Ax(k)$, where $A \in \overline{\mathbb{R}}_{\max}^{n\times n}$, the sojourn time matrix $T_{xx}(\cdot,\mu)$ follows the dynamics*

$$T_{xx}(k+1,\mu) = (AT_{xx}(k,\mu)) \phi A = A(T_{xx}(k,\mu) \phi A) \ ,$$

provided that $T_{xx}(\cdot,\mu)$ never assumes infinite values. More generally, the following inequalities always hold true

$$T_{xx}(k+1,\mu) \geq (AT_{xx}(k,\mu)) \phi A \geq A(T_{xx}(k,\mu) \phi A) \ .$$

Proof We have

$$\begin{aligned}
T_{xx}(k+1,\mu) &= x(k+1) \phi x(k+1-\mu) = (Ax(k)) \phi (Ax(k-\mu)) \\
&= ((Ax(k)) \phi x(k-\mu)) \phi A && \text{by (f.9),} \\
&\geq (A(x(k) \phi x(k-\mu))) \phi A && \text{by (f.12),} \\
&\geq A((x(k) \phi x(k-\mu)) \phi A) && \text{by (f.12).}
\end{aligned}$$

The two inequalities become equalities in the case when $T_{uv}(k,\mu)$ has finite entries only. Indeed, in $\overline{\mathbb{R}}_{\max}$, the only counterexamples to equality in (f.12) are the cases when ε and/or $\top$ are involved: for example, $\varepsilon = \varepsilon \otimes (\varepsilon \phi \varepsilon) < (\varepsilon \otimes \varepsilon) \phi \varepsilon = \top$. ■

The following result provides insight into how correlations are transformed by linear systems.

Theorem 6.44 (Nondecreasing correlation principle) *Consider a (MIMO) shift-invariant system with (matrix) impulse response $H \in \mathcal{S}$ and two inputs signals u and v with their corresponding outputs y and z, respectively. Then*

$$y \phi z \ \geq \ (v \backslash\!\!\!\circ u)(H \phi H) \ , \tag{6.20}$$

$$z \backslash\!\!\!\circ y \ \geq \ (v \backslash\!\!\!\circ u) \bigwedge_{i,j} H_{ij} \phi H_{ij} \ . \tag{6.21}$$

Proof Observe first that, for all i, j,

$$(u \phi v)_{ij}(k) = \bigwedge_l \left(u_i(l) \phi v_j(l-k)\right) = \bigwedge_l \left(v_j(l-k) \backslash\!\!\!\circ u_i(l)\right)$$

because $\otimes$ is commutative for scalars. Using this equality for $i = j$ and the obvious fact that $(u \mathbin{\circ\!\!\!/} v)_{ij} \geq \varepsilon$ for $i \neq j$, we have that $u \mathbin{\circ\!\!\!/} v \geq (v \mathbin{\circ\!\!\!\backslash} u)e$, where e is the identity matrix. Then, we have

$$
\begin{aligned}
y \mathbin{\circ\!\!\!/} z &= (Hu) \mathbin{\circ\!\!\!/} (Hv) & \\
&= ((Hu) \mathbin{\circ\!\!\!/} v) \mathbin{\circ\!\!\!/} H & \text{by (f.9),} \\
&\geq (H(u \mathbin{\circ\!\!\!/} v)) \mathbin{\circ\!\!\!/} H & \text{by (f.12),} \\
&\geq ((v \mathbin{\circ\!\!\!\backslash} u)H) \mathbin{\circ\!\!\!/} H & \text{as explained above,} \\
&\geq (v \mathbin{\circ\!\!\!\backslash} u)(H \mathbin{\circ\!\!\!/} H) & \text{by (f.12).}
\end{aligned}
$$

This proves (6.20). Inequality (6.21) is obtained easily from (6.20) and (4.82). ■

Remark 6.45

1. Since $H \mathbin{\circ\!\!\!/} H \geq e$, by (f.6), Inequality (6.21) implies that $z \mathbin{\circ\!\!\!\backslash} y \geq v \mathbin{\circ\!\!\!\backslash} u$, which means that, in the SISO case, the correlation of output signals is not less than the correlation of inputs.

2. For autocorrelations, (6.20) becomes $y \mathbin{\circ\!\!\!/} y \geq (u \mathbin{\circ\!\!\!\backslash} u)(H \mathbin{\circ\!\!\!/} H) \geq H \mathbin{\circ\!\!\!/} H$ since $(u \mathbin{\circ\!\!\!\backslash} u) \geq e$. This is a second correlation principle, which states that the autocorrelation of outputs is not less than the intrinsic correlation $H \mathbin{\circ\!\!\!/} H$ of the system. ■

Theorem 6.44 suggests the importance of quotients of the form $A \mathbin{\circ\!\!\!/} A$. Theorem 4.59 and Corollary 4.69 gave an algebraic characterization of these quotients.

6.6.2 Stability and Stabilization

In this subsection we are concerned with what we call the internal stability of systems. The discussion will be limited to systems modeling timed event graphs. This notion of internal stability means that there is no accumulation of tokens in places or, dually, that the sojourn times of tokens remain finite. Let us start our discussion on stability by studying the relation between the asymptotic slopes of functions and their correlations.

Definition 6.46 (Asymptotic slope) *Let $h \in \acute{\overline{\mathbb{R}}}_{\max}(\acute{\gamma})$ be represented by $\acute{e}(p \oplus q\gamma^{\nu}(a\gamma^{r})^{*})$ (see Remark 6.33; without loss of generality, we assume that a is nonnegative). Then, the asymptotic slope of $h \in \acute{\overline{\mathbb{R}}}_{\max}(\acute{\gamma})$, denoted $\mathrm{sl}_{\infty}(h)$, is defined by the ratio a/r.*

As observed in Remark 6.33, the ratio a/r is independent of the particular representation of this type which was chosen for h. Note the difference between the slope introduced in the proof of Theorem 6.32 and this asymptotic slope: in the context of $\acute{\overline{\mathbb{R}}}_{\max}(\acute{\gamma})$, the former would be the maximum ratio $a(n)/n$ among the monomials $a(n)\acute{\gamma}^{n}$ appearing in h; the latter is the limit of such ratios when n goes to infinity.

Theorem 6.47 *Given a realization of some $h \in \overline{\mathbb{R}}_{\max}(\acute{\gamma})$ by an event graph with 'internal state' x, for any rational input (dater) u such that $u(k) = \varepsilon, \forall k < 0$, the corresponding dater x is also rational and such that $x(k) = \varepsilon, \forall k < 0$. The following equivalence holds true:*

$$\{(A): \quad \forall i,j,\ (R_{xx})_{ij} \neq \varepsilon\} \quad \Leftrightarrow \quad \{(B): \quad \forall i,j,\ \mathrm{sl}_\infty(x_i) = \mathrm{sl}_\infty(x_j)\} \ .$$

Proof The case of zero slopes must be handled separately. Suppose that for some i and j, $\mathrm{sl}_\infty(x_i) \geq \mathrm{sl}_\infty(x_j)$ and that, moreover, $\mathrm{sl}_\infty(x_i) > 0$. Then it is easy to see that there exists a shift γ^μ such that $x_i \geq \gamma^\mu x_j$. Therefore, for all $k \in \mathbb{Z}$, $x_i(k) \geq x_j(k-\mu)$ and $(R_{xx})_{ij}(\mu) \geq e$. Consequently, if $\mathrm{sl}_\infty(x_i) = \mathrm{sl}_\infty(x_j) > 0$, $(R_{xx})_{ij} > \varepsilon$ and (A) holds true. If $\mathrm{sl}_\infty(x_i) = \mathrm{sl}_\infty(x_j) = 0$, x_i and x_j are then polynomials, that is, they can be (minimally) represented by a finite number of coefficients not equal to ε. In this case, it is easy to conclude this part of the proof by remembering that $\varepsilon \not\phi \varepsilon = \top$.

Conversely, if (B) does not hold, that is, there exists a pair (i,j) such that $\mathrm{sl}_\infty(x_j) > \mathrm{sl}_\infty(x_j)$, then whatever $\mu \in \mathbb{Z}$, $x_i(k)$ increases to infinity strictly faster than $x_j(k-\mu)$ when $k \to +\infty$. Hence, for all $\mu \in \mathbb{Z}$, $\bigwedge_k x_i(k) \not\phi x_j(k-\mu) = \varepsilon$ (the $\wedge$ is obtained as a limit when $k \to +\infty$) and (A) is contradicted. ∎

Definition 6.48 (Internal stability) *When the equivalent conditions (A) and (B) hold true* for all inputs *of the type described in Theorem 6.47, we say that the realization is internally stable.*

Remark 6.49 Owing to Remark 6.42 (point 4), in the situation of Definition 6.47, and if all daters x_i remain finite, one can obtain an upper bound for the sojourn times of tokens in any internal path of the event graph (using appropriate shifts μ). The condition that the x_i remain finite is satisfied if the inputs remain finite (no indefinite 'starving' of tokens at the input transitions) and if the system has no deadlocks. A deadlock would be revealed by infinite asymptotic slopes for the daters associated with transitions belonging to the deadlocked circuits.

However, even for internally stable event graphs with finite inputs and no deadlocks, there might happen that tokens incur unbounded sojourn times: this typically occurs in places which are located immediately downstream of u-transitions (sources), when one uses inputs which are too fast with respect to the potential throughput of the event graph (that is, when the asymptotic slopes of these inputs are *less* than the common value of the asymptotic slopes of the x_i). ∎

Corollary 6.50 *Given a rational impulse response h and a realization described by a triple of matrices*[4] *(A, B, C), this realization is internally stable if and only*

$$\forall i,j \ , \quad (H \not\phi H)_{ij} \neq \varepsilon \ ,$$

where $H \stackrel{\text{def}}{=} \acute{e}(\gamma A)^* B$.

[4]See Definition 6.37, except that minimality is not required here.

Proof The condition is sufficient because $x = Hu$ and $x \phi x \geq H \phi H$ by Remark 6.45 (point 2). Conversely, since $(H \phi H)_{ij} = \bigwedge_l H_{il} \phi H_{jl}$ and if $(H \phi H)_{ij} = \varepsilon$, then there exists l_o such that $H_{il_o} \phi H_{jl_o} = \varepsilon$. The system does not satisfy the requirement of Definition 6.48 for the input $u_{l_o} = \acute{e}$, $u_l = \varepsilon, l \neq k$. ■

Theorem 6.51 *If the internal subgraph of an event graph (that is, the subgraph obtained by deleting input and output transitions together with the arcs connecting them to other transitions) is strongly connected, then this system is internally stable.*

Proof Indeed if the internal subgraph is strongly connected, for any pair of internal nodes (i, j), there exists a path ρ from j to i containing μ_ρ tokens (μ_ρ may be equal to 0). Then we have $x_i(k) \geq \alpha_\rho \otimes x_j(k - \mu_\rho)$, where α_ρ is the sum of the holding times of the places in the path ρ (i.e. $\alpha_\rho = |\rho|_\mathrm{w}$; $\alpha_\rho > \varepsilon$, indeed $\alpha_\rho \geq e$). Therefore $(T_{xx})_{ij}(k, \mu_\rho) \geq t_\rho$ for all k. This holds for any input and Definition 6.48 is satisfied. ■

When a given (open-loop) event graph is not internally stable, we consider the problem of obtaining this property by 'closing the loop' between inputs and outputs. By this we mean that u will be obtained from y by $u = Fy \oplus v$, where F is a 'feedback' matrix of appropriate dimensions with entries in $\overline{\mathbb{R}}_{\max}(\acute{\gamma})$, and v is the new input (of the same dimension as u). The situation is depicted in Figure 6.5 from which it appears that (at least some components of) u, respectively y, do no

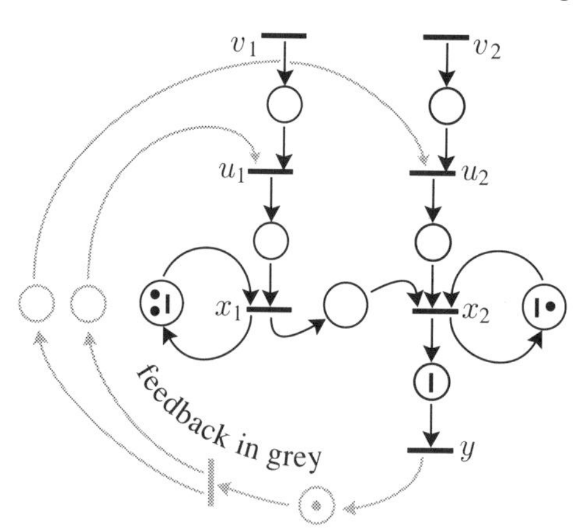

Figure 6.5: An unstable timed event graph with a stabilizing feedback

longer correspond to sources, respectively sinks. The feedback should in general be dynamic in the sense that F should indeed contain terms $a(n)\acute{\gamma}^n$ with $n \geq 1$ and $a(n) \geq e$, in order to avoid deadlocks in the closed-loop system. A term of this type in F_{ij} means that there exists a path from y_j to u_i (in grey in the figure) with a total number of n tokens in the initial marking and a total holding time of $a(n)$ time units.

The stabilization of event graphs by output feedback requires the introduction of the following notions.

Definition 6.52

Structural Controllability *An event graph is structurally controllable if every internal transition can be reached by a path from at least one input transition.*

Structural Observability *An event graph is structurally observable if, from every internal transition, there exists a path to at least one output transition.*

Theorem 6.53 (Feedback stabilization) *Any structurally controllable and observable event graph can be made internally stable by output feedback.*

Proof The idea of the proof is to fulfill the sufficient condition of strong connectedness (mentioned in Theorem 6.51) for the internal subgraph of the closed-loop graph. Under the assumptions of structural controllability and observability of the open-loop system, it should not be difficult to see that this sufficient condition can indeed be satisfied if an effective feedback connection is established from any output to any input. ■

Of course, one can imagine more refined strategies in order to attempt to minimize the number of feedback links so introduced. Obviously, input transitions which are upstream of *several* m.s.c.s.'s of the internal subgraph must be preferably used and a similar remark applies to output transitions.

Example 6.54 The timed event graph represented in Figure 6.5 is not internally stable in the open-loop configuration. For instance, if tokens are input through u_1 at the rate of 2 tokens per time unit, then tokens accumulate indefinitely in the place between x_1 and x_2 since the throughput of x_2 is limited to one token per time unit, whereas x_1 can process tokens at the given input rate. On the other hand, the system can be stabilized by the feedback shown in Figure 6.5 (grey lines). ■

6.6.3 Loop Shaping

In the previous subsection we saw how to obtain an internally stable system by closing the loop between input and output, provided the system is structurally controllable and observable. However, this operation creates new circuits whereas it preserves the circuits already existing in the open-loop system. Therefore, the maximum cycle mean may only increase when passing from the open-loop to the closed-loop system, which means that the throughput (inverse of the maximum cycle mean) may only be worse, resulting in a loss of performance.

The newly created circuits traverse the feedback arcs. If any such circuit, say ζ, happens to be critical, it suffices to increase the number of tokens in the corresponding feedback path in such a way that the cycle mean $|\zeta|_w/|\zeta|_t$ ceases to be critical. This reasoning justifies the following theorem which improves the previous one.

Theorem 6.55 *Any structurally controllable and observable event graph can be made internally stable by output feedback without altering its original open-loop throughput.*

Another more algebraic view on this problem can be explained in the simple case of a SISO system. Let h be its (rational) impulse response. The open-loop throughput is $1/\mathrm{sl}_\infty(h)$. If one uses the feedback law $u = fy \oplus v$, the closed-loop system is $y = (hf)^* hv$. Then it can be proved that if $f = \gamma^\mu$, there exists μ large enough such that $\mathrm{sl}_\infty\left((h\gamma^\mu)^* h\right) = \mathrm{sl}_\infty(h)$.

An interesting question is to determine the *minimum* number of tokens (which may represent costly resources practically) such that a desired throughput is achieved. This problem is discussed in [62].

6.7 NOTES

This chapter is based on the two articles [111] and [112]. The idea of extending the application of the max-plus algebra to continuous systems was proposed by R. Nikoukhah during a Max-Plus' working group meeting. It is quite natural once one realizes that time-invariant max-plus linear systems indeed perform sup-convolutions of the inputs with their impulse responses. Continuous Petri nets have also been studied in [14] and [99].

Formal and numerical transfer functions are isomorphic in conventional algebra, and therefore they are not always clearly distinguished in the literature on system theory. The situation is quite different in the max-plus context. The terminology 'transfer function' was reserved for the Fenchel transform of the impulse response in this chapter.

In the literature on optimization, the idea of considering dynamic systems based on vector sums of convex objects appeared from time to time but with no connection to the modeling of synchronization mechanisms.

The characterization of rational impulse responses in terms of periodicity was given for the first time in [41]. A program for symbolic computation based on this periodic characterization of rational systems has been developed by S. Gaubert. It is called MAX [62]. An analogous notion of periodicity exists in the Petri net literature [36].

The second-order theory developed in the second part has two origins: the first stems from [112], the second from [4]. The first is concerned with finding a max-plus equivalent of the autocorrelation of a process, the second with describing the recurrent equation of differences. The application to stability provides another view on the stabilization by feedback described for the first time in [41]. The nondecreasing correlation principle was found by S. Gaubert.

The interesting problem of the optimization of the number of tokens involved in the loop shaping issue has been solved in [62] but was not discussed here.

PART IV

Stochastic Systems

CHAPTER 7

Ergodic Theory of Event Graphs

7.1 INTRODUCTION

The main practical concerns of this chapter are the construction of the stationary regime of stochastic event graphs and the conditions on the statistics of the holding times under which such a stationary regime exists. The basis for the analysis is the set of equations which govern the evolution of daters, established in Chapter 2. In §7.6 we will see that this construction also allows us to determine the stationary regime of the marking process.

The main tool for addressing these problems is ergodic theory: the existence problem is stated in terms of a 'random eigenpair problem' which generalizes the eigenpair problem formulation of Chapter 3 in the deterministic case, and which can be seen as the $\mathbb{R}_{\max}$-analogue of that of Oseledeç's multiplicative ergodic theorem in conventional algebra.

Section 7.2 focuses on a simple one-dimensional nonautonomous example. This example is the Petri net analogue of the classical $G/G/1$ queue. Most of the basic probabilistic tools to be used in this chapter are introduced through this simple example. These tools are based on the probabilistic formalism of [6]. More advanced probabilistic material, and in particular the ergodic theorems which are used or referred to in the chapter, are gathered in §7.7.

Section 7.3 gives the basic first-order theorems which indicate how the daters grow in such a stochastic framework. The growth rates given in these first-order theorems are shown to be the $\mathbb{R}_{\max}$-analogues of Lyapunov exponents in conventional algebra, and generalizations of cycle times in the deterministic case.

Second-order theorems are concerned with the construction of the eigenpairs. This construction is based on the analysis of ratios of daters (in the $\mathbb{R}_{\max}$ sense). This second-order theory is first presented for multidimensional nonautonomous systems in §7.4. It is shown that under appropriate statistical assumptions, this type of systems admits a unique stationary regime which is reached in finite time, regardless of the initial condition, provided the 'Lyapunov exponents' of the m.s.c.s.'s are less than the asymptotic rate of the input. Section 7.5 focuses on the autonomous case. We provide a simple and natural condition for the uniqueness and the *reachability* of the stationary regime.

Throughout the chapter, we will consider two levels of abstraction.

- The first level is that of stochastic Petri nets, for which we will use the notation of Chapter 2, and for which the reference dioid will be $\mathbb{R}_{\max}$. This level will provide examples and will require a particular attention because of the difficulties related to the initial conditions.

- The second one is that of linear algebra in a stochastic context. The discussions will rely upon the notion of residuation introduced in Chapter 4. We will try to consider general dioids, although most of the practical results which we have at this stage are limited to $\mathbb{R}_{\max}$.

For each section of the chapter, we will try to indicate which level is considered.

7.2 A SIMPLE EXAMPLE IN $\mathbb{R}_{\max}$

7.2.1 The Event Graph

We first consider a simple example of an event graph with a single input: the event graph has two transitions q_1 and q_2, two places p_1 and p_2, and the following topology: p_2 is a recycling of q_2, $\pi(p_1) = q_1$, $\sigma(p_1) = q_2$, and q_1 is an input transition with input sequence $\{u(k)\}_{k\geq 1}$. The initial marking has one token in p_i, with lag time w_i, $i = 1, 2$, so that the set $\mathcal{Q}'$ of transitions followed by at least one place with nonzero initial marking is $\{q_2\}$ (see Figure 7.1). The holding times in p_1 are all

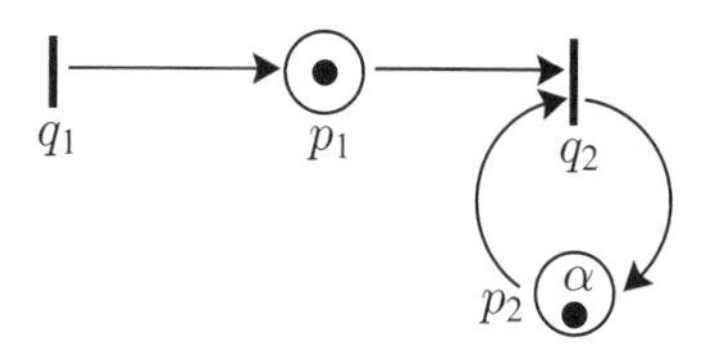

Figure 7.1: A simple example

zero, whereas those in p_2 are given by the sequence $\{\alpha(k)\}$. Observe that $M = 1$ and that both $|\mathcal{Q}'|$ and $|\mathcal{I}|$ are equal to 1. Accordingly, the matrices $\overline{A}(k, k-1)$ and $\overline{B}(k, k-1)$ in (2.38) are one-dimensional:

$$\overline{A}(k, k-1) = \big(\alpha(k)\big) \ , \quad \overline{B}(k, k-1) = (e) \ .$$

Let $A(k) = \alpha(k+1)$. Equation (2.38) reads

$$x(k+1) = A(k)x(k) \oplus u(k) \oplus v(k+1) \ , \quad k \geq 0 \ , \tag{7.1}$$

where $v(1) = w_1 \oplus w_2$ and $v(k) = \varepsilon$ for $k \neq 1$. In this equation the continuation for $(u(0), x(0))$ is $(u(0), x(0)) = (\varepsilon, \varepsilon)$. The input is weakly compatible if $u(1) \geq e$. The initial lag times are weakly compatible if

$$w_2 \leq \alpha(1) \ , \quad w_1 \leq e \quad \text{and} \quad w_1 \oplus w_2 \geq e \ . \tag{7.2}$$

Since each transition is followed by at most one place with a nonzero initial marking, any weakly compatible initial condition is compatible, so that we can rewrite the preceding equation as

$$x(k+1) = A(k)x(k) \oplus u(k) \ , \quad k \geq 0 \ , \tag{7.3}$$

provided we now take $x(0) = w_2 \circ\!\!\!/ \alpha(1)$ and $u(0) = w_1$.

Remark 7.1 With the statistical assumptions described in the next subsection, this system is very close to the FIFO $G/G/1/\infty$ queue. The G/G part states that both the ratios of the input sequence (see (7.4) below) and the holding times form general stationary sequences. The $1/\infty$ part states that there is a single server and that there is an infinite buffer in front of the server. This system is also known as the *producer-consumer* system in theoretical computer science. The input u_1 into transition q_1 features the external input stream of customers, p_1 is the infinite buffer which stores customers to be served, q_2 features the single server, and the holding times in p_2 represent the service times. ∎

7.2.2 Statistical Assumptions

The statistical assumptions are as follows: the firing times are 0, as well as the holding times in p_1. The holding times $\alpha(k), k \geq 1$, (or equivalently $A(k)$) and the ratios $\{U(k)\}_{k\geq 1}$ of the input sequence, where

$$U(k) \stackrel{\text{def}}{=} u(k+1) \circ\!\!\!/ u(k) \ , \tag{7.4}$$

form two jointly stationary and ergodic sequences on some probability space $(\Omega, \mathbb{F}, \mathbb{P})$. This assumption can be seen as the stochastic generalization of the case of periodic input considered in Chapter 3: the constant ratios of the input and the constant holding times in Chapter 3 are now replaced by stationary ratios. Whenever needed, we will stress the fact that $A(k)$, $U(k)$ and $w \stackrel{\text{def}}{=} (w_1, w_2)$ are random variables, namely measurable functions from Ω into $\mathbb{R}$, by writing $A(k;\omega)$, $U(k;\omega)$ and $w(\omega)$ instead of $A(k)$, $U(k)$, and w, respectively. Observe that this is tantamount to using the same notation for a function and for the value which it takes at a particular point. The context should always allow the reader to decide what is meant.

Before going further in the analysis of the system, we comment on the statistical framework, and on what will be meant by joint stationarity and ergodicity of the two sequences $\{A(k)\}$ and $\{U(k)\}$ throughout this chapter and the next one.

Definition 7.2 (θ-shift) *The mapping $\theta : \Omega \to \Omega$ is a shift operator on $(\Omega, \mathbb{F}, \mathbb{P})$ if it is bijective and measurable from Ω onto itself, and if it is such that the probability law $\mathbb{P}$ is left invariant by θ, namely $\mathbb{E}[f] = \mathbb{E}[f \circ \theta]$, for all measurable and integrable functions $f : \Omega \to \mathbb{R}$, where $\mathbb{E}$ denotes the mathematical expectation with respect to $\mathbb{P}$.*

> By convention, the composition operator '$\circ$' has the highest priority in all formulæ. For instance, $f \circ gh$ means $(f \circ g)h$.

Definition 7.3 (θ-stationarity) *We say that a sequence of $\mathbb{R}$-valued random variables $\{a(k;\omega)\}_{k\in\mathbb{Z}}$ defined on $(\Omega, \mathbb{F}, \mathbb{P})$ is θ-stationary if the relation*

$$a(k;\omega) = a(0; \theta^k(\omega)) \tag{7.5}$$

holds for all $k \geq 0$, where θ^k is the composition of θ by itself k times: $\theta^{k+1} = \theta^k \circ \theta$, and $\theta^0 = I_\Omega$, the identity.

Remark 7.4 Another way of stating the preceding definition consists in requiring that $a(0)\circ\theta^k = \gamma^{-k}a(0)$, for all $k \in \mathbb{Z}$, where γ is the backward shift operator on the numbering of sequences which was defined in §5.3.2. ■

In the present example, we will assume that the data of the problem, namely both sequences $\{A(k)\}$ and $\{U(k)\}$, are θ-stationary. We immediately obtain from this and from (7.5) that for all integers m, the relation

$$\begin{aligned}\mathbb{E}[h(A(0), U(0), \ldots, A(k), U(k))] \\ = \mathbb{E}[h(A(m), U(m), \ldots, A(m+k), U(m+k))]\end{aligned}$$

holds for all measurable functions $h : \mathbb{R}^{2(k+1)} \to \mathbb{R}$ such that the expectation exists. This is a natural property to expect from joint stationarity indeed. Starting from this assumption, we will then be interested in proving that other quantities associated with the event graph also satisfy the θ-stationarity property.

Similarly, the joint ergodicity of the sequences $\{A(k)\}$ and $\{U(k)\}$ is obtained when assuming that θ is $\mathbb{P}$-*ergodic*:

Definition 7.5 (Ergodic shift) *The shift θ is said to be ergodic if the almost sure (a.s.) limit*

$$\lim_{k\to\infty} \frac{1}{k}\sum_{l=1}^{k} f\circ\theta^l = \mathbb{E}[f] \quad a.s. \tag{7.6}$$

holds for all measurable and integrable functions $f : \Omega \to \mathbb{R}$.

Owing to (7.5), the last property implies in particular that

$$\lim_{k\to\infty} \frac{1}{k}\sum_{l=1}^{k} A(l) = \mathbb{E}[A(0)] \text{ a.s.} \quad \text{and} \quad \lim_{k\to\infty} \frac{1}{k}\sum_{l=1}^{k} U(l) = \mathbb{E}[U(0)] \text{ a.s.},$$

provided $A(0)$ and $U(0)$ are integrable, which corresponds to the conventional meaning of the ergodicity of both sequences. The joint ergodicity becomes more apparent from the formula

$$\lim_{k\to\infty} \left(\bigotimes_{l=1}^{k} h\left(A(l), U(l)\right) \right)^{1/k} = \mathbb{E}[h(A(0), U(0))] \quad \text{a.s.}$$

for all measurable functions $h : \mathbb{R}^2 \to \mathbb{R}$ such that the expectation exists, which is also a direct consequence of our definition.

A measurable set $\mathcal{A}$ of $\mathbb{F}$ (an 'event') is said to be θ-invariant if the indicator function of $\mathcal{A}$, which will be denoted $1_{\{\mathcal{A}\}}$, satisfies the relation $1_{\{\mathcal{A}\}}\circ\theta = 1_{\{\mathcal{A}\}}$. We will often make use of Birkhoff's pointwise ergodic theorem (see [20]).

Theorem 7.6 (Birkhoff) *The shift operator θ is ergodic if and only if the only sets of the σ-algebra $\mathbb{F}$ which are θ-invariant are the trivial sets $\emptyset$ and Ω.*

Example 7.7 (Canonical probability space) In the particular example which we consider, the data consist of the two sequences $\{A(k)\}$ and $\{U(k)\}$. A concrete example of such a shift operator is provided by the translation operator γ^{-1} on the *canonical space* of the two sequences, which is defined as follows:

- Ω is the space of bi-infinite sequences of the form $\ldots, z(-2), z(-1), z(0), z(1), z(2), \ldots$, where $z(k) = (s(k), t(k)) \in \mathbb{R}^2$ for all $k \in \mathbb{Z}$;
- $\mathbb{F}$ is the σ-algebra generated by the coordinate mappings $e_k(\omega), k \in \mathbb{Z}$, where $e_k(\omega) = z(k)$;
- $\mathbb{P}$ is a probability measure on the measurable space $(\Omega, \mathbb{F})$.

On this probability space, we can then take

$$\theta(\ldots, z(-1), z(0), z(1), \ldots) = \gamma^{-1}(\ldots, z(-1), z(0), z(1), \ldots) ,$$

that is $e_k(\theta(\omega)) = e_{k+1}(\omega)$ for all $k \in \mathbb{Z}$. Within this framework, the θ-stationarity assumption boils down to the assumption that the probability law $\mathbb{P}$ of the two sequences is left invariant by γ.

If $A(k; \omega)$ denotes the first component of $e_k(\omega)$, and $U(k; \omega)$ the second one, we obtain that (7.5) is indeed satisfied by both sequences. ∎

Remark 7.8 If we consider a sequence of random variables, say $\{b(k; \omega)\}$, defined on this canonical probability space, which is different from the coordinate process, it is *not* true in general that $b(0) \circ \theta^k = b(k)$. It is clear that $b(0; \theta(\omega)) = b(0; \gamma^{-1}(\omega))$, but in general, it is not true that $b(0; \gamma^{-1}(\omega)) = \gamma^{-1} b(0; \omega)$ because the translation operator which is used at the right-hand side of the last relation has nothing to do with the specific one used at the left-hand side, which operates on the sequences of Ω. For instance, take $b(k; \omega) = kA(k; \omega)$. We have $b(0) \circ \theta^k = 0$, which clearly differs from $kA(k)$ unless $A(\cdot) = 0$. ∎

7.2.3 Statement of the Eigenvalue Problem

We can rewrite the equations governing this system as

$$\begin{cases} u(k+1) &= U(k)u(k) , \\ x(k+1) &= A(k)x(k) \oplus u(k) , \end{cases}$$

or equivalently as

$$X(k+1) = D(k)X(k) , \quad k \geq 0 , \tag{7.7}$$

where

$$X(k) = \begin{pmatrix} u(k) \\ x(k) \end{pmatrix} \quad \text{and} \quad D(k) = \begin{pmatrix} U(k) & \varepsilon \\ e & A(k) \end{pmatrix} .$$

The variables $A(k)$ and $U(k)$ are assumed to be integrable random variables, defined on a probability space $(\Omega, \mathbb{F}, \mathbb{P}, \theta)$, and such that

$$A(k) = A \circ \theta^k , \quad U(k) = U \circ \theta^k , \quad k \in \mathbb{Z} ,$$

where $A \stackrel{\text{def}}{=} A(0)$ and $U \stackrel{\text{def}}{=} U(0)$. Under this assumption, $D(k) = D{\circ}\theta^k$, with $D = D(0)$.

In the deterministic setting, we looked for periodic regimes in terms of eigenpairs associated with the $\mathbb{R}_{\max}$ matrix describing the dynamics of the event graph (see §3.7). In the stochastic setting defined above, we state the problem as the following 'random eigenpair' problem.

> Can we find an eigenvector $X = (X_1, X_2)$, normalized in such a way that $X_1 = e$, and an eigenvalue λ, which are both random variables defined on $(\Omega, \mathbb{F}, \mathbb{P}, \theta)$, and such that
>
> $$DX = \lambda X{\circ}\theta \ ? \tag{7.8}$$

In view of the specific form of D, this eigenpair property reads

$$\left.\begin{array}{rcl} \lambda & = & U \ , \\ e \oplus AX_2 & = & UX_2{\circ}\theta \ , \end{array}\right\} \tag{7.9}$$

so that the only true unknown is X_2. Assume that the above eigenpair is finite; whenever we take $x(0) = X_2$ and $u(0) = X_1 = e$ in Equation (7.7), we obtain

$$\left.\begin{array}{rcl} u(1) & = & U \ , \\ x(1) & = & e \oplus AX_2 \ . \end{array}\right\} \tag{7.10}$$

From (7.9) and (7.10), we see that

$$x(1) - u(1) = X_2{\circ}\theta + U - U = (x(0) - u(0)){\circ}\theta \ .$$

More generally, we prove in the same way that for all $k \geq 0$,

$$x(k) - u(k) = (x(0) - u(0)){\circ}\theta^k \ .$$

Therefore, if the above eigenpair problem has a finite solution, we can find an initial condition such that the random variables $x(k) - u(k)$ are stationary.

Let us show that for this initial condition the marking process is also stationary: let $N^+(k)$ denote the number of tokens in p_1 at the epoch when transition q_2 fires for the k-th time, namely at $x(k)$. Within our setting, $N^+(k)$ is a random variable. For instance $N^+(1)$ is given by the following expression (see §2.5.6):

$$N^+(1) = \sum_{h=1}^{\infty} 1_{\{x(1) \geq u(h)\}} = \sum_{h=1}^{\infty} 1_{\{X_2{\circ}\theta \geq \sum_{l=1}^{h-1} U{\circ}\theta^l\}} \ ,$$

where $\sum_1^0 = 0$ by convention. Similarly, when using the convention $\sum_2^1 = 0$, we obtain

$$\begin{aligned} N^+(2) &= \sum_{h=2}^{\infty} 1_{\{x(2) \geq u(h)\}} = \sum_{h=2}^{\infty} 1_{\{X_2{\circ}\theta^2 \geq \sum_{l=2}^{h-1} U{\circ}\theta^l\}} \\ &= \sum_{h=2}^{\infty} 1_{\{\Delta{\circ}\theta \geq \sum_{l=1}^{h-1} U{\circ}\theta^l\}}{\circ}\theta = N^+(1){\circ}\theta \ , \end{aligned}$$

and more generally $N^+(k+1) = N^+(1){\circ}\theta^k$.

7.2.3.1 Solution of the Eigenpair Problem

For this classical example, it is customary to take

$$\Delta \stackrel{\text{def}}{=} X_2 \circ \theta U$$

as unknown, rather than X_2, mainly because this new unknown is nonnegative; indeed it is immediately seen from (7.9) that

$$\Delta \circ \theta = e \oplus \left(A \circ \theta \not{} U \Delta \right) = e \oplus F \Delta \ , \tag{7.11}$$

where

$$F \stackrel{\text{def}}{=} A \circ \theta \not{} U \ .$$

The construction of a solution to (7.11) is based on a backward construction which is common in ergodic theory, and which will be used on several occasions in this chapter. The backward process $\Delta(k), k \geq 0$, associated with (7.11) is the random process on the probability space $(\Omega, \mathbb{F}, \mathbb{P})$ defined by $\Delta(0) = e$ and

$$\begin{aligned} \Delta(k+1) &= \left(e \oplus F \Delta(k) \right) \circ \theta^{-1} \\ &= e \oplus F \circ \theta^{-1} \Delta(k) \circ \theta^{-1} \ , \qquad k \geq 0 \ . \end{aligned} \tag{7.12}$$

We will return to the physical interpretation of this process in the next subsection.

A nice property of the backward process is that $\Delta(k) = \Delta(k;\omega)$ is nondecreasing in k for all ω. This is obtained by induction: it is true that $\Delta(1) \geq \Delta(0) = e$; assuming that $\Delta(k) \geq \Delta(k-1)$, we obtain from (7.12) that

$$\begin{aligned} \Delta(k+1) \circ \theta &= e \oplus F \Delta(k) \\ &\geq e \oplus F \Delta(k-1) = \Delta(k) \circ \theta \ . \end{aligned}$$

Let Δ be the a.s. limit of $\Delta(k)$ as k goes to ∞ (the a.s. limit exists because $\Delta(k)$ is nondecreasing for all ω). The random variable Δ may be finite or infinite. In both cases we obtain that Δ satisfies (7.11) by letting k go to ∞ in (7.12).

7.2.3.2 Finiteness of the Eigenvector

The main result of this subsection is the following theorem.

Theorem 7.9 (Stability condition) *If* $\mathbb{E}[A] < \mathbb{E}[U]$, *then the eigenvector* $X = \left(e, (\Delta \not{} U) \circ \theta^{-1} \right)$ *is* $\mathbb{P}$*-a.s. finite.*

Since U is a.s. finite (U is assumed to have a finite mean), it follows from the very definition of Δ that X is a.s. finite if and only if Δ is a.s. finite. The event $\{\Delta = \infty\}$ is θ-invariant. Indeed, if $\Delta(\omega) = \infty$, then $\Delta(\theta(\omega)) = \infty$, in view of (7.11) and of the assumption that $U(\omega) < \infty$ a.s. Similarly, $\Delta(\theta(\omega)) = \infty$ implies $\Delta(\omega) = \infty$, since $A(\omega) < \infty$ a.s. Therefore, in view of the ergodic assumption, $\mathbb{P}[\Delta = \infty]$ is either 0 or 1: either Δ is finite with probability 1, or it is infinite with probability 1 (see Theorem 7.6).

Lemma 7.10 (Backward star) *The following relation holds:*

$$\Delta(k) = \bigoplus_{l=0}^{k} \bigotimes_{h=1}^{l} F \circ \theta^{-h} , \tag{7.13}$$

where the $\otimes$-product over an empty set (when $l = 0$) is e by convention.

Proof The proof is by induction on k. The relation holds true for $k = 0$ since both sides are equal to e. Assume the relation holds up to some $k \geq 0$. Then using (7.12), we obtain

$$\begin{aligned} \Delta(k+1) &= F \circ \theta^{-1} \otimes \left(\bigoplus_{l=0}^{k} \bigotimes_{h=1}^{l} F \circ \theta^{-h} \right) \circ \theta^{-1} \oplus e \\ &= \left(\bigoplus_{l=1}^{k+1} \bigotimes_{h=1}^{l} F \circ \theta^{-h} \right) \oplus e = \bigoplus_{l=0}^{k+1} \bigotimes_{h=1}^{l} F \circ \theta^{-h} , \end{aligned}$$

where we used the distributivity of $\otimes$ with respect to $\oplus$ and the associativity and commutativity of $\oplus$ to pass from the first expression to the second in the last equation. ■

Remark 7.11 The property that $\Delta(k)$ is nondecreasing, which was already shown in the preceding subsection, is obvious from (7.13), since this relation shows that $\Delta(k)$ consists of the maximum of an increasing set of random variables. ■

Proof of Theorem 7.9 If $\mathbb{E}[F] < 0$ (or equivalently $\mathbb{E}[A] < \mathbb{E}[U]$), from the pointwise ergodic theorem we obtain that the a.s. limit

$$\lim_{k \to \infty} \left(\bigotimes_{h=1}^{k} F \circ \theta^{-h} \right)^{1/k} = \frac{1}{k} \sum_{h=1}^{k} (A \circ \theta^{-h+1} - U \circ \theta^{-h}) = \mathbb{E}[A - U] < 0$$

holds (we used the obvious property that θ is $\mathbb{P}$-ergodic if and only if θ^{-1} is $\mathbb{P}$-ergodic). Therefore,

$$S(k) \stackrel{\text{def}}{=} \sum_{h=1}^{k} (A \circ \theta^{-h} - U \circ \theta^{-h})$$

tends to $-\infty$ a.s. as k goes to ∞, which in turn implies that $S(k) < 0$ for all k greater than a finite random integer L. Hence $\Delta(k)$ is a.s. finite in view of (7.13), since it is the maximum of an a.s. finite number L of finite random variables. ■

Remark 7.12 A partial converse of the preceding result is the following: if $\mathbb{E}[A] > \mathbb{E}[U]$, then no $\mathbb{P}$-a.s. finite solution of (7.11) exists. To prove this, it is enough to show that $\Delta = \infty$, $\mathbb{P}$-a.s., and that Δ is the least nonnegative solution of (7.11). The latter is proved by induction. If we start with a nonnegative solution Ξ of (7.11) for which $\Xi \geq \Delta(0) = e$, it is easily checked that $\Xi \geq \Delta(k)$ implies $\Xi \geq \Delta(k+1)$ (this is true because $\Delta(k+1)\circ\theta = e \oplus F\Delta(k) \leq e \oplus F\Xi = \Xi\circ\theta$). As for the proof of $\Delta = \infty$, $\mathbb{P}$-a.s., it follows from the fact that $S(l)$ then tends to ∞ a.s. as l goes to ∞. This in turn implies that $\Delta(k)$ tends to infinity as well, in view of (7.13). ■

Remark 7.13 The random variable Δ may be finite and nonintegrable. A simple example of this situation is provided by the $M/G/1$ case (namely $u(k)$ is the k-th epoch of a Poisson process and $\{\alpha(k)\}$ is an independent i.i.d. sequence), whenever the service times $\alpha(k)$ have infinite second moments (see [46]). ■

7.2.4 Relation with Event Graph

This section focuses on the relation between the eigenpair which was constructed in the previous section and the stochastic event graph which motivated our preliminary example. Consider the 'ratios' $\delta(k) \stackrel{\text{def}}{=} x(k+1)\phi u(k) = x(k+1) - u(k)$, $k \geq 0$. By using (7.3), we obtain

$$\begin{aligned} x(k+2) - u(k+1) &= \max\big(A(k+1) + x(k+1), u(k+1)\big) - u(k+1) \\ &= \max\big(A(k+1) + x(k+1) - u(k+1), 0\big) \\ &= \max\big(A(k+1) + \delta(k) - U(k), 0\big) \ , \end{aligned}$$

which corresponds to the $\mathbb{R}_{\max}$ relation

$$\delta(k+1) = e \oplus F(k)\delta(k) \ , \quad k \geq 0 \ , \tag{7.14}$$

where the initial condition $\delta(0)$ is given by the relation

$$\delta(0) \stackrel{\text{def}}{=} (A(0)x(0) \oplus u(0))\phi u(0) = (w_2 \oplus w_1)\phi w_1 \ , \tag{7.15}$$

and $F(k) = A(k+1)\phi U(k) = F\circ\theta^k$, $k \geq 0$. When making use of Assumption (7.2), we obtain

$$\delta(0) = w_2 \phi w_1 \geq e \ , \tag{7.16}$$

for all weakly compatible initial lag times w. This lower bound is achievable whenever $w_1 = w_2 = e$.

In what follows we will emphasize the dependence on the initial condition by adding a second optional argument to the δ function: for instance, $\delta(k; z)$ will denote the value of $\delta(k)$ whenever $\delta(0) = z$. Of special interest to us will be the sequence $\{\delta(k; e)\}$ defined by (7.14), and by the initial condition $\delta(0; e) = e$.

It is immediately checked by induction that

$$\Delta(k) = \delta(k; e)\circ\theta^{-k} \ , \tag{7.17}$$

which shows that $\delta(k; e)$ and $\Delta(k)$ have the same probability law. Indeed, this is true for $k = 0$, and assuming it is true for some $k \geq 0$, we obtain from (7.12) that

$$\begin{aligned} \Delta(k+1) &= \big(e \oplus F(\delta(k; e)\circ\theta^{-k})\big)\circ\theta^{-1} \\ &= \big(e \oplus F(k)\delta(k; e)\big)\circ\theta^{-k-1} \\ &= \delta(k+1; e)\circ\theta^{-k-1} \ , \end{aligned}$$

where we used the property that $F = F(k)\circ\theta^{-k}$. Therefore, the random variable $\Delta(k)$ stands for the value of $\delta(k;e)$, when replacing the sequence $U(0)$, $U(1)$,... by $U(-k)$, $U(1-k)$,..., and the sequence of holding times $A(0)$, $A(1)$,... by $A(-k)$, $A(1-k)$,..., respectively.

Remark 7.14 Another interpretation is as follows: we go backward in time and we define the continuation of $u(k)$ by

$$u(k) = u(1) - \sum_{l=k}^{0} U(l) \ , \quad k \leq 0 \ .$$

If we assume that the holding time of the k-th token in p_2 is $\alpha(k)$, and that the entrance of the k-th token in p_1 takes place at time $u(k)$, for all $k \in \mathbb{Z}$, we can then interpret $\Delta(k)$ as the value of $x(1) \not/ u(0)$ given that the value of $x(-k+1) \not/ u(-k)$ is e. ■

Remark 7.15 Assume Δ is a.s. finite. If we take the initial lag times such that

$$w_2 = \Delta w_1 \ , \tag{7.18}$$

then $\delta(0) = \Delta$, so that $\delta(1)$ is equal to $\Delta\circ\theta$, in view of (7.16); more generally we have $\delta(k) = \Delta\circ\theta^k$ for all $k \geq 0$. In words, we found initial lag times which make the ratio process $\delta(k)$ θ-stationary. Observe, however, that these initial lag times are not weakly compatible in general. For instance, if we take $w_1 = u(0) = e$ (w_1 and w_2 are only defined through (7.18) up to an additive constant), and if $\Delta > \alpha(1)$ with a positive probability, then (7.18) shows that the compatibility relation $w_2 \leq \alpha(1)$ cannot hold almost surely. ■

Remark 7.16 If the ratios $\delta(k)$ are finite and θ-stationary, the θ-stationarity of the ratios $x(k+1) \not/ x(k)$ is easily obtained from the following relation

$$\frac{x(k+1)}{x(k)} = \frac{x(k+1)}{u(k)} \frac{u(k)}{u(k-1)} \frac{u(k-1)}{x(k)} \ .$$

If $x(k+1) \not/ u(k) = \Delta\circ\theta^k$, since $u(k+1) \not/ u(k) = U\circ\theta^k$, we then have

$$x(k+1) \not/ x(k) = \left(\Delta\circ\theta^k \not/ U\circ\theta^{k-1}\right) \not/ \Delta\circ\theta^{k-1} \ .$$

Therefore

$$x(k+1) \not/ x(k) = \left\{\left(\Delta \not/ U\circ\theta^{-1}\right) \not/ \Delta\circ\theta^{-1}\right\} \circ\theta^k \ .$$

■

Remark 7.17 By computing the star operation forward (or more directly by using (7.17) and (7.13)), we obtain the expression

$$\delta(k;e) = \bigoplus_{l=0}^{k} \bigotimes_{h=k-l}^{k-1} F(h) \ . \tag{7.19}$$

The sequence $\{\delta(k;e)\}$ is not monotone in general. ■

7.2.5 Uniqueness and Coupling

The main result of this section is the following theorem.

Theorem 7.18 *If the stability condition* $\mathbb{E}[A] < \mathbb{E}[U]$ *is satisfied, there exists a unique finite random eigenvalue* λ *and a unique finite random eigenvector* $X = (X_1, X_2)$, *with* $X_1 = e$, *such that (7.8) holds. In addition, for all finite random initial conditions* $X(0) = (X_1(0), X_2(0))$ *with* $X_1(0) = e$, *there exists a finite integer-valued random variable* K *such that, for* $k \geq K$,

$$X(k+1) = D(k)D(k-1)\dots D(1)DX(0) = \lambda(k)\lambda(k-1)\dots\lambda(1)\lambda(0)X\circ\theta^{k+1} , \tag{7.20}$$

where X *and* λ *are defined as above.*

The main tool for proving this theorem is the notion of coupling.

Definition 7.19 (Coupling) *The random sequence* $\{W(k)\}_{k\geq 0}$ *defined on the probability space* $(\Omega, \mathbb{F}, \mathbb{P})$ couples in finite time *(or simply couples) with the stationary sequence generated by the random variable* V *if there exists a finite integer-valued random variable* K *such that*

$$W(k) = V\circ\theta^k, \quad \forall k \geq K .$$

We also say that the sequence $\{V\circ\theta^k\}$ is reached by coupling by the sequence $\{W(k)\}$. Coupling implies weak convergence: more precisely, if $\{W(k)\}$ couples with the sequence generated by V, then $W(k)$ converges weakly to V as k goes to ∞. (see [6, Chapter 2]). We start with the following lemma which deals with stochastic event graphs.

Lemma 7.20 *Assume that* $\mathbb{E}[A] < \mathbb{E}[U]$. *Then for all finite and compatible initial lag times* $w = (w_1, w_2)$, *there exists a positive integer* $H(w;\omega)$ *such that for all* $k \geq H(w)$, $\delta(k; z) = \delta(k; e)$, *where* $z = z(w;\omega)$ *is the initial condition defined in (7.15).*

Proof We first prove that $H(w) = H'(w)$, where

$$H'(w) \stackrel{\text{def}}{=} \inf\{k \geq 0 \mid \delta(k; z) = \delta(k; e)\} .$$

In words, after the first time when $\delta(k; z)$ and $\delta(k; e)$ meet, their paths are identical forever. The proof is by induction: if for some ω, $\delta(k; z; \omega) = \delta(k; e; \omega)$, then from (7.14) we obtain that $\delta(k+1; z; \omega) = \delta(k+1; e; \omega)$.

It is easily checked by induction on k that for all weakly compatible initial lag times $\delta(k; z) \geq \delta(k; e) \geq e$, for all $k \geq 0$.

Assume that the statement of the theorem does not hold. Then the paths $\delta(k; z)$ and $\delta(k; e)$ never meet with a positive probability, so that the event

$$\mathcal{A} = \{\delta(k; z) > \delta(k; e) \geq e, \forall k \geq 0\}$$

has a positive probability. For all ω in $\mathcal{A}$, we obtain

$$\begin{aligned} \delta(k;z) &= e \oplus F(k-1)\delta(k-1;z) \\ &= F(k-1)\delta(k-1;z) \ , \end{aligned}$$

for all $k \geq 0$. Therefore, if $\mathcal{A}$ has a positive probability, the relation

$$\delta(k;z) = z \otimes \bigotimes_{l=0}^{k-1} F{\circ}\theta^l$$

holds with a positive probability. Owing to the ergodic assumption, $\sum_{l=0}^{k-1} F{\circ}\theta^l$ tends to $-\infty$ a.s. if $\mathbb{E}[A] < \mathbb{E}[U]$. Therefore, under the assumption $\mathbb{E}[A] < \mathbb{E}[U]$, the last relation readily implies that $\delta(k;z) \to -\infty$ when $k \to \infty$, with a positive probability, which is impossible since $\delta(k;z) \geq 0$. ■

The general coupling property for the ratio process of the event graph is summarized in the following lemma.

Lemma 7.21 *Let w be an arbitrary finite and compatible initial lag time vector. The sequence $\{\delta(k;z)\}$ couples with the sequence generated by the random variable Δ, so that $\delta(k;\delta(0;w))$ converges weakly to Δ when k tends to ∞. If $\mathbb{E}[A] > \mathbb{E}[U]$, then $\delta(k;z)$ converges a.s. to ∞ when k tends to ∞. More precisely,*

$$\lim_k \left(\delta(k,z)\right)^{1/k} = \mathbb{E}[A \not{o} U] > e \quad a.s.$$

Proof From Lemma 7.20, there exists a finite integer $H = H(z) > 0$ such that for all $k \geq H$, $\delta(k;z) = \delta(k;e)$ a.s. Using again Lemma 7.20, we obtain another finite integer $H' = H(\Delta)$ such that for all $k \geq H'$, $\Delta{\circ}\theta^k = \delta(k;\Delta) = \delta(k;e)$ a.s. Hence, for all $k \geq \max\{H, H'\}$, $\Delta{\circ}\theta^k = \delta(k;z)$ a.s. As for the case $\mathbb{E}[A] > \mathbb{E}[U]$, we should use the bound $\delta(k;z) \geq \delta(k;e)$ and the fact that

$$\lim_k \left(\delta(k;e)\right)^{1/k} = \mathbb{E}[A] \not{o} \mathbb{E}[U] > e \ ,$$

which follows from (7.13), to prove that $\lim \delta(k;z) = \infty$ a.s. ■

Corollary 7.22 *If $\mathbb{E}[A] < \mathbb{E}[U]$, then Δ is the unique finite solution of (7.11).*

Proof The uniqueness of the possible stationary regimes follows from the coupling property: if Ξ is another stationary regime, namely a finite solution of (7.11), then we first obtain $\Xi{\circ}\theta \geq e$ a.s. from the fact that Ξ satisfies (7.11), so that Ξ is necessarily nonnegative. In addition, from the coupling property we obtain that $\Xi{\circ}\theta^k = \delta(k;\Xi) = \delta(k;e) = \delta(k;\Xi) = \Delta{\circ}\theta^k$, for all $k \geq \max\{H(\Delta), H(\Xi)\} < \infty$. Therefore $\Delta = \Xi$. ■

Proof of Theorem 7.18 The existence part is established in §7.2.3.1–7.2.3.2. For λ and X as in the theorem, we must have $\lambda = U$ and $X_2{\circ}\theta U$ necessarily satisfies

(7.11). Therefore, $X_2 \circ \theta U = \Delta$ in view of Corollary 7.22. The last property of the theorem is a mere rephrasing of Lemma 7.21, once we notice that the coupling of $\{\delta(k)\}$ with $\{\Delta \circ \theta^k\}$ implies that of $\{x(k) \not/ u(k)\}$ with $\{X_2 \circ \theta^k\}$. In fact, the last assertion is only proved for initial conditions such that $\delta(0) \geq e$ (see the proof of the Lemma 7.21); the extension to more general finite initial conditions is obtained in the same way. ■

Remark 7.23 The only difficulty in the preceding eigenpair problem lies in finding X_2, or equivalently Δ. In the case when the sequences $\{A(k)\}$ and $\{U(k)\}$ are both i.i.d. (independent and identically distributed) and mutually independent, the problem of finding the distribution function of X_2 is solved using Wiener-Hopf factorization [46]. ■

7.2.6 First-Order and Second-Order Theorems

The aim of what follows is primarily to extend Theorem 7.18 to more general classes of matrices D. Of particular interest to us will be matrices which correspond to certain types of autonomous and nonautonomous event graphs, like those introduced in Chapter 2.

The results generalizing the eigenpair property of Theorem 7.18 will be referred to as second-order theorems, because they are concerned with ratios of the state variables. These theorems can be seen as $\mathbb{R}_{\max}$-instances of *multiplicative ergodic theorems* (see Theorem 7.108).

In what follows, the constants which characterize the growth rates of the state variables $x_j(k)$, and which generalize those in Theorem 7.24 below, will be referred to as Lyapunov exponents; these theorems will be called first-order or rate theorems. We conclude the section with the first-order theorem associated with our simple example.

Let (e_1, e_2) denote the following vectors of $\mathbb{R}^2$: $e_1 = (e, \varepsilon)$ and $e_2 = (\varepsilon, e)$.

Theorem 7.24 *The growth rate of $X(k)$ is characterized by the relations*

$$\begin{aligned}\lim_{k\to\infty} X_1(k)^{1/k} &= \lim_{k\to\infty} \left(e_1 D(k)D(k-1)\dots D(1)D(0)X(0)\right)^{1/k} \\ &= \mathbb{E}[U] \quad a.s.,\end{aligned}$$

and

$$\begin{aligned}\lim_{k\to\infty} X_2(k)^{1/k} &= \lim_{k\to\infty} \left(e_2 D(k)D(k-1)\dots D(1)D(0)X(0)\right)^{1/k} \\ &= \mathbb{E}[A] \oplus \mathbb{E}[U] \quad a.s.,\end{aligned}$$

regardless of the (finite) initial condition $X(0)$.

Proof The first assertion of the theorem is trivial. As for the second, we have

$$\lim_k \left(x(k)\right)^{1/k} = \lim_k \left(u(k-1)\right)^{1/k} \lim_k \left(x(k) \not/ u(k-1)\right)^{1/k}$$

$$= \mathbb{E}[U] \lim_k \left(x(k) \not/ u(k-1)\right)^{1/k} .$$

If $\mathbb{E}[A] > \mathbb{E}[U]$, then Lemma 7.21 implies that

$$\lim_k \left(x(k) \not/ u(k-1)\right)^{1/k} = \mathbb{E}[A] \not/ \mathbb{E}[U] .$$

If $\mathbb{E}[A] < \mathbb{E}[U]$, we obtain from the coupling property of Lemma 7.21 that

$$\begin{aligned} \lim_k \left(x(k) \not/ u(k-1)\right)^{1/k} &= \lim_k \left(\Delta \circ \theta^k\right)^{1/k} \\ &= \lim_k \left(\bigoplus_{i=1}^{k} (\Delta \not/ \Delta \circ \theta^{-1}) \circ \theta^i \right)^{1/k} . \end{aligned} \tag{7.21}$$

If Δ is integrable in addition to being finite, so is $\Delta \not/ \Delta \circ \theta^{-1}$, and we therefore have $\mathbb{E}\left[\Delta \not/ \Delta \circ \theta^{-1}\right] = e$; thus Birkhoff's Theorem and (7.21) immediately imply that

$$\lim_k \left(x(k) \not/ u(k-1)\right)^{1/k} = \mathbb{E}\left[\Delta \not/ \Delta \circ \theta^{-1}\right] = e .$$

Even if Δ is not integrable (which may happen even in this simple case, see Remark 7.13), the random variable $\Delta \not/ \Delta \circ \theta^{-1}$ is integrable as can be seen when using the following bounds obtained from (7.11):

$$F \circ \theta^{-1} \leq \Delta \not/ \Delta \circ \theta^{-1} \leq A \circ \theta^{-1} .$$

Therefore, in this case too, when using (7.21), we also obtain that

$$\lim_k \left(x(k) \not/ u(k-1)\right)^{1/k} = \mathbb{E}\left[\Delta \not/ \Delta \circ \theta^{-1}\right] ,$$

from Birkhoff's Theorem. We now prove that

$$\mathbb{E}\left[\Delta - \Delta \circ \theta^{-1}\right] = 0 , \tag{7.22}$$

which implies that in this case too $\lim_k \left(x(k)\right)^{1/k} = \mathbb{E}[U]$. In order to prove (7.22), observe that

$$\left|\min(\Delta, t) - \min(\Delta \circ \theta^{-1}, t)\right| \leq \left|\Delta - \Delta \circ \theta^{-1}\right| ,$$

for all $t \in \mathbb{R}^+$. Thus, from the Lebesgue dominated convergence theorem, we obtain that

$$\begin{aligned} 0 &= \lim_{t \to \infty} \mathbb{E}\left[\min(\Delta, t) - \min(\Delta \circ \theta^{-1}, t)\right] \\ &= \mathbb{E}\left[\lim_{t \to \infty} (\min(\Delta, t) - \min(\Delta \circ \theta^{-1}, t))\right] = \mathbb{E}\left[\Delta - \Delta \circ \theta^{-1}\right] . \end{aligned}$$

If $\mathbb{E}[A] = \mathbb{E}[U]$, either Δ is a.s. finite, in which case the preceding method applies, or it is a.s. infinite. We will consider the case $\Delta = \infty$ a.s. later on (see Theorem the:sum); the results is that

$$\lim_k \left(x(k)\right)^{1/k} = \mathbb{E}[U]$$

in this case too. ■

The proof of the following lemma is contained in the proof of the preceding theorem, and will be important in what follows.

Lemma 7.25 *If* Δ *is a finite (not necessarily integrable) random variable such that* $\Delta - \Delta{\circ}\theta^{-1}$ *is integrable, namely* $\mathbb{E}\left[\left|\Delta - \Delta{\circ}\theta^{-1}\right|\right] < \infty$, *where* $|x|$ *denotes conventional absolute value in* $\mathbb{R}$, *then* $\mathbb{E}\left[\Delta - \Delta{\circ}\theta^{-1}\right] = 0$.

7.3 FIRST-ORDER THEOREMS

7.3.1 Notation and Statistical Assumptions

Let $\mathcal{D}$ be a general dioid. For $A \in \mathcal{D}^{p\times q}$, let

$$|A|_{\oplus} \stackrel{\text{def}}{=} \bigoplus_{i=1}^{p}\bigoplus_{j=1}^{q} A_{ij} \tag{7.23}$$

and

$$|A|_{\wedge} \stackrel{\text{def}}{=} \bigwedge_{i=1}^{p}\bigwedge_{j=1}^{q} A_{ij} \ . \tag{7.24}$$

We will often use the following properties.

Lemma 7.26 *For all pairs of matrices* (A, B) *such that the product* AB *is well defined, we have*

$$\left.\begin{array}{rcl} |AB|_{\oplus} & \leq & |A|_{\oplus}\, |B|_{\oplus} \ , \\ |AB|_{\oplus} & \geq & |A|_{\wedge}\, |B|_{\oplus} \ , \\ |AB|_{\oplus} & \geq & |A|_{\oplus}\, |B|_{\wedge} \ , \end{array}\right\} \tag{7.25}$$

and

$$\left.\begin{array}{rcl} |AB|_{\wedge} & \geq & |A|_{\wedge}\, |B|_{\wedge} \ , \\ |AB|_{\wedge} & \leq & |A|_{\wedge}\, |B|_{\oplus} \ , \\ |AB|_{\wedge} & \leq & |A|_{\oplus}\, |B|_{\wedge} \ , \end{array}\right\} \tag{7.26}$$

where $\leq$ *is the order associated with* $\oplus$ *in* $\mathcal{D}$.

Proof Since $A_{ik} \leq |A|_{\oplus}$ for all i, k,

$$\bigoplus_{i,j}\bigoplus_{k} A_{ik}B_{kj} \leq |A|_{\oplus}\left(\bigoplus_{j,k} B_{kj}\right) = |A|_{\oplus}\, |B|_{\oplus} \ .$$

The proof of the other formulæ is similar. ■

The equation of interest in this section is

$$x(k+1) = A(k)x(k) \ , \tag{7.27}$$

where $A(k)$, respectively $x(k)$, is a random square matrix, respectively a random column vector, with entries taking their values in $\mathcal{D}$. We will stress the dependence of $x(k)$ on the initial condition by writing $x(k; x_0)$.

The random variables $A(k), k \in \mathbb{Z}$, and the initial condition x_0 are assumed to be defined on a common probability space $(\Omega, \mathbb{F}, \mathbb{P}, \theta)$, where θ is a shift which leaves $\mathbb{P}$ invariant, and is ergodic, with

$$A(k) = A \circ \theta^k \ , \quad k \in \mathbb{Z} \ . \tag{7.28}$$

Most of the section will be devoted to the case when $\mathcal{D}$ is $\mathbb{R}_{\max}$. Within this setting, $x(k)$ will be an $\mathfrak{n}$-dimensional column vector and $A(k)$ an $\mathfrak{n} \times \mathfrak{n}$ matrix. In this case, each entry of A is either a.s. equal to ε or nonnegative, and each diagonal entry of A is nonnegative. We start with a few examples in this dioid.

7.3.2 Examples in $\mathbb{R}_{\max}$

7.3.2.1 Example 1: Autonomous Event Graphs

Consider the evolution equation of a FIFO and autonomous stochastic event graph in its standard form, as given in Equation (2.31). If the initial condition of this event graph is compatible, (2.31) is of the type (7.27). In addition assume that the holding times $\alpha_i(k), p_i \in \mathcal{P}$, and the initial lag times of this event graph are random variables defined on a common probability space $(\Omega, \mathbb{F}, \mathbb{P}, \theta)$, and that the sequence $\{\alpha_i(k)\}$ is θ-stationary, i.e.

$$\alpha_i(k) = \alpha_i \circ \theta^k \ , \quad k \in \mathbb{Z} \ , \quad p_i \in \mathcal{P} \ ,$$

where α_i is finite, nonnegative and integrable. Then it easily checked that the matrices $\widetilde{A}(k)$ in (2.31) satisfy the θ-stationarity property and that each entry of $\widetilde{A}$ is either a.s. equal to ε or nonnegative and integrable. In view of the FIFO assumption, it is always true that $x_j(k+1) \geq x_j(k)$, so that the diagonal entry $\widetilde{A}_{jj}(k)$ can be assumed to satisfy the bound $\widetilde{A}_{jj}(k) \geq e$ without loss of generality. Therefore, under the foregoing statistical assumptions, any FIFO and autonomous stochastic event graph with compatible initial condition satisfies an evolution equation which falls into the framework considered above. Conversely, as was pointed out in §2.5.4, we can also view any equation of the type (7.27) as the standard evolution equation of an event graph with compatible initial condition and where the initial marking is $(0, 1)$-valued.

7.3.2.2 Example 2: Nonautonomous Event Graphs

Similarly, consider the evolution equation of a FIFO nonautonomous stochastic event graph in its standard form (2.39). If the initial condition is compatible, this equation then reads

$$\widetilde{x}(k+1) = \widetilde{A}(k)\widetilde{x}(k) \oplus \widetilde{B}(k)\widetilde{u}(k) \ . \tag{7.29}$$

If we define $X(k)$ to be the following $M(|\mathrm{Q}|+|\mathrm{I}|)$-dimensional vector and $A(k)$ to be the following matrix:

$$X(k) = \begin{pmatrix} \widetilde{u}(k) \\ \widetilde{x}(k) \end{pmatrix}, \quad A(k) = \begin{pmatrix} \widetilde{U}(k) & \varepsilon \\ \widetilde{B}(k) & \widetilde{A}(k) \end{pmatrix}, \tag{7.30}$$

where $\widetilde{U}(k)$ is the diagonal matrix with entries

$$\widetilde{U}_{jj}(k) = \widetilde{u}_j(k+1) \not/ \widetilde{u}_j(k) \ , \quad q_j \in \mathrm{I} \ , \tag{7.31}$$

then it is immediate that (2.39) can also be rewritten as

$$X(k+1) = A(k)X(k) \ , \quad k \geq 1 \ . \tag{7.32}$$

This transformation is tantamount to viewing each input transition j as a recycled transition where the holding times of the recycling place are given by the sequence $\left\{\widetilde{U}_{jj}(k)\right\}$. If the holding times $\alpha_i(k)$ and the *inter-input times* $U_{jj}(k)$ satisfy the θ-stationarity conditions

$$\alpha_i(k) = \alpha_i \circ \theta^k \ , \quad p_i \in \mathfrak{P} \ , \quad U_{jj}(k) = U_{jj} \circ \theta^k \ , \quad q_j \in \mathrm{I} \ , \quad k \in \mathbb{Z} \ ,$$

where the random variables α_i and U_{jj} are positive and integrable, then the matrices $A(k)$ satisfy the θ-stationarity condition (7.28) and the additional conditions mentioned above. Hence, the framework described at the beginning of this section also covers the nonautonomous case, provided we make additional θ-stationarity assumptions on the inter-input times.

7.3.3 Maximal Lyapunov Exponent in $\mathbb{R}_{\max}$

We assume that the nonnegative entries of A are all integrable. Under this condition the sequence $\{x(k; x_0)\}$ defined by (7.27) converges to ∞ a.s. in a way which is quantified by the following theorem.

Theorem 7.27 *There exists a constant $e \leq \mathfrak{a} < \top = \infty$ such that, for all finite initial conditions x_0, the a.s. limit*

$$\lim_{k\to\infty} |x(k;x_0)|_{\oplus}^{1/k} = \lim_{k\to\infty} |A(k-1)A(k-2)\ldots A(k)A(0)x_0|_{\oplus}^{1/k} = \mathfrak{a} \quad \textit{a.s.} \tag{7.33}$$

holds. If the initial condition is integrable, in addition we have

$$\lim_{k\to\infty} \mathbb{E}\left[|x(k;x_0)|_{\oplus}^{1/k}\right] = \lim_{k\to\infty} \mathbb{E}\left[|x(k;x_0)|_{\oplus}\right]^{1/k} = \mathfrak{a} \ . \tag{7.34}$$

Proof By induction, we obtain that $|x(k;e)|_{\oplus}$ is integrable for all $k \geq 0$ (using the integrability assumptions together with the fact that $\max(a,b) \leq |a|+|b|$, for a and b in $\mathbb{R}$). Therefore we have

$$e \leq \mathbb{E}\left[|x(k;e)|_{\oplus}\right] < \top \ , \quad \forall k \geq 0 \ .$$

Let

$$\xi_{m,m+k} = |x(k;e)|_{\oplus} \circ \theta^m \;, \quad m \in \mathbb{Z} \;, \quad k \geq 0 \;. \tag{7.35}$$

Since

$$|x(k,e)|_{\oplus} = \left|A \circ \theta^{k-1} \ldots Ae\right|_{\oplus} = \left|A \circ \theta^{k-1} \ldots A\right|_{\oplus} \;,$$

we obtain from Lemma 7.26 that for all $k \geq 1$, and all $0 \leq p \leq k$,

$$\left|A \circ \theta^{k-1} \ldots A \circ \theta^p A \circ \theta^{p-1} \ldots A\right|_{\oplus} \circ \theta^m \leq \left|A \circ \theta^{k-1} \ldots A \circ \theta^p\right|_{\oplus} \circ \theta^m \left|A \circ \theta^{p-1} \ldots A\right|_{\oplus} \circ \theta^m \;,$$

that is, $\xi_{m,m+k} \leq \xi_{m,m+p} + \xi_{m+p,m+k}$, so that $\xi_{m,m+k}$ is a nonnegative and integrable subadditive process. From Kingman's Theorem on subadditive ergodic processes (see Theorem 7.106), we obtain

$$\lim_{k \to \infty} (\xi_{0k})^{1/k} = \lim_{k \to \infty} \mathbb{E}\left[(\xi_{0k})^{1/k}\right] = \mathfrak{a} \quad \text{a.s.},$$

for some constant $\mathfrak{a} < \infty$, which concludes the proof for $|x(k;e)|_{\oplus}$. From the relation $x(k) = A(k-1) \ldots A(0)x_0$ and from (7.25), we obtain the immediate bounds

$$|x(k;e)|_{\oplus} |x_0|_{\wedge} \leq |x(k;x_0)|_{\oplus} \leq |x(k;e)|_{\oplus} |x_0|_{\oplus} \;, \quad k \geq 0 \;, \quad \forall x_0 \text{ finite} \;.$$

Therefore

$$|x(k;e)|_{\oplus}^{1/k} |x_0|_{\wedge}^{1/k} \leq |x(k;x_0)|_{\oplus}^{1/k} \leq |x(k;e)|_{\oplus}^{1/k} |x_0|_{\oplus}^{1/k} \;, \tag{7.36}$$

for all $k \geq 0$. Property (7.33) follows immediately when letting k go to ∞. If, in addition, x_0 is integrable, we first prove by induction that $x(k;x_0)$ is integrable for all $k \geq 0$. We can hence take expectations in (7.36) and use the fact that $\lim_{k \to \infty} \mathbb{E}\left[(\xi_{0k})^{1/k}\right] = \mathfrak{a}$ to obtain (7.34). ■

Remark 7.28 Certain representations of stochastic event graphs considered in Chapter 2, such as the representation of Corollary 2.62 for instance, involve initial conditions with ε entries, for which Theorem 7.27 cannot be applied directly. However, it is easy to check that one can replace these entries by appropriate finite entries without altering the value of $x(\cdot)$. ■

Remark 7.29 It will also be useful to know when the constant $\mathfrak{a}$ is strictly positive. A sufficient condition for this is that there exists at least a circuit of the precedence graph of A and two nodes i_0 and j_0 in this circuit such that $\mathbb{E}\left[A_{j_0 i_0}(k)\right] > e$. Under this condition the positiveness of $\mathfrak{a}$ is obtained from the bound

$$x_{j_0}(k\mathfrak{n}) \geq A_{j_0 i_0}(k\mathfrak{n} - 1) x_{i_0}((k-1)\mathfrak{n}) \;.$$

This in turn implies

$$\mathbb{E}\left[x_{j_0}(k\mathfrak{n};e)\right] \geq k\mathbb{E}\left[A_{j_0 i_0}(k)\right] = kC \;,$$

with $C > 0$, which implies that $\mathfrak{a} > e$. Note that in the stochastic event graph setting, this condition is tantamount to having a circuit of the event graph with at least one place with a positive mean holding time. ■

7.3.4 The Strongly Connected Case

The framework is that of the previous section. Let $\mathcal{G}(A)$ denote the precedence graph of the square matrix A (see §2.3). Although matrix A depends on ω, the assumption that its entries are either a.s. equal to ε or a.s. finite implies that $\mathcal{G}(A)$ is either a.s. strongly connected or a.s. nonstrongly connected (or equivalently, either A is a.s. irreducible or it is a.s. nonirreducible). In this subsection we assume that we are in the former case.

Remark 7.30 The assumption that A is irreducible and the assumption that the diagonal entries of A are different from ε imply that A is aperiodic (see Definition 2.15 and the theorem which follows this definition). More precisely, the matrix

$$G(k) \stackrel{\text{def}}{=} A(k+\mathfrak{n}-1)A(k+\mathfrak{n}-2)\dots A(k) \ , \quad k \in \mathbb{Z} \ , \tag{7.37}$$

is such that $G_{ij}(k) \geq e$ for all pairs $(i,j) \in \{1,\dots,\mathfrak{n}\}^2$. ■

We know from the preceding subsection that $|x(k)|_\oplus$ grows like $\mathfrak{a}^k$. In fact, in the case considered here, each individual state variable $x_j(k)$ has the same growth rate, as shown in the following lemma.

Corollary 7.31 *If matrix A is irreducible, then for all finite initial conditions x_0, and for all $j = 1,\dots,\mathfrak{n}$, we have*

$$\lim_{k\to\infty} \left(x_j(k;x_0)\right)^{1/k} = \mathfrak{a} \quad a.s., \tag{7.38}$$

where $\mathfrak{a}$ is the maximal Lyapunov exponent of Theorem 7.27. If the initial condition is integrable, we also have

$$\lim_{k\to\infty} \mathbb{E}\left[\left(x_j(k;x_0)\right)^{1/k}\right] = \mathfrak{a} \ . \tag{7.39}$$

Proof From Remark 7.30, we obtain that $x_j(k;x_0) \geq x_i(k-\mathfrak{n};x_0)$ for all $i,j = 1,\dots,\mathfrak{n}$, and $k > \mathfrak{n}$. The property (7.38) follows then from the bounds

$$|x(k-\mathfrak{n})|_\oplus \leq x_j(k) \leq |x(k)|_\oplus \ , \quad \forall j = 1,\dots,\mathfrak{n} \ , \tag{7.40}$$

and from Theorem 7.27. ■

Corollary 7.32 *Under the foregoing assumptions, if A is irreducible, the a.s. limits*

$$\lim_{k\to\infty} |x(k)|_\wedge^{1/k} = \mathfrak{a} \quad a.s. \tag{7.41}$$

and

$$\lim_{k\to\infty} |A(k-1)\dots A(1)A(0)|_\wedge^{1/k} = \mathfrak{a} \quad a.s. \tag{7.42}$$

hold.

Proof Equation (7.41) follows from (7.40). As for the second relation, it is immediate that

$$\limsup_k |A(k-1)\dots A(1)A(0)|_{\wedge}^{1/k} \leq \mathfrak{a} \quad \text{a.s.}$$

In addition, we have

$$\begin{aligned}\big(A(k-1+\mathfrak{n})A(k-2+\mathfrak{n})\dots A(1)A(0)\big)_{ji} \\ = \bigoplus_{l=1}^{\mathfrak{n}} \Big[\big(A(k-1+\mathfrak{n})A(k-2+\mathfrak{n})\dots A(\mathfrak{n}+1)A(\mathfrak{n})\big)_{jl} \\ \otimes \big(A(\mathfrak{n}-1)A(\mathfrak{n}-2)\dots A(1)A(0)\big)_{li}\Big] .\end{aligned}$$

In view of Remark 7.30, this implies

$$\begin{aligned}\big(A(k-1+\mathfrak{n})A(k-2+\mathfrak{n})\dots A(1)A(0)\big)_{ji} \\ \geq \bigoplus_{l=1}^{\mathfrak{n}} (A(k-1+\mathfrak{n})A(k-2+\mathfrak{n})\dots A(\mathfrak{n}+1)A(\mathfrak{n}))_{jl} \\ = |x(k,e)\circ\theta^n|_{\oplus} .\end{aligned}$$

Thus the a.s. limit

$$\liminf_k (A(k-1)\dots A(1)A(0))_{ji}^{1/k} \geq \mathfrak{a} \quad \text{a.s.}$$

holds as a direct consequence of Theorem 7.27. ∎

Remark 7.33 Thus, in the case of a strongly connected stochastic event graph, all transitions have the same asymptotic firing rate; the constant $\mathfrak{a}$ is also called the *cycle time* of the strongly connected event graph. Its inverse $\mathfrak{a}^{-1}$ is often called its *throughput*. ∎

7.3.5 General Graph

Consider the decomposition of $\mathcal{G}(A)$ into its m.s.c.s.'s (§2.2). For the same reasons as above, the number N_A of its m.s.c.s.'s and their topologies are nonrandom. We will use the notations of §2.2 for the m.s.c.s.'s and the reduced graph $(\overline{\mathcal{V}}, \overline{\mathcal{E}})$ of $\mathcal{G}(A)$. The reduced graph is acyclic and connected (provided the precedence graph is connected, which will be assumed in the what follows). Remember that a m.s.c.s. $(\mathcal{V}_n, \mathcal{E}_n)$ is said to be a source subgraph if node n of the reduced graph has no predecessors, and that it is said to be a nonsource subgraph otherwise.

Remark 7.34 In the particular case when the equation of interest is that of a nonautonomous event graph of the form (7.32), each recycled transition associated with an input transition will be seen as a source subgraph. ∎

When there is no ambiguity, the notation π, π^* and π^+ will be used to represent the usual sets of predecessor nodes in the *reduced graph*. Without loss of generality, the numbering of the nodes is assumed to be compatible with the graph in the sense that $(m,n) \in \overline{\mathcal{E}}$ implies $m < n$. In particular, the source subgraphs are numbered $\{1,\ldots,N_0\}$. For all $1 \leq n \leq N_A$, we will make use of the restrictions $A_{(n)(m)}$, $x_{(n)}$, $A_{(\leq n)(\leq m)}$, $x_{(\leq n)}$, etc. defined in Notation 2.5.

The maximal Lyapunov exponent associated with the matrix $A_{(n)}(k) \stackrel{\text{def}}{=} A_{(n)(n)}(k)$ (respectively $A_{(\leq n)} \stackrel{\text{def}}{=} A_{(\leq n)(\leq n)}$ or $A_{(<n)} \stackrel{\text{def}}{=} A_{(<n)(<n)}$) will be denoted $\mathfrak{a}_{(n)}$ (respectively $a_{(\leq n)}$ or $a_{(<n)}$).

Observe that in general, $x_{(n)}(k)$ does not coincide with the solution of the evolution equation

$$y(k+1) = A_{(n)}(k)y(k) \ , \quad k \geq 0 \ ,$$

with initial condition $y(0) = x_{(n)}(0)$. However, the sequence $\{x_{(\leq n)}(k)\}$ (respectively $\{x_{(<n)}(k)\}$) is the solution of the evolution equation

$$\begin{aligned} x_{(\leq n)}(k+1) &= A_{(\leq n)}(k)x_{(\leq n)}(k) \ , \\ \text{(respectively} \quad x_{(<n)}(k+1) &= A_{(<n)}(k)x_{(<n)}(k)) \ , \quad k \geq 0 \ , \end{aligned}$$

with initial condition $x_{(\leq n)}(0)$ (respectively $x_{(<n)}(0)$).

Lemma 7.35 *For all finite initial conditions, the following a.s. limits hold:*

$$\lim_{k\to\infty} \left|x_{(n)}(k)\right|_{\oplus}^{1/k} = \mathfrak{a}_{(\leq n)} \quad a.s., \tag{7.43}$$

and

$$\lim_{k\to\infty} \left(x_j(k)\right)^{1/k} = \mathfrak{a}_{(\leq n)} \quad a.s., \quad \forall j \in \mathcal{V}_n \ . \tag{7.44}$$

If the initial condition is integrable, we also have

$$\lim_{k\to\infty} \mathbb{E}\left[\left|x_{(n)}(k)\right|_{\oplus}^{1/k}\right] = \mathfrak{a}_{(\leq n)} \tag{7.45}$$

and

$$\lim_{k\to\infty} \mathbb{E}\left[\left(x_j(k)\right)^{1/k}\right] = \mathfrak{a}_{(\leq n)} \ , \quad \forall j \in \mathcal{V}_n \ . \tag{7.46}$$

Proof It is obvious from the definition that $\left|x_{(n)}(k)\right|_{\oplus} \leq \left|x_{(\leq n)}(k)\right|_{\oplus}$, so that

$$\liminf_k \left|x_{(n)}(k)\right|_{\oplus}^{1/k} \leq \mathfrak{a}_{(\leq n)} \ .$$

When using the fact that there exists a path of length less than $\mathfrak{n}$ from h to j in $\mathcal{G}(A)$, for all $j \in \mathcal{V}_n$ and $h \in \bigcup_{m\in\pi^*(n)} \mathcal{V}_m$, together with the assumption on the diagonal entries of A, we obtain the following bound from (7.27):

$$x_j(k+1) \geq \bigoplus_{\{h\in\mathcal{V}_m, m\in\pi^*(n)\}} x_h(k-\mathfrak{n}) \ , \quad \forall j \in \mathcal{V}_n \ ,$$

provided $k \geq \mathfrak{n}$. Therefore $\left|x_{(n)}(k+1)\right|_{\oplus} \geq \left|x_{(\leq n)}(k-\mathfrak{n})\right|_{\oplus}$, for $k \geq \mathfrak{n}$, so that

$$\limsup_{k} \left|x_{(n)}(k)\right|_{\oplus}^{1/k} \geq \mathfrak{a}_{(\leq n)} \quad \text{a.s.}$$

The proof of the individual a.s. limits in (7.44) follows the same lines as in Corollary 7.31. In the integrable case, the proof of the convergence of the expectations is immediate. ■

Owing to the acyclic nature of the reduced graph, the vector $x_{(n)}(k)$ satisfies the equation

$$x_{(n)}(k+1) = A_{(n)}(k)x_{(n)}(k) \oplus s(n,k+1) \ , \tag{7.47}$$

where

$$s(n,k+1) \stackrel{\text{def}}{=} A_{(n)(<n)}(k)x_{(<n)}(k) \ . \tag{7.48}$$

Equation (7.47) is the basis for proving the following property.

Theorem 7.36 *The constant $\mathfrak{a}_{(\leq n)}$, which characterizes the growth rate of the variables $x_j(k)$, $j \in \mathcal{V}_n$, is obtained from the constants $\mathfrak{a}_{(m)}$, $1 \leq m \leq n$, by the relation*

$$\mathfrak{a}_{(\leq n)} = \bigoplus_{m \in \pi^*(n)} \mathfrak{a}_{(m)} \ . \tag{7.49}$$

Proof We first prove that

$$\lim_{k\to\infty} |s(n,k)|_{\oplus}^{1/k} = \mathfrak{a}_{(<n)} \quad \text{a.s.,} \tag{7.50}$$

for all $N_0 < n \leq N$. From (7.48), we obtain that

$$|s(n,k+1)|_{\oplus} \leq |A(k)|_{\oplus} \left(\bigoplus_{m \in \pi^+(n)} \left|x_{(m)}(k)\right|_{\oplus} \right) ,$$

so that

$$|s(n,k+1)|_{\oplus}^{1/k} \leq |A(k)|_{\oplus}^{1/k} \left|x_{(<n)}(k)\right|_{\oplus}^{1/k} \ . \tag{7.51}$$

The integrability assumption on A implies that

$$\lim_{k} |A(k)|_{\oplus}^{1/k} = e \quad \text{a.s.}$$

(using the same technique as in the proof of Theorem 7.24). Letting k go to ∞ in (7.51) then implies

$$\liminf_{k\to\infty} |s(n,k)|_{\oplus}^{1/k} \leq \mathfrak{a}_{(<m)} \quad \text{a.s.}$$

By using the same type of arguments as in Lemma 7.35, from (7.48) we obtain that

$$|s(n,k+1)|_{\oplus} \geq \left|x_{(<n)}(k-\mathfrak{n})\right|_{\oplus} \quad \text{a.s.,}$$

which in turn implies

$$\limsup_{k\to\infty} |s(n,k)|_{\oplus}^{1/k} \geq \mathfrak{a}_{(<m)} \quad \text{a.s.}$$

This concludes the proof of (7.50).

It is clear from (7.48) that $\left|x_{(n)}(k+1)\right|_{\oplus} \geq |s(n,k+1)|_{\oplus}$, so that necessarily $\left|x_{(n)}(k+1)\right|_{\oplus}^{1/k} \geq |s(n,k+1)|_{\oplus}^{1/k}$, and hence $\mathfrak{a}_{(\leq n)} \geq \mathfrak{a}_{(<n)}$. Owing to the individual limits of Equation 7.44, for all $j \in \mathcal{V}_n$, $(x_{(n)})_j(k) \sim \mathfrak{a}^k_{(\leq n)}$, whereas $|s(n,k)|_{\oplus} \sim \mathfrak{a}^k_{(<n)}$, so that if $\mathfrak{a}_{(\leq n)} > \mathfrak{a}_{(<n)}$, then there exists a finite integer-valued random variable K such that

$$A_{(n)}(k)x_{(n)}(k) \geq s(n,k) \ , \quad \forall k \geq K \ .$$

Accordingly, Equation (7.48) reads

$$x_{(n)}(k+1) = A_{(n)}(k)x_{(n)}(k) \ ,$$

for $k \geq K$. Let $y(k;x_0)$ denote the solution of the equation

$$y(k+1) = A_{(n)}(k)y(k) \ , \quad k \geq 0 \ ,$$

with initial condition $y(0) = (x_0)_{(n)}$. On $\{K = h\}$, we have

$$\begin{aligned} x_{(n)}(k) &= A_{(n)}(k)\ldots A_{(n)}(h)x_{(n)}(h) \\ &= (A_{(n)}(k-h)\ldots A_{(n)}(0)x_{(n)}(h)\circ\theta^{-h})\circ\theta^h \\ &= y(k-h; x_{(n)}(h)\circ\theta^{-h})\circ\theta^h \ , \end{aligned}$$

for all $k \geq h$. Thus, on the event $\{K = h\}$

$$\lim_k \left|x_{(n)}(k)\right|_{\oplus}^{1/k} = \lim_k \left|y(k-h; x_{(n)}(h)\circ\theta^{-h})\right|_{\oplus}^{1/k} \circ\theta^h = \mathfrak{a}_{(n)} \quad \text{a.s.,} \tag{7.52}$$

where we used the a.s. convergence result of Theorem 7.27 applied to matrix $A_{(n)}(k)$. Since K is finite, $\bigcup_h \{K = h\} = \Omega$, so that

$$\lim_k \left(x_j(k)\right)^{1/k} = \mathfrak{a}_{(n)} \quad \text{a.s.,} \quad \forall j \in \mathcal{V}_n \ .$$

Therefore, $\mathfrak{a}_{(\leq n)} \geq \mathfrak{a}_{(<n)}$, and $\mathfrak{a}_{(\leq n)} > \mathfrak{a}_{(<n)}$ implies $\mathfrak{a}_{(\leq n)} = \mathfrak{a}_{(n)}$, that is, $\mathfrak{a}_{(\leq n)} = \mathfrak{a}_{(<n)} \oplus \mathfrak{a}_{(n)}$. The proof of (7.49) is obtained from the last relation by an immediate induction on n. ■

Example 7.37 (Acyclic fork-join networks of queues) Consider the stochastic event graph of Figure 7.2. This example features an acyclic fork-join queuing network which is characterized by an acyclic graph $\widehat{\mathcal{G}} = (\widehat{\mathcal{V}}, \widehat{\mathcal{E}})$, with nodes $\{0, 1, \ldots, \mathfrak{n}\}$. In this graph, $\widehat{\pi}(j)$ will denote the set of predecessors of node j and $\widehat{\sigma}(j)$ the set of its successors. This graph has a single source node denoted 0.

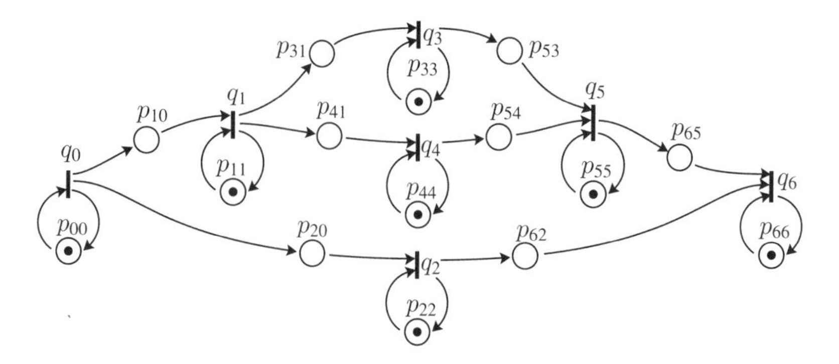

Figure 7.2: An acyclic fork-join network of queues

With this graph, we associate a FIFO stochastic event graph, for which we use the conventional notation. The set of transitions is $\{q_0, q_1, \ldots, q_\mathfrak{n}\}$; each transition q_j is recycled, with associated place p_{jj}, $0 \leq j \leq \mathfrak{n}$; in addition, a place p_{ji} is associated with each pair of nodes $0 \leq i, j \leq \mathfrak{n}$ such that $i \in \widehat{\pi}(j)$. Transition q_0, which generates the input to the queuing network, is a recycled transition such that $\sigma(q_0) = \{p_{00}, p_{i0}\}_{i \in \widehat{\sigma}(0)}$, and $\pi(q_0) = \{p_0\}$. Transition q_j, which represents queue j of the queuing network, admits the set $\{p_{jj}, p_{ji}\}_{i \in \widehat{\pi}(j)}$ as a predecessor set and the set $\{p_{jj}, p_{ij}\}_{i \in \widehat{\sigma}(j)}$ as a successor set.

If $\widehat{\sigma}(j)$ has two or more elements, one says that there is a 'fork' from queue j to the queues of $\widehat{\sigma}(j)$. As a result of this fork, when a departure takes place from queue j, this creates simultaneous arrivals into successor queues. Similarly, when $\widehat{\pi}(j)$ has two or more elements, one says that there is a 'join'. Clearly, the effect of a join in queue j is to synchronize the outputs of the queues of $\widehat{\pi}(j)$.

Let $\alpha_j(k)$ denote the holding times in p_{jj} and $U(k)$ denote those in p_{00}, and assume that all the other holding times are zero. For instance, the matrix $A(k)$ associated with the autonomous event graph of Figure 7.2 is characterized by the formula

$$A(k-1) = \begin{pmatrix} \alpha_0(k) & e & e & \varepsilon & \varepsilon & \varepsilon & \varepsilon \\ \varepsilon & \alpha_1(k) & \varepsilon & e & e & \varepsilon & \varepsilon \\ \varepsilon & \varepsilon & \alpha_2(k) & \varepsilon & \varepsilon & \varepsilon & e \\ \varepsilon & \varepsilon & \varepsilon & \alpha_3(k) & \varepsilon & e & \varepsilon \\ \varepsilon & \varepsilon & \varepsilon & \varepsilon & \alpha_4(k) & e & \varepsilon \\ \varepsilon & \varepsilon & \varepsilon & \varepsilon & \varepsilon & \alpha_5(k) & e \\ \varepsilon & \varepsilon & \varepsilon & \varepsilon & \varepsilon & \varepsilon & \alpha_6(k) \end{pmatrix}$$

It is easily checked that $N_A = \mathfrak{n} + 1$, and that each m.s.c.s. consists of exactly one transition, so that the firing rate of transition j is simply

$$\mathfrak{a}_{(\leq j)} = \bigoplus_{i \in \widehat{\pi}^*(j)} \mathbb{E}[\alpha_j] \ .$$

■

7.3.6 First-Order Theorems in Other Dioids

The rate theorem (Theorem 7.27) which was established in the preceding sections for $\oplus = \max$ and $\otimes = +$ is essentially based on the following two ingredients:

- the relation
$$|AB|_{\oplus} \leq |A|_{\oplus} \; |B|_{\oplus} \; ; \tag{7.53}$$
- Kingman's subadditive ergodic theorem which relies on the fact that $\otimes = +$.

The first result holds in a general dioid (see Lemma 7.26).

Example 7.38 For instance, if $\mathcal{D} = \mathbb{R}_{\min}$, the first relation of (7.25) translates into the property $|AB|_{\oplus} \geq |A|_{\oplus} \; |B|_{\oplus}$ (where $\geq$ denotes here the conventional ordering in $\mathbb{R}$), since $\leq$ corresponds to the 'reverse' of $\leq_{\min}$. ∎

In order to extend the second result to a general dioid, we need a 'sub-$\otimes$ ergodic theorem' stating that any $\mathcal{D}$-valued random sequence $\{a_{mn}\}_{m<n\in\mathbb{N}}$ which satisfies the conditions
$$a_{mn} \leq a_{mp} \otimes a_{pn} \; , \quad \forall m < p < n \; ,$$
and
$$a_{m,m+k} = a_{0k} \circ \theta^m \; , \quad \forall m, k \in \mathbb{N} \; ,$$
is such that
$$\exists \lim_{k\to\infty} (a_{0k})^{1/k} = a \quad \text{a.s.},$$
where a is some constant (the meaning of the limit will not be discussed precisely here). For instance, such a theorem will follow from Kingman's subadditive ergodic theorem if $\otimes$ is $+$ or $\times$. Thus, under the same type of statistical assumptions as in Theorem 7.27, we can prove the existence of maximal Lyapunov exponents for linear systems of the type (7.27) in $\mathbb{R}_{\min}$.

7.4 SECOND-ORDER THEOREMS; NONAUTONOMOUS CASE

7.4.1 Notation and Assumptions

In this section the dioid $\mathcal{D}$ under consideration is general. The basic equation of interest is the evolution equation
$$x(k+1) = A(k)x(k) \oplus B(k)u(k) \; , \quad k \geq 0 \; , \tag{7.54}$$
where $x(k)$, $u(k)$ and $A(k)$ all belong to $\mathcal{D}$.

The sequences $\{A(k), B(k)\}_{k\in\mathbb{Z}}$ and $\{u(k)\}_{k\geq 0}$ are assumed to be given, as well as the initial condition x_0. The sequence $\{u(k)\}$ is assumed to be finite and nondecreasing in k. Let $U(k)$ be defined by
$$U(k) = \frac{u(k+1)}{u(k)} \; , \quad k \geq 0 \; . \tag{7.55}$$

7.4.2 Ratio Equation in a General Dioid

As in the preliminary example, we define the *ratio process*, which consists of the sequence

$$\delta(k) = \frac{x(k+1)}{u(k)} \ , \quad k \geq 0 \ . \tag{7.56}$$

The first aim of this subsection is to derive an evolution equation for this ratio process.

Theorem 7.39 (Ratio Equation) *The variables $\delta(k)$ satisfy the inequalities*

$$\delta(k+1) \geq A(k+1)\delta(k)U_-(k) \oplus B(k+1)U_+(k)U_-(k) \ , \quad k \geq 0 \ , \tag{7.57}$$

where the initial condition is $\delta(0) = \delta_0$ is equal to $x(1) \oslash u(0)$, and where

$$U_-(k) \stackrel{\text{def}}{=} \frac{u(k)}{u(k+1)} \ , \qquad U_+(k) \stackrel{\text{def}}{=} U(k) = \frac{u(k+1)}{u(k)} \ . \tag{7.58}$$

Proof By successively using formulæ (f.2) and (f.12) of Table 4.1, we obtain

$$\begin{aligned} \frac{x(k+1)}{u(k-1)} &= \frac{A(k)x(k) \oplus B(k)u(k)}{u(k-1)} \\ &\geq \frac{A(k)x(k)}{u(k-1)} \oplus \frac{B(k)u(k)}{u(k-1)} \\ &\geq A(k)\frac{x(k)}{u(k-1)} \oplus B(k)\frac{u(k)}{u(k-1)} \ . \end{aligned}$$

From (f.5), we obtain

$$u(k-1) \geq \frac{u(k-1)}{u(k)} u(k) \ ,$$

so that

$$\begin{aligned} \frac{x(k+1)}{u(k-1)} &\leq \frac{x(k+1)}{\left(u(k-1) \oslash u(k)\right) u(k)} \\ &= \frac{x(k+1) \oslash u(k)}{u(k-1) \oslash u(k)} \ , \end{aligned}$$

where we used (f.9) in order to obtain the last relation. By using the notation $U_-(k)$ and $U_+(k)$ defined in the lemma, we finally obtain

$$\frac{\delta(k)}{U_-(k-1)} \geq A(k)\delta(k-1) \oplus B(k)U_+(k-1) \ , \quad k \geq 1 \ .$$

Therefore,

$$\begin{aligned} \frac{\delta(k)}{U_-(k-1)} U_-(k-1) \ \geq \ & A(k)\delta(k-1)U_-(k-1) \\ & \oplus B(k)U_+(k-1)U_-(k-1) \ , \quad k \geq 1 \ , \end{aligned}$$

and the proof is completed since the left-hand side of the last expression is less than $\delta(k)$ (owing to (f.5)). ■

In what follows, we will concentrate on the least solution of Inequalities (7.57).

Lemma 7.40 *The least solution of (7.57) is also the solution of the set of equations*

$$\delta(k+1) = A(k+1)\delta(k)U_-(k) \oplus B(k+1)U_+(k)U_-(k) \ , \quad k \geq 0 \ , \tag{7.59}$$

with initial condition $\delta(0) = x(1) \phi u(0)$.

Proof The minimum element of the set $\{x \geq a\}$ being a, the result of the lemma follows from an immediate induction. ■

Example 7.41 Let the dioid of scalars be $\mathbb{R}_{\max}$ and let x and u be vectors in this dioid, of dimension $\mathfrak{n}$ and $\mathfrak{m}$ respectively. Taking Remark 4.80 into account, the above calculations are still valid. A more direct derivation of the same result is obtained by subtracting $u_i(k)$ from the j-th line of (7.54); we directly obtain

$$\begin{aligned}\delta_{ji}(k+1) &= \bigoplus_{l=1}^{\mathfrak{n}} \left(A_{jl}(k+1)\left(x_l(k+1) \phi u_i(k+1)\right)\right) \\ &\quad \oplus \bigoplus_{l=1}^{\mathfrak{m}} \left(B_{jl}(k+1)\left(u_l(k+1) \phi u_i(k+1)\right)\right) \ .\end{aligned}$$

By using the property that $a \phi a = e$ for all $a \neq \varepsilon \in \mathbb{R}$, it is immediately verified that under the finiteness assumptions which were made on $u(k)$, this can be rewritten as

$$\begin{aligned}\delta_{ji}(k+1) &= \bigoplus_{l=1}^{\mathfrak{n}} \bigoplus_{p=1}^{\mathfrak{m}} \left(A_{jl}(k+1)\left(x_l(k+1) \phi u_p(k)\right)\left(u_p(k) \phi u_i(k+1)\right)\right) \\ &\quad \oplus \bigoplus_{l=1}^{\mathfrak{m}} \bigoplus_{p=1}^{\mathfrak{m}} \left(B_{jl}(k+1)\left(u_l(k+1) \phi u_p(k)\right)\left(u_p(k) \phi u_i(k+1)\right)\right) \ ,\end{aligned}$$

which is a mere rephrasing of the matrix relation (7.59). ■

7.4.3 Stationary Solution of the Ratio Equation

All the data are now assumed to be defined on a common probability space $(\Omega, \mathbb{F}, \mathbb{P}, \theta)$, with the usual assumptions on θ. The variables $A(k)$ and $B(k)$ are assumed to be θ-stationary, with $A(k) = A \circ \theta^k$ and $B(k) = B \circ \theta^k$. The dioid $\mathcal{D}$ is assumed to be complete. In addition to this, it is assumed that the sequence of ratios $\{U_+(k)\}$ and $\{U_-(k)\}$ defined in (7.58) are such that

$$U_+(k) = U_+ \circ \theta^k \ , \quad U_-(k) = U_- \circ \theta^k \ , \quad k \in \mathbb{Z} \ .$$

As in the preliminary example, we are interested in the possibility of making the ratios $\delta(k)$ stationary. For this, we will use a backward construction which generalizes the construction of the preliminary example.

Definition 7.42 (Backward process) *The backward process associated with the least solution of (7.57) is the sequence of variables $\{\Delta(k)\}_{k\geq 0}$ defined as the solution of the set of relations*

$$\begin{aligned} \Delta(0)\circ\theta &= C\ , \\ \Delta(k+1)\circ\theta &= A\circ\theta\Delta(k)U_-\oplus C\ , \quad k\geq 0\ , \end{aligned} \tag{7.60}$$

where $U_+ \stackrel{\text{def}}{=} U_+(0)$, $U_- \stackrel{\text{def}}{=} U_-(0)$ and $C \stackrel{\text{def}}{=} B\circ\theta U_+U_-$.

Lemma 7.43 *The sequence $\{\Delta(k)\}$ is coordinatewise nondecreasing in k with respect to the order in $\mathcal{D}$.*

Proof It is clear that $\Delta(1)\geq C\circ\theta^{-1} = \Delta(0)$. Assume now that for some $k\geq 1$, $\Delta(k)\geq\Delta(k-1)$. Then

$$\begin{aligned} \Delta(k+1)\circ\theta &= A\circ\theta\Delta(k)U_-\oplus C \\ &\geq A\circ\theta\Delta(k-1)U_-\oplus C = \Delta(k)\circ\theta\ , \end{aligned}$$

where we used the fact that the mapping $x\mapsto AxB\oplus C$ is isotone. ■

Lemma 7.44 *The random variable $\Delta = \bigoplus_{k\geq 0}\Delta(k)$ satisfies the relation*

$$\Delta\circ\theta = A\circ\theta\Delta U_-\oplus C\ . \tag{7.61}$$

Proof The sum Δ belongs to $\mathcal{D}$ because this dioid is complete. Since the mapping $x\mapsto AxB\oplus C$ is l.s.c., we also have

$$\begin{aligned} \Delta\circ\theta &= \bigoplus_{k\geq 0}\left(A\circ\theta\Delta(k)U_-\oplus C\right) \\ &= A\circ\theta\left(\bigoplus_{k\geq 0}\Delta(k)\right)U_-\oplus C \\ &= A\circ\theta\Delta U_-\oplus C\ . \end{aligned}$$

■

Lemma 7.45 *In the case when $A\in\mathcal{D}^{n\times n}$, $U+$ and $U_-\in\mathcal{D}^{m\times m}$, and C and $\Delta\in\mathcal{D}^{n\times m}$, if the diagonal elements of A are greater than or equal to e and U_- has finite entries, then the event $\mathcal{B} = \{|\Delta|_\oplus = \top\}$ is of probability either 0 or 1.*

Proof Owing to the assumption that the diagonal elements of A are a.s. finite, $|\Delta|_\oplus = \top$ implies $|A\circ\theta\Delta|_\oplus = \top$, which in turn implies that $|A\circ\theta\Delta U_-|_\oplus = \top$ (since U_- has all its entries a.s. finite). Therefore, $|\Delta|_\oplus = \top$ implies that

$$|\Delta\circ\theta|_\oplus \geq |A\circ\theta\Delta U_-|_\oplus = \top\ .$$

Thus, the measurable set $\mathcal{B}$ is such that the indicator functions $1_{\mathcal{B}}\circ\theta^k$ are nondecreasing in k. This is enough to ensure that $\mathcal{B}$ is of probability 0 or 1; indeed, when using the ergodicity assumption on the shift θ, and the nondecreasingness property, we obtain

$$\mathbb{P}[\mathcal{B}] = \lim_{k\to\infty} \frac{1}{k} \sum_{l=1}^{k} 1_{\mathcal{B}}\circ\theta^l \geq 1_{\mathcal{B}} \quad \text{a.s.},$$

so that if $\mathbb{P}[\mathcal{B}] > 0$, then the indicator function $1_{\mathcal{B}}$ is equal to 1 a.s. (an indicator function which is positive is necessarily equal to 1). ■

The following expansion of the backward process generalizes the star operation of Lemma 7.10.

Lemma 7.46 *The relation*

$$\Delta(k) = \bigoplus_{l=0}^{k} \left(\bigotimes_{h=1}^{l} A(-h+1) \right) C(-l-1) \left(\bigotimes_{h=1}^{l} U_-(h-l-1) \right) \tag{7.62}$$

holds, where the $\otimes$-product over an empty set (when $l = 0$) is equal to e by convention, and $C(k) = B(k+1)U_+(k)U_-(k)$.

As in our previous example, the nondecreasingness of the sequence $\{\Delta(k)\}$ becomes transparent from this formula.

The main question we are now interested in consists in determining the conditions under which the limiting value Δ is a.s. finite. The answer to this question is based on ergodic theory arguments, and is therefore dependent on the specific dioid which is considered. We will concentrate on the $\mathbb{R}_{\max}$ case for the rest of this section.

7.4.4 Specialization to $\mathbb{R}_{\max}$

7.4.4.1 Statistical Assumptions

In this subsection the underlying scalar dioid is $\mathbb{R}_{\max}$. In view of the results of the preceding subsection and of those of the preliminary example, the most natural statistical assumptions would consist in taking

- $U_+(k) = U_+\circ\theta^k$ (which implies that $U_-(k) = U_-\circ\theta^k$);
- U_+ integrable (which implies U_- integrable).

We will rather take the weaker assumptions

- $\{U_+(k)\}$ couples with $\{U_+\circ\theta^k\}$, where all the entries of U_+ are a.s. finite (which implies that $\{U_-(k)\}$ couples with $\{U_-\circ\theta^k\}$, where U_- has finite entries);
- $(U_+)_{ii}$ integrable for all $i = 1, \ldots, \mathfrak{n}$.

The motivations for taking these weaker assumptions will be commented upon in the next subsection.

Remark 7.47 Let $\mathfrak{u}_i \stackrel{\text{def}}{=} \mathbb{E}[(U_+)_{ii}]$. Since $u_i(k) = u_i(0) \bigotimes_{l=0}^{k-1} (U_+)_{ii}(l),\ k \geq 1$, from the coupling assumption and the assumption $\mathbb{E}\left[(U_+)_{ii}\right] = \mathfrak{u}_i$, we obtain that

$$\lim_{k\to\infty} \left(u_i(k)\right)^{1/k} = \mathfrak{u}_i \quad \text{a.s.,} \quad \forall i = 1, \dots, \mathfrak{m} \ . \tag{7.63}$$

This can only be compatible with the assumption that $U_+(k)$ couples with a stationary and finite sequence if $\mathfrak{u}_i = \mathfrak{u}_j$ for all $i, j = 1, \dots, \mathfrak{m}$ (since $u_i(k) - u_j(k)$ cannot couple with a finite stationary sequence if $\mathfrak{u}_i \neq \mathfrak{u}_j$). Therefore, a direct conclusion of our assumptions is that

$$\lim_{k\to\infty} |u(k)|_{\oplus}^{1/k} = \lim_{k\to\infty} |u(k)|_{\wedge}^{1/k} = \mathfrak{u} \quad \text{a.s.} \tag{7.64}$$

More general cases with $\mathbb{E}[(U_+)_{ii}]= \mathfrak{u}_i$ and $\mathfrak{u}_i \neq \mathfrak{u}_j$ for some pair (i, j), are of limited practical interest, as shown by the following example. ■

Example 7.48 Matrix A has a single m.s.c.s., and its Lyapunov exponent $\mathfrak{a}$ is such that $\mathfrak{a} < \mathfrak{u}_i < \mathfrak{u}_j$. In view of Theorem 7.36, we then have

$$x_l(k) \sim (\mathfrak{a} \oplus \mathfrak{u}_i \oplus \mathfrak{u}_j)^k = \mathfrak{u}_j^k \ ,$$

for all $l = 1, \dots, \mathfrak{n}$. Therefore, $x_l(k) - u_i(k)$ tends a.s. to ∞ for all l as k goes to ∞. Thus, in such a situation, some of the ratios necessarily become infinite. ■

We conclude this section with an algebraic interpretation of the assumptions on U_+, and a statement of the $\mathbb{R}_{\max}$-eigenvalue problem.

Lemma 7.49 *Let $V(k), k \geq 0$ be a sequence of $\mathfrak{m}\times\mathfrak{m}$ matrices. The two conditions*

- $V(k) = v(k+1) \phi v(k),\ k \geq 0,$ *where $v(k),\ k \geq 0$ is a sequence of finite $\mathfrak{m}$-dimensional vectors,*
- $V(k) = V{\circ}\theta^k,\ k \geq 0,$

are equivalent to the existence of a unique finite $\mathbb{R}^{\mathfrak{m}}$ eigenvector y, with $y_1 = e$, and a unique eigenvalue $\beta \in \mathbb{R}$ such that $Vy = \beta y{\circ}\theta$ and $V = (\beta y{\circ}\theta) \phi y$.

Proof We first show that under the first two conditions, there exists a unique pair of finite vectors (y, z) such that $y_1 = e$ and $V = z \phi y$. We have $V = v(1) \phi v(0) = z \phi y$, where $y_i \stackrel{\text{def}}{=} v_i(0) \phi v_1(0)$ and $z_i \stackrel{\text{def}}{=} v_i(1) \phi v_1(0)$. We have $y_1 = e$; let us show that y and z are uniquely defined from V: from the very definition of ϕ, we have

$$V_{ij} = z_i \phi y_j \ .$$

By taking $j = 1$ in the last relation, we see that z is uniquely defined from V, since $z_i = V_{i1}$; therefore $y_j = z_i \not/ V_{ij}$ does not depend on i and is uniquely determined from V.

Let β be the random variable $\beta = z_1$. We have $z = \beta w$, where $w_i = z_i \not/ \beta = v_i(1) \not/ v_1(1)$. We now conclude the proof of the eigenpair property by showing that $w = y{\circ}\theta$. We have $V{\circ}\theta = v(2) \not/ v(1) = v \not/ w$, where $v_i \stackrel{\text{def}}{=} v_i(2) \not/ v_1(1)$. Since $w_1 = e$, the above uniqueness property shows that $w = y{\circ}\theta$ indeed.

Conversely, if we assume that $V = (\beta y{\circ}\theta) \not/ y$, for some finite (β, y), we obtain $V(0) = v(1) \not/ v(0)$ with $v(0) = y$ and $v(1) = \beta y{\circ}\theta$. More generally, we have $V(k) = v(k+1) \not/ v(k)$, for all $k \geq 1$, when taking $v(k) = \beta^k \ldots \beta y{\circ}\theta^k$. ■

Under the assumptions listed above, the system of interest can be rewritten as

$$\left.\begin{array}{rcl} u(k+1) & = & U_+(k)u(k) \ , \\ x(k+1) & = & A(k)x(k) \oplus B(k)u(k) \ , \end{array}\right\} \tag{7.65}$$

or equivalently as

$$X(k+1) = D(k)X(k) \ , \quad k \geq 0 \ , \tag{7.66}$$

where

$$X(k) = \begin{pmatrix} u(k) \\ x(k) \end{pmatrix}$$

and

$$D(k) = \begin{pmatrix} U_+(k) & \varepsilon \\ B(k) & A(k) \end{pmatrix} .$$

In view of the preceding lemma, the assumptions on $D(k)$ can be summarized as follows: the matrices $D(k)$ couple in finite time with a stationary sequence $\{D{\circ}\theta^k\}$, where

$$D = \begin{pmatrix} U_+ & \varepsilon \\ B & A \end{pmatrix} ;$$

the matrix U_+ is such that

$$U_+ = \frac{\lambda u{\circ}\theta}{u} \ , \tag{7.67}$$

where (λ, u) are uniquely defined (u is a finite random vector $u \in \mathbb{R}^m$ with $u_1 = e$ and λ is a nonnegative and finite random variable).

The problem of interest is then similar to the random eigenpair problem of §7.2.3: can we continue the random eigenpair property $U_+ u = \lambda u{\circ}\theta$, which follows from (7.67), to the following eigenpair property of D:

$$DX = \lambda X{\circ}\theta \ ? \tag{7.68}$$

Remark 7.50 The assumption that $(U_+)_{ij}(k)$ couples with a stationary sequence for *all* $i, j = 1, \ldots, m$, which is equivalent to the eigenpair property of (7.67), is necessary for the second-order theorems of the following subsections; in particular, if this property is only satisfied by the diagonal terms $(U_+)_{ii}(k)$ (like for instance in the formulation (7.30) of the evolution equation), then these stability theorems do not hold, as shown by Example 7.98. ■

7.4.4.2 *Example: Nonautonomous Event Graphs*

We know from Corollary 2.82 that a nonautonomous FIFO stochastic event graph with recycled transitions and with a compatible initial condition in its standard form satisfies an equation of the form (7.54) in $\mathbb{R}_{\max}$. We make the following stationarity assumptions:

- the holding times $\alpha_i(k)$, $p_i \in \mathcal{P}$, $k \in \mathbb{Z}$, are θ-stationary and integrable;
- the ratios $U_{ij}(k) \stackrel{\text{def}}{=} u_i(k+1) \not{/} u_j(k)$, q_i, $q_j \in \mathcal{I}$, $k \in \mathbb{N}$, are finite and couple in finite time with a θ-stationary sequence $\{U_{ij} \circ \theta^k\}$, which satisfies the integrability and rate conditions mentioned above.

As a direct consequence of the first assumption, the sequences $\{A(k)\}$ and $\{B(k)\}$ are both θ-stationary and the entries of these matrices which are not a.s. equal to ε are integrable. In particular, the diagonal entries of $A(k)$ are a.s. nonnegative and integrable owing to the assumptions that the transitions are all recycled.

Remark 7.51 We now list a few motivations for the assumptions on $U_+(k)$, which will become more apparent in §7.4.4.5. In (7.54), we would like to be able to take the input vector $u(k) \in \mathbb{R}^{\mathfrak{m}}$ equal to the output of some other stochastic event graph (incidentally, this is the most practical way of building input vectors which satisfy the conditions of §7.4.4.1). For instance, consider the vector $(u(k), x(k+1)) \in \mathbb{R}^2$, associated with Equation (7.3), as an output signal of the system analyzed in the preliminary example. The first assumption (coupling of $U_+(k)$ with $U_+ \circ \theta^k$) is motivated by the following observations:

- Even in the stable case, the output process of a stochastic event graph is usually not such that $U_+(k)$ is θ-stationary from $k = 0$ on. For instance, if we take the specific vector mentioned above as an input vector of Equation (7.54), with $\mathfrak{m} = 2$, we know from the preliminary example that the corresponding sequences $\{U_+(k)\}$ and $\{U_-(k)\}$ couple with stationary sequences, provided appropriate rate conditions are satisfied.
- More generally, for stochastic event graphs, the assumption that $\{U_+(k)\}$ is θ-stationary may not be consistent with the assumption that the initial condition is compatible.

As for the second assumption (integrability of the diagonal terms only), we also know from the preliminary example that the nondiagonal entries of U_+ or U_- are not integrable in general (see Remark 7.13). ■

7.4.4.3 *Finiteness of the Stationary Solution of the Ratio Equation; Strongly Connected Case*

Matrix A associated with (7.54) is assumed to have a strongly connected precedence graph. Let $\mathfrak{a}$ denote the maximal Lyapunov exponent associated with A (see Corollary 7.31). Let

$$\xi(k; x_0) \stackrel{\text{def}}{=} |A(k) \dots A(2)A(1)\delta(0; x_0)U_-(0)U_-(1) \dots U_-(k-1)|_{\oplus} \ , \qquad (7.69)$$

where x_0 is a finite random initial condition in $\mathbb{R}^{n\times m}$.

Lemma 7.52 *For all finite initial conditions* $x_0 \in \mathbb{R}^{n\times m}$,

$$\lim_{k\to\infty} \left(\xi(k, x_0)\right)^{1/k} = \mathfrak{a} \not/ \mathfrak{u} \quad a.s. \tag{7.70}$$

Proof The property stated in (7.70) is similar to the one proved in Theorem 7.27. Indeed, from (7.25) we obtain that

$$|\xi(k, x_0)|_\oplus \le |A(k)\dots A(2)A(1)|_\oplus \; |\delta(0; x_0)|_\oplus \; |U_-(0)U_-(1)\dots U_-(k-1)|_\oplus \; .$$

We know from Theorem 7.27 that

$$\lim_k |A(k)A(k-1)\dots A(2)A(1)|_\oplus^{1/k} = \mathfrak{a} \quad \text{a.s.}$$

As for the product $U_-(0)U_-(1)\dots U_-(k-1)$, we cannot apply the same theorem because the entries of $U_-(k)$ are not assumed to be integrable anymore, and because the θ-stationarity is replaced by the coupling assumption. However, owing to the specific form of $U_-(k)$, and to the fact that the scalar dioid is $\mathbb{R}_{\max}$, the relation

$$U_-(0)\dots U_-(k-1) = \frac{u(0)}{u(k)}$$

holds, so that

$$\lim_k |U_-(0)\dots U_-(k-1)|_\oplus^{1/k} = e \not/ \mathfrak{u} \quad \text{a.s.} \; ,$$

in view of (7.63). This immediately implies that

$$\limsup_k \left(\xi(k; x_0)\right)^{1/k} \le \mathfrak{a} \not/ \mathfrak{u} \quad \text{a.s.}$$

On the other hand, (7.25) and (7.26) imply that

$$\xi(k; x_0) \ge |(A(k-1)\dots A(0))|_\oplus \; |\delta(0; x_0)|_\wedge \left(|u(0)|_\wedge \not/ |u(k)|_\oplus\right) \; .$$

By using this inequality together with the above a.s. limits, we finally obtain

$$\liminf_k \left(\xi(k; x_0)\right)^{1/k} \ge \mathfrak{a} \not/ \mathfrak{u} \quad \text{a.s.}$$

■

Let

$$\zeta(k) \stackrel{\text{def}}{=} \left|A \dots A \circ \theta^{-k+1} C \circ \theta^{-k-1} U_- \circ \theta^{-k} \dots U_- \circ \theta^{-1}\right|_\oplus \; , \tag{7.71}$$

where C was defined in Definition 7.42. The following lemma is very similar to the preceding one, although it is somewhat more difficult to prove.

Lemma 7.53 *Under the foregoing assumptions*

$$\lim_{k\to\infty} (\zeta(k))^{1/k} = \mathfrak{a} \not{} \mathfrak{u} \quad a.s. \tag{7.72}$$

Proof When using the same arguments as in Theorem 7.27 and Corollary 7.32 (applied to the shift θ^{-1}, which is also ergodic), we obtain that for all i, j,

$$\begin{aligned} \lim_k \left(\left(A \ldots A \circ \theta^{-k+1} \right)_{ij} \right)^{1/k} &= \lim_k \left| A \ldots A \circ \theta^{-k+1} \right|_{\oplus}^{1/k} \\ &= \lim_k \mathbb{E} \left[\left| A \ldots A \circ \theta^{-k+1} \right|_{\oplus}^{1/k} \right] = \mathfrak{a}' \ , \end{aligned}$$

for some positive and finite constant $\mathfrak{a}'$ (the first two limits are understood in the a.s. sense). Since

$$\mathbb{E} \left[\left| A \ldots A \circ \theta^{-k+1} \right|_{\oplus}^{1/k} \right] = \mathbb{E} \left[\left| A \circ \theta^{k-1} \ldots A \right|_{\oplus}^{1/k} \right] \ ,$$

we necessarily have $\mathfrak{a} = \mathfrak{a}'$. Therefore, for all i, j,

$$\lim_k \left(\left(A \ldots A \circ \theta^{-k+1} \right)_{ij} \right)^{1/k} = \mathfrak{a} \quad \text{a.s.} \tag{7.73}$$

We now show that we also have

$$\lim_k \left(\left(U_- \circ \theta^{-k} \ldots U_- \circ \theta^{-1} \right)_{ij} \right)^{1/k} = e \not{} \mathfrak{u} \quad \text{a.s.,} \tag{7.74}$$

for all $i, j = 1, \ldots, \mathfrak{m}$. For this, we will use the specific forms of U_+ and U_-, which imply that there exists a unique pair of vectors $(b(0), b(1))$ in $(\mathbb{R}^{\mathfrak{m}})^2$ such that $b_1(0) = e$ and

$$U_+ = \frac{b(1)}{b(0)} \ , \quad U_- = \frac{b(0)}{b(1)}$$

(see Lemma 7.49). Let $\{b(k)\}_{k\geq 1}$ be the sequence of $\mathbb{R}^{\mathfrak{m}}$-valued vectors defined by the relations

$$b(k) = U_+ \circ \theta^{k-1} b(k-1) = U_+ \circ \theta^{k-1} \ldots U_+ \circ \theta^1 b(1) \ , \quad k \geq 2 \ .$$

We have

$$U_+ \circ \theta^k = \frac{b(k+1)}{b(k)} \ , \qquad U_- \circ \theta^k = \frac{b(k)}{b(k+1)} \ , \qquad k \geq 0 \ . \tag{7.75}$$

This implies

$$\begin{aligned} \left(U_- \circ \theta^{-k} \ldots U_- \circ \theta^{-1} \right)_{ij} &= \left(U_- \ldots U_- \circ \theta^{k-1} \right)_{ij} \circ \theta^{-k} \qquad (7.76) \\ &= b_i(0) \circ \theta^{-k} \not{} b_j(k) \circ \theta^{-k} \ . \qquad (7.77) \end{aligned}$$

Therefore for $i = j$

$$\left(U_-\circ\theta^{-k}\dots U_-\circ\theta^{-1}\right)_{ii} = \bigotimes_{l=-1}^{-k}(e\not/U_{ii})\circ\theta^l \ ,$$

so that the ergodicity of θ^{-1} and the assumption $\mathbb{E}[U_{ii}] = \mathfrak{u}$ show that (7.74) holds at least for $i = j$. In view of (7.76), we have

$$\left(U_-\circ\theta^{-k}\dots U_-\circ\theta^{-1}\right)_{ij} = \left(U_-\circ\theta^{-k}\dots U_-\circ\theta^{-1}\right)_{jj} b_i(0)\circ\theta^{-k}\not/b_j(0)\circ\theta^{-k} \ .$$

We conclude the proof of (7.74) by showing that

$$\lim_k \left(b_i(0)\circ\theta^{-k}\right)^{1/k} = e \quad \text{a.s.} \ , \tag{7.78}$$

for all $i = 1,\dots,\mathfrak{n}$ (this is immediate if the entries of U_- or U_+ are integrable, but we have not assumed this). We know from Lemma 7.49 that

$$b(0)\circ\theta = b(1)\not/(b_1(1)) \ . \tag{7.79}$$

Therefore, for all $i = 1,\dots,\mathfrak{n}$, the relation

$$\frac{b_i(0)\circ\theta}{b_i(0)} = \frac{(b_i(1)\not/b_i(0))}{(b_1(1)\not/b_1(0))} = \frac{(U_+)_{ii}}{(U_+)_{11}}$$

holds, which shows that $b_i(0)\circ\theta\not/b_i(0)$ and hence $b_i(0)\circ\theta^{-1}\not/b_i(0)$ are integrable. This implies that $\mathbb{E}[b_i(0)\circ\theta^{-1}\not/b_i(0)] = e$ (see Lemma 7.25). Since

$$b_i(0)\circ\theta^{-k} = b_i(0)\bigotimes_{h=0}^{-k+1}\left(b_i(0)\circ\theta^{-1}\not/b_i(0)\right)\circ\theta^h \ ,$$

this and Birkhoff's Theorem imply that

$$\begin{aligned}\lim_k \left[b_i(0)\circ\theta^{-k}\right]^{1/k} &= \lim_k \left[b_i(0)\bigotimes_{h=0}^{-k+1}\left(b_i(0)\circ\theta^{-1}\not/b_i(0)\right)\circ\theta^h\right]^{1/k} \\ &= \lim_k \left(\bigotimes_{h=0}^{-k+1}\left(b_i(0)\circ\theta^{-1}\not/b_i(0)\right)\circ\theta^h\right)^{1/k} \\ &= \mathbb{E}\left[b_i(0)\circ\theta^{-1}\not/b_i(0)\right] = e \ . \end{aligned} \tag{7.80}$$

Finally, since $C = B\circ\theta U_+U_- = B\circ\theta b(1)\not/b(1)$, we obtain from (7.79) that

$$C = B\circ\theta\frac{b(1)}{b(1)} = B\circ\theta\frac{b(1)\not/b_1(1)}{b(1)\not/b_1(1)} = \left(B\frac{b(0)}{b(0)}\right)\circ\theta \ . \tag{7.81}$$

Then either the j-th line of B is ε, and $C_{ji}(k) = \varepsilon$ for all i and k, or it is different from ε and (7.81), the integrability assumption on the non ε elements of B, and (7.78) imply that

$$\lim_{k\to\infty} \left(C_{ji}\circ\theta^{-k}\right)^{1/k} = e \quad \text{a.s. }, \tag{7.82}$$

for all i. The proof of the lemma is concluded from (7.73), (7.74) and (7.82). ■

Remark 7.54 When using the notation of the preceding theorem, the initial condition for the backward recurrence should be

$$\Delta(0) = \frac{Bb(0)}{b(0)} \ .$$

If $U_+(k)$ is stationary, then for all x_0, the initial condition δ_0 of the ratio process satisfies the bound

$$\delta_0 = \frac{A(0)x_0 \oplus B(0)u(0)}{u(0)} \geq \frac{B(0)u(0)}{b(0)} = \frac{Bb(0)}{b(0)} = C\circ\theta^{-1} \ .$$

■

Remark 7.55 If A is not strongly connected, the proof of Lemma 7.53 allows one to conclude that

$$\limsup_{k\to\infty} (\zeta(k))^{1/k} \leq \mathfrak{a} \not\!/ \mathfrak{u} \quad \text{a.s. }, \tag{7.83}$$

where $\mathfrak{a}$ is the maximal Lyapunov exponent of A. ■

In the following theorem, which holds regardless of the strong connectedness of A, $\mathfrak{a}$ is the maximal Lyapunov exponent of A.

Theorem 7.56 *If $\mathfrak{a} < \mathfrak{u}$, then $|\Delta|_\oplus < \infty$ a.s., and there exists an initial condition $\delta(0)$ such that the solution $\delta(k)$ of the ratio equation (7.59) forms a stationary and ergodic process.*

Proof From Lemma 7.45, either $|\Delta(k)|_\oplus$ tends to ∞ a.s., or $\Delta(k)$ tends to Δ with $|\Delta|_\oplus < \top$. Assume we are in the first case. Then in view of (7.62),

$$\limsup_{k\to\infty} \zeta(k) = \infty \quad \text{a.s.,}$$

which contradicts (7.83) if $\mathfrak{a} < \mathfrak{u}$. Therefore $\mathfrak{a} < \mathfrak{u}$ implies that $|\Delta|_\oplus < \infty$ a.s., and hence $\Delta < \infty$ a.s. Taking $\delta(0) = \Delta$ makes $\delta(k)$ stationary in view of (7.61). ■

Remark 7.57 If we return to the initial system (7.54), we may ask whether this system has an initial condition $(x(0), u(0))$ which renders the ratio process $\delta(k)$ stationary. The answer to this question is equivalent to the existence of a solution to the equation

$$\frac{Ax(0) \oplus Bu(0)}{u(0)} = \Delta \ ,$$

where the unknowns are $x(0)$, and $u(0)$, and where Δ is the random variable defined in Lemma 7.44. We will not pursue this line of thought since we will see that the stationary regime $\{\Delta \circ \theta^k\}$ is actually reached by coupling regardless of the initial condition $\delta(0)$. ■

Lemma 7.58 *In the strongly connected case, if $\mathfrak{a} > \mathfrak{u}$, the variables $\Delta(k)$ all converge to ∞ a.s.*

Proof In view of (7.62), $\mathfrak{a} > \mathfrak{u}$ implies $|\Delta(k)|_{\oplus}$ tends to ∞ and hence $|\Delta|_{\oplus} = \infty$. From (7.61), we obtain

$$\Delta \circ \theta^n \geq G \circ \theta \Delta H \ ,$$

where

$$G \stackrel{\text{def}}{=} \bigotimes_{l=1}^{n} A \circ \theta^l \ , \quad H \stackrel{\text{def}}{=} \bigotimes_{l=0}^{n-1} U_- \circ \theta^l \ .$$

In view of our assumptions on A, the matrix $G(k)$ defined in (see (7.37)) is such that $G_{ij} \geq e$ for all $i, j = 1, \ldots, n$. This together with $|\Delta|_{\oplus} = \top$ a.s. implies $\Delta_j \circ \theta^n = \top$ a.s. for all $j = 1, \ldots, n$ (since U_- is a.s. finite). Therefore $\mathfrak{a} > \mathfrak{u}$ implies $\Delta_j = \top$ a.s. for all $j = 1, \ldots, n$. ■

Corollary 7.59 *The random variable Δ is the least stationary solution of (7.61).*

Proof Starting with a solution Ξ of (7.61), we necessarily have $\Xi \geq C \circ \theta^{-1} = \Delta(0)$. It is easily checked that $\Xi \geq \Delta(k)$ implies $\Xi \geq \Delta(k+1)$ (the proof is essentially the same as for the preliminary example). Therefore $\Xi \geq \Delta$. ■

Hence if $\mathfrak{a} > \mathfrak{u}$, there is no finite stationary regime for ratios of the type $x_j(k) \not/ u_i(k)$.

Remark 7.60 A few remarks are in order concerning stochastic event graphs. If there exist initial lag times $x(0)$ and $u(0)$ which make the ratio process stationary, these lag times are not compatible in general (see Remark 7.15).

- Nothing general can be said about the critical case $\mathfrak{u} = \mathfrak{a}$. As in queuing theory, it may happen that Δ is finite or infinite, depending on higher order statistics. For instance, if the holding times are deterministic, the variable Δ is finite (see Chapter 3). In the case of i.i.d. exponentially distributed holding times, it will be infinite.

- It is not always true that the variables Δ are integrable. Simple counterexamples can be found in queuing theory. For instance, the stationary waiting times in the $GI/GI/1$ queue (see §7.2) fall in this category of ratios and are only integrable under certain specific conditions on the second moments of the holding (service and interarrival) times.

- Assume that the constant $\mathfrak{a}$ is fixed; what is the minimal value of $\mathfrak{u}$ for which the system can be stabilized? The preceding remarks show that the threshold is for $\mathfrak{u} = \mathfrak{a}$, the cycle time of the strongly connected component. In other words, the minimal asymptotic rate of the input for which the system can be stabilized, is equal to the cycle time of the system with an infinite supply of tokens in each of the inputs. This result can then be seen as a generalization to Petri nets of a result known as Lavenberg's Theorem [83] in queuing theory. ■

7.4.4.4 Coupling; Strongly Connected Case

Theorem 7.61 *Under the statistical assumptions of §7.4.4.1, if $\mathfrak{a} < \mathfrak{u}$, there exists a unique random matrix Δ such that for all finite initial conditions $\delta(0)$, the random variables $\{\delta(k;\delta(0))\}$ couple in finite time with the sequence $\{\Delta\circ\theta^k\}$, regardless of the initial conditions provided they are finite. If $\mathfrak{a} > \mathfrak{u}$, then for all $j = 1,\ldots,\mathfrak{n}$, and $i = 1,\ldots,\mathfrak{m}$, $\delta_{ji}(k;\delta(0))$ converges a.s. to ∞ when k tends to ∞.*

The proof is based on the following lemma.

Lemma 7.62 *Assume that $\mathfrak{a} < \mathfrak{u}$. Then for any finite initial condition $\delta(0)$, there exists a positive integer $K(\delta(0))$ such that for all $k \geq K(\delta(0))$, $\delta(k;\delta(0)) = \delta(k; C\circ\theta^{-1})$.*

Proof The proof is by contradiction. Assume that $\delta(k;\delta(0)) \neq \delta(k;C\circ\theta^{-1})$ for all $k \geq 0$. This means that for any fixed $k \geq 1$, there exists a pair of integers (j_k, i_k), with $1 \leq j_k \leq \mathfrak{n}$ and $1 \leq i_k \leq \mathfrak{m}$, such that $\delta_{j_k i_k}(k;z_1) > \delta_{j_k i_k}(k;z_2)$, where z_1 is either z or $C\circ\theta^{-1}$ and z_2 is the other one. In view of (7.59), we necessarily have $\delta_{j_k i_k}(k;z_1) = \big(A(k)\delta(k-1;z_1)U_-(k-1)\big)_{j_k i_k}$, for if it were not the case, we would have

$$\delta_{j_k i_k}(k;z_1) = \delta_{j_k i_k}(k;z_2) = \big(B(k)U_+(k-1)U_-(k-1)\big)_{j_k i_k}$$

(in $\mathbb{R}_{\max}$, $a \oplus b \neq a$ implies $a \oplus b = b$). This in turn implies the existence of a pair of integers (j_{k-1}, i_{k-1}), with $1 \leq j_{k-1} \leq \mathfrak{n}$ and $1 \leq i_{k-1} \leq \mathfrak{m}$, such that

$$\delta_{j_k i_k}(k;z_1) = A_{j_k j_{k-1}}(k)\delta_{j_{k-1} i_{k-1}}(k-1;z_1)U_-(k-1)_{i_{k-1} i_k} \ .$$

It is easy to see that necessarily $\delta_{j_{k-1} i_{k-1}}(k-1;z_1) > \delta_{j_{k-1} i_{k-1}}(k-1;z_2)$. If this were not true, we would then have

$$\begin{aligned}
\delta_{j_k i_k}(k;z_1) &= A_{j_k j_{k-1}}(k)\delta_{j_{k-1} i_{k-1}}(k-1;z_1)U_-(k-1)_{i_{k-1} i_k} \\
&\leq A_{j_k j_{k-1}}(k)\delta_{j_{k-1} i_{k-1}}(k-1;z_2)U_-(k-1)_{i_{k-1} i_k} \\
&\leq \bigoplus_{p=1}^{\mathfrak{n}} \bigoplus_{q=1}^{\mathfrak{m}} A_{j_k p}(k)\delta_{pq}(k-1;z_2)U_-(k-1)_{q i_k} \\
&\quad \oplus \big(B(k)U_+(k-1)U_-(k-1)\big)_{j_k i_k} \\
&= \delta_{j_k i_k}(k;z_2) \ ,
\end{aligned}$$

which would contradict the definition of j_k and i_k.

More generally, by using the same argument iteratively, we can find a sequence of pairs $\{(j_{k-l}, i_{k-l})\}_{l=1,2,\ldots,k}$ such that for all l in this range,

$$\delta_{j_{k-l+1}i_{k-l+1}}(k-l+1;z_1) = A(k-l+1)_{j_{k-l+1}j_{k-l}}\delta(k-l;z_1)_{j_{k-l}i_{k-l}}U_-(k-l)_{i_{k-l}i_{k-l+1}} .$$

Therefore, there exists a sequence of pairs such that

$$\delta_{j_k i_k}(k;z_1) = \bigotimes_{l=1}^{k} A_{j_{k-l+1}j_{k-l}}(k-l+1)\delta_{j_0 i_0}(0;z_1) \bigotimes_{l=0}^{k-1} U_-(l)_{i_l i_{l+1}} ,$$

which implies that $\delta_{j_k i_k}(k;z_1) \le \xi(k;z_1)$, where $\xi(k;z_1)$ is defined by (7.69). On the other hand, we know from (7.59) that $\delta(k;z_1) \ge B(k)U_+(k-1)U_-(k-1)$; thus for some pair (j_k, i_k) such that the line $B_{j_k \cdot}$ has some nonvanishing entries,

$$\left(\xi(k;z_1)\right)^{1/k} \ge \left[\left(B(k)U_+(k-1)U_-(k-1)\right)_{j_k i_k}\right]^{1/k} . \tag{7.84}$$

Owing to (7.70), $\xi(k;z_1)^{1/k} \to \mathfrak{a} \not/ \mathfrak{u}$ when $k \to \infty$. Similarly, our assumptions on the rates (7.63) and on the integrability of the entries of B imply that

$$\begin{aligned}
&\lim_{k\to\infty} \left(B(k)U_+(k-1)U_-(k-1)\right)_{j_k i_k}^{1/k} \\
&\quad = \lim_{k\to\infty} \bigoplus_l \left(B_{j_k l}(k)\right)^{1/k} \left(\left(u_l(k)\right)^{1/k} \not/ \left(u_{i_k}(k)\right)^{1/k}\right) \\
&\quad = \bigoplus_l \left(\lim_{k\to\infty} \left(B_{j_k l}(k)\right)^{1/k}\right) \left(\lim_{k\to\infty} \left(u_l(k)\right)^{1/k} \not/ \lim_{k\to\infty} \left(u_{i_k}(k)\right)^{1/k}\right) \\
&\quad = e \quad \text{a.s.}
\end{aligned}$$

By letting k go to ∞ in (7.84), we finally obtain $\mathfrak{a} \not/ \mathfrak{u} \ge e$, where the contradiction is obtained, since we have assumed that $\mathfrak{a} \not/ \mathfrak{u} < e$. ■

The proof of Theorem 7.61 is obtained from the last lemma using the same arguments as in the preliminary example (see Lemma 7.21).

Corollary 7.63 *Under the stability condition $\mathfrak{a} < \mathfrak{u}$, Equation (7.61) admits a unique finite solution.*

Proof Given the coupling property of Theorem 7.61, the proof is the same as the one of Corollary 7.22. ■

Remark 7.64 The coupling of the ratios $x(k+1) \not/ u(k)$ with a finite θ-stationary sequence implies the coupling of other ratios like $x(k+1) \not/ x(k)$ with a θ-stationary and finite sequence. In order to see this, write

$$\frac{x(k+1)}{x(k)} = \delta(k)U_-(k)\delta_-(k) ,$$

where $\delta_-(k)$ is the matrix with entries $(\delta_-(k))_{ij} = e \not{/} \delta_{ji}(k)$. The coupling of $\{\delta(k)\}$ with a stationary sequence implies that of $\{\delta_-(k)\}$, which in turn implies that of $\{x(k+1) \not{/} x(k)\}$, provided $\{U_-(k)\}$ couples with a stationary sequence too. ■

Example 7.65 (Blocking after service) Consider a network of $\mathfrak{n}$ machines in series. The first machine has an infinite input buffer and is fed by an external arrival stream of items. There are no intermediate buffers between machine j and machine $j+1$, $1 \leq j \leq \mathfrak{n}-1$, and an item having completed its service in machine j is blocked there as long as machine $j+1$ is not empty.

It is easily checked that this mechanism is adequately described by the timed event graph of Figure 7.3. The input transition q_0 is associated with the input

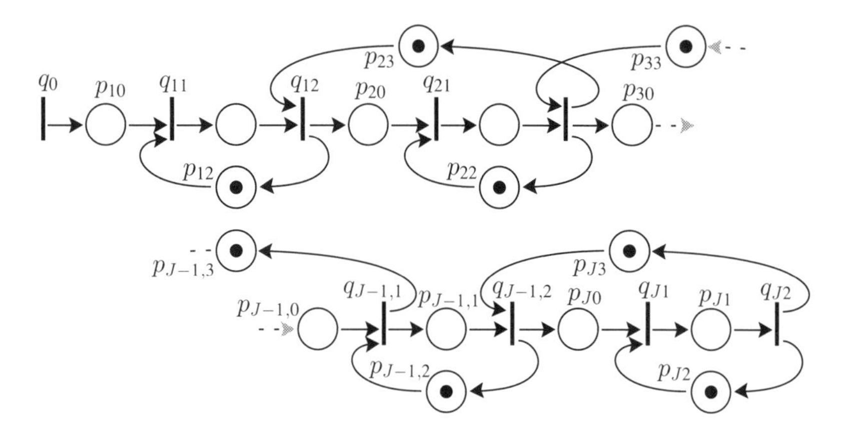

Figure 7.3: Blocking after service

function $u(k)$, with inter-input times $U(k)$; transitions q_{j1} and q_{j2} are associated with the behavior of machine j, with $j = 1, \ldots, \mathfrak{n}$. Place p_{j0}, which precedes q_{j1}, represents the device for transporting items from machine $j-1$ to machine j. The holding times associated with these places (namely the transportation times) will be assumed to be identically equal to e. Transition q_{j1} represents the admittance to machine j, and q_{j2} the exit out of machine j. The holding times in p_{j1} will be denoted $\alpha_j(k)$ and represent the processing time of the k-th item to enter machine j. Finally, the feedback arc which contains $p_{j+1,3}$ forbids one to send an item from machine j to machine $j+1$ if the latter still has an item (this is the blocking mechanism). Similarly, the feedback arc which contains p_{j2} prevents an item from entering machine j if this one contains another item. The two places p_{j2} and p_{j3} are assumed to have zero holding times.

All these variables are defined on the probability space $(\Omega, \mathbb{F}, \mathbb{P}, \theta)$, as well as the compatible initial lag times $v_j \in \mathbb{R}$, where v_j is both the initial lag time of the token in p_{j2}, and the one of the token in p_{j3}.

It is easily checked that the state space can be reduced to the set of transitions $\mathcal{Q}' = \{q_{12}, \ldots, q_{\mathfrak{n}2}\}$, and that the corresponding canonical form of the evolution equation of Corollary 2.82 reads

$$\widetilde{x}(k+1) = \widetilde{A}(k)\widetilde{x}(k) \oplus \widetilde{B}(k)\widetilde{u}(k) ,$$

where the $\mathfrak{n} \times \mathfrak{n}$ matrix $\widetilde{A}(k)$ is given by the relation

$$\widetilde{A}_{ji}(k) = \begin{cases} \bigotimes_{l=i}^{j} \alpha_l(k+1) & \text{for } i = 1, \ldots, j \; ; \\ e & \text{for } i = j+1 \; ; \\ \varepsilon & \text{for } i = j+2, \ldots, \mathfrak{n} \; , \end{cases}$$

whereas the $\mathfrak{n} \times 1$ matrix $\widetilde{B}(k)$ is defined by $\widetilde{B}_{j1}(k) = \bigotimes_{l=1}^{j} \alpha_l(k+1)$, and $\widetilde{u}(k) = u(k+1)$. The precedence graph is strongly connected, and ratios of the form $\widetilde{x}_j(k+1) \not/ \widetilde{u}(k)$ with $j = 1, \ldots, \mathfrak{n}$, admit a unique stationary regime if $\mathfrak{u} = \mathbb{E}[U] > \mathfrak{a}$, where $\mathfrak{a}$ is the cycle time associated with matrices $A(k)$. The ratio $\widetilde{x}_j(k+1) \not/ \widetilde{u}(k)$ here represents the delay between the arrival of an item in the network and the time at which it starts being processed by machine j. If this rate condition is fulfilled, the unique stationary regime is reached with coupling regardless of the initial lag times. ■

7.4.4.5 *More on Coupling; General Case*

The notations and the statistical assumptions of the present section are those of §7.4.4.3. The only difference concerns the precedence graph associated with A, which is not supposed to be strongly connected here. We assume that A has N_A m.s.c.s.'s. As in the preceding subsections, the basic process of interest is the ratio process

$$\eta(k) \stackrel{\text{def}}{=} \frac{X(k+1)}{X(k)} , \tag{7.85}$$

where $\{X(k)\}$ is the sequence defined in (7.66).

For the proof of the next theorem, it will be convenient to use Equation (7.66). By construction, the number of m.s.c.s.'s of D is $N_D = N_A + 1$.

Remark 7.66 The $\delta(k)$ process defined in (7.56) is the ratio of the state variables $x(k+1)$ and the input $u(k)$, whereas the ratio $\eta(k)$ defined above is $\begin{pmatrix} u'(k+1) & x'(k+1) \end{pmatrix}' \not/ \begin{pmatrix} u'(k) & x'(k) \end{pmatrix}'$. The restriction of the latter matrix to the set of coordinates $(>1), (1)$ coincides with $\delta(k)$. ■

Theorem 7.67 *If the Lyapunov exponents $\mathfrak{a}_{(n)}$ of Theorem 7.36 satisfy the condition*

$$\bigoplus_{n=1}^{N_A} \mathfrak{a}_{(n)} < \mathfrak{u} , \tag{7.86}$$

then there exists a unique finite random $(\mathfrak{n}+\mathfrak{m}) \times (\mathfrak{n}+\mathfrak{m})$ matrix E such that the sequence $\{\eta(k)\}$, defined in (7.85), couples in finite time with a uniquely defined *stationary and ergodic process $\{E\circ\theta^k\}$, regardless of the initial condition. If*

$$\bigoplus_{n=1}^{N_A} \mathfrak{a}_{(n)} > \mathfrak{u} \ , \tag{7.87}$$

let n_0 be the first $n = 1, \ldots, N_A$, such that $\mathfrak{a}_{(n)} > \mathfrak{u}$. Then all ratios of the form $\eta_{ji}(k), j \in \mathcal{V}_{n_0}, i \in \mathcal{V}_m, m < n_0$, tend to ∞ a.s., for all finite initial conditions.

Proof We prove by induction on $n = 1, \ldots, N_D$, that under the rate condition (7.86), the matrices

$$\frac{X_{(\leq n)}(k+1)}{X_{(\leq n)}(k)} \ ,$$

couple in finite time with a uniquely defined and finite θ-stationary process, and that the stationary regime is such that

$$\mathbb{E}[\left(X_{(\leq n)}(k+1)\right)_i \not/ \left(X_{(\leq n)}(k)\right)_i] = \mathfrak{u} \ , \quad \forall i \ .$$

This induction assumption is satisfied for the source m.s.c.s. of D, namely for $n = 1$, which corresponds to the set of input variables, because of the assumptions which were made on the input functions. It is also satisfied for all source m.s.c.s.'s of A, namely for all $n = 2, \ldots, N_0 + 1$ (the numbering we adopt here for m.s.c.s.'s is that relative to D), in view of Theorem 7.61 and Remark 7.64.

Assume that the induction assumption is true up to rank $n - 1$, where $n - 1 \geq N_0 + 1$ and that $\mathfrak{a}_{(n)} < \mathfrak{u}$. From (7.66) we obtain

$$X_{(n)}(k+1) = D_{(n),(n)}(k)X_{(n)}(k) \oplus D_{(n),(\leq n-1)}(k)X_{(\leq n-1)}(k) \ ,$$

where the notation should be clear in view of our conventions. From the assumption that the ratios of $X_{(\leq n-1)}(k+1) \not/ X_{(\leq n-1)}(k)$ couple in finite time with θ-stationary sequences, and that the rates of the coordinates of $X_{(\leq n-1)}(k)$ are all $\mathfrak{u}$, we obtain from Theorem 7.61 that the ratios $X_{(n)}(k+1) \not/ X_{(\leq n-1)}(k)$ also couple with a finite and stationary sequence. This immediately implies that the sequence $\left\{X_{(\leq n)}(k+1) \not/ X_{(\leq n)}(k)\right\}$ also couples with a θ-stationary sequence.

For all $i \in \mathcal{V}_n$ and in the stationary regime, $\mathbb{E}[X_i(k+1) \not/ X_i(k)] = \mathfrak{u}$. This follows from the rate property of Theorem 7.36, which implies that

$$\lim_{k\to\infty} \left(X_i(k)\right)^{1/k} = \mathfrak{u} \quad \text{a.s.}, \tag{7.88}$$

for all $i \in \mathcal{V}_n$, and from the following simple argument: the variables $X_i(k+1) \not/ X_i(k)$ are nonnegative (the diagonal elements of D are all assumed to be greater than e), and couple with a θ-stationary sequence $\{\Delta_{ii}\circ\theta^k\}$; therefore, either the random variable Δ_{ii} is integrable, or $\mathbb{E}[\Delta_{ii}] = \infty$. Since

$$X_i(k) = X_i(0) \bigotimes_{l=1}^{k} \left(X_i(l) \not/ X_i(l-1)\right) \ ,$$

if $\mathbb{E}[\Delta_{ii}] = \infty$, Birkhoff's Theorem implies that

$$\begin{aligned}\lim_k \left(X_i(k)\right)^{1/k} &= \lim_k \left(\bigotimes_{l=1}^{k} X_i(l) \not/ X(l-1)_i\right)^{1/k} \\ &= \lim_k \left(\bigotimes_{l=K}^{k} \Delta_{ii} \circ \theta^l\right)^{1/k} = \infty \quad \text{a.s.,}\end{aligned}$$

where K is the coupling time of $X_i(k+1) \not/ X_i(k)$ with its stationary regime. This is in contradiction with (7.88). Therefore Δ_{ii} is integrable and necessarily of mean value $\mathfrak{u}$.

The uniqueness property follows from Corollary 7.63. ■

Remark 7.68 The only general statement which can be made concerning integrability is as follows: ratios of the form $x_i(k+1) \not/ x_i(k)$ are always integrable, whereas ratios of the form $x_j(k+1) \not/ x_i(k)$, where i and j belong to different m.s.c.s.'s, say $i \in \mathcal{V}_m$ and $j \in \mathcal{V}_n$ with $m \neq n$, may be finite and nonintegrable. ■

7.4.5 Multiplicative Ergodic Theorems in $\mathbb{R}_{\max}$

We are now in a position to answer the eigenpair problem associated with Equation (7.54) and stated in (7.68). The notations concerning the m.s.c.s.'s of A and D are those of Theorem 7.67.

Theorem 7.69 *Under the statistical assumptions of §7.4.4.1, if the stability condition $\bigoplus_{n=1}^{N_A} \mathfrak{a}_{(n)} < \mathfrak{u}$ is satisfied, there exist a unique finite random eigenvalue λ and a unique finite random eigenvector $X \in \mathbb{R}^{\mathfrak{m}+\mathfrak{n}}$, with $X_1 = e$, and such that (7.68) holds.*

Proof Owing to (7.67), we know that the restriction $D_{(1),(1)} = U_+$ of D to the first $\mathfrak{m}$ coordinates, satisfies the eigenpair property

$$D_{(1),(1)} u = \lambda u \circ \theta \ ,$$

for a uniquely defined finite eigenpair (λ, u), with $u_1 = e$. Therefore, the first $\mathfrak{m}$ coordinates of X and λ are uniquely defined, and in particular $X_{(1)} = u$.

Let Δ be the $\mathfrak{n} \times \mathfrak{m}$ random matrix of Lemma 7.44 (the finiteness of Δ is obtained from Theorem 7.56). Let y be the $\mathbb{R}^{\mathfrak{n}}$ vector defined by

$$y \circ \theta \stackrel{\text{def}}{=} (\Delta u) \not/ \lambda \ , \tag{7.89}$$

so that $\Delta u = \lambda y \circ \theta$. When multiplying (7.61) by u, we obtain

$$\lambda y \circ \theta = A \Delta \circ \theta^{-1} U_- \circ \theta^{-1} u \oplus B (U_+ U_-) \circ \theta^{-1} u \ . \tag{7.90}$$

In view of the definition of U_+, we have

$$U_+ U_- = \frac{\lambda u \circ \theta}{u} \frac{u}{\lambda u \circ \theta} = \frac{u \circ \theta}{u \circ \theta} \ .$$

Therefore $(U_+U_-)\circ\theta^{-1}u = u$. Similarly,

$$\Delta\circ\theta^{-1}U_-\circ\theta^{-1}u = \Delta\circ\theta^{-1}\frac{u\circ\theta^{-1}}{\lambda\circ\theta^{-1}u}u = y \ .$$

Thus, the right-hand side of (7.90) reads $Ay \oplus Bu$, and we conclude the proof of the existence part by taking $X_{(>1)} = y$.

The uniqueness property follows from the following observations: if X and X' are two different eigenvectors, they can only differ through their last coordinates (i.e. the coordinates corresponding to (>1)), since u and λ are uniquely defined. Then there exist two different vectors y and y' in $\mathbb{R}^n$ such that

$$\lambda y\circ\theta = Ay \oplus Bu \ , \quad \lambda y'\circ\theta = Ay' \oplus Bu \ .$$

By defining

$$\Delta \stackrel{\text{def}}{=} \lambda\frac{y\circ\theta}{u} \ , \quad \Delta' \stackrel{\text{def}}{=} \lambda\frac{y'\circ\theta}{u} \ ,$$

we obtain two different finite matrices which are easily shown to satisfy Equation (7.61). Since this equation has a unique finite solution, we reach a contradiction.

■

We conclude this subsection by the following rewriting of the coupling property of Theorem 7.67, which holds whenever $U_+(k)$ is stationary, and which generalizes Theorem 7.18.

Theorem 7.70 *If the stability condition $\bigoplus_{n=1}^{N_A} \mathfrak{a}_{(n)} < \mathfrak{u}$ is satisfied, and if the matrices $U_+(k)$ are θ-stationary, then for all finite random initial conditions $X(0)$ with $X_1(0) = e$, a finite integer-valued random variable K exists, such that for all $k \geq K$,*

$$X(k+1) = D(k)\ldots D(1)D(0)X(0) = \lambda\circ\theta^k\ldots\lambda\circ\theta\lambda X\circ\theta^{k+1} \ , \tag{7.91}$$

where (X, λ) is the eigenpair of Theorem 7.69.

Proof Owing to the coupling property of Theorem 7.67, we have that $X(k) \not{e} X_1(k)$ couples in finite time with $X\circ\theta^k$. We obtain (7.91) by $\otimes$-multiplying the equality

$$X(k) \not{e} X_1(k) = X\circ\theta^k \ ,$$

which holds for k greater than the coupling time K, by

$$\lambda\circ\theta^{k-1}\ldots\lambda = \frac{X_1(k)}{X_1(k-1)}\cdots\frac{X_1(1)}{X_1(0)} \ ,$$

and by using the assumption that $X_1(0) = e$. ■

7.5 SECOND-ORDER THEOREMS; AUTONOMOUS CASE

The equation of interest in this section is

$$x(k+1) = A(k)x(k) \ , \quad k \geq 0 \ , \tag{7.92}$$

with initial condition x_0, where $x(k)$ and $A(k)$ have their entries in a dioid $\mathcal{D}$.

7.5.1 Ratio Equation

As in the preceding sections, the basic process of interest is the ratio process[1]

$$\delta(k; x_0) = \frac{x(k+1; x_0)}{x(k; x_0)} \ . \tag{7.93}$$

The aim of this subsection is to determine the condition under which this process admits a stationary regime where

$$\delta(k) = \delta \circ \theta^k \ , \quad k \geq 0 \ , \tag{7.94}$$

and to quantify the nature of the convergence of $\delta(k; x_0)$ to this regime.

Lemma 7.71 *The state variables $\delta(k; x_0)$ satisfy the inequalities*

$$\delta(k+1; x_0) \geq \frac{A(k+1)\delta(k)}{A(k)} \ , \quad k \geq 0 \ , \tag{7.95}$$

with the initial condition

$$\delta(0; x_0) = \frac{A(0)x(0)}{x(0)} \ . \tag{7.96}$$

Proof We have

$$\begin{aligned} \delta(k+1) &= \frac{x(k+2)}{x(k+1)} = \frac{A(k+1)x(k+1)}{A(k)x(k)} \\ &= \frac{\left(A(k+1)x(k+1)\right) \oslash x(k)}{A(k)} \\ &\geq \frac{A(k+1)\left(x(k+1) \oslash x(k)\right)}{A(k)} \ , \end{aligned}$$

where we successively used (f.9) and (f.12) in the second and the third relations. ■

Example 7.72 Consider the case of matrices with entries in $\mathbb{R}_{\max}$; matrix $A(k)$ is such that

$$A_{ij}(k) \neq \varepsilon \ , \quad \forall i, j = 1, \ldots, n \ , \quad k \geq 0 \ .$$

[1] We will use the same symbol $\delta(\cdot)$ to represent this ratio process and the one defined in Equation (7.56) in the nonautonomous case; the context should help determining which one is meant; in this section, the optional argument of $\delta(k)$ will be x_0.

In $\mathbb{R}_{\max}$, if A and B are finite $\mathfrak{n} \times \mathfrak{n}$ matrices, the $\mathfrak{n} \times \mathfrak{n}$ matrix $X = A \not\circ B$ is given by the relation

$$X_{ij} = \bigwedge_{k=1}^{\mathfrak{n}} A_{ik} \not\circ B_{jk} \ , \quad \forall i, j = 1, \ldots, \mathfrak{n} \tag{7.97}$$

(see Equation (4.82)). Using this relation and the finiteness assumption on $A(k)$, we directly obtain that for all $k \geq 1$, $i, j = 1, \ldots, \mathfrak{n}$,

$$\begin{aligned}
\delta_{ji}(k+1; x_0) &= x_j(k+2; x_0) \not\circ \left(\bigoplus_{g=1}^{\mathfrak{n}} A_{ig}(k) x_g(k; x_0) \right) \\
&= \bigwedge_{g=1}^{\mathfrak{n}} \left(x_j(k+2; x_0) \not\circ \left(A_{ig}(k) x_g(k; x_0) \right) \right) \\
&= \bigwedge_{g=1}^{\mathfrak{n}} \bigoplus_{h=1}^{\mathfrak{n}} \left(A_{jh}(k+1) \delta_{hg}(k; x_0) \right) \not\circ A_{ig}(k) \\
&= \bigwedge_{g=1}^{\mathfrak{n}} \left(A(k+1) \delta(k; x_0) \right)_{jg} \not\circ A_{ig}(k) \\
&= \left(\frac{A(k+1)\delta(k; x_0)}{A(k)} \right)_{ji} .
\end{aligned}$$

■

The following lemma is proved exactly in the same way as Lemma 7.40.

Lemma 7.73 *The least solution of (7.95) is also the solution of the set of equations*

$$\delta(k+1; x_0) = \frac{A(k+1)\delta(k; x_0)}{A(k)} , \tag{7.98}$$

with the initial condition $\delta(0; x_0) = (A(0)x(0)) \not\circ x(0)$.

7.5.2 Backward Process

The statistical assumptions on $A(k)$ are those of §7.3.1. The dioid $\mathcal{D}$ is assumed to be complete.

Theorem 7.74 *Under the above assumption, Equation (7.98) has a* θ*-stationary subsolution in the sense that there exists a random variable* Δ *satisfying the relation*

$$\Delta \circ \theta \leq \frac{A \circ \theta \Delta}{A} . \tag{7.99}$$

This subsolution is a solution if the right division $x \mapsto x \not\circ b$ *is l.s.c.*

The proof is based on the backward process defined by

$$\Delta(0) = A \ , \qquad \Delta(k+1)\circ\theta = \frac{A\circ\theta\Delta(k)}{A} \ , \quad k \geq 0 \ . \tag{7.100}$$

Lemma 7.75 *The sequence $\{\Delta(k)\}$ is nondecreasing.*

Proof By using (f.6), we obtain

$$\Delta(1)\circ\theta = \frac{A\circ\theta A}{A} \geq A\circ\theta \ ,$$

so that $\Delta(1) \geq \Delta(0)$. Assume $\Delta(k) \geq \Delta(k-1)$, for $k \geq 1$. Then

$$\begin{aligned}\Delta(k+1)\circ\theta &= \frac{A\circ\theta\Delta(k)}{A} \\ &\geq \frac{A\circ\theta\Delta(k-1)}{A} \\ &= \Delta(k)\circ\theta \ ,\end{aligned}$$

where we used the isotony of the mapping $x \mapsto (ax)\not\phi b$. ■

Proof of Theorem 7.74 From the preceding lemma, we obtain that the sum

$$\Delta \stackrel{\text{def}}{=} \bigoplus_{k\geq 0} \Delta(k) \tag{7.101}$$

exists since $\mathcal{D}$ is assumed to be complete. By summing up the equations in (7.100) and by using the nondecreasingness of $\Delta(k)$ and the fact that the mapping $x \mapsto (ax)\not\phi b$ is isotone, we directly obtain

$$\Delta = \bigoplus_{k\geq 1} \Delta(k)\circ\theta = \bigoplus_{k\geq 0} \frac{A\circ\theta\Delta(k)}{A} \leq \frac{A\circ\theta\Delta}{A} \ .$$

If the mapping $x \mapsto (ax)\not\phi b$ is l.s.c., there is equality in the last relation. A sufficient condition for this is that the right division operator $x \mapsto x\not\phi b$ be l.s.c. ■

The subsolution Δ satisfies the following extremal property.

Lemma 7.76 *The random variable Δ is less than or equal to any solution of (7.99) greater than or equal to A.*

Proof Let Ξ be an arbitrary solution of (7.99) such that $\Xi \geq A = \Delta(0)$. If $\Xi \geq \Delta(k)$, we obtain

$$\Xi\circ\theta = \frac{A\circ\theta\Xi}{A} \geq \frac{A\circ\theta\Delta(k)}{A} \geq \Delta(k+1)\circ\theta \ .$$

Thus, $\Xi \geq \Delta$. ■

As in the nonautonomous case, the question of finiteness of the minimal (sub)solution Δ will only be addressed in specific dioids. However, we have the following general bound.

Lemma 7.77 *For all $k \geq 1$,*

$$\Delta(k)\circ\theta \leq \frac{A\circ\theta A \ldots A\circ\theta^{-k+1}}{A \ldots A\circ\theta^{-k+1}} \ .$$

Proof The bound clearly holds for $k = 1$, in view of the definition. Assume it is true for $k \geq 1$. Then

$$\begin{aligned}
\Delta(k+1)\circ\theta &= \frac{A\circ\theta\Delta(k)}{A} \\
&\leq \frac{A\circ\theta\left(\left(AA\circ\theta^{-1}\ldots A\circ\theta^{-k}\right) \circ\!\!\!/ \left(A\circ\theta^{-1}\ldots A\circ\theta^{-k}\right)\right)}{A} \\
&\leq \frac{\left(A\circ\theta A\circ\theta^{-1}\ldots A\circ\theta^{-k}\right) \circ\!\!\!/ \left(A\circ\theta^{-1}\ldots A\circ\theta^{-k}\right)}{A} \\
&= \frac{A\circ\theta A\circ\theta^{-1}\ldots A\circ\theta^{-k}}{AA\circ\theta^{-1}\ldots A\circ\theta^{-k}} \ ,
\end{aligned}$$

where we used (f.12) and (f.9) to obtain the last two relations. ■

7.5.3 From Stationary Ratios to Random Eigenpairs

In this subsection we make use of the following inequalities which hold in any dioid $\mathcal{D}$.

Lemma 7.78 *For all $x \in \mathcal{D}$,*

$$x \leq \frac{e}{x \circ\!\!\!\backslash e} \quad \text{and} \quad x \leq \frac{e}{e \circ\!\!\!/ x} \ . \tag{7.102}$$

Proof When taking $x = e$ and $a = x$ in Formula (f.5) of Table 4.1, we obtain $x(x \circ\!\!\!\backslash e) \leq e$, which immediately implies the first relation in view of the definition of the residuation of right multiplication. The second relation is obtained in the same way. ■

The results of this subsection are concerned with Equation (7.92) in a general dioid $\mathcal{D}$, where $A(k) = A\circ\theta^k$.

Theorem 7.79 *Assume that the ratio equation associated with (7.92) admits a stationary subsolution Δ in $\mathcal{D}$ such that*

$$\Delta\circ\theta \leq \frac{A\circ\theta\Delta}{A} \ , \tag{7.103}$$

and such that

$$\Delta = \frac{Ax}{x} \ , \tag{7.104}$$

for some x in $\mathcal{D}$. Then there exists a right super-eigenpair (λ, X) such that

$$AX \geq X\circ\theta\lambda \ . \tag{7.105}$$

Proof Let $y = Ax$. From (7.103), (f.12) and (f.9), we obtain

$$\Delta\circ\theta \leq \frac{A\circ\theta(y \circ\!\!\!/\, x)}{A} \leq \frac{(A\circ\theta y)\circ\!\!\!/\, x}{A} = \frac{A\circ\theta y}{Ax} = \frac{A\circ\theta y}{y} \ .$$

Therefore $\Delta\circ\theta y \leq A\circ\theta y$, that is, $(y\circ\theta) \circ\!\!\!/\, (x\circ\theta) y \leq A\circ\theta y$. By using (7.102), we obtain $x\circ\theta \leq (e \circ\!\!\!/\, x\circ\theta) \circ\!\!\!\backslash\, e$, so that

$$A\circ\theta y \geq \frac{y\circ\theta}{(e \circ\!\!\!/\, x\circ\theta) \circ\!\!\!\backslash\, e} y \geq \frac{y\circ\theta}{e} \frac{e}{x\circ\theta} y \ ,$$

where we used (f.11) in order to obtain the last relation. Since $x \circ\!\!\!/\, e = x$, we finally obtain

$$A\circ\theta y \geq y\circ\theta \left(\frac{e}{x\circ\theta}\right) y \ .$$

When taking

$$X = (Ax)\circ\theta^{-1} \tag{7.106}$$

and

$$\lambda = \frac{e}{x} X \ , \tag{7.107}$$

we directly obtain (7.105). ■

Remark 7.80 The terminology which is used in the preceding theorem comes from the case when the dioid of interest is a matrix dioid associated with some scalar dioid $\mathcal{D}$. In this case, it is easily checked from (4.82) that if A is $n \times n$ and $x(k)$ $n \times 1$, then λ, defined in (7.107), is a scalar and X, defined in (7.106), is an $n \times 1$ matrix. ■

7.5.4 Finiteness and Coupling in $\mathbb{R}_{\max}$; Positive Case

In this subsection, $\mathcal{D} = \mathbb{R}_{\max}$. By using (7.97), it is easily checked that the right-division operator of matrices with finite entries in $\mathbb{R}_{\max}$ is l.s.c., so that the subsolution Δ of Theorem 7.74 is a solution. The statistical assumptions are those of the previous section. In addition, we assume that A is *positive* in the sense that $A_{ij}(k) \geq e$ for all $i, j = 1 \ldots, n$. More general conditions will be considered in the next subsection.

7.5.4.1 *Finiteness of the Minimal Stationary Solution*

We start with a few preliminary lemmas.

Lemma 7.81 *For all initial conditions x_0, and all $k \geq 0$, $\delta(k; x_0)$ satisfies the bounds*

$$A(k) \leq \delta(k; x_0) \ , \tag{7.108}$$

and for all $k \geq 1$,

$$|\delta(k; x_0)|_\oplus \leq |A(k)|_\oplus \ |A(k-1)|_\oplus \ . \tag{7.109}$$

Proof Assume that for some $k \geq 0$, $\delta(k;x_0) \geq A(k)$; this is true for $k = 0$ in view of (7.96) and (f.6) which imply that

$$\delta(0) = \frac{A(0)x_0}{x_0} \geq A(0) \ .$$

Then

$$\delta(k+1;x_0) = \frac{A(k+1)\delta(k;x_0)}{A(k)} \geq \frac{A(k+1)A(k)}{A(k)} \geq A(k+1) \ ,$$

where we successively used the isotony of the mapping $x \mapsto (ax) \circ\!\!\!/ b$ and (f.6). This completes the proof of the lower bound.

As for the upper bound, let $\delta_-(k;x_0)$ be the matrix

$$\delta_-(k;x_0) = \frac{x(k)}{x(k+1)} \ , \quad k \geq 0 \ .$$

For all $k \geq 1$, we have

$$\begin{aligned} \delta_-(k;x_0) &= \frac{x(k)}{A(k)A(k-1)x(k-1)} = \frac{x(k) \circ\!\!\!/ x(k-1)}{A(k)A(k-1)} \\ &= \frac{\delta(k-1;x_0)}{A(k)A(k-1)} \geq \frac{A(k-1)}{A(k)A(k-1)} \ , \end{aligned}$$

where we successively used (f.9) and the lower bound on $\delta(k)$. Since we are in $\mathbb{R}_{\max}$, this reads (see Equation (4.82))

$$x_i(k) \circ\!\!\!/ x_j(k+1) \geq \bigwedge_{l=1}^{n} A_{il}(k-1) \circ\!\!\!/ \left(A(k)A(k-1)\right)_{jl} \ ,$$

which can be rewritten as

$$x_j(k+1) \circ\!\!\!/ x_i(k) \leq \bigoplus_{l=1}^{n} \left(A(k)A(k-1)\right)_{jl} \circ\!\!\!/ A_{il}(k-1) \ .$$

Since $A_{il}(k-1) \geq e$ for all i,l, we finally obtain

$$\begin{aligned} |\delta(k)|_\oplus &\leq \bigoplus_{j,l=1}^{n} \left(A(k)A(k-1)\right)_{jl} \\ &= |A(k)A(k-1)|_\oplus \leq |A(k)|_\oplus \, |A(k-1)|_\oplus \ , \end{aligned}$$

which concludes the proof of the upper bound. ∎

Lemma 7.82 *For all $k \geq 0$, the random variable $\Delta(k)$ satisfies the bounds*

$$A \leq \Delta(k) \ , \tag{7.110}$$

and

$$|\Delta(k)|_\oplus \leq |A|_\oplus \left|A \circ \theta^{-1}\right|_\oplus \ . \tag{7.111}$$

Proof The fact that $\Delta(k) \geq A$ is clear since $\Delta(k)$ is nondecreasing and $\Delta(0) = A$. In order to prove the upper bound in (7.111), we first establish the property that for all x_0,

$$\Delta(k) \leq \delta(k; x_0)\circ\theta^{-k} \ , \quad k \geq 0 \ . \tag{7.112}$$

The proof is by induction; the property holds for $k = 0$, in view of (7.108) considered for $k = 0$. By assuming it holds for some $k \geq 0$, we then obtain

$$\begin{aligned} \Delta(k+1)\circ\theta &= \frac{A\circ\theta\Delta(k)}{A} \leq \frac{A\circ\theta\delta(k; x_0)\circ\theta^{-k}}{A} \\ &= \frac{A\circ\theta^{k+1}\delta(k; x_0)}{A\circ\theta^k} \circ\theta^{-k} = \delta(k+1; x_0)\circ\theta^{-k} \ , \end{aligned}$$

which concludes the proof of (7.112). From (7.112) and (7.109), we immediately obtain that

$$|\Delta(k)|_\oplus \leq |\delta(k; x_0)|_\oplus \circ\theta^{-k} \leq |A|_\oplus \left|A\circ\theta^{-1}\right|_\oplus \ , \quad \forall k \geq 0 \ .$$

■

By putting together the results obtained in this section, we see that the nondecreasing and bounded sequence $\{\Delta(k)\}$ necessarily converges to a finite and integrable limit.

Theorem 7.83 *In $\mathbb{R}_{\max}$, if A is integrable and such that $A_{ij} \geq e$ for all i, j, the equation*

$$\Delta\circ\theta = \frac{A\circ\theta\Delta}{A} \tag{7.113}$$

always admit the limiting value Δ of the sequence $\{\Delta(k)\}$ as a finite and integrable solution. Any other solution of (7.113) is bounded from below by Δ.

However, the situation is slightly different from the one encountered in the nonautonomous case: in particular, the equality in law $\Delta(k) = \delta(k)\circ\theta^{-k}$ has no reason to hold here (because it is not true in general that we can take $\delta(0) = \Delta(0)$). Therefore, nothing ensures a priori that $\delta(k)$ converges weakly to Δ as k goes to ∞; the only thing which we know from Lemma 7.76 is that any stationary regime of the ratio process is bounded from below by Δ. The conditions under which this minimal solution can be a weak limit for the ratio process are the main focus of the following subsections.

7.5.4.2 Reachability

Definition 7.84 (Direct reachability) *A stationary solution Δ of (7.113) is directly reachable if there exists an initial condition x_0 for which the ratio process defined in (7.93) coincides with the stationary process $\{\Delta\circ\theta^k\}$, in the sense that*

$$\delta(k; x_0) = \Delta\circ\theta^k \ , \quad k \geq 0 \ .$$

Lemma 7.85 *The stationary solution* Δ *of (7.113) is reachable if and only if the system of equations*

$$\Delta = \frac{Ax}{x} \tag{7.114}$$

has a finite solution $x \in \mathbb{R}^n$ *with* $x_1 = e$. *If such a solution exists, it is unique.*

Proof If such a solution exists, the ratio process (7.98) can then be made stationary by adopting $x_0 = x$ as the initial condition (see (7.96)). It is clear that there is no loss of generality in assuming that $x_1 = e$. We now prove that (7.114) has at most one finite solution with $x_1 = e$. Indeed, for all n-dimensional column vectors a and b with finite entries in $\mathbb{R}_{\max}$, the relation

$$\frac{a \oslash b}{a \oslash b} = \frac{b}{b}$$

holds (use (4.82) repeatedly). Therefore,

$$x = \left(\frac{x}{x}\right)_{\cdot 1} = \left(\frac{Ax \oslash x}{Ax \oslash x}\right)_{\cdot 1} = \left(\frac{\Delta}{\Delta}\right)_{\cdot 1} ,$$

so that x is uniquely defined from Δ. ■

Remark 7.86 Using the construction in the preceding proof, it is easy to check that the set

$$\left\{\omega \;\middle|\; \exists x(\omega) \in \mathbb{R}^n : x_1(\omega) = 0, \Delta(\omega) = \frac{A(\omega)x(\omega)}{x(\omega)}\right\}$$

can be rewritten as

$$\left\{\omega \;\middle|\; \Delta(\omega) = \frac{A(\omega)\left(\Delta(\omega) \circ\!\!\!\!\setminus \Delta(\omega)\right)_{\cdot 1}}{\left(\Delta(\omega) \circ\!\!\!\!\setminus \Delta(\omega)\right)_{\cdot 1}}\right\} , \tag{7.115}$$

where its measurability becomes more apparent. ■

Similarly, a stationary solution Δ of (7.113) is said to be *reachable by coupling* if there exists an initial condition x_0 and a finite random variable K such that $\delta(k; x_0) = \Delta \circ \theta^k$, $k \geq K$. The aim of the following subsections is to give sufficient conditions under which the minimal stationary solution of Theorem 7.74 satisfies the above reachability properties.

7.5.4.3 Conditions for Reachability

For $k \geq 0$, let $\mathcal{A}(k)$ denote the event

$$\mathcal{A}(k) \stackrel{\text{def}}{=} \left\{\omega \;\middle|\; \exists x \in \mathbb{R}^n : \Delta(k) = \frac{Ax}{x}\right\} , \tag{7.116}$$

where $\Delta(k)$ is the nondecreasing backward process defined in (7.100), and let $\mathcal{A}$ be the event

$$\mathcal{A} \stackrel{\text{def}}{=} \left\{ \omega \;\middle|\; \exists x \in \mathbb{R}^n : \Delta = \frac{Ax}{x} \right\} , \tag{7.117}$$

where Δ is the a.s. limit of $\Delta(k)$. The following notation will be used in what follows: for all events $\mathcal{B}$, $\mathcal{B}\circ\theta$ will denote the set $\mathcal{B}\circ\theta = \{\omega \in \Omega \mid 1_{\mathcal{B}}\circ\theta(\omega) = 1\}$.

Lemma 7.87 *For all $k \geq 0$, $\mathcal{A}(k)$ is included in $\mathcal{A}(k+1)\circ\theta$ and $\mathcal{A}$ is included in $\mathcal{A}\circ\theta$.*

Proof For $\omega \in \mathcal{A}(k)$, there exists a finite random vector x such that $\Delta(k) = Ax \circ\!\!\!/\, x$. Hence,

$$\begin{aligned} \Delta(k+1)\circ\theta &= \frac{A\circ\theta\Delta(k)}{A} = \frac{A\circ\theta\left(Ax \circ\!\!\!/\, x\right)}{A} \\ &= \frac{(A\circ\theta Ax) \circ\!\!\!/\, x}{A} = \frac{A\circ\theta Ax}{Ax} \qquad \text{(from (f.12) and (f.9))} \\ &= \frac{A\circ\theta\left(\Delta(k)\right)_{.1}}{\left(\Delta(k)\right)_{.1}} , \end{aligned}$$

so that $\theta(\omega)$ belongs to $\mathcal{A}(k+1)$ (take $x = \left(\Delta(k)\right)_{.1}\circ\theta^{-1}$). The proof of the second inclusion is similar. ∎

Lemma 7.88 *If $\mathbb{P}[\mathcal{A}(k)] > 0$ for some k, then $\limsup_{k\to\infty} \mathcal{A}(k) = \Omega$ a.s.*

Proof If $\mathbb{P}[\mathcal{A}(k)] > 0$ for some $k \geq 0$, the ergodic assumption implies that $\lim_{h\to\infty} \left(\sum_{m=1}^{h} 1_{\{\mathcal{A}(k)\}}\circ\theta^{-m}\right)/h = \mathbb{P}[\mathcal{A}(k)] > 0$ a.s., so that necessarily $\limsup_{m\to\infty} \mathcal{A}(k)\circ\theta^{-m} = \Omega$ a.s. From Lemma 7.87, we obtain by an immediate induction that for all $m \geq 0$, $\mathcal{A}(k+m) \supseteq \mathcal{A}(k)\circ\theta^{-m}$. Thus, the last relation implies that $\limsup_{k\to\infty} \mathcal{A}(k) = \Omega$ a.s. ∎

Lemma 7.89 *If $\mathbb{P}[\mathcal{A}(k)] > 0$ for some k, then $\mathbb{P}[\mathcal{A}] = 1$.*

Proof Let $\mathcal{H}$ be the subset of $\mathbb{R}^{n\times n}$ defined by

$$\mathcal{H} = \left\{ \Delta \in \mathbb{R}^{n\times n} \;\middle|\; \exists x \in \mathbb{R}^n : \Delta = \frac{Ax}{x} \right\} .$$

When using (7.115), we obtain the equivalent representation

$$\mathcal{H} = \left\{ \Delta \in \mathbb{R}^{n\times n} \;\middle|\; \Delta = \frac{A\left(\Delta \circ\!\!\!\backslash \Delta\right)_{.1}}{\left(\Delta \circ\!\!\!\backslash \Delta\right)_{.1}} \right\} ,$$

from which it is immediate that $\mathcal{H}$ is a closed subset of $\mathbb{R}^{n\times n}$. Owing to Lemma 7.88, if $P[\mathcal{A}(k)] > 0$ for some k, then for almost all $\omega \in \Omega$ there exists a sequence of integers $k_n \uparrow \infty$ such that $\omega \in \mathcal{A}(k_n)$ or equivalently such that $\Delta(k_n) \in \mathcal{H}$ for all $n \geq 1$. Since $\mathcal{H}$ is closed, the a.s. limit Δ of $\Delta(k_n)$ when n goes to ∞ is also in $\mathcal{H}$, so that $\omega \in \mathcal{A}$. ∎

Let $h^\star$ be a fixed integer such that $1 \leq h^\star \leq \mathfrak{n}$, and let $\mathcal{B}$ be the event

$$\mathcal{B} = \{A{\circ}\theta_{\cdot h^\star} A_{h^\star\cdot} = A{\circ}\theta A\} \ , \tag{7.118}$$

or equivalently,

$$\mathcal{B} = \left\{ A_{jh^\star}{\circ}\theta A_{h^\star i} = \bigotimes_{h=1}^{\mathfrak{n}} A_{jh}{\circ}\theta A_{hi} \ , \quad \forall i,j = 1,\ldots,\mathfrak{n} \right\} \ . \tag{7.119}$$

Theorem 7.90 *If there exists $h^\star$ such that $\mathbb{P}[\mathcal{B}] > 0$, the stationary regime defined by Δ is directly reachable.*

Proof If a is an $\mathfrak{n}$-dimensional row (respectively column) vector with finite entries in $\mathbb{R}_{\max}$, then we can use the group structure of $\otimes$ to write

$$a = \frac{e}{a \backslash e} \quad \left(\text{respectively } \ a = \frac{e}{e / a}\right) \ ,$$

where e is 1×1. If c is a scalar, we have $c = e/(e/c)$. On $\mathcal{B}$, we therefore have

$$\begin{aligned} \Delta(1){\circ}\theta &= \frac{A{\circ}\theta A}{A} = \frac{A{\circ}\theta_{\cdot h^\star} A_{h^\star\cdot}}{A} = \frac{A{\circ}\theta_{\cdot h^\star}\left(e/\left(A_{h^\star\cdot} \backslash e\right)\right)}{A} \\ &= \frac{A{\circ}\theta_{\cdot h^\star} / \left(A_{h^\star\cdot} \backslash e\right)}{A} = \frac{A{\circ}\theta_{\cdot h^\star}}{A\left(A_{h^\star\cdot} \backslash e\right)} \ . \end{aligned}$$

In these equalities, we used (f.12) in order to obtain the fourth equality, and (f.9) in order to obtain the last one. Both are equalities because $\mathcal{D} = \mathbb{R}_{\max}$ and the entries of the matrices which are dealt with are finite.

Let $Z = A\left(A_{h^\star\cdot} \backslash e\right)$. On $\mathcal{B}$, we have

$$A{\circ}\theta Z = A{\circ}\theta A \frac{e}{A_{h^\star\cdot}} = A{\circ}\theta_{\cdot h^\star} A_{h^\star\cdot} \frac{e}{A_{h^\star\cdot}} = A{\circ}\theta_{\cdot h^\star} \ .$$

Therefore, on $\mathcal{B}$,

$$\Delta(1){\circ}\theta = \frac{A{\circ}\theta_{\cdot h^\star}}{Z} = \frac{A{\circ}\theta Z}{Z} \ ,$$

so that $\mathcal{B} \subset \mathcal{A}(1){\circ}\theta$. The proof is immediately concluded from Lemma 7.89. ∎

Remark 7.91 Consider the particular case of an autonomous event graph for which the random variables $\alpha_j(k)$ are mutually independent. It is easily checked that a sufficient condition for $\mathbb{P}[\mathcal{B}] > 0$ is that there exists one transition in $\mathcal{Q}'$ such that all the places which follow it have holding times with an infinite support. As we will see in Example 7.92, weaker conditions, like for instance having one transition followed by at least one place with infinite holding times, may be enough to ensure this property.

As in the nonautonomous case, there is no reason for the initial condition, the existence of which is proved in Theorem 7.90, to be compatible. ■

Example 7.92 (Manufacturing blocking) Consider a closed cyclic network of n machines. There are no intermediate buffers between machines. The migration of items is controlled by manufacturing blocking, as defined in Example 7.65: when an item is finished in machine j, $0 \leq j \leq n-1$, it enters machine $s(j) = (j+1) \bmod n$ if $s(j)$ is empty. Otherwise it is blocked in j until $s(j)$ is empty. In this example, all machine indices are understood to be modulo n.

A network of this type, with $n = 3$, is described by the timed event graph of Figure 7.4. The interpretation of the various types of places and transitions is

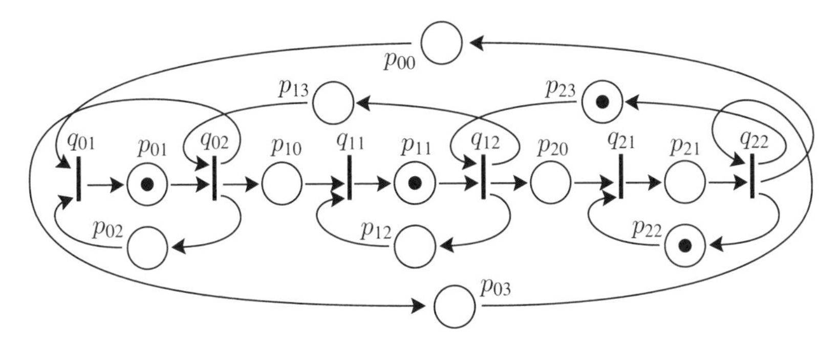

Figure 7.4: Closed manufacturing blocking

the same as in Example 7.65. The only nonzero holding times are those in p_{j1}, $j = 0, \ldots, n$, which will be denoted $\alpha_j(k)$, representing the service time of the k-th item in machine j. Let μ_j denote the initial number of items in p_{j1}. It is assumed that $\mu_j = 0$ or 1, that

$$0 < \sum_{j=0}^{n-1} \mu_j < n ,$$

and that if $\mu_j = 0$, then there is an initial token in place p_{j2} and in place p_{j3}.

The state variables are the firing times of the transitions followed by at least one place with a nonzero initial marking, namely

$$x_j(k) = \begin{cases} x_{j1} & \text{if } \mu_j = 1 ; \\ x_{j2} & \text{if } \mu_j = 0 . \end{cases}$$

The initial lag times are assumed to be compatible, and are given under the form of the vector $z \in \left(\mathbb{R}^+\right)^n$, where z_j represents the lag time of the initial item in machine j if $\mu_j = 1$, and the lag times of the two tokens in p_{j2} and p_{j3} otherwise. In the first case, z_j can be seen as the epoch when machine j starts working. In the second one, z_j is the time when some past workload of machine j has been completed.

Let $r(j)$ be the number of machines to the right of j such that $\mu_j = 1$, plus 1 and let $l(j)$ be the number of machines to the left of j such that $\mu_j = 0$, plus 1. It is easily checked that the $\mathfrak{n} \times \mathfrak{n}$ matrix $A(k)$ is given by the relation

$$A_{ji}(k) = \begin{cases} \bigotimes_{h=i}^{j-\mu_j} \alpha_h(k+1) & \text{if } i \in \{j - l(j), \dots, j-1\} \ ; \\ \alpha_i(k+1) & \text{if } i \in \{j, \dots, j + r(j) - 1\} \text{ and} \\ & i \notin \{j - l(j), \dots, j-1\} \ ; \\ e & \text{if } i = j + r(j) \text{ and} \\ & i \notin \{j - l(j), \dots, j-1\} \ ; \\ \varepsilon & \text{otherwise,} \end{cases}$$

that the precedence graph is strongly connected, and that under the condition of Theorem 7.90, the ratios of the form $x_j(k+1) \not/ x_i(k)$, $i, j = 0, \dots, \mathfrak{n} - 1$, admit a stationary regime which is reached with coupling regardless of the initial condition.

For the example of Figure 7.4, $A(k)$ reads

$$A(k) = \begin{pmatrix} \alpha_0(k+1) & \alpha_1(k+1)\alpha_2(k+1) & \alpha_2(k+1) \\ \alpha_0(k+1) & \alpha_1(k+1) & e \\ \alpha_0(k+1) & \alpha_1(k+1)\alpha_2(k+1) & \alpha_2(k+1) \end{pmatrix} . \tag{7.120}$$

Matrix A being positive, we can directly apply Theorems 7.90 and 7.93. The condition $\mathbb{P}[\mathcal{B}] > 0$ is satisfied if the random variables $\alpha_j(k)$ are mutually independent and if one of the distribution functions S_0 and S_1 has an infinite support, where $S_j(t) = \mathbb{P}[\alpha_j \leq t]$, $t \in \mathbb{R}^+$. ■

7.5.4.4 *Coupling*

The main result of this subsection is the following theorem.

Theorem 7.93 *If there exists $h^\star$ such that $\mathbb{P}[\mathcal{B}] > 0$, the stationary sequence $\{\Delta \circ \theta^k\}$ is the unique stationary solution of (7.98). For any initial condition x_0, the sequence $\{\delta(k; x_0)\}$ couples with this stationary sequence.*

Proof Let $C(k)$, $k \geq 1$, be the event

$$C(k) = \{A(k+1)x(k+1) = A_{\cdot h^\star}(k+1)x_{h^\star}(k+1)\} \ .$$

We first prove that on the event $C(k)$, the relation

$$\delta(k+1) = \frac{A_{\cdot h^\star}(k+1)}{A(k)\left(\delta_{h^\star \cdot}(k) \backslash e\right)} \tag{7.121}$$

holds. On $C(k)$, we have indeed

$$\delta(k+1) = \frac{A(k+1)x(k+1)}{A(k)x(k)} = \frac{A_{\cdot h^\star}(k+1)x_{h^\star}(k+1)}{A(k)x(k)}$$

$$
\begin{aligned}
&= \frac{A_{\cdot h^*}(k+1)\left(e \oslash \left(e \oslash x_{h^*}(k+1)\right)\right)}{A(k)x(k)} = \frac{A_{\cdot h^*}(k+1)}{A(k)x(k)\left(e \oslash x_{h^*}(k+1)\right)} \\
&= \frac{A_{\cdot h^*}(k+1)}{A(k)\left(x(k) \oslash x_{h^*}(k+1)\right)} = \frac{A_{\cdot h^*}(k+1)}{A(k)\left(\delta_{h^*\cdot}(k) \obackslash e\right)} \ .
\end{aligned}
$$

Therefore, on $\mathcal{C}(k)$,

$$
\begin{aligned}
\delta(k+2) &= \frac{A(k+2)x(k+2)}{x(k+2)} = A(k+2)\frac{x(k+2)}{x(k+2)} \\
&= A(k+2)\frac{\delta_{\cdot 1}(k+1)}{\delta_{\cdot 1}(k+1)} = A(k+2)\frac{A_{\cdot h^*}(k+1)}{A_{\cdot h^*}(k+1)} \ ,
\end{aligned}
$$

where the last relation follows from (7.121). This last formula shows that on the event $\mathcal{C}(k)$, regardless of x_0,

$$
\delta(k+2; x_0) = \phi(A(k+1), A(k+2)) \ , \tag{7.122}
$$

where ϕ is a measurable function which we will not need in explicit form.

If we can show that for all k, $\mathcal{D}\circ\theta^k \subset \mathcal{C}(k)$, where $\mathcal{D}$ is an event of positive probability, then Equation (7.122) implies that $\delta(k)\circ\theta^{-k}$ couples with a uniquely defined finite stationary sequence. This result is a direct consequence of Borovkov's renovating events theorem (Theorem 7.107 shows that $\mathcal{C}(k)$ is a renovating event of length 2). The second step of the proof consists in showing that $\mathcal{B}\circ\theta^k \subset \mathcal{C}(k)$. In order to do so, we first prove that

$$
\mathcal{B} \subset \left\{ A\frac{e}{A_{h^*\cdot}} \le \frac{A\circ\theta_{\cdot h^*}}{A\circ\theta} \right\} \ . \tag{7.123}
$$

The property $A\circ\theta A = A\circ\theta_{\cdot h^*}A_{h^*\cdot}$ implies that $A\circ\theta A\left(A_{h^*\cdot} \obackslash e\right) = A\circ\theta_{\cdot h^*}$, which in turn implies that $A\left(A_{h^*\cdot} \obackslash e\right) \le \left(A\circ\theta \obackslash A\circ\theta_{\cdot h^*}\right)$, where we used the very definition of left division in order to obtain the last implication. This immediately implies (7.123).

We are now in a position to conclude the proof by showing that

$$
\mathcal{B}\circ\theta^k \subset \mathcal{C}(k) \ . \tag{7.124}
$$

Inequality (7.109) implies

$$
\frac{x(k+1)}{x(k+1)} = \frac{x(k+1)}{A(k)x(k)} = \frac{\delta(k)}{A(k)} \ge \frac{A(k)}{A(k)} \ .
$$

Therefore, for all h,

$$
\frac{x_h(k+1)}{x(k+1)} \ge \frac{A_{h\cdot}(k)}{A(k)} \ ,
$$

or, equivalently,

$$
\frac{x(k+1)}{x_h(k+1)} \le A(k)\frac{e}{A_{h\cdot}(k)} \ . \tag{7.125}
$$

On the event $\mathcal{B}\circ\theta^k$

$$A(k+1)A(k) = A_{\cdot h^\star}(k+1)A_{h^\star\cdot}(k) \ .$$

By using (7.123) and (7.125), we therefore obtain that on this event,

$$\frac{x(k+1)}{x_{h^\star}(k+1)} \le A(k)\frac{e}{A_{h^\star\cdot}(k)} \le \frac{A_{\cdot h^\star}(k+1)}{A(k+1)} \ .$$

From the very definition of left division, the last relation implies that

$$A(k+1)\frac{x(k+1)}{x_{h^\star}(k+1)} \le A_{\cdot h^\star}(k+1) \ ,$$

so that

$$A(k+1)x(k+1) \le A_{\cdot h^\star}(k+1)x_{h^\star}(k+1) \ .$$

which concludes the proof of (7.124). Therefore, there exists a stationary sequence of *renovating events* of length 2. The coupling property is then a direct application of Borovkov's Theorem. ∎

7.5.5 Finiteness and Coupling in $\mathbb{R}_{\max}$; Strongly Connected Case

The assumptions of this subsection are the same as in §7.5.4, except for the positiveness assumption which is replaced by the assumption that the precedence graph of A has a deterministic topology (namely the entries which are equal to ε with a positive probability are a.s. equal to ε) and is strongly connected. We also make the usual assumption that the diagonal entries are nonvanishing. Under these assumptions, the matrices

$$E(k) \stackrel{\text{def}}{=} A(\mathfrak{n}k+\mathfrak{n}-1)A(\mathfrak{n}k+\mathfrak{n}-2)\ldots A(\mathfrak{n}k+1)A(\mathfrak{n}k) \ , \quad k \in \mathbb{Z} \ ,$$

are such that $E_{ij}(k) \ge e$ for all pairs (i,j) (this follows from Remark 7.30 and from the fact that $E(k) = G(\mathfrak{n}k)$. In this subsection, it will also be assumed that the shift

$$\Theta \stackrel{\text{def}}{=} \theta^{\mathfrak{n}}$$

is ergodic (the ergodicity of θ does not grant the ergodicity of $\theta^k, k > 1$, in general). Observe that $E(k) = E\circ\Theta^k$, where $E \stackrel{\text{def}}{=} E(0)$.

Let $X(k) \in \mathbb{R}^{\mathfrak{n}}$ be defined by the relation

$$X(k) = x(\mathfrak{n}k) \ , \quad k \ge 0 \ . \tag{7.126}$$

It is easily checked from (7.92) that the state variables $X(k)$ satisfy the relation

$$X(0) = x(0) \ , \qquad X(k+1) = E(k)X(k) \ , \quad k \ge 0 \ . \tag{7.127}$$

Following our usual notation, we will stress the dependence on the initial condition $x_0 \stackrel{\text{def}}{=} x(0)$ by writing $X(k; x_0)$ when needed.

Theorem 7.94 *Under the assumption that there exists $h^\star$ such that*

$$\mathbb{P}\left[E\circ\Theta_{\cdot h^\star}E_{h^\star\cdot} = E\circ\theta E\right] > 0 \ , \tag{7.128}$$

the ratios $\delta(k) = x(k+1)\not\circ x(k)$ also admit a stationary regime $\{\delta\circ\theta^k\}$. This stationary regime is unique, integrable and directly reachable. Whatever the initial condition x_0, $\delta(k;x_0)$ couples with it in finite time.

Proof Under Assumption (7.128), the ratio process $X(k+1)\not\circ X(k)$ couples with a stationary regime $\Delta\circ\Theta^k$, which is directly reachable (Theorems 7.90 and 7.93). Therefore, the equation

$$\Delta = \frac{Ex}{x} \tag{7.129}$$

has a unique solution satisfying the condition $x_1 = e$. From the very definition, taking x as the initial condition makes the ratios

$$\Delta(k;x) \stackrel{\text{def}}{=} \frac{x((k+1)\mathfrak{n};x)}{x(k\mathfrak{n};x)}$$

stationary in k, and more precisely such that $\Delta(k;x) = \Delta\circ\Theta^k$, $k \geq 0$. Therefore, the ratios $x((k+1)\mathfrak{n}+1;x)\not\circ x(k\mathfrak{n}+1;x)$ are stationary in k, as can be seen when writing them as

$$\begin{aligned}
\frac{x((k+1)\mathfrak{n}+1;x)}{x(k\mathfrak{n}+1;x)} &= \frac{A((k+1)\mathfrak{n})x((k+1)\mathfrak{n})}{A(k\mathfrak{n})x(k\mathfrak{n})} \\
&= \frac{A((k+1)\mathfrak{n})x((k+1)\mathfrak{n})\not\circ x(k\mathfrak{n})}{A(k\mathfrak{n})x(k\mathfrak{n})} \\
&= \frac{A((k+1)\mathfrak{n})\Delta(k,x)}{A(k\mathfrak{n})x(k\mathfrak{n})} \ ,
\end{aligned}$$

and when using the stationarity of $\Delta(k;x)$. But this ratio process is the one generated by the event graph when taking $\{A(k)\circ\theta\}_{k\geq 0}$ as the timing sequence, and y as the initial condition, where $y = x(1;x)\not\circ x_1(1;x)$. In view of the uniqueness property mentioned in Theorem 7.93 we immediately obtain that $x(\mathfrak{n}+1;x)\not\circ x(1;x) = \Delta\circ\theta$. Since $y_1 = e$, this in turn implies that $y = x\circ\theta$, owing to the uniqueness property mentioned in Lemma 7.85.

We show that the ratio process $\delta(k;x)$ satisfies (7.94). We have

$$\begin{aligned}
\delta(1;x) &= \frac{A(1)x(1)}{x(1)} = A(1)\frac{x(1)}{x(1)} \\
&= A(1)\frac{x(1)\not\circ x_1(1)}{x(1)\not\circ x_1(1)} = A\circ\theta\frac{x}{x}\circ\theta = \frac{Ax}{x}\circ\theta \\
&= \delta(0;x)\circ\theta \ ,
\end{aligned} \tag{7.130}$$

so that $\delta(k;x)$ satisfies (7.94) for $k=1$. In addition, $\delta(k)$ satisfies the equation

$$\delta(k+1) = \frac{A(k+1)x(k+1)}{A(k)} = \frac{A(k+1)\delta(k)}{A(k)} \ .$$

From this relation and (7.130), we prove by an immediate induction that $\delta(k)$ satisfies (7.94) for all $k \geq 0$.

One proves in the same way that the coupling of the ratios $\Delta(k)$ with a uniquely defined stationary process implies the same property for $\delta(k)$. The integrability property follows from the integrability of x and of the finite entries of A and from the relation $\delta(0; x) = Ax \not/ x$. ∎

Remark 7.95 If we replace $\mathfrak{n}$ in (7.126) by another integer $\mathfrak{n}'$, such that

- $(G(\mathfrak{n}'k)_{ij} \geq e$ for all $i, j = 1, \ldots, \mathfrak{n}$ (this condition is satisfied for all $\mathfrak{n}' \geq \mathfrak{n}$);
- $\theta^{\mathfrak{n}'}$ is ergodic,

then the whole construction is unchanged. As a consequence, whenever the variables A associated with $\mathfrak{n}$ do not satisfy the reachability and coupling conditions of Theorems 7.90 and 7.93, we still have the option to test this condition on the variables A' associated with $\mathfrak{n}'$. ∎

7.5.6 Finiteness and Coupling in $\mathbb{R}_{\max}$; General Case

The framework is the same as in §7.5.5, but A is not supposed to have a strongly connected precedence graph anymore. The notations concerning the decomposition of the precedence graph into m.s.c.s.'s are those of the end of §7.3.5. In particular, we will number the source subgraphs $1, \ldots, N_0$ and the nonsource ones $N_0 + 1, \ldots, N$.

We know from §7.5.5 that the stationary regime of the ratio process of the source subgraphs can be constructed using the techniques developed there. The only remaining problem consists in the construction of the stationary regime of nonsource subgraphs.

Consider first the case when the reduced graph has a single source, namely $N_0 = 1$, and assume that it satisfies the assumption of Theorem 7.94. Then we obtain from this theorem that the ratios $x_{(1)}(k+1) \not/ x_{(1)}(k)$ couple in finite time with a stationary, ergodic and integrable process $\delta_{(1)} \circ \theta^k$, which satisfies the property $\mathbb{E}\left[\left(\delta_{(1)}\right)_{ii}\right] = \mathfrak{a}_{(1)}$, where $\mathfrak{a}_{(1)}$ is the maximal Lyapunov exponent associated with $A_{(1)}$. The same technique as in the general nonautonomous case (see §7.4.4.5) allows us to prove the following theorem.

Theorem 7.96 *If $A_{(1)}$ satisfies the assumptions of Theorem 7.94, and if the condition $\bigoplus_{n=2}^{N} \mathfrak{a}_{(n)} < \mathfrak{a}_{(1)}$ holds true, then a unique finite random matrix δ exists such that the ratio process $\delta(k) = x(k+1) \not/ x(k)$ couples in finite time with the stationary and ergodic process $\delta \circ \theta^k$, regardless of the initial condition. If $\bigoplus_{n=2}^{N} \mathfrak{a}_{(n)} > \mathfrak{a}_{(1)}$, let n_0 be the first $n \in \{2, \ldots, N\}$ such that $\mathfrak{a}_{(n)} > \mathfrak{a}_{(1)}$. Then all ratios of the form $\delta_{ji}(k)$, $j \in \mathcal{V}_{n_0}$, $i \in \mathcal{V}_m$, $m \in \pi^+(n_0)$ tend to ∞ a.s. for all initial conditions.*

Remark 7.97 Nothing general can be said with respect to the critical case, namely when $\bigoplus_{n \geq 2} \mathfrak{a}_{(n)} = \mathfrak{a}_{(1)}$ (e.g. queuing theory). ∎

Consider now the case when the reduced graph has several sources, namely $N_0 > 1$. If the sources have different cycle times, it is clear that some of the ratios of the processes $x_j(k)$, $j \in \mathcal{V}_n$, $n = 1, \dots, N_0$, can neither be made stationary nor couple with a stationary sequence. Even if all these m.s.c.s.'s have the Lyapunov exponents, nothing general can be said about the stationarity of the variables $\delta_{ij}(k)$ for $j \in \mathcal{V}_n$, $i \in \mathcal{V}_m$, $m, n = 1, \dots, N_0$, $m \neq n$, as exemplified in the following simple situation.

Example 7.98 Consider a timed event graph with three recycled transitions q_1, q_2 and t and five places p_1, p_2, p'_1, p'_2 and r. Place p_i (respectively r) is the place associated with the recycling of q_i, $i = 1, 2$ (respectively t) and p'_i is the place connecting q_i to t (see Figure 7.5). Within the terminology of Example 7.37,

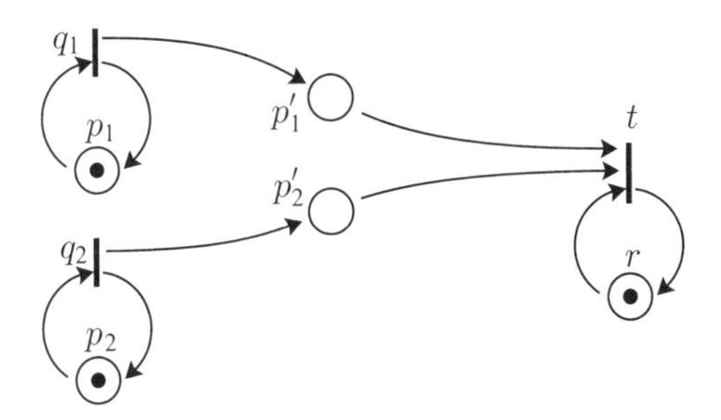

Figure 7.5: Counterexample

this system is a *join* queue with one server (transition t) and with two sources (transitions q_1 and q_2). This example can be seen as the simplest assembly problem in manufacturing: engines are produced by q_1 and car bodies by q_2, whereas t is the assembly machine. With our terminology, we have two source m.s.c.s.'s $\mathcal{G}_i$, with $\mathcal{V}_i = \{q_i\}$ and $\mathcal{E}_i = (q_i, q_i)$, $i = 1, 2$, and one nonsource subgraph $\mathcal{G}_3$, with $\mathcal{V}_3 = \{t\}$ and $\mathcal{E}_3 = (t, t)$. Assume the holding times in r, p'_1 and p'_2 are zero and that the holding times in p_1 and p_2 are mutually independent i.i.d. sequences $\{\alpha_1(k)\}$ and $\{\alpha_2(k)\}$ with common mean λ. If the variables $\alpha_1(k)$ and $\alpha_2(k)$ are deterministic, the ratios $x_1(k+1) \not{/} x_2(k)$ (with obvious notation) are stationary and finite whatever the initial condition. However, if the two sequences are made of exponentially distributed random variables with parameter λ, these ratios form a null recurrent Markov chain on $\mathbb{R}$ which admits no invariant measure with finite mass, so that they cannot be made stationary.

This example can also be formulated as a nonautonomous system with a two-dimensional input vector $(u_1(k), u_2(k))$ whenever we remove the recycling place p_i associated with transitions q_i, and replace it by an input function $u_i(k)$, $i = 1, 2$. In this formulation we see that in the exponential case (i.e. $u_1(k)$ and $u_2(k)$ are the epochs of two *independent* Poisson processes with the same intensity), the matrices $U_+(k)$ and $U_-(k)$ do not satisfy the assumptions of §7.4.4.1 (although the diagonal terms of these matrices are stationary and integrable, the nondiagonal terms do not couple with finite stationary processes). We see that in this case, our second-order theorems do not apply. ■

7.5.7 Multiplicative Ergodic Theorems in $\mathbb{R}_{\max}$

We will limit ourselves to the case when A is positive (see §7.5.4).

Theorem 7.99 *If the event* $\mathcal{B} = \{\omega \mid \exists h^\star : A\circ\theta A = A\circ\theta_{\cdot h^\star} A_{h^\star\cdot}\}$ *has a positive probability, then there exists a unique finite eigenpair* $\{\lambda, X\}$, *with* $X_1 = e$ *and such that* $AX = \lambda X\circ\theta$. *This eigenpair is integrable, and*

$$\mathbb{E}[\lambda] = \mathfrak{a} \ , \tag{7.131}$$

where $\mathfrak{a}$ *is the maximal Lyapunov exponent of* A. *In addition, the following coupling property takes place: for all finite initial conditions* $x(0) = x_0$, *with* $(x_0)_1 = e$, *there exists a finite integer-valued random variable* K *such that, for all* $k \geq K$,

$$\frac{x(k+1; x_0)}{x(k; x_0)} = \left(\lambda \frac{X\circ\theta}{X}\right)\circ\theta^k \ .$$

Proof We know from Theorem 7.90 that under the above assumptions, Equation (7.105) has a unique finite solution Δ for which (7.105) is satisfied with equality, and such that (7.104) has a solution. When specializing the formulæ of Theorem 7.79 to vectors and matrices with entries in $\mathbb{R}_{\max}$ as considered here, it is easily checked that whenever the subsolution Δ of (7.103) is finite and is a solution, then the super-eigenvalue inequality of (7.105) becomes an eigenvalue equality, so that there exists a pair (λ, X) satisfying the eigenpair property (7.68).

First it is easily checked that if (X, λ) is a finite eigenpair, then $(AX)\not/X = (\lambda X\circ\theta)\not/X$ is a solution of the ratio equation (7.113). Under the foregoing assumptions, this equation has a unique solution Δ. Since the equation $\Delta = (Ax)\not/x$ has a unique solution x such that $x_1 = e$ (Lemma 7.85), the eigenvector X is uniquely defined if we decide that $X_1 = e$. The same property holds for λ since $\lambda(X\circ\theta)\not/X = \Delta$. Property (7.131) follows from the relation

$$x(k+1; X) = \lambda\circ\theta^k \ldots \lambda X\circ\theta^{k+1} \ ,$$

which implies that

$$\left(x_1(k+1; X)\right)^{1/k} = \left(\bigotimes_{h=0}^{k} \lambda\circ\theta^h\right)^{1/k} \ .$$

The result follows immediately from the pointwise ergodic theorem and from Corollary 7.31. ■

7.6 STATIONARY MARKING OF STOCHASTIC EVENT GRAPHS

In this section we consider a stochastic event graph, with all its transitions recycled, and where the places in the recycling all have positive holding times. We return to the notation of conventional algebra.

Definition 7.100 (Stable place) *A place of the event graph is said to be stable if the number of tokens in this place at time t (the marking at time t), converges weakly to a finite random variable when time goes to ∞. The event graph is said to be stable if all the places are stable.*

The aim of this section is to determine the conditions under which the event graph is stable and to construct the stationary regime of the marking process, under the usual stationarity and ergodicity on the holding times.

Remark 7.101 Let $\mathcal{P}^0$ be the subset of places connecting two transitions belonging to the same strongly connected subgraph, and $\mathcal{P}^1$ be the subset of places connecting transitions which belong to no circuit. The marking of a place in $\mathcal{P}^0$ is bounded, and the only places which can possibly have an infinite marking when time goes to ∞ are those of $\mathcal{P}^1$ (see Chapter 2). ■

Pick some place p_i in $\mathcal{P}$ and let $q_j = \pi(p_i)$, and $q_l = \sigma(p_i)$. Assume that there exists an initial condition x_0 such that $x_j(0; x_0) = 0$, and such that the ratios $x(k+1; x_0) \not/ x(k; x_0)$ are stationary and ergodic (the conditions for such an initial condition to exist are given in Theorems 7.67 and 7.96, respectively, for the nonautonomous and the autonomous case). Since the sequence

$$b(k) \stackrel{\text{def}}{=} x(k; x_0) \not/ x_j(k; x_0) \ , \quad k \geq 0 \ ,$$

is stationary and ergodic, it can be continued to a bi-infinite stationary and ergodic sequence by the relation $b(k) \stackrel{\text{def}}{=} b \circ \theta^k$, $k \in \mathbb{Z}$, where $b = b(0)$. A similar continuation also holds for the sequence

$$d(k) \stackrel{\text{def}}{=} \delta_{jj}(k; x_0) \stackrel{\text{def}}{=} d \circ \theta^k \ , \quad k \geq 0 \ .$$

Because of the assumption that q_j is recycled and with positive holding times in the recycling, $d > 0$. Therefore, we can consider the sequence $\{d(k)\}$ as the stationary inter-event times of a point process defined on $(\Omega, \mathbb{F}, \mathbb{P}, \theta)$.

Definition 7.102 *We define $\mathcal{N}$ as the marked point process on $(\Omega, \mathbb{F}, \mathbb{P}, \theta)$ with inter-event times sequence $\{d(k)\}_{k \in \mathbb{Z}}$ and with the $\mathbb{R}^{|\mathcal{Q}|}$-valued mark sequence $\left\{ \left(b_h(k) \right)_{q_h \in \mathcal{Q}} \right\}_{k \in \mathbb{Z}}$. Namely, the k-th point of $\mathcal{N}$ is*

$$t(k) \stackrel{\text{def}}{=} \begin{cases} x_j(k; x_0) & \text{for } k \geq 0 \ ; \\ \sum_{h=k}^{-1} -d(k) & \text{for } k < 0 \ , \end{cases}$$

and its mark is $\{b_h(k), q_h \in \mathcal{Q}\}$.

The interarrival times and the marks being θ-stationary, this point process is stationary (in its so-called Palm version). Owing to our assumptions, $\mathcal{N}$ has a finite intensity and no double points.

Let $T(k) = (T_1(k), \ldots, T_{\mathfrak{n}}(k))$, where $\mathfrak{n} = |\mathcal{Q}|$, be the sequence

$$T_h(k) \stackrel{\text{def}}{=} t(k) + b_h(k) \ , \quad q_h \in \mathcal{Q} \ , \quad k \in \mathbb{Z} \ ,$$

and let N_i^- be the random variable

$$N_i^- = \sum_{k \le 0} 1_{\{T_l(k+\mu_i)>0\}} \ , \tag{7.132}$$

where $q_l = \sigma(p_i)$. This variable is a.s. finite. Indeed, $T_l(k)$ satisfies the relations

$$\lim_{k \to \infty} \frac{T_l(k)}{k} = c > 0 \ ,$$

where c is a positive constant. Therefore $\{T_l(k)\}$ is an increasing sequence such that $\lim_{k \to -\infty} T_l(k) = -\infty$ a.s. Hence there exists a finite integer-valued random variable H such that $T_l(k) \le 0$ for all $k \le -H$.

Theorem 7.103 *Under the assumptions of Theorem 7.67 (respectively 7.96), if $\mathfrak{a}_{(n)} < \mathfrak{u}$, for all $n = 1, \dots, N$, (respectively $\mathfrak{a}_{(n)} < \mathfrak{a}_{(1)}$, for all $n = 2, \dots, N$), where N denotes the number of m.s.c.s.'s of the event graph, then the event graph is stable whatever the initial condition, and the marking in place p_i at arrival epochs converges weakly to the random variable N_i^-. Conversely, if n_0 is the first $n = 1, \dots, N$, such that $\mathfrak{a}_{\{n\}} > \mathfrak{u}$ (respectively the first $n = 2, \dots, N$, such that $\mathfrak{a}_{(n)} > \mathfrak{a}_{(n)}$), then the places connecting the transitions of $\mathcal{Q}_m \cup \mathcal{I}$ (respectively $\mathcal{Q}_m$), $m < n_0$, to transitions of $\mathcal{Q}_{n_0}$ are all unstable whatever the initial condition.*

Proof Let $N_i^-(k)$ be the number of tokens in p_i just after time $x_j(k)$, $k \ge 1$, where $q_j = \pi(p_i)$. From (2.42) we obtain that

$$N_i^-(k) = \sum_{h=1}^{k+\mu_i} 1_{\{x_l(h)>x_j(k)\}} = \sum_{h=0}^{k+\mu_i} 1_{\{x_l(k+\mu_i-h)>x_j(k)\}} \ . \tag{7.133}$$

We first prove the last assertion of the theorem. Assume that p_i is the place connecting a transition of $\mathcal{Q}_m \cup \mathcal{I}$ (respectively $\mathcal{Q}_m$) to a transition of $\mathcal{Q}_{n_0}$. Owing to the property that

$$\lim_k \frac{x_l(k) - x_j(k)}{k} = \mathfrak{a}_{(n_0)} - \mathfrak{u} > 0 \ ,$$

$$\left(\text{respectively} \quad \lim_k \frac{x_l(k) - x_j(k)}{k} = \mathfrak{a}_{(n_0)} - \mathfrak{a}_{(1)} > 0 \right) \ ,$$

and to the increasingness of the sequences $\{x_l(k)\}$ and $\{x_j(k)\}$, we obtain that, for all H, there exits K such that, for all $k \ge K$ and $h = 1, \dots, H$, $x_l(k-h) - x_j(k) \ge 0$. It follows immediately from this that $N_i^-(k) \ge H$ for $k \ge K$. Therefore, $N_i^-(k)$ tends to ∞ a.s.

We now prove the first part. We know that the ratios of $x_h(k)$, $q_h \in \mathcal{Q}$, couple with their stationary regime in a finite random time K. This implies that for all fixed h, the sequence $\{x_l(k + \mu_i - h) - x_j(k)\}$ couples with a stationary process. More precisely, for all $k \ge K + h$, and $h > \mu_i$, $x_l(k + \mu_i - h) - x_j(k) = -\rho_l(\mu_i - h) \circ \theta^k$, where

$$\rho_l(k) \stackrel{\text{def}}{=} \sum_{n=-1}^{k} d \circ \theta^n - b_l \circ \theta^{-k} = T_j(0) - T_l(k) \ , \quad k < 0 \ ,$$

in view of the uniqueness of the stationary regimes of the ratios. Define $H = \inf\{k \mid k \geq K, x_l(h) - x_j(k) < 0 \ , \quad \forall h = 1, \dots, K\}$. This H is a.s. finite since K is finite and $x_j(k)$ tends to ∞ a.s. Therefore,

$$N_i^-(k) = \sum_{1 \leq h \leq k-K} 1_{\{x(k+\mu_i-h)_l - x_j(k) > 0\}} = \sum_{1 \leq h \leq k-K} 1_{\{-\rho_l(\mu_i - h) \circ \theta^k > 0\}} \ ,$$

for all $k \geq H$. On the other hand,

$$N_i^- \circ \theta^k = \sum_{0 \leq h} 1_{\{T_l(k+\mu_i-h) - T_j(k) > 0\}} = \sum_{0 \leq h} 1_{\{-\rho_l(\mu_i-h) \circ \theta^k > 0\}} \ .$$

Since $T_j(k)$ tends to ∞ as k goes to ∞, we obtain that there exists an L such that

$$\sum_{k-h \leq K} 1_{\{T_l(k+\mu_i-h) - T_j(k) > 0\}} = 0 \ ,$$

for all $k \geq L$. Therefore, $N_i^-(k) = N_i^- \circ \theta^k$ for $k \geq \max(H, L)$, and the stationary regime of the marking process is reached with coupling, regardless of the initial condition. ■

Remark 7.104 This stationary regime is unique, owing to the uniqueness of the stationary regime of the ratio process. ■

Remark 7.105 The preceding construction gives the Palm probability of the number of tokens in p_i at arrival epochs. The stationary distribution of the number of tokens in p_i in 'continuous time' is then obtained via the Palm inversion formula (see [5, p. 17]). ■

7.7 APPENDIX ON ERGODIC THEOREMS

The aim of this section is to state in a concise form a few basic ergodic theorems (in conventional algebra) which are either used or referred to in this chapter. The basic data are a probability space $(\Omega, \mathbb{F}, \mathbb{P})$ on which a shift operator θ is defined (see Definition 7.2). This shift is assumed to be stationary and ergodic.

Theorem 7.106 (Kingman's subadditive ergodic theorem) *Let $\xi_{m,n}$, $m > n \in \mathbb{Z}$ be an integrable random process on $(\Omega, \mathbb{F}, \mathbb{P})$ such that*

$$\xi_{m,m+p} = \xi_{0,p} \circ \theta^m \ , \quad \forall m \in \mathbb{Z} \ , \quad \forall p > 0 \quad \textit{(stationarity)} \ ,$$

and

$$\xi_{m,n} \leq \xi_{m,p} + \xi_{p,n} \ , \quad \forall m < p < n \quad \textit{(subadditivity)} \ .$$

Assume in addition that there exists a positive constant A such that $\mathbb{E}[\xi_{0,p}] \geq -Ap$, for all $p > 0$. Then there exists a constant γ such that the following two equations hold:

$$\lim_{p \to \infty} \frac{\xi_{0,p}}{p} = \gamma \quad \textit{a.s.}, \qquad \lim_{p \to \infty} \frac{\mathbb{E}[\xi_{0,p}]}{p} = \gamma \ .$$

For the proof, see [75], [76].

Theorem 7.107 (Borovkov's renovating events theorem) *Let $\{u(k)\}$ be a θ-stationary $\mathbb{R}^{\mathfrak{n}}$-valued sequence of random variables defined on $(\Omega, \mathbb{F}, \mathbb{P})$. Let $\{x(k)\}$ be the $\mathbb{R}^K$-valued sequence of random variables defined by the recurrence relation*

$$x(k+1) = a(x(k), u(k)) \quad , \quad k \geq 0 \ , \tag{7.134}$$

where a is a continuous mapping $\mathbb{R}^K \times \mathbb{R}^{\mathfrak{n}} \to \mathbb{R}^K$, and by the random initial condition $x(0)$. The event $\mathfrak{A}(k) \in \mathbb{F}$ is said to be a renovating event of length $m \geq 1$ and of associated function $\phi : \mathbb{R}^{m\mathfrak{n}} \to \mathbb{R}^K$ if, on $A(k)$, the relation

$$x(k+m) = \phi\left(u(k), \ldots, u(k+m-1)\right)$$

holds. If the random process $x(k)$ admits a sequence $\{\mathfrak{A}(k)\}$ of renovating events, all of length m and associated function ϕ, such that $\mathfrak{A}(k) = \mathfrak{A}(0)\circ\theta^k$, $\forall k \geq 0$, and $\mathbb{P}[\mathfrak{A}(0)] > 0$, then, the sequence $\{x(k)\circ\theta^{-k}\}$ converges a.s. to a finite random variable z, which does not depend upon the initial condition $x(0)$. The sequence $\{z\circ\theta^k\}$ is a finite solution of (7.134), and the sequence $\{x(k)\}$ couples with it in finite time for all finite initial conditions.

For the proof, see [26], [6].

For any matrix A, let $|A|$ denote its operator norm, namely

$$|A| = \sup_{\|x\|=1} \|Ax\| \ ,$$

where $\|x\|$ denotes the Euclidean norm of vector x.

Theorem 7.108 (Oseledeç's multiplicative ergodic theorem) *Let $\{A(k)\}$ be a sequence of $\mathfrak{n} \times \mathfrak{n}$ random matrices with nonnegative entries, defined on the probability space $(\Omega, \mathbb{F}, \mathbb{P})$. Assume that $A(k) = A(0)\circ\theta^k$, for all $k \in \mathbb{Z}$, and that $\mathbb{E}\left[\max\left(\log(|A(0)|), 0\right)\right] < \infty$. Then there exists a constant γ (the maximal Lyapunov exponent of the sequence) such that*

$$\lim_{k\to\infty} \frac{1}{k} \log\left(|A(k)\ldots A(1)|\right) = \gamma \ , \quad a.s.$$

In addition, there exists a random eigenspace $V(\omega)$ of dimension d constant, $d \leq \mathfrak{n}$, such that $A(1)V = V\circ\theta$ and such that for all random vectors x in V,

$$\lim_{k\to\infty} \frac{1}{k} \log\left(\|A(k)\ldots A(1)x\|\right) = \gamma \ , \quad a.s.$$

Whenever $d = 1$, there exists an eigenpair $\{\lambda, X\}$ such that $A(0)X = \lambda X\circ\theta$ and $\mathbb{E}[\lambda] = \gamma$.

In fact, Oseledeç's Theorem gives the existence of other eigenvalues as well. Our statement of this theorem is limited to the maximal eigenvalue and its associated eigenspace (see [106], [45]).

7.8 NOTES

The preliminary example of §7.2 was first analyzed by R.M. Loynes in 1962 [87]. The probabilistic formalism introduced in §7.2 is that developed for queues by P. Brémaud and one of the coauthors in [6]. The sections on the relationship between stochastic event graphs and $\mathbb{R}_{\max}$-multiplicative ergodic theory (§7.3–7.6) are mainly based on [11], [13]. As to the writing of this book, this approach provides a more or less systematic way for analyzing nonautonomous systems. The situation is somewhat less satisfactory in the autonomous case: in particular, only the case when the eigenspace associated with the maximal exponent has dimension 1 was considered. This practically covers cases with 'sufficiently random' entries of A, as shown by the results of §7.5; however, we know from the analysis of Chapter 3 that an eigenspace of dimension 1 is rarely sufficient to handle the case of deterministic systems. Autonomous deterministic systems can fortunately be addresses via the spectral methods of Chapter 3. However, some systems are neither deterministic nor random enough to satisfy the conditions of §7.4. Filling up this 'theoretical gap' between purely deterministic and sufficiently random systems is clearly tantamount to understanding the structure of the eigenspace associated with the maximal exponent when this eigenspace is of dimension greater than 1.

CHAPTER 8

Computational Issues in Stochastic Event Graphs

8.1 INTRODUCTION

This chapter gathers miscellaneous results pertaining to the computation of the cycle times and the stationary regimes of stochastic event graphs. The existence and uniqueness of these two quantities are discussed in Chapter 7: the cycle time of a stochastic event graph is the maximal Lyapunov exponent associated with the matrices $\widetilde{A}(k)$ of its standard equation, and stationary regimes correspond to stochastic eigenpairs of $\widetilde{A}$.

Section 8.2 focuses on monotonicity properties of daters and counters considered as functions of the data (e.g. firing and holding times, initial marking, topology of the graph, etc.). These results lead to the derivation of a lower bound for the cycle time, which is based on the results of Chapter 3 concerning the deterministic case. It is also shown that the throughput is a concave function of the initial marking, provided that the firing and holding times satisfy appropriate statistical properties.

Section 8.3 is concerned with the relationship between stochastic event graphs and a class of age-dependent branching processes. Large deviation techniques are used to provide an estimate for the cycle time, which is also shown to be an upper bound.

The last section contains miscellaneous computational results which can be obtained in the Markovian case. Whenever the firing and the holding times have discrete distribution functions with finite support, simple sufficient conditions for the ratio process to have a finite state space Markov chain structure are given. In the continuous and infinite support case, partial results on functional equations satisfied by the stationary distribution functions are provided. These results are then used for computing the distribution of the stationary regime.

The sections of this chapter can be read (almost) independently. Each section has its own prerequisites: basic properties of stochastic orders for §8.2 (see [123]); notions of branching processes and of large deviations ([3]) in §8.3; elementary Markov chain theory in §8.4. Throughout the whole chapter, the scalar dioid of reference is $\mathbb{R}_{\max}$, unless otherwise specified.

8.2 MONOTONICITY PROPERTIES

8.2.1 Notation for Stochastic Ordering

Let x and $x^\dagger$ be $\mathbb{R}^n$-valued random variables. Three classical stochastic ordering relations between x and $x^\dagger$ will be considered in this section.

Notation 8.1

Stochastic ordering $\leq_{\text{st}}$: $x \leq_{\text{st}} x^\dagger$ if $\mathbb{E}[f(x)] \leq \mathbb{E}[f(x^\dagger)]$, for all nondecreasing functions $f : \mathbb{R}^n \to \mathbb{R}$.

Convex ordering $\leq_{\text{cx}}$: $x \leq_{\text{cx}} x^\dagger$ if $\mathbb{E}[f(x)] \leq \mathbb{E}[f(x^\dagger)]$, for all convex functions $f : \mathbb{R}^n \to \mathbb{R}$.

Increasing convex ordering $\leq_{\text{icx}}$: $x \leq_{\text{icx}} x^\dagger$ if $\mathbb{E}[f(x)] \leq \mathbb{E}[f(x^\dagger)]$, for all convex and nondecreasing functions $f : \mathbb{R}^n \to \mathbb{R}$. ■

Let $x = \{x(1), \ldots, x(k), \ldots\}$ (respectively $x(\cdot) = x(t), t \in \mathbb{R}^+$) and $x^\dagger = \{x^\dagger(1), \ldots, x^\dagger(k), \ldots\}$ (respectively $x^\dagger(\cdot) = \{x^\dagger(t)\}_{t \in \mathbb{R}^+}$) be two $\mathbb{R}^n$-valued stochastic sequences (respectively processes) defined on the probability space $(\Omega, \mathbb{F}, \mathbb{P})$. The sequence $x^\dagger$ is said to dominate x (respectively the process $x^\dagger(\cdot)$ dominates $x(\cdot)$) for one of the above ordering relations, say $\leq_{\text{st}}$, which is denoted $x \leq_{\text{st}} x^\dagger$ (respectively $x(\cdot) \leq_{\text{st}} x^\dagger(\cdot)$), if all corresponding finite dimensional distributions compare for this ordering.

For basic properties of these orderings, see §8.5.

8.2.2 Monotonicity Table for Stochastic Event Graphs

The basic model of this section is a live autonomous event graph, where all transitions are assumed to be recycled. The nonautonomous case leads to similar results and will not be considered in this chapter. The notation and basic definitions concerning stochastic event graphs are those of Chapter 2 and Chapter 7. The following concise notation will be used:

Data

Firing times: $\beta(k)$ denotes the vector $\beta_j(k), j = 1, \ldots, |\mathcal{Q}|$, and β the sequence $\{\beta(k)\}$.

Holding times: $\alpha(k)$ denotes the vector $\alpha_i(k), i = 1, \ldots, |\mathcal{P}|$, and α the sequence $\{\alpha(k)\}$.

Timing sequence: $\eta(k)$ denotes the vector $(\beta(k), \alpha(k))$ and η the sequence $\{\eta(k)\}$.

Initial marking: μ denotes the vector $\mu_i, i = 1 \ldots, |\mathcal{P}|$.

State Variables With each dater sequence $\{x_j(k)\}_{k \in \mathbb{N}}$ we associate the $\mathbb{N}$-valued counter function $x_j(t)$, $t \in \mathbb{R}^+$, defined by the relation

$$x_j(t) = \sup\{k \mid x_j(k) \leq t\} \ .$$

Daters: $x(k)$ denotes the vector $x_j(k), j = 1, \dots |\mathcal{Q}|$, and x the sequence $\{x(k)\}$.

Counters: $x(t)$ denotes the vector $x_j(t), j = 1, \dots |\mathcal{Q}|$, $t \in \mathbb{R}$, and $x(\cdot)$ the function $x(t)$.

The aim of the following sections is to prove various monotonicity properties of the state variables and of their asymptotic characteristics considered as functions of the data. By asymptotic characteristics, we mean the cycle time $\mathfrak{a}$ of the event graph and its (conventional) inverse τ, the throughput. A typical question is as follows: if one replaces one of the data, say μ or β, by $\mu^\dagger$ or $\beta^\dagger$ respectively, where the new data are greater than the initial ones for some partial ordering, what result do we obtain on the various state variables? The main properties along these lines are summarized in Table 8.1. The reader should refer to the following subsections in order to obtain the specific assumptions under which the reported monotonicity properties hold. These assumptions are not always the most general ones under which these properties hold. For instance, we have tried to avoid the intricate issues associated with the initial conditions by choosing assumptions leading to short proofs, although most of the properties of the table extend to more general initial conditions.

Table 8.1: Monotonicity of the state variables

Data	Variation of data	Daters	Cycle time	Counters	Throughput
μ	$\mu \le \mu^\dagger$	$x \ge x^\dagger$	$\mathfrak{a} \ge \mathfrak{a}^\dagger$	$x(\cdot) \le x^\dagger(\cdot)$	$\tau \le \tau^\dagger$
$\mathcal{G}$	$\mathcal{P} \subset \mathcal{P}^\dagger$ $\mathcal{Q} \subset \mathcal{Q}^\dagger$ $\mathcal{E} \subset \mathcal{E}^\dagger$	$x \le x^\dagger$	$\mathfrak{a} \le \mathfrak{a}^\dagger$	$x(\cdot) \ge x^\dagger(\cdot)$	$\tau \ge \tau^\dagger$
η	$\eta \le_{\text{st}} \eta^\dagger$	$x \le_{\text{st}} x^\dagger$	$\mathfrak{a} \le \mathfrak{a}^\dagger$	$x(\cdot) \ge_{\text{st}} x^\dagger(\cdot)$	$\tau \ge \tau^\dagger$
	$\eta \le_{\text{cx}} \eta^\dagger$	$x \le_{\text{icx}} x^\dagger$	$\mathfrak{a} \le \mathfrak{a}^\dagger$		$\tau \ge \tau^\dagger$
	$\eta \le_{\text{icx}} \eta^\dagger$	$x \le_{\text{icx}} x^\dagger$	$\mathfrak{a} \le \mathfrak{a}^\dagger$		$\tau \ge \tau^\dagger$

8.2.3 Properties of Daters

8.2.3.1 Stochastic Monotonicity

Monotonicity with respect to the Timing Sequence In this paragraph, we assume that the entrance times are all equal to e (see Remark 2.75). Let $\eta^\dagger$ be another timing sequence associated with the same event graph (i.e. the topology and the initial marking of the event graph are unchanged and each timing variable is replaced by the corresponding dagger variable) and let $x^\dagger$ be the resulting daters. We first compare the sequences $x^\dagger$ and x, whenever the timing sequences can be compared for some integral ordering.

Theorem 8.2 *If $\eta \leq_{\mathrm{st}} \eta^\dagger$, then $x \leq_{\mathrm{st}} x^\dagger$.*

Proof The initial condition $\widetilde{x}(0)$ is untouched by the transformation of the variables, in view of the definition given in Remark 2.75. The matrices $\widetilde{A}(k)$ in (2.31) are nondecreasing functions of the variables α, β (see Equations (2.15), (2.28) and the definition of $\widetilde{A(k)}$). Therefore, from (8.40), we obtain that

$$\{\widetilde{x}(0), \widetilde{A}(0), \widetilde{A}(1), \ldots\} \leq_{\mathrm{st}} \{\widetilde{x}^\dagger(0), \widetilde{A}^\dagger(0), \widetilde{A}^\dagger(1), \ldots\} \ .$$

We now use the canonical representation (2.31) to represent the evolution equations of interest as recursions of the type (8.43), where the mapping a is coordinatewise nondecreasing and such that the sequences defined in (8.44) satisfy the relation $\{\xi(k)\} \leq_{\mathrm{st}} \{\xi^\dagger(k)\}$. The proof is then concluded from Theorem 8.60. ∎

In the next theorem, the timing variables are assumed to be integrable.

Theorem 8.3 *If $\eta \leq_{\mathrm{icx}} \eta^\dagger$, then $x \leq_{\mathrm{icx}} x^\dagger$.*

Proof The entries of matrices $\widetilde{A}(k), k \geq 1$, are nondecreasing and convex functions of the variables α, β. So, the assumption $\eta \leq_{\mathrm{icx}} \eta^\dagger$ and (8.41) imply that

$$\{\widetilde{x}(0), \widetilde{A}(0), \widetilde{A}(1), \ldots\} \leq_{\mathrm{icx}} \{\widetilde{x}^\dagger(0), \widetilde{A}^\dagger(0), \widetilde{A}^\dagger(1), \ldots\} \ .$$

Since the mapping a of the preceding proof is nondecreasing and convex, the result immediately follows from Theorem 8.62. ∎

Remark 8.4 Assume that the holding and firing times are all integrable. Then, it follows from (8.42) that the 'deterministic version' of the event graph Γ, with firing times $\overline{\beta}_j(k) = \mathbb{E}\left[\beta_j(k)\right]$ and holding times $\overline{\alpha}_i(k) = \mathbb{E}\left[\alpha_i(k)\right]$, leads to a sequence of daters $\{\overline{x}(k)\}$ which is a lower bound of $\{x(k)\}$ in the $\leq_{\mathrm{icx}}$ sense. In particular, since the daters are integrable under these assumptions, it follows from Lemma 8.59 that for all $k \geq 1$ and $q_j \in \mathcal{Q}$, $\mathbb{E}\left[x_j(k)\right] \geq \overline{x}_j(k)$. ∎

Example 8.5 For example, by applying Theorem 8.3 to the cyclic queuing network with finite buffers of Example 7.92, one obtains that the departure times from the queues are $\leq_{\mathrm{icx}}$-nondecreasing functions of the service times. ∎

Monotonicity with respect to the Initial Marking Here, the discussion will be limited to the case when all initial lag times are equal to e (which is a compatible initial condition). It will be convenient to use (2.19) as the basic evolution equation. Under our assumption on the initial lag times, this equation reads

$$x_j(k) = \bigoplus_{\{i \in \pi^q(j) | k > \mu_i\}} \beta_{\pi^p(i)}(k - \mu_i)\alpha_i(k)x_{\pi^p(i)}(k - \mu_i) \oplus e \ , \tag{8.1}$$

for $k \geq 1$, where $x_j(k) = \varepsilon$ if $k \leq 0$. Consider the same event graph as above, but with the initial marking $\mu_i^\dagger, p_i \in \mathfrak{P}$, in place of μ_i, and let $\{x^\dagger(k)\}$ denote the corresponding dater sequence.

Theorem 8.6 *Under the assumption that all the initial lag times are equal to e, if $\mu_i \leq \mu_i^\dagger$, $\forall p_i \in \mathfrak{P}$, then the coordinatewise inequality $x \geq x^\dagger$ holds.*

Proof Owing to the assumption that all transitions are recycled and to the preceding convention concerning the continuation of $x(k)$ for $k \leq 0$, it is easy to check that

$$x_j(k) \geq \beta_j(k-1)x_j(k-1) \ , \quad \forall j = 1, \ldots, \mathfrak{n}, \quad k \in \mathbb{Z} \ .$$

Therefore

$$x_j(k) \geq x_j(k-1) \ , \quad \forall j = 1, \ldots, \mathfrak{n}, \quad k \in \mathbb{Z} \ ,$$

and

$$\beta_j(k)x_j(k) \geq \beta_j(k-1)x_j(k-1) \ , \quad \forall j = 1, \ldots, \mathfrak{n}, \quad k \in \mathbb{Z} \ . \tag{8.2}$$

We prove that $x_j(k) \leq x_j^\dagger(k)$ for all $j = 1, \ldots, \mathfrak{n}$, and $k \geq 1$. The proof is by induction on (j, k). Since the event graph is assumed to be live, the numbering of the transitions can be chosen in such a way that for all $(j,k), j = 1, \ldots, \mathfrak{n}$, $k \geq 1$, the variables $x_i(l)$ which are found at the right-hand side of (8.1) are always such that either $l < k$ or $l = k$, but $i < j$. Therefore, there exists a way of numbering the transitions such that the daters $x_j(k)$ can be computed recursively in the order

$$x_1(1), x_2(1), \ldots, x_{\mathfrak{n}}(1), x_1(2), \ldots, x_{\mathfrak{n}}(2), \ldots, x_1(k), \ldots, x_{\mathfrak{n}}(k), \ldots .$$

Assume that the property holds up to (j, k) excluding this point (it holds for $(1, 1)$ since with our assumptions, we necessarily have $x_1(1) = x_1^\dagger(1) = e$). Then, we have

$$\begin{aligned}
x_j^\dagger(k) &= \bigoplus_{\{i \in \pi^q(j) | k > \mu_i^\dagger\}} \beta_{\pi^p(i)}(k - \mu_i^\dagger)\alpha_i(k)x^\dagger_{\pi^p(i)}(k - \mu_i^\dagger) \oplus e \\
&\leq \bigoplus_{\{i \in \pi^q(j) | k > \mu_i^\dagger\}} \beta_{\pi^p(i)}(k - \mu_i)\alpha_i(k)x^\dagger_{\pi^p(i)}(k - \mu_i) \oplus e \\
&\leq \bigoplus_{\{i \in \pi^q(j) | k > \mu_i^\dagger\}} \beta_{\pi^p(i)}(k - \mu_i)\alpha_i(k)x_{\pi^p(i)}(k - \mu_i) \oplus e \\
&\leq \bigoplus_{\{i \in \pi^q(j) | k > \mu_i\}} \beta_{\pi^p(i)}(k - \mu_i)\alpha_i(k)x_{\pi^p(i)}(k - \mu_i) \oplus e \\
&= x_j(k) \ ,
\end{aligned}$$

where we successively used the monotonicity property (8.2), the induction assumption, and finally the fact that we sum up with respect to a larger set. ■

Example 8.7 As an application of this result, one obtains the monotonicity of departure times in closed cyclic networks with blocking, as a function of the population and the buffer sizes. ■

Stochastic Monotonicity with respect to the Topology Consider two event graphs with associated graphs $\Gamma = ((\mathcal{P} \cup \mathcal{Q}), \mathcal{E})$ and $\Gamma^\dagger = ((\mathcal{P}^\dagger \cup \mathcal{Q}^\dagger), \mathcal{E}^\dagger)$, where

$$\mathcal{P} \subseteq \mathcal{P}^\dagger \ , \qquad \mathcal{Q} \subseteq \mathcal{Q}^\dagger \ , \qquad \mathcal{E} \subseteq \mathcal{E}^\dagger \ . \tag{8.3}$$

The event graph $\Gamma^\dagger$ is such that the initial marking, the initial lag times, the firing and holding times of those places and transitions which belong both to Γ and $\Gamma^\dagger$ are the same. Let $x^\dagger$ denote the sequence of daters of $\Gamma^\dagger$, and $[x^\dagger]$ the restriction of $x^\dagger$ to the set of transitions which belong to Q and $Q^\dagger$.

Theorem 8.8 *Under the foregoing assumptions, $x \leq x^\dagger$ coordinatewise.*

Proof The proof is based on Equation (2.19) and is again by induction on (j, k). Assume that the property $x_i(l) \leq x_i^\dagger(l)$ holds up to (j, k) excluding this point. The point $(1, 1)$ is not necessarily the first one to compute in the total order associated with $\Gamma^\dagger$, but we are sure that all the places present in Γ preceding q_1 are present in $\Gamma^\dagger$ too, and with the same number of initial tokens and the same lag times, so that the property necessarily holds for (1,1). Then, denoting $\pi^\dagger$ the predecessor function in $\Gamma^\dagger$, we obtain the following relation for all $q_j \in \mathcal{Q}$:

$$\begin{aligned}
x_j^\dagger(k) &= \bigoplus_{\{i \in \pi^{\dagger,q}(j) | k > \mu_i^\dagger\}} \beta^\dagger_{\pi^{\dagger,p}(i)}(k - \mu_i^\dagger)\alpha_i^\dagger(k) x^\dagger_{\pi^{\dagger,p}(i)}(k - \mu_i^\dagger) \\
&\quad \oplus \bigoplus_{\{i \in \pi^{\dagger,q}(j) | k \leq \mu_i^\dagger\}} w_i^\dagger(k) \\
&\geq \bigoplus_{\{i \in \pi^{q}(j) | k > \mu_i^\dagger\}} \beta^\dagger_{\pi^{p}(i)}(k - \mu_i^\dagger)\alpha_i^\dagger(k) x^\dagger_{\pi^{p}(i)}(k - \mu_i^\dagger) \\
&\quad \oplus \bigoplus_{\{i \in \pi^{q}(j) | k \leq \mu_i^\dagger\}} w_i^\dagger(k) \\
&= \bigoplus_{\{i \in \pi^{q}(j) | k > \mu_i\}} \beta_{\pi^{p}(i)}(k - \mu_i)\alpha_i(k) x^\dagger_{\pi^{p}(i)}(k - \mu_i) \\
&\quad \oplus \bigoplus_{\{i \in \pi^{q}(j) | k \leq \mu_i\}} w_i(k) \\
&\geq \bigoplus_{\{i \in \pi^{q}(j) | k > \mu_i\}} \beta_{\pi^{p}(i)}(k - \mu_i)\alpha_i(k) x_{\pi^{p}(i)}(k - \mu_i) \\
&\quad \oplus \bigoplus_{\{i \in \pi^{q}(j) | k \leq \mu_i\}} w_i(k) \\
&= x_j(k) \ ,
\end{aligned}$$

where we successively used the assumptions (8.3), the assumption that the initial condition and the firing and holding times of the nodes of $\Gamma \cap \Gamma^\dagger$ are the same, and finally the induction assumption. ∎

8.2.4 Properties of Counters

In this subsection we assume that tokens incur no holding times in places (so that the only timing variables come from the firing times of the transitions) and that there is at most one place between two subsequent transitions q_i and q_j, which will be denoted (j, i), with initial marking $\mu(j, i) \in \mathbb{N}$. It will also be assumed that the initial lag times are all equal to e (which leads to a compatible initial condition). Under these assumptions, the evolution equation (2.19) in Chapter 2 reads

$$x_j(k) = \bigoplus_{i \in p(j)} \beta_i(k - \mu(j,i))\, x_i(k - \mu(j,i)) \oplus e \ , \quad k \geq 1 \ ,$$

with initial condition $x_j(k) = \varepsilon$ for all $k \leq 0$. In this relation, $p(j) \stackrel{\text{def}}{=} \pi^p(\pi^q(j))$.

8.2.4.1 Evolution Equations

Let $x_j(t)$ (respectively $y_j(t)$) denote the number of firings which transition j initiated (respectively completed) by time $t, t \in \mathbb{R}^+$. Without loss of generality, we assume that both $x_j(t)$ and $y_j(t)$ are right continuous.

Remark 8.9 The mappings $x_j(k) : \mathbb{N} \to \mathbb{R}$ and $x_j(t) : \mathbb{R} \to \mathbb{N}$ are related by the formulae

$$x_j(t) = \sup\{k \mid x_j(k) \leq t\} \ , \tag{8.4}$$

$$x_j(k) = \inf\{t \mid x_j(t) \geq k\} \ . \tag{8.5}$$

When using the definitions of §4.4, we see that the isotone and l.s.c. mapping $x_j(k) : \mathbb{N} \to \mathbb{R}$ admits the u.s.c. mapping $x_j(t) : \mathbb{R} \to \mathbb{N}$ as its residual; similarly, the isotone and u.s.c. mapping $x_j(t) : \mathbb{R} \to \mathbb{N}$ admits the l.s.c. mapping $x_j(k) : \mathbb{N} \to \mathbb{R}$ as its dual residual. ■

In this subsection it is assumed that the firing times are all strictly positive.

Theorem 8.10 *The random variables $x_j(t)$ and $y_j(t)$, $1 \leq j \leq \mathfrak{n}$, $t \geq 0$, satisfy the following evolution equations:*

$$x_j(t) = \min_{i \in p(j)} \left(y_i(t) + \mu(j,i)\right) \ , \tag{8.6}$$

$$y_j(t) = \int_0^t 1_{\{\beta_j(x_j(u)) \leq t-u\}}\, x_j(\mathrm{d}u) \ , \tag{8.7}$$

where, for all $j = 1, \ldots, \mathfrak{n}$, $y_j(0) = 0$ and $x_j(t) = 0, \forall t < 0$.

Proof By time t, transition j initiated exactly as many firings as the minimum over $i \in p(j)$ of the number of tokens which entered place (j, i) by time t (including the initial tokens). Since a place is preceded by exactly one transition, the number of tokens which entered place (j, i) by t equals $\mu(j, i)$ plus the number of firings which transition i completed by time t. The Stieljes integral in (8.7) is a compact way of writing the sum

$$\sum_{k=0}^{\infty} 1_{\{x_j(k)+\beta_j(k)\le t\}} \ .$$

■

In the deterministic case, (8.7) takes the simpler form $y_j(t) = x_j(t-\beta_j)$, which leads back to the following familiar result.

Corollary 8.11 *If the firing times are deterministic (i.e. $\beta_j(k) = \beta_j$), then (8.6) and (8.7) reduce to the $\mathbb{R}_{\max}$ equation*

$$x_j(t) = \bigwedge_{i\in p(j)} \mu(j,i)x_i(t-\beta_i) \ , \tag{8.8}$$

where $x_i(t) = e$ for $t < 0$.

8.2.4.2 Stochastic Monotonicity

From (8.4), one obtains that for any fixed n-tuple $t_1 < \ldots < t_n$ in $\mathbb{R}^+$, the vector $(x(t_1),\ldots,x(t_n))$ is a nonincreasing function of x. Therefore, each $\le_{\text{st}}$-monotonicity property of the sequence x with respect to $\le_{\text{st}}$ yields a dual stochastic monotonicity property of $x(\cdot)$ (see Table 8.1).

8.2.4.3 Concavity with respect to the Initial Marking

Throughout this subsection it is assumed that the sequences $\{\beta_j(k)\}_k$ are mutually independent in j.

Theorem 8.12 *If the random variables $\beta_j(k)$ are i.i.d., with exponential distribution of parameter λ_j, then, for any $t \ge 0$, and any $1 \le j \le \mathfrak{n}$, $x_j(t)$ and $y_j(t)$ are stochastically increasing and concave (see Definition 8.63) in the initial marking $\mu \in \mathbb{N}^{|\mathfrak{E}|}$.*

Proof Let $\{b_j(n)\}_{n=1}^{\infty}$, $1 \le j \le \mathfrak{n}$, be mutually independent sequences of i.i.d. random variables where $b_j(n)$ is exponentially distributed with parameter λ_j. Let $t_0, t_1, t_2, \ldots, t_n, \ldots$ be the times defined by $t_0 = 0$ and

$$t_n = t_{n-1} + \min_{1\le j\le \mathfrak{n}} b_j(n) \ , \quad n \ge 1 \ ,$$

and let $\chi_j(n)$ be the indicator function

$$\chi_j(n) = 1_{\{t_n = t_{n-1}+b_j(n)\}} \ .$$

Let $\Gamma^\dagger$ be an event graph with the same topology and initial marking as Γ, and with the following dynamics (which differs from the dynamics defined in Chapter 2): in $\Gamma^\dagger$, for all transitions j enabled at time t_n^+, the residual firing time of j at time t_n^+ (namely the time which elapses between t_n and the completion of the ongoing firing of transition j) is resampled and taken equal to $b_j(n+1)$.

- If $\chi_j(n+1) = 1$ for a transition j which belongs to the set of transitions enabled at time t_n^+ in $\Gamma^\dagger$, then transition j is fired at time t_{n+1}^-, which defines a new set of enabled transitions at time t_{n+1}^+ by the usual token production and consumption rule.
- If $\chi_{j'}(n+1) = 1$ for a transition j' which is not enabled at time t_n^+ in $\Gamma^\dagger$, nothing happens as far as the marking process is concerned.

For each transition j which is enabled at t_{n+1}^+ (either still ongoing or newly enabled), one resamples a *new* residual firing time equal to $b_j(n+2)$, etc.

For $\Gamma^\dagger$ defined above, it is easily checked that the variables $x_j^\dagger(t)$ (respectively $y_j^\dagger(t)$) representing the number of firings of transition j initiated (respectively completed) by time t, satisfy the following equations:

$$\begin{aligned} y_j^\dagger(0) &= 0 \ , \\ y_j^\dagger(t) &= y_j^\dagger(t_n) \stackrel{\text{def}}{=} Y_j(n) \ , \quad t_n \le t < t_{n+1} \ , \\ x_j^\dagger(t) &= x_j^\dagger(t_n) \stackrel{\text{def}}{=} X_j(n) \ , \quad t_n \le t < t_{n+1} \ , \end{aligned} \tag{8.9}$$

and

$$X_j(n) = \bigwedge_{i \in p(j)} \left(Y_i(n)\mu(i,j)\right) \ , \tag{8.10}$$

$$Y_i(n+1) = \left(Y_i(n)\chi_i(n+1)\right) \wedge X_i(n) \ , \tag{8.11}$$

for all $n = 0, 1, 2, \ldots$. Equation (8.10) is obtained in the same way as (8.6). In order to obtain Equation (8.11) observe that

$$Y_i(n+1) \le Y_i(n) + \chi_i(n+1)$$

(equality holds if i is enabled at time t_n^+), and

$$\begin{aligned} i \text{ enabled at } t_n^+ &\Leftrightarrow Y_i(n) = X_i(n) - 1 \ ; \\ i \text{ not enabled at } t_n^+ &\Leftrightarrow Y_i(n) = X_i(n) \end{aligned}$$

(because of the recycling of transition i, there is at most one uncompleted firing initiated on this transition). Equation (8.11) is obtained from the following observation: either i is not enabled at t_n^+, and the smaller term in the right-hand side of (8.11) is $X_i(n)$, or i is enabled and $Y_i(n+1) = Y_i(n) + \chi_i(n+1)$.

It is now immediate to prove by induction that, for all realizations of the random variables $b_j(k)$, the state variables $X_j(n)$ and $Y_j(n)$, $1 \le j \le \mathfrak{n}, n \ge 0$, are nondecreasing and concave functions of μ. The variables $x_j^\dagger(t)$ and $y_j^\dagger(t)$ satisfy the same property in view of (8.9) and of the fact that the variables $b(n)$ do not depend upon μ. Thus, if μ and μ' are initial markings such that $\nu = \rho\mu + (1-\rho)\mu'$ is in $\mathbb{N}^{|\mathcal{P}|}$ for some real parameter $\rho \in (0,1)$, we have, with obvious notations,

$$x_j^\dagger(t;\nu) \ge \rho x_j^\dagger(t;\mu) + (1-\rho)x_j^\dagger(t;\mu') \ ,$$

for all ω, t and j, with a similar result for y.

Owing to the memoryless property of the exponential distribution, the counters $x_j^\dagger(t)$ and $y_j^\dagger(t)$, $1 \leq j \leq \mathfrak{n}$, are equal in distribution to $x_j(t)$ and $y_j(t)$, respectively. Therefore, under appropriate integrability assumptions,

$$\mathbb{E}\left[x_j(t;\nu)\right] \geq \rho\mathbb{E}\left[x_j(t;\mu)\right] + (1-\rho)\mathbb{E}\left[x_j(t;\mu')\right] ,$$

so that $x_j(t)$ (and $y_j(t)$) are stochastically increasing and concave in μ indeed. ■

We now define the class of PERT-exponential distribution functions, which will allow us to generalize Theorem 8.12.

Definition 8.13 (Stochastic PERT graph) *A stochastic PERT graph is a connected, directed, acyclic and weighted graph with a single source node and a single sink node, where the weights are random variables associated with nodes.*

There is no loss of generality in the assumption that only nodes are weighted; one can equivalently weight arcs or both arcs and nodes. In any weighted directed acyclic graph, the path with maximal weight is called the *critical path.*

Definition 8.14 (PERT-exponential distribution function) *The distribution function of a random variable X is of PERT-exponential type if X can be expressed as the weight of the critical path of a stochastic PERT graph $\mathcal{G}$ where the weights of the nodes are mutually independent random variables with exponential distribution functions.*

Notation 8.15 Such a distribution function will be denoted $F(\mathcal{G},\lambda)$, where $\mathcal{G}$ is the underlying graph and $\lambda = (\lambda_1,\ldots,\lambda_{|\mathcal{G}|})$, where λ_i is the parameter of the exponential distribution associated with node i in $\mathcal{G}$ (we will assume that the source and sink nodes are numbered 1 and $|\mathcal{G}|$ respectively). ■

Definition 8.16 (Log-concave functions) *A function $f : \mathbb{R}^n \to \mathbb{R}^+$ is* log-concave *if for all $x, y \in \mathbb{R}^n$ and $0 < \rho < 1$, the inequality $f(\rho x+(1-\rho)y) \geq f^\rho(x)f^{(1-\rho)}(y)$ holds.*

Theorem 8.17 *PERT-exponential distribution functions are* log-concave.

For a proof, see [10].

Theorem 8.18 *If the firing times of a stochastic event graph Γ are all mutually independent, and if for all transitions j, $1 \leq j \leq \mathfrak{n}$, the firing times $\beta_j(k)$ are i.i.d. random variables with PERT-exponential distribution function, then, for all $t \geq 0$, and all $1 \leq j \leq \mathfrak{n}$, $x_j(t)$ and $y_j(t)$ are stochastically increasing and concave in the initial marking μ.*

Proof Let $F(\mathcal{G}^j,\lambda^j)$ be the distribution function associated with transition q_j of Γ. Let $n^j \stackrel{\text{def}}{=} |\mathcal{G}^j|$.

For all j, consider the stochastic event graph Γ^j defined from the PERT graph $\mathcal{G}^j$ as follows: with each node i of $\mathcal{G}^j$, we associate a transition q_i^j in Γ^j; similarly, to each arc in $\mathcal{G}^j$ corresponds an arc in Γ^j and a place on this arc. The initial marking and the holding times of each of these places are zero. The firing times of transition q_i^j are i.i.d. random variables with exponential distribution function of parameter λ_i^j.

We now construct a stochastic event graph $\Gamma^\dagger$ which is defined from Γ and $\Gamma^j, j = 1, \ldots, \mathfrak{n}$, as follows: for all j, $1 \leq j \leq \mathfrak{n}$, we 'replace' transition q_j in Γ by the event graph Γ^j; all the places of Γ are kept in $\Gamma^\dagger$ together with their initial marking; we take $\pi^\dagger(q_1^j)$ equal to the set $\pi(q_j)$ and $\sigma^\dagger(q_{n^j}^j)$ equal to $\sigma(q_j)$, for all $j = 1, \ldots, \mathfrak{n}$; finally, we add a new feedback arc from transition $q_{n^j}^j$ to transition q_1^j, for all j; the place on this arc is assigned one token in its initial marking and zero holding and lag times. This transformation is depicted in Figure 8.1. If the number

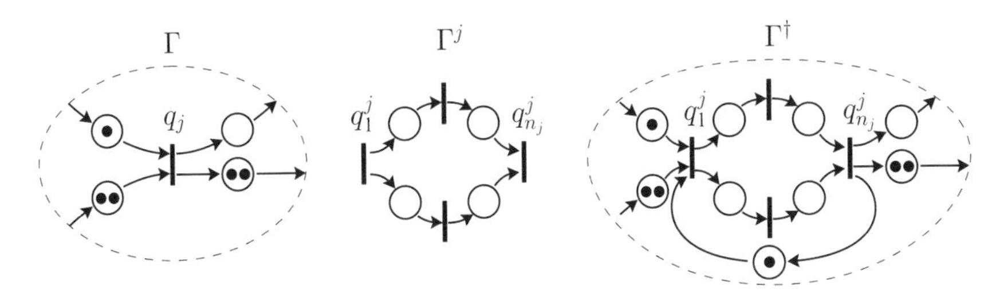

Figure 8.1: Transformation of an event graph with PERT-exponential firing times

of firings of transition q_1^j initiated (respectively completed) by time t in $\Gamma^\dagger$ is denoted $x_{j,1}^\dagger(t)$ (respectively $y_{j,1}^\dagger(t)$), then an immediate coupling argument shows that

$$x_j(t) =_{\text{st}} x_{j,1}^\dagger(t) \ , \quad y_j(t) =_{\text{st}} y_{j,1}^\dagger(t) \ , \quad \forall t \geq 0 \ , \quad \forall 1 \leq j \leq \mathfrak{n} \ ,$$

where the symbol $=_{\text{st}}$ denotes equivalence in law.

Let $\mu^\dagger$ denote the initial marking of $\Gamma^\dagger$. Applying Theorem 8.12 to $\Gamma^\dagger$ implies that for all $t \geq 0$, and all $1 \leq j \leq \mathfrak{n}$, $x_{j,1}^\dagger(t)$ and $y_{j,1}^\dagger(t)$ are stochastically increasing and concave in the initial marking $\mu^\dagger \in \mathbb{N}^{|\mathcal{Q}^\dagger|}$. Consequently, $x_j(t)$ and $y_j(t)$ are stochastically increasing and concave in the initial marking $\mu \in \mathbb{N}^{|\mathcal{Q}|}$, $t \geq 0$, $1 \leq j \leq \mathfrak{n}$. ■

Remark 8.19 It is easy to see that PERT-exponential distribution functions include Erlang distribution functions as a special case. Therefore one can approximate step functions with PERT-exponential distributions. Theorem 8.18 can be shown to hold when some of the firing times are deterministic, by using some adequate limiting argument. In the particular case when all firing times are deterministic and integer-valued, one can also prove the concavity of the counters by an immediate induction argument based on Formula (8.8). ■

8.2.5 Properties of Cycle Times

Throughout this subsection we suppose that the sequences of holding and firing times satisfy the joint stationary and ergodic assumptions of §7.3 and are integrable. We also assume that the event graph under consideration is strongly connected and its cycle time (see §7.3.4) is denoted $\mathfrak{a}$, so that the following a.s. limits hold:

$$\lim_{k\to\infty} \left|(x(k))^{1/k}\right|_{\oplus} = \lim_{k\to\infty} \mathbb{E}\left[|x(k)|_{\oplus}\right]^{1/k} = \mathfrak{a} \quad \text{a.s.,} \tag{8.12}$$

provided that the initial lag times are integrable. The throughput of the event graph will be denoted $\tau \overset{\text{def}}{=} \mathfrak{a}^{-1}$.

8.2.5.1 *First-Order Theorems for Counters*

Theorem 8.20 *For a strongly connected stochastic event graph satisfying the foregoing assumptions, the following a.s. limits hold:*

$$\lim_{t\to\infty} |x(t)|_{\wedge}^{1/t} = \lim_{t\to\infty} |x(t)|_{\oplus}^{1/t} = \lim_{t\to\infty} (x_j(t))^{1/t} = \tau \quad \textit{a.s.} \tag{8.13}$$

for all $1 \le j \le \mathfrak{n}$.

Proof For all $k \geq 1$ and $1 \le j \le \mathfrak{n}$, we have

$$x_j(t) = k \ , \quad \text{for } t \text{ in the interval} \quad x_j(k) \le t < x_j(k+1) \ ,$$

which implies that

$$\frac{x_j(k)}{k} \le \frac{t}{x_j(t)} < \frac{x_j(k+1)}{k} \ , \quad \text{for } t \text{ in} \quad x_j(k) \le t < x_j(k+1) \ .$$

When letting t or k go to ∞ in the last relation and when using (8.12), we obtain

$$\lim_{t\to\infty} (x_j(t))^{1/t} = \tau \quad \text{a.s.}$$

The proofs for $|x(t)|_{\wedge}$ and $|x(k)|_{\oplus}$ follow immediately. ■

Remark 8.21 In the particular case of a deterministic event graph, $\mathfrak{a}$ is given by the formula

$$\mathfrak{a} = \max_{\zeta} \frac{|\zeta|_{\mathrm{w}}}{|\zeta|_{\mathrm{t}}} \ ,$$

where ζ ranges over the set of circuits of Γ, $|\zeta|_{\mathrm{w}}$ is the sum of the firing and holding times in circuit ζ and $|\zeta|_{\mathrm{t}}$ is the number of tokens in the initial marking of ζ (see (3.7)). ■

The remainder of this section focuses on the derivation of monotonicity, convexity properties of cycle time and throughput.

8.2.5.2 Bounds on Cycle Times

The stochastic comparison properties obtained in §8.2.3 admit the following corollaries.

Corollary 8.22 *If both sequences* $\{\alpha(k), \beta(k)\}$ *and* $\{\alpha^\dagger(k), \beta^\dagger(k)\}$ *satisfy the above stationarity, ergodicity and integrability assumptions, and the assumptions of Theorem 8.2 (respectively Theorem 8.3), then the associated cycle times* $\mathfrak{a}$ *and* $\mathfrak{a}^{dagger}$ *are such that* $\mathfrak{a} \leq \mathfrak{a}^\dagger$.

Proof If the initial condition is integrable, the relation

$$\mathbb{E}\left[|x(k)|_\oplus\right] \leq \mathbb{E}\left[|x^\dagger(k)|_\oplus\right] ,$$

holds for all $k \geq 1$, as a direct consequence of Theorem 8.2 (respectively Theorem 8.3) because the function $x \mapsto f(x) = x$ is nondecreasing (and convex). Dividing this inequality by k and letting k go to ∞ yield the result in view of (8.12). ■

Remark 8.23 The preceding result extends immediately to the non strongly connected case. ■

The observation which was made in Remark 8.4 allows one to provide a general lower bound for the cycle time $\mathfrak{a}$ as shown by the following corollary.

Corollary 8.24 *Under the assumptions of Corollary 7.31, the cycle time* $\mathfrak{a}$ *of the stochastic event graph satisfies the bound*

$$\mathfrak{a} \geq \max_\zeta \frac{\mathbb{E}\left[|\zeta|_{\mathrm{w}}\right]}{|\zeta|_{\mathrm{t}}} .$$

where $\mathbb{E}\left[|\zeta|_{\mathrm{w}}\right]$ *denotes the mathematical expectation of the sum of holding and firing times in the circuit* ζ.

The right-hand side of the last expression is also the cycle time of a deterministic event graph with the same topology as the initial one, but with the firing and holding times replaced by their mean values. Under the above statistical assumptions, one obtains the following corollary in the same way.

Corollary 8.25 *Under the assumptions of Theorem 8.6 (respectively Theorem 8.8)* $\mathfrak{a} \geq \mathfrak{a}^\dagger$ *(respectively* $\mathfrak{a} \leq \mathfrak{a}^\dagger$*), where* $\mathfrak{a}^\dagger$ *is the cycle time of the event graph* $\Gamma^\dagger$ *which is considered in this theorem.*

Example 8.26 For instance, the throughput in queuing networks with blocking, open or closed, is stochastically decreasing in the service times (as a consequence of the property of line 4 in the last column of Table 8.1), and increasing in the buffer sizes and in the customer population (line 1), regardless of the statistical assumptions. ■

8.2.5.3 *Concavity with respect to the Initial Marking*

The concavity properties established in §8.2.4 for counters together with relation (8.13) readily imply a related concavity property of the throughput with respect to the initial marking.

Corollary 8.27 *Under the assumptions of §8.2.4.3, if the firing times are mutually independent and if in addition the firing times $\{\beta_j(k)\}$ are i.i.d. with PERT-exponential distribution functions, then the throughput τ is increasing and concave in the initial marking $\mu \in \mathbb{N}^{|\mathcal{E}|}$.*

Example 8.28 For instance, the throughput in queuing networks with blocking and with PERT-exponential service times is stochastically decreasing in the service times, and increasing and concave in the buffer sizes or the customer population. ■

8.2.6 Comparison of Ratios

Certain ratios of the state process $x(k)$, and hence the marking in the corresponding places, also exhibit interesting stochastic ordering properties. Roughly speaking, the places in question are those which do not belong to a strongly connected component of the event graph, which corresponds to the set of places with a marking which is not structurally bounded. The properties of interest are established through simple examples.

8.2.6.1 *Assumptions*

We consider an event graph with several strongly connected subgraphs. We assume that this event graph is in its canonical form, namely all places have exactly one token in the initial marking (see Remark 2.77). Let $\mathcal{Q}_{(n)}$ be the set of transitions corresponding to one of the subgraphs, where n is not a source node in the reduced graph. The evolution equation (7.47) of Chapter 7 reads

$$x_{(n)}(k+1) = A_{(n)(n)}(k)x_{(n)}(k) \oplus A_{(n)(<n)}(k)x_{(<n)}(k) \ , \quad k \geq 1 \ .$$

Let

$$\delta(k) \stackrel{\text{def}}{=} \frac{x_{(n)}(k+1)}{x_{(<n)}(k)} \ ,$$

$$U_+(k) \stackrel{\text{def}}{=} \frac{x_{(<n)}(k+1)}{x_{(<n)}(k)} \ , \qquad U_-(k) \stackrel{\text{def}}{=} \frac{x_{(<n)}(k)}{x_{(<n)}(k+1)} \ .$$

Then, the ratio process $\delta(k)$ satisfies the equation

$$\delta(k+1) = A(k+1)\delta(k)U_-(k) \oplus B(k+1)U_+(k)U_-(k) \ , \quad k \geq 0 \ , \qquad (8.14)$$

where $A(k) = A_{(n)(n)}(k)$ and $B(k) = A_{(n)(<n)}(k)$ (see §7.4.2).

8.2.6.2 Stochastic Ordering Results

Let $v(k) = (A(k+1), B(k+1), U_-(k))$ and $y(k) = \delta(k)$. Equation (8.14) can be rewritten as $y(k+1) = a(y(k), v(k))$.

Lemma 8.29 *The mapping $a(\cdot)$ satisfies the assumptions of Theorem 8.61.*

Proof The nondecreasingness of $y \mapsto a(y, v)$ is obvious. As for the convexity of the mapping $(y, v) \mapsto a(y, v)$, observe that $a(y(k), v(k))$ is given as the maximum of two functions, $A(k+1)\delta(k)U_-(k)$ and $B(k+1)U_+(k)U_-(k)$, so that it is sufficient to prove that each of these functions has the desired convexity property, which is clear for the first one. In order to prove the convexity property for the second function, we rewrite the entries of $U_-(k)U_+(k)$ as

$$\left(U_+(k)U_-(k)\right)_{ij} = U_-(k)_{1j} \not/ U_-(k)_{1i} \enspace ,$$

so that the entries of the second function can be rewritten as

$$\begin{aligned} \left(B(k+1)U_+(k)U_-(k)\right)_{ij} &= \bigoplus_l B_{il}(k+1)U_-(k)_{1j} \not/ U_-(k)_{1l} \\ &= U_-(k)_{1j} \left(\bigoplus_l B_{il}(k+1) \not/ U_-(k)_{1l} \right) \enspace . \end{aligned}$$

Since the mapping $(B(k+1), U_-(k)) \mapsto B_{il}(k+1) \not/ U_-(k)_{1l}$ is convex (it is linear in the conventional sense), each of these entries is the sum of two convex functions in the variables $U_-(k), B(k+1)$, which concludes the proof. ■

As a direct consequence of Theorem 8.61, we obtain the following result.

Corollary 8.30 *If one replaces the sequence $\{\delta(0), A(k), B(k), U_-(k)\}$ by a sequence $\left\{\delta^\dagger(0), A^\dagger(k), B^\dagger(k), U_-^\dagger(k)\right\}$ such that*

$$\{\delta(0), A(k), B(k), U_-(k)\} \geq_{\mathrm{cx}} \left\{\delta^\dagger(0), A^\dagger(k), B^\dagger(k), U_-^\dagger(k)\right\} \enspace ,$$

then the resulting ratio sequence $\delta^\dagger$ is such that $\delta \geq_{\mathrm{icx}} \delta^\dagger$.

Interesting applications of this property arise when the firing and holding times of the event graph are all mutually independent, so that the two sequences of random variables $\{A(k)\}$ and $\{B(k), U_-(k)\}$ of the preceding theorem are also mutually independent. For example, when applying the result of Corollary 8.30 to the sequences

$$B^\dagger(k) = \mathbb{E}\left[B(k)\right] \enspace , \quad U_-^\dagger(k) = \mathbb{E}\left[U_-(k)\right] \enspace , \quad A^\dagger(k) = A(k) \enspace ,$$

we obtain

$$\{\delta(k)\} \geq_{\mathrm{icx}} \left\{\delta^\dagger(k)\right\} \enspace , \tag{8.15}$$

provided the initial conditions which are chosen for both systems satisfy the assumption of the corollary. In particular, we obtain the relation

$$x_j(k+1) \not{} x_i(k) \geq_{\text{icx}} \delta^\dagger_{ji}(k) \ ,$$

for all $q_j \in Q_{(n)}$ and $q_i \in \pi^2(q_j) \cap \{Q \setminus Q_{(n)}\}$. The random variable $\delta^\dagger_{ji}(k)$ can be interpreted as the ratio $x^\dagger_j(k+1) \not{} u^\dagger_i(k)$ of a nonautonomous event graph $\Gamma^\dagger$ with the same topology as $Q_{(n)}$, with a $|Q_{(<n)}|$-dimensional input $u^\dagger(k)$ and with the evolution equation

$$x^\dagger(k+1) = A^\dagger(k)x^\dagger(k) \oplus B^\dagger(k)u^\dagger(k) \ .$$

The input process $u^\dagger(k)$ is determined by the relations

$$U^\dagger_-(k) = \frac{u^\dagger(k)}{u^\dagger(k+1)}$$

and by the initial condition $u^\dagger(0)$. This second event graph is 'simpler' than the previous one in that the influence of the predecessors of $Q_{(n)}$ is captured by the first moments of the variables $U_-(k)$ and $B(k)$ only.

Another example of application of Corollary 8.30 consists in choosing

$$B^\dagger(k) = \mathbb{E}\left[B(k)\right] \ , \quad U^\dagger_-(k) = \mathbb{E}\left[U_-(k)\right] \ , \quad A^\dagger(k) = \mathbb{E}[A(k)] \ .$$

With such a definition, we always have $v \geq_{cx} v^\dagger$, which leads to a comparison result between the ratio process of a stochastic event graph and that of a deterministic one (for which one can use the results of Chapter 3).

The conditions under which a stationary solution of (8.14) and its †-counterpart exist, are given in Theorem 7.96. Let us assume that these conditions are satisfied for both systems, so that one can construct the stationary marking in the initial event graph and in $\Gamma^\dagger$. One can then apply Little's formula ([77]) and (8.15) to derive the following bound on the stationary marking N_i in a place $p_i = \pi(q_j)$, where $q_j \in Q_{(n)}$ and $\pi(p_i) \notin Q_{(n)}$.

$$\mathbb{E}[N_i] = \frac{\mathbb{E}\left[\delta_{j,\pi^p(i)}\right]}{\mathfrak{a}_1} \geq \frac{\mathbb{E}\left[\delta^\dagger_{j,\pi^p(i)}\right]}{\mathfrak{a}_1} \ .$$

In this relation, $\mathfrak{a}_1$ is the cycle time of transitions q_j and $q_{\pi^p(i)}$ (these cycle times must coincide since the place is assumed to be stable). The real number $\mathbb{E}\left[\delta_{j,\pi^p(i)}\right]$ represents the average time spent by a token in p_i in the stationary regime (the time spent by the k-th token of p_i in this place is $x_j(k) - x_{\pi^p(i)}(k-1)$ indeed).

Similarly, under the preceding independence assumptions, it is easily seen that the variables $\delta(k)$ are stochastically increasing and convex in $\{A_{(n)(n)}(k)\}$.

8.3 EVENT GRAPHS AND BRANCHING PROCESSES

This section focuses on the derivation of bounds and estimates for cycle times of strongly connected stochastic event graphs with i.i.d. firing times. We use association properties satisfied by partial sums of the firing times in order to prove that the daters can be compared for $\leq_{\text{st}}$ with the last birth in a multitype branching process, the structure of which is determined from the characteristics of the event graph. Classical large deviation estimates are then used to compute the growth rate of this last birth epoch, following the method developed in [19]. This allows one to derive a computable upper bound for the cycle time, which is exemplified on tandem queuing networks with communication blocking.

8.3.1 Statistical Assumptions

The assumptions are those of §8.2.4. In addition to this, we assume that the sequences $\{\beta_j(k)\}_{k=1}^{+\infty}, j = 1, \ldots, \mathfrak{n}$, are mutually independent sequences of i.i.d. nonnegative and integrable random variables defined on a common probability space $(\Omega, \mathbb{F}, \mathbb{P})$, and that the initial number of tokens in any place is at most 1 (this last assumption introduces no loss of generality, see Remark 2.77).

We know from Chapter 2 that whenever the event graph under consideration is live, it is possible to rewrite its equation as

$$x(k) = A(k)x(k-1) \ , \quad k \geq 1 \ , \tag{8.16}$$

where matrix $A(k)$ is defined as follows:

$$A_{jj'}(k) = \bigoplus_{\{(j'=i_0,i_1,i_2\ldots,i_{h-1},i_h=j)\in S(j',j,1)\}} \beta_{j'}(k-1) \otimes \left(\bigotimes_{m=1}^{h-1} \beta_{i_m}(k) \right) \ , \tag{8.17}$$

with the usual convention if the set $S(j', j, 1)$ is empty (see Remark 2.69). It is assumed that the event graph under consideration is strongly connected, and that the initial condition $x(0)$ is equal to e (since we are only interested in determining the cycle time, this last assumption introduces no loss of generality). The following theorem is based on the notion of association (see Definition 8.64).

8.3.2 Statistical Properties

Lemma 8.31 *Under the foregoing statistical assumptions, $\{A_{ij}(k), x_j(k), i, j = 1, \ldots, \mathfrak{n}, k \geq 0\}$ forms a set of associated random variables (see Definition 8.64).*

Proof The independence assumption on the firing times implies that the random variables $\{A(k)\}$ are associated since they are obtained as increasing functions of associated random variables (see (8.17)). The result for $x_j(k)$ follows immediately from (8.16) and Theorem 8.67. ■

Lemma 8.32 *For all $j_0, j_1, \ldots, j_h \in \{1, \ldots, \mathfrak{n}\}$, the random variables $A_{j_{k+1}j_k}(k)$, $k = 0, \ldots, h$, are mutually independent.*

Proof In view of (8.17), the random variables $A_{j_{k+1}j_k}(k)$ can all be written in the form

$$A_{j_{k+1}j_k}(k) = \phi_k(\beta_{l_1}(k), \ldots, \beta_{l_{p_k}}(k), \beta_{l'_1}(k-1)) \ ,$$

for some indices $l_1, \ldots, l'_1$. Since the random variables $\{\beta_j(k)\}_{j,k}$ are mutually independent, it is enough to show that the arguments of the functions ϕ_k, $k = 0, \ldots, h$, are disjoint to prove the property.

The only situation where these sets of arguments could fail being disjoint is for two adjacent terms $A_{j_{k+1}j_k}(k)$ and $A_{j_{k+2}j_{k+1}}(k+1)$ having one argument of the type $\beta_j(k)$ in common. The only such argument in $A_{j_{k+2}j_{k+1}}(k+1)$ is $\beta_{j_{k+1}}(k)$. Assume that this is also an argument of $A_{j_{k+1}j_k}(k)$. Then, there exists a circuit crossing j_{k+1} and j_{k+1} with zero initial marking in all the places of the circuit, which contradicts the liveness assumption. ■

8.3.3 Simple Bounds on Cycle Times

Let $\mathfrak{a}$ be the cycle time of $A(k)$. Since we assumed strong connectedness, we have

$$\lim_{k\to\infty} \mathbb{E}\left[x_j(k)\right]^{1/k} = \lim_{k\to\infty} (x_j(k))^{1/k} = \mathfrak{a} \quad \text{a.s.,} \quad \forall j = 1, \ldots, \mathfrak{n} \ . \tag{8.18}$$

Let

- N be the maximal degree of the transitions which are followed by at least one place with a nonzero initial marking (the degree of a node is the number of arcs incident with this node).
- b be a random variable which is a $\leq_{\text{st}}$ upper bound of each of the random variables $A_{ij}(0)$, namely

$$A_{ij}(0) \leq_{\text{st}} b \ , \quad \forall i, j = 1, \ldots, \mathfrak{n} \ .$$

- $b(z)$ be the Laplace transform[1] $\mathbb{E}\left[\exp(zb)\right]$, which is assumed to be finite in a neighborhood of $z = 0$.
- $M(x)$ be the Cramer-Legendre transform of the distribution function of the random variable b, namely

$$M(x) = \inf_{z\in\mathbb{R}} \left(\log(b(z)) - zx\right) \ .$$

The present section is devoted to the proof of the following result.

Theorem 8.33 *Let $\gamma = \inf\{x \mid x > \mathbb{E}[b], M(x)+\log(N) < 0\}$. Under the foregoing assumptions, the cycle time of the event graph admits the upper bound $\mathfrak{a} \leq \gamma$.*

We start with two preliminary lemmas.

---

[1]This is not the usual Laplace transform which would read $\mathbb{E}\left[\exp(-zb)\right]$.

Lemma 8.34 *For all* $\epsilon > 0$, *and for all* $j = 1, \dots, \mathfrak{n}$,

$$\lim_{k\to\infty} \mathbb{P}\left[\frac{x_j(k)}{k} < \mathfrak{a} + \epsilon\right] = 1 \ ,$$

and

$$\lim_{k\to\infty} \mathbb{P}\left[\frac{x_j(k)}{k} < \mathfrak{a} - \epsilon\right] = 0 \ .$$

Proof The property follows immediately from the fact that a.s. convergence implies convergence in probability and from (8.18). ∎

Lemma 8.35 *If* $c \in \mathbb{R}$ *is such that*

$$\lim_{k\to\infty} \mathbb{P}\left[x_j(k) - kc \le 0\right] = 1 \ , \tag{8.19}$$

for some $j = 1, \dots, \mathfrak{n}$, *then* $c \ge \mathfrak{a}$.

Proof Under the assumption (8.19),

$$\lim_{k\to\infty} \mathbb{P}\left[\frac{x_j(k)}{k} \le c\right] = 1 \ ,$$

so that we cannot have $c = \mathfrak{a} - \epsilon$ for some $\epsilon > 0$, in view of Lemma 8.34. Therefore, $c \ge \mathfrak{a}$. ∎

Proof of Theorem 8.33 Under our assumptions, Equation (8.16) reads $x(k+1) = A(k) \otimes x(k)$ with the initial condition $x(0) = e$ and it is easily checked by induction that

$$x_j(k) = \bigoplus_{j_0,\dots,j_{k-1}\in\{i,\dots,\mathfrak{n}\}} \bigotimes_{h=0}^{k-1} A_{j_{h+1},j_h}(h) \ , \tag{8.20}$$

where $j_k = j$. Therefore,

$$\mathbb{P}\left[x_j(k) - ck \le 0\right] = \mathbb{P}\left[\max_{j_0,\dots,j_{k-1}\in\{i,\dots,\mathfrak{n}\}} \sum_{h=0}^{k-1} C_{j_{h+1},j_h}(h) \le 0\right] ,$$

where $C_{ij}(k) \stackrel{\text{def}}{=} A_{ij}(k) - c$.

For k fixed, Lemma 8.31 implies that the variables $\sum_{h=0}^{k-1} C_{j_{h+1},j_h}(h)$, where $j_0, \dots, j_{k-1}$ vary over the set $\{1, \dots, \mathfrak{n}\}^k$, are associated. Therefore, from Lemma 8.66,

$$\mathbb{P}\left[\max_{j_0,\dots,j_{k-1}\in\{i,\dots,\mathfrak{n}\}} \sum_{h=0}^{k-1} C_{j_{h+1}j_h}(h) \le 0\right] \ge \prod_{j_0,\dots,j_{k-1}\in\{1,\dots,\mathfrak{n}\}} \mathbb{P}\left[\sum_{h=0}^{k-1} C_{j_{h+1},j_h}(h) \le 0\right] .$$

Since the random variables $A_{j_{h+1},j_h}(h)$ are independent (see Lemma 8.32), and $\leq_{\mathrm{st}}$-bounded from above by b, we have

$$\mathbb{P}\left[\sum_{h=0}^{k-1} C_{j_{h+1},j_h}(h) \leq 0\right] \geq \mathbb{P}\left[\sum_{h=0}^{k-1} (b(h) - c) \leq 0\right] ,$$

where $\{b(h)\}$ is a sequence of i.i.d. random variables with the same distribution function as b. Now, Chernoff's Theorem ([3]) implies

$$\mathbb{P}\left[\sum_{h=0}^{k-1} b(h) > ck\right] = \exp\left(M(c)k + \mathrm{o}(k)\right) ,$$

for all $c > \mathbb{E}[b]$, so that

$$\mathbb{P}\left[x_j(k) - ck \leq 0\right] \geq \left(1 - \exp(M(c)k + \mathrm{o}(k))\right)^{C_j(k)} ,$$

where $C_j(k)$ denotes the number of paths $j_0, \dots, j_{k-1}$ which satisfy the property $\sum_{h=0}^{k-1} C_{j_{h+1},j_h}(h) \neq -\infty$. Therefore, it is enough to have the limit

$$C_j(k) \exp(kM(c)) \to 0$$

when k goes to ∞, in order to obtain

$$\lim_{k\to\infty} \mathbb{P}\left[x_j(k) - ck \leq 0\right] = 1 . \tag{8.21}$$

Clearly, the bound $C_j(k) \leq N^k$ holds, so that a sufficient condition for (8.21) to hold is $M(c) + \log(N) < 0$. In other words, for $c > \mathbb{E}[b]$ such that $M(c) + \log N < 0$, (8.21) holds, so that $c \geq \mathfrak{a}$ in view of Lemma 8.35. ■

In fact, we proved the following and more general result.

Corollary 8.36 *If* $\log(C_j(k)) = Ck + \mathrm{o}(k)$, *then* $\mathfrak{a} \leq \inf\{c \mid M(c) + C < 0\}$.

Example 8.37 (Blocking queues in tandem) Consider the example of Figure 8.2, which represents a line of processors with blocking before service, also called communication blocking in the exponential case. Let $\mathfrak{n}$ denote the number of processors, each of which is represented by a transition. In Figure 8.2, $\mathfrak{n}$ equals 4. The first processor (on the left of the figure) has an infinite buffer of items to serve. Between two successive processors, the buffer is of capacity one (which is captured by the fact that there are two tokens in any of the upper circuits originating from a processor). The processors are single servers with a FIFO discipline (which is captured by the lower circuit associated with each transition). It is assumed that all transitions have exponentially distributed firing times with parameter 1. In this example, we have $N = 3$, $b(z) = (1 - z)^{-1}$. The Cramer-Legendre transform of $b(z)$ is given by

$$M(x) = \inf_{z\in[0,1)} \left(-zx - \log(1 - z)\right) .$$

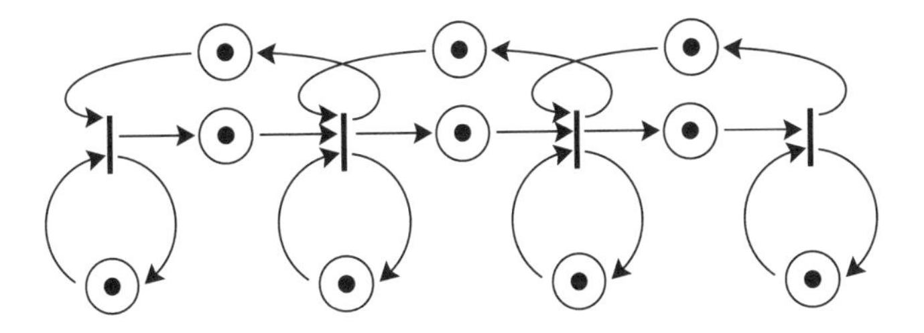

Figure 8.2: Communication blocking: 4 nodes, 1 buffer

The derivative of the function $-zx - \log(1-z)$ with respect to z vanishes for $z = 1 - x^{-1}$ and this point is a minimum. Therefore

$$M(x) = 1 - x + \log(x) \ .$$

As a direct application of Theorem 8.33, we obtain

$$\mathfrak{a} \leq \inf\{x \mid 1 - x + \log(x) + \log(3) < 0\} \ ,$$

which provides the following uniform bound in $\mathfrak{n}$: $\mathfrak{a} \leq 3.33$. In other words, the throughput of the systems is always greater than 0.3, regardless of the number of processors.

If we apply Theorem 8.36 using the following more precise estimate of $C_j(k)$

$$\frac{1}{k} \log C_j(k) = 1 + 2\cos\left(\frac{\pi}{\mathfrak{n}+1}\right) + \mathrm{o}(k) \leq 3 \tag{8.22}$$

(see below for its proof), we obtain

$$\mathfrak{a} \leq \inf\left\{x \;\middle|\; 1 - x + \log(x) < \log\left(1 + 2\cos\left(\frac{\pi}{5}\right)\right)\right\} \simeq 3.09 \ .$$

If the service times are Erlang-3 with mean 1, namely if $b(z) = (3/(3-z))^3$, in the same way, from Theorem 8.33, we obtain that

$$\mathfrak{a} \leq \inf\{x \mid 3(x - 1 - \log(x)) > \log(3)\} \simeq 2.11 \ ,$$

which corresponds to a throughput greater than 0.48.

Proof of Formula (8.22) Let K denote the adjacency matrix of the precedence graph of $A(0)$, namely the $\mathfrak{n} \times \mathfrak{n}$ matrix such that $K_{i,j} = 1$ if $i \in \pi(j)$, and 0 otherwise. For $\mathfrak{n} = 4$, we obtain for instance

$$K = \begin{pmatrix} 1 & 1 & 0 & 0 \\ 1 & 1 & 1 & 0 \\ 0 & 1 & 1 & 1 \\ 0 & 0 & 1 & 1 \end{pmatrix} .$$

It is easily checked by induction that $K_{i,j}^k$ counts the number of paths of length k from i to j. Let P be the substochastic matrix defined by $P = K/3$, and let λ_n denote the Perron-Frobenius eigenvalue associated with P. From the irreducibility of P, we obtain

$$C_j(k) = \sum_{i=1}^{n} K_{i,j}^k = \mathrm{O}(3\lambda_n)^k \ .$$

In order to evaluate λ_n, we introduce the Markov chain Z_k with substochastic transition matrix P and uniform initial measure. We then have

$$\mathbb{P}\left[Z_k = 1\right] = \mathrm{O}(\lambda_n)^k \ .$$

We can now evaluate $\mathbb{P}\left[Z_k = j\right]$ using the recurrence relations

$$\begin{aligned} \mathbb{P}\left[Z_{k+1} = j\right] &= \mathbb{P}\left[Z_k = j-1\right]/3 + \mathbb{P}\left[Z_k = j\right]/3 + \mathbb{P}\left[Z_k = j+1\right]/3 \ , \\ & \qquad\qquad 1 < j < n \ , \\ \mathbb{P}\left[Z_{k+1} = 1\right] &= \mathbb{P}\left[Z_k = 1\right]/3 + \mathbb{P}\left[Z_k = 2\right]/3 \ , \\ \mathbb{P}\left[Z_{k+1} = n\right] &= \mathbb{P}\left[Z_k = n-1\right]/3 + \mathbb{P}\left[Z_k = n\right]/3 \ . \end{aligned} \tag{8.23}$$

Let

$$P(x, y) = \sum_{k=0}^{\infty} \sum_{j=1}^{n} x^k y^j \mathbb{P}\left[Z_k = j\right] \ .$$

From (8.23), we obtain

$$P(x, y) = \frac{G(y) - F(x)(1 + y^{n+1})}{3 - x(y + 1 + y^{-1})} \ , \tag{8.24}$$

where $F(x)$ is the function

$$F(x) = x \sum_{k=0}^{\infty} \mathbb{P}\left[Z_k = 1\right] x^k \ ,$$

and

$$G(y) = 1 + y + y^2 + \ldots + y^n \ .$$

The denominator of (8.24) vanishes for $x = x(y) = 3/(y + 1 + y^{-1}) \leq 1$. Therefore, we necessarily have

$$F(x) = \frac{G(y)}{1 + y^{n+1}} \ ,$$

for $x = x(y)$. The poles of $F(x)$ are for $y^{n+1} = -1$, namely for

$$y(l) = \exp\left(\frac{i\pi + 2il\pi}{n+1}\right), \quad l = 0, \ldots, n \ .$$

We have

$$x(y(l)) = \frac{3}{1 + 2\cos\left(\frac{\pi(1+2l)}{n+1}\right)} \ ,$$

the smallest of which is for $l = 0$. Therefore, from classical theorems on generating functions,

$$\mathbb{P}[Z_k = 1] = \mathrm{O}(x(y(0))^{-k}) \ ,$$

Or, equivalently,

$$\lambda_n = \frac{1 + 2\cos\left(\frac{\pi}{n+1}\right)}{3} \ ,$$

which in turn implies

$$C_1(k) = 1 + 2\cos\left(\frac{\pi}{n+1}\right) \ .$$

■

8.3.4 General Case

8.3.4.1 Multitype Branching Processes

The class of age-dependent multitype branching processes considered in this section is a special case of those considered in [19]. There are n types; the branching process is characterized by a family of integer-valued random processes

$$Z_{ji}^{kl}(t) \ ; \quad t \in \mathbb{R}^+; \quad i, j = 1, \ldots, n \ ; \quad k, l = 1, 2, \ldots \ ,$$

where

- the $n \times n$ matrices $Z^{kl}(\cdot)$ are i.i.d. for $l, k = 1, 2, \ldots$;
- the variables $Z_{ji}^{kl}(\cdot)$ are mutually independent in i, j, and this for all $l, k = 1, 2, \ldots$.

Index k refers to generation levels from an initial generation called 1, and index l is used to count individuals of a given type within a generation level. If the branching process is initiated by a single generation-1 individual of type j, born at time 0, this individual gives birth to a total of $Z_{ji}^{11}(\infty)$ generation-2 individuals of type i, one at each jump time of $Z_{ji}^{11}(\cdot)$. Once a generation-k individual of type i is born, it is assigned an integer l, different from all the integers assigned to already born individuals of the same generation and type (for instance, the individuals of generation 2 and type i can be numbered $1, \ldots, Z_{ji}^{11}(\infty)$). Then, the random function $Z_{ih}^{kl}(t)$ is used to determine the number of generation-$(k+1)$ individuals of type h born from the latter individual in less than $t \in \mathbb{R}^+$ after its own birth.

Let $T_{ji}^{(k)}(t) \in \mathbb{N}$ denote the total number of generation-k individuals of type i born by time t in a branching process initiated at time 0 from a single generation-1 individual of type j. Let $F_{ji}(t)$ be the monotonic function defined by the relation

$$F_{ji}(t) = \mathbb{E}\left[Z_{ji}^{11}(t)\right] \ , \quad t \in \mathbb{R}^+ \ ,$$

and let $\Phi(z)$ be the $\mathfrak{n} \times \mathfrak{n}$ matrix with entries

$$\Phi_{ji}(z) = \int_0^\infty \exp(zt)\, F_{ji}(\mathrm{d}t) \ .$$

We assume that there exists a real neighborhood of 0 where the matrix $\Phi(z)$ is finite.

Lemma 8.38 (Biggins) *Under the above assumptions,*

$$\mathbb{E}\left[\int_0^\infty \exp(zt)\, T_{ji}^{(k)}(\mathrm{d}t)\right] = \Phi_{ji}^k(z) \ , \tag{8.25}$$

where Φ^k denotes the k-th power of Φ.

Proof Let $\mathbb{F}_k$ denote the σ-field of the events up to the k-th generation. Owing to the independence assumptions, we obtain the vector relation

$$\mathbb{E}\left[\int_0^\infty \exp(zt)\, T_j^{(k+1)}(\mathrm{d}t) \;\middle|\; \mathbb{F}_k\right] = \left(\int_0^\infty \exp(zt)\, T_j^{(k)}(\mathrm{d}t)\right) \Phi(z) \ ,$$

where $T_j^{(k)(\cdot)}$ denotes the vector $(T_{j1}^{(k)}(\cdot), \dots, T_{j\mathfrak{n}}^{(k)}(\cdot))$. By taking expectations in the last expression, we obtain (8.25). ∎

8.3.4.2 Comparison between Event Graphs and Branching Processes

Consider now the following specific age-dependent branching process associated with the stochastic event graph under consideration:

- there are as many types as there are transitions followed by at least one place with a nonzero initial marking, namely $\mathfrak{n}$;
- the random vector $Z_j^{11}(t)$ is defined through its probability law by the relation $Z_{ji}^{11}(t) =_{\mathrm{st}} 1_{i \in p(j)} 1_{A_{ji}(0) \le t}$, for all $i, j \in \{1, \dots, \mathfrak{n}\}$. This fully defines the probability law of the matrices $Z^{kl}(\cdot)$ in view of the independence assumptions.

Observe that, for this specific branching process, an individual of type j gives birth to at most one individual of type i, for all i, j.

Let $\widehat{x}_j(k)$ be the epoch of the latest birth of all generation-k individuals ever born in the above branching process, when this one is initiated by an individual of type j at time 0.

Lemma 8.39 *Under the foregoing statistical assumptions, for all $j \in \{1, \dots, \mathfrak{n}\}$ and $k \ge 1$,*

$$x_j(k) \le_{\mathrm{st}} \widehat{x}_j(k) \ ,$$

provided $x(0) = e$.

Proof From the definition of $x(k)$, for all $t \in \mathbb{R}^+$,

$$\mathbb{P}[x_j(k) \le t] = \mathbb{P}\left[\bigoplus_{j_0,\dots,j_{k-1}\in\{1,\dots,n\}} \bigotimes_{h=0}^{k-1} A_{j_{h+1},j_h}(h) \le t \right] ,$$

where $j_k = j$. Therefore, the association property of Lemma 8.31 implies that

$$\begin{aligned} \mathbb{P}\left[x_j(k) \le t\right] &\ge \prod_{j_0,\dots,j_{k-1}\in\{i,\dots,n\}} \mathbb{P}\left[\sum_{h=0}^{k-1} A_{j_{h+1},j_h}(h) \le t\right] \\ &= \prod_{\substack{j_0,\dots,j_{k-1}\in\{i,\dots,n\} \\ j_h\in p(j_{h+1})}} \mathbb{P}\left[\sum_{h=0}^{k-1} A_{j_{h+1},j_h}(h) \le t\right] . \end{aligned} \tag{8.26}$$

Now, from its very definition, the event $\{x_j(k) \le t\}$ can be written as

$$\bigcap_{\substack{j_0,\dots,j_{k-1}\in\{i,\dots,n\} \\ j_h\in p(j_{h+1})}} \left\{\sum_{h=0}^{k-1} \overline{A}_{j_{h+1},j_h}(h) \le t\right\} ,$$

where the random variables $\overline{A}_{j_{h+1},j_h}$ are all mutually independent, and where $\overline{A}_{j_{h+1},j_h}$ has the same probability law as A_{j_{h+1},j_h}. Since the random variables in the right-hand side of (8.26) are also mutually independent (see Lemma 8.32), the latter expression coincides with $\mathbb{P}\left[\widehat{x}_j(k) \le t\right]$. ■

8.3.4.3 *Upper Bounds for Cycle times*

Whenever the integrals defining the entries of $\Phi(z)$ converge, this matrix is positive; its Perron-Frobenius eigenvalue ([61]) is denoted $\phi(z)$. Let $M(x)$ be the Cramer-Legendre transform of $\phi(z)$:

$$M(x) = \inf_{z>0} \left(\log(\phi(z)) - zx\right) .$$

It is well known that $M(x)$ is decreasing for $x \ge 0$ (see [3]). Let γ be defined by

$$\gamma = \inf\{x \mid M(x) < 0\} .$$

Theorem 8.40 *Under the foregoing statistical assumptions, the cycle time* $\mathfrak{a}$ *of the event graph is such that*

$$\mathfrak{a} \le \gamma . \tag{8.27}$$

Proof We first prove that

$$\limsup_k \frac{\widehat{x}_j(k)}{k} \le \gamma \quad \text{a.s.} \tag{8.28}$$

Let $v(z)$ be the right eigenvector associated with the maximal eigenvalue $\phi(z)$. From (8.25), we obtain that

$$\left\langle \mathbb{E}\left[\int_0^\infty \exp(zt)\, T_j^{(k)}(\mathrm{d}t)\right], v(z) \right\rangle = \phi^k(z) v_j(z) \;,$$

so that

$$\left\langle \mathbb{E}\left[\int_0^\infty \exp(zt)\, T_j^{(k)}(\mathrm{d}t)\right], 1 \right\rangle \leq \phi^k(z) v_j(z) u(z) \;, \tag{8.29}$$

where $u(z) = (\min_i v_i(z))^{-1}$ ($v(z)$ is strictly positive due to the Perron-Frobenius theorem). Now, since $\widehat{x}_j(k) = \sup\{t \mid \exists i = 1, \ldots, \mathfrak{n}, T_{ji}^{(k)}(t) = 0\}$, we have

$$\begin{aligned} \left\langle \mathbb{E}\left[\int_0^\infty \exp(zt)\, T_j^{(k)}(\mathrm{d}t)\right], 1 \right\rangle &= \sum_{i=1}^{\mathfrak{n}} \mathbb{E}\left[\int_0^\infty \exp(zt)\, T_{ji}^{(k)}(\mathrm{d}t)\right] \\ &\geq \mathbb{E}\left[\exp(z\widehat{x}_j(k))\right] \;. \end{aligned} \tag{8.30}$$

In addition, for $z \geq 0$,

$$\mathbb{P}\left[\frac{\widehat{x}_j(k)}{k} \geq c\right] \leq \mathbb{E}\left[\exp\left(z\left(\frac{\widehat{x}_j(k)}{k} - c\right)\right)\right] .$$

This, plus (8.29) and (8.30), in turn imply

$$\lim_k \frac{1}{k} \log \mathbb{P}\left[\widehat{x}_j(k) \geq kc\right] \leq \inf_{z>0}(\log(\phi(z)) - zc) = M(c) \;.$$

Therefore, for all c such that $M(c) < 0$, $\sum_{k\geq 1} \mathbb{P}\left[\widehat{x}_j(k) \geq kc\right] < \infty$, so that the Borel-Cantelli Lemma immediately implies (8.28).

From Lemma 8.39, for all bounded and nondecreasing functions f,

$$\mathbb{E}\left[f\left(\frac{x_j(k)}{k}\right)\right] \leq \mathbb{E}\left[f\left(\frac{\widehat{x}_j(k)}{k}\right)\right] \;,$$

for all $j = 1, \ldots, \mathfrak{n}$.

In view of (8.18), from the Lebesgue dominated convergence theorem we obtain that, for f bounded and continuous,

$$\lim_k \mathbb{E}\left[f\left(\frac{x_j(k)}{k}\right)\right] = f(\mathfrak{a}) \;.$$

In addition, for f continuous, monotonic, nondecreasing and bounded, we also have

$$\begin{aligned} \limsup_k \mathbb{E}\left[f\left(\frac{\widehat{x}_j(k)}{k}\right)\right] &\leq \mathbb{E}\left[\limsup_k f\left(\frac{\widehat{x}_j(k)}{k}\right)\right] = \mathbb{E}\left[f\left(\limsup_k \frac{\widehat{x}_j(k)}{k}\right)\right] \\ &\leq f(\gamma) \;, \end{aligned}$$

where we successively used Fatou's lemma, the monotonicity and continuity of f, and finally (8.28). Therefore $f(\mathfrak{a}) \leq f(\gamma)$, for all nondecreasing, continuous and bounded f, which immediately implies (8.27). ■

Observe that in the particular case when all non $-\infty$ entries of $A(1)$ have the same distribution characterized by the function $b(z)$, the eigenvalue of interest is precisely $\phi(z) = b(z)C$, where C is the Perron-Frobenius eigenvalue of the adjacency matrix associated with the matrix A, namely the maximal eigenvalue of matrix $\Phi(0)$.

Example 8.41 (Blocking queues with transportation times) The example is that of the line of processors described previously, but with a deterministic transportation time between processors. The associated event graph is obtained from that of Figure 8.2 by replacing each buffer by two buffers connected by a transportation transition with deterministic firing times δ as shown in Figure 8.3. The statistical

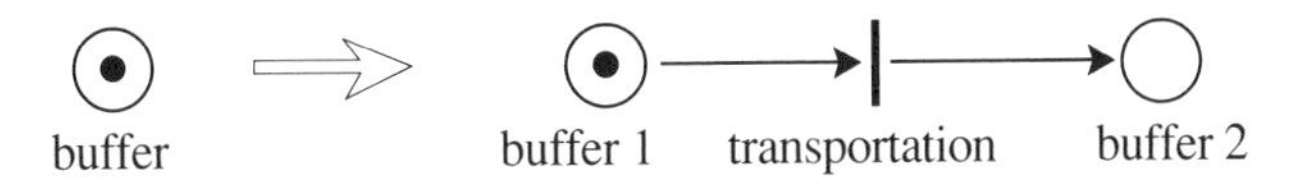

Figure 8.3: Blocking with transportation times

assumptions concerning the firing times of transitions associated with processors are those of the previous example. In this example, we have

$$\Phi(z) = \begin{pmatrix} \frac{1}{1-z} & \frac{1}{1-z} & 0 & 0 \\ \frac{\exp(\delta z)}{1-z} & \frac{1}{1-z} & \frac{1}{1-z} & 0 \\ 0 & \frac{\exp(\delta z)}{1-z} & \frac{1}{1-z} & \frac{1}{1-z} \\ 0 & 0 & \frac{\exp(\delta z)}{1-z} & \frac{1}{1-z} \end{pmatrix} .$$

The Perron-Frobenius eigenvalue of this matrix is

$$\phi(z) = \frac{1}{1-z}\left(1 + 2\exp\left(\frac{\delta z}{2}\right)\cos\left(\frac{\pi}{n+1}\right)\right) \tag{8.31}$$

(the proof is similar to that of the previous case). The technique is then the same as above for deriving the upper bound γ. The lower bounds $\mathfrak{a}^+$ given in the following arrays are those obtained by convex ordering following the method indicated in Theorem 8.24. For $n = 4$, one obtains the following array at the left-hand side. For n large, the lower bound is unchanged, and we obtain the following upper bound indicated in the array at the right-hand side.

δ	0	1	2	3
γ	3.1	3.3	3.7	4.2
$\mathfrak{a}^+$	1	1.5	2	2.5

δ	0	1	2	3
γ	3.3	3.6	4.0	4.4
$\mathfrak{a}^+$	1	1.5	2	2.5

■

Remark 8.42 Better upper bounds can be derived when considering the constant γ_l, $l \geq 0$, associated with the n-type, age-dependent branching process of probability law

$$Z_{ji}^{11}(t) =_{\mathrm{st}} 1_{\{-\infty<(A(0)A(1)\ldots A(l))_{ji}\leq t\}} \ .$$

The constant γ referred to in Theorem 8.40 corresponds to γ_0. It can be shown that the sequence $\{\gamma_l\}$ decreases to a limit when l tends to ∞. ∎

Remark 8.43 Lower bounds on cycle times based on convex ordering were discussed in the previous section. We have shown how to derive upper bounds based on large deviations in the present section. Since the stability region of a non strongly connected event graph is obtained by comparing the cycle times of its strongly connected components (see Theorems 7.69 and 7.96), these two bounding methods also provide a way to analyze the stability region of this class of systems. ∎

8.4 MARKOVIAN ANALYSIS

8.4.1 Markov Property

The evolution equation studied in this section is

$$x(k+1) = A(k)x(k) \ , \quad k = 0, 1, 2, \ldots, \tag{8.32}$$

with initial condition $x_0 \in \mathbb{R}^n$.

Theorem 8.44 *If the matrices $A(k)$ are i.i.d. and independent of the initial condition x_0, the sequence $\{z(k)\} \stackrel{\mathrm{def}}{=} \{x(k) \not/ x_1(k)\}$ forms an $\mathbb{R}^n$-valued Markov chain.*

Proof The Markov property follows immediately from the relation

$$z(k+1) = \frac{x(k+1)}{x_1(k+1)} = \frac{A(k)x(k)}{(A(k)x(k))_1} = \frac{A(k)z(k)}{(A(k)z(k))_1} \ ,$$

and from the independence assumptions (see Theorem 8.68). ∎

There is no general theory available for computing the invariant measure of this Markov chain. The following sections will therefore focus on simple examples. These examples are obtained either from specific problems described in Chapter 1 or from simplifying mathematical assumptions[2] on the structure of matrices $A(k)$. The quantities of interest are

$$\lim_{k\to\infty} \mathbb{E}\left[x_i(k+1) \not/ x_i(k)\right] \ ,$$

for an arbitrary i, which coincides with the Lyapunov exponent of m.s.c.s. $[i]$, and the distribution of the stationary ratios.

[2]These examples are defined through (8.32) as stochastic $\mathbb{R}_{\max}$-linear systems; they do not necessarily have an interpretation in terms of FIFO stochastic event graphs, as defined in Chapter 2.

8.4.2 Discrete Distributions

Example 8.45 Consider the case when $x \in \mathbb{R}^2$ and when matrix $A(k)$ is one of the following two matrices:

$$\begin{pmatrix} 3 & 7 \\ 2 & 4 \end{pmatrix}, \quad \begin{pmatrix} 3 & 5 \\ 2 & 4 \end{pmatrix},$$

each with probability 1/2. This example was also mentioned in §1.3. Starting from an arbitrary x_0-vector, say $x_0 = \begin{pmatrix} 0 & 2 \end{pmatrix}'$, we will set up the reachability tree of all possible normalized states. This is indicated in Table 8.2 which gives the list of

Table 8.2: Transitions of the Markov chain

Initial state	$A_{12} = 7$		$A_{12} = 5$	
$n_1 = \begin{pmatrix} 0 & 2 \end{pmatrix}'$	n_2	9	n_3	7
$n_2 = \begin{pmatrix} 0 & -3 \end{pmatrix}'$	n_4	4	n_3	3
$n_3 = \begin{pmatrix} 0 & -1 \end{pmatrix}'$	n_2	6	n_3	4
$n_4 = \begin{pmatrix} 0 & -2 \end{pmatrix}'$	n_2	5	n_3	3

state transitions of $z(k)$ (the normalization here means that the first component of the normalized state is always 0), together with the corresponding value of $(A(k)z(k))_1$ (the normalization factor).

In order to obtain a concise notation, the different normalized state vectors are denoted $n_i, i = 1, \ldots$. The table is obtained in the following way. The initial state is $n_1 \stackrel{\text{def}}{=} \begin{pmatrix} 0 & 2 \end{pmatrix}'$. From there, two states can be reached in one step, depending on the value of $A(0)$: $\begin{pmatrix} 0 & -3 \end{pmatrix}'$ and $\begin{pmatrix} 0 & -1 \end{pmatrix}'$. Both normalized states are added to the list and denoted n_2 and n_3, respectively. The normalization factors are 9 and 7, respectively. When taking n_2 as initial state, two states can be reached: $\begin{pmatrix} 0 & -2 \end{pmatrix}'$ and $\begin{pmatrix} 0 & -1 \end{pmatrix}'$. Only the first of these normalized states is new; it is added to the list and called n_4, and so on. For the current example, it turns out that there exist four different states (see Table 8.2).

From this table, one directly notices that the system never returns to n_1. Hence this state is transient. In fact, the Markov chain has a single recurrence class, which consists of the three states n_2, n_3 and n_4: from the definition of $A(k)$, we obtain

$$\begin{aligned} z_2(k+1) &= x_2(k+1) - x_1(k+1) \\ &= \max\left(2 + x_1(k), 4 + x_2(k)\right) - \max\left(3 + x_1(k), A_{12}(k) + x_2(k)\right) , \end{aligned}$$

where $A_{12}(k)$ is equal to either 7 or 5. Rewriting the right-hand side results in

$$z_2(k+1) = \max\left(0, 2 + z_2(k)\right) - \max\left(1, A_{12}(k) - 2 + z_2(k)\right) .$$

Whatever integer value we assume for $z_2(k)$, $z_2(k+1)$ can only assume one of the values -1, -2 or -3. The transition matrix of the restriction of this Markov chain to this recurrence class is

$$\begin{pmatrix} 0 & 1/2 & 1/2 \\ 1/2 & 1/2 & 1/2 \\ 1/2 & 0 & 0 \end{pmatrix} .$$

The stationary distribution of this chain is easily calculated to be

$$\mu(n_2) = 1/3 \ , \quad \mu(n_3) = 1/2 \ , \quad \mu(n_4) = 1/6 \ .$$

The average cycle time is then

$$\mu(n_2)(4\mu(A_1) + 3\mu(A_2)) + \mu(n_3)(6\mu(A_1) + 4\mu(A_2)) \\ + \mu(n_4)(5\mu(A_1) + 3\mu(A_2)) = 13/3 \ .$$

■

The crucial feature in this method is that the number of different normalized state vectors is finite. We now give a few theorems which provide simple sufficient conditions for the finiteness of the state space within this context.

Theorem 8.46 *Consider the $\mathfrak{n}$-dimensional equation (8.32). Assume that for all entries $A_{ij}(k)$ there exist finite real numbers $\underline{A}_{ij}$ and $\overline{A}_{ij}$ such that*

$$\mathbb{P}[\underline{A}_{ij} \le A_{ij}(k) \le \overline{A}_{ij}] = 1 \ , \quad \forall k \ge 0 \ .$$

Suppose that $z(0)$ is finite. Then, for $k = 1, 2, \ldots$, all elements $z_j(k)$ of the Markov chain are bounded and we have

$$\min_{1 \le i \le n} (\underline{A}_{ji} - \overline{A}_{1i}) \le z_j(k) \le \max_{1 \le i \le n} (\overline{A}_{ji} - \underline{A}_{1i}) \ , \quad j = 2, \ldots, \mathfrak{n} \ . \tag{8.33}$$

Proof We have $z_j(0) = x_j(0) - x_1(0)$, which is finite. From the definition of z it follows that

$$\begin{aligned} z_j(1) &= \left(A_{j1}(0) \oplus A_{j2}(0) z_2(0) \oplus \cdots \oplus A_{j\mathfrak{n}}(0) z_{\mathfrak{n}}(0)\right) \\ &\quad \not\circ \left(A_{11}(0) \oplus A_{12}(0) z_2(0) \oplus \cdots \oplus A_{1\mathfrak{n}}(0) z_{\mathfrak{n}}(0)\right) \ . \end{aligned}$$

Let $q \in \{1, \ldots, \mathfrak{n}\}$ be such that

$$A_{jq}(0) z_q(0) = A_{j1}(0) \oplus A_{j2}(0) z_2(0) \oplus \cdots \oplus A_{j\mathfrak{n}}(0) z_{\mathfrak{n}}(0) \ ,$$

and let $r \in \{1, \ldots, \mathfrak{n}\}$ be such that

$$A_{1r}(0) z_r(0) = A_{11}(0) \oplus A_{12}(0) z_2(0) \oplus \cdots \oplus A_{1\mathfrak{n}}(0) z_{\mathfrak{n}}(0) \ .$$

Then,

$$\begin{aligned} z_j(1) &= A_{jq}(0)z_q(0) - A_{1r}(0)z_r(0) &\geq A_{jr}(0)z_r(0) - A_{1r}(0)z_r(0) \\ &= A_{jr}(0) - A_{1r}(0) &\geq \underline{A}_{jr} - \overline{A}_{1r} \ , \end{aligned}$$

whereas on the other hand,

$$\begin{aligned} z_j(1) &= A_{jq}(0)z_q(0) - A_{1r}(0)z_r(0) &\leq A_{jq}(0)z_q(0) - A_{1q}(0)z_q(0) \\ &= A_{jq}(0) - A_{1q}(0) &\leq \overline{A}_{jq} - \underline{A}_{1q} \ . \end{aligned}$$

The property extends immediately to $z(k)$, $k \geq 1$. ■

Remark 8.47 Theorem 8.46 can straightforwardly be generalized to matrices $A(k)$ which are such that, for some l, $\bigotimes_{k=0}^{l} A(k)$ has all its entries bounded from below and from above. ■

The preceding theorem admits the following obvious corollary.

Corollary 8.48 *If the matrices $A(k)$ are i.i.d. with integer-valued entries which satisfy the conditions of Theorem 8.46, and if all entries of $z(0)$ are finite, integer-valued, and independent of matrices $A(k)$, then the Markov chain $z(k)$ has a finite state space.*

Remark 8.49 It is possible that under certain conditions, bounds exist which are better than those given in Theorem 8.46, as shown by the following two-dimensional example with

$$\mathbb{P}[A_{ij}(k) = 0] = \mathbb{P}[A_{ij}(k) = 1] = 1/2 \ , \qquad \text{except for } i = j = 1 \ ;$$
$$\mathbb{P}[A_{11}(k) = 1] = \mathbb{P}[A_{11}(k) = 2] = 1/2 \ .$$

Then the greatest lower bound and least upper bound of the random variables are

$$\underline{A}_{11} = 1 \ ; \quad \overline{A}_{11} = 2 \ ; \quad \underline{A}_{ij} = 0 \ ; \quad \overline{A}_{ij} = 1 \ .$$

According to Theorem 8.46, we have

$$-2 = \min(0 - 1, 0 - 2) \leq z(k) \leq \max(1 - 0, 1 - 1) = 1 \ .$$

In the integer-valued case, it follows from this that the state space of $z(k)$ is given by the set $\{-2, -1, 0\}$. Hence, in this case, $z(k)$ will not achieve the upper bound of Theorem 8.46 with positive probability. ■

This theorem can easily be extended to include rational values.

Corollary 8.50 *If all entries of $A(k)$ are rational-valued and satisfy the conditions of Theorem 8.46 for all k a.s. and if all entries of $z(0)$ are rational, then the state space of the Markov chain remains finite.*

We now give an example in which the number of elements in the state space does depend on the actual values of the random variables, but in which on the other hand the rationality does not play a role.

Example 8.51 Consider (8.32), with $n = 2$, and where the random variables $A_{ij}(k)$ have the following support

$$A_{11} = 0 \text{ or } 1 \ , \quad A_{12} = 0 \text{ or } \alpha \ , \quad A_{21} = 1 \ , \quad A_{22} = 0 \text{ or } \alpha \ .$$

Let $\alpha > 1$ and let all possible outcomes have positive probabilities. For this two-dimensional system, the Markov chain reduces to $z(k) = x_2(k) - x_1(k) \in \mathbb{R}$. Using Theorem 8.46, one obtains that $-\alpha \leq z(k) \leq \alpha$. From Corollaries 8.48 and 8.50, we know that if α is rational, then the Markov chain has a finite state space (at least for a proper choice of $z(0)$).

Depending on the value of α, the recurrent state space of $z(k)$ can be determined. For all $\alpha > 1$, $z(k)$ can assume the following six states with positive probability:

$$0 \ , \quad 1 \ , \quad 1 - \alpha \ , \quad \alpha \ , \quad \alpha - 1 \ , \quad -\alpha \ .$$

For $\alpha \geq 2$, these are the only values $z(k)$ can assume. For $1 < \alpha < 2$, the following states are also in the state space:

$$2 - \alpha \ , \quad 2 - 2\alpha \ , \quad 2\alpha - 2 \ , \quad -1 \ .$$

For $3/2 \leq \alpha < 2$ the state space consists of just these ten states. But for $1 < \alpha < 3/2$ the following values can also be assumed:

$$3 - 2\alpha \ , \quad 3 - 3\alpha \ , \quad \alpha - 3 \ , \quad \alpha - 2 \ .$$

Again, for $4/3 \leq \alpha < 3/2$, the state space consists of the given fourteen states. But, for $1 < \alpha < 4/3$, four other states are also possible, resulting in eighteen states, whereas for $1 < \alpha < 5/4$ again four new states are possible, etc. We see that if α comes closer to one, the number of states increases (stepwise). But for any value of α the total number of states remains finite. Also for $\alpha = 1$ the number of states is finite (in fact, the state space is then equal to {-1,0,1}). Also, for all values of α (both rational and irrational) within a certain interval, the number of elements of the state space of the Markov chain $z(k)$ is the same. ■

Example 8.52 Consider the following six-dimensional case:

$$A(k) = \begin{pmatrix} \varepsilon & e & \varepsilon & \varepsilon & \varepsilon & \varepsilon \\ \varepsilon & \varepsilon & e & \varepsilon & e & \varepsilon \\ \varepsilon & \varepsilon & \varepsilon & e & \varepsilon & \varepsilon \\ e & \varepsilon & \varepsilon & \varepsilon & e & \varepsilon \\ \varepsilon & \varepsilon & \varepsilon & \varepsilon & \varepsilon & e \\ a(k) & \varepsilon & e & \varepsilon & \varepsilon & \varepsilon \end{pmatrix} ,$$

with

$$\mathbb{P}[a(k) = e] = \mathbb{P}[a(k) = 1] = 1/2 \ .$$

The matrix is irreducible since the graph corresponding to the matrix is strongly connected. But it turns out that, in this case, the state space of the Markov chain becomes infinite. This can be seen as follows. From a state $\begin{pmatrix} l & e & l & e & l & e \end{pmatrix}'$ the following states are possible:

$$\begin{pmatrix} e & l & e & l & e & l \end{pmatrix}' \quad \text{and} \quad \begin{pmatrix} e & l & e & l & e & l+1 \end{pmatrix}' .$$

After the last state, the state

$$\begin{pmatrix} l+1 & e & l+1 & e & l+1 & e \end{pmatrix}'$$

can be reached with positive probability, and from this state the state

$$\begin{pmatrix} e & l+1 & e & l+1 & e & l+2 \end{pmatrix}'$$

is possible, etc. ■

Example 8.53 (Railway Traffic Example Revisited) In §1.2.6 a railway system in a metropolitan area was studied. This example will be studied in greater detail now. The three railway stations S_1, S_2 and S_3 are connected by a railway system as indicated in Figure 1.10. The railway system consists of two inner circles, along which the trains run in opposite direction, and of three outer circles. The model which describes the departure times of the nine trains is $x(k+1) = A_1 x(k)$. Two other models were described in §1.2.6 as well, depending on other routing schemes of the trains. These models were characterized by the transition matrices A_2 and A_3 respectively. The elements of these matrices were e, ε or s_{ij}, the latter referring to the traveling time from station S_i to station S_j. The quantity s_{ii} refers to the traveling time of the outer circle connected to station S_i. It is assumed that all s_{ij}-quantities are equal to 1, except for s_{31}. The latter quantity is random and is either 1 or 2. Each time a train runs from S_3 to S_1 there is a probability p, $0 \leq p \leq 1$, that the train will be delayed, i.e. $s_{31} = 2$ rather than $s_{31} = 1$. Thus matrices A_i become k-dependent and will be denoted $A_i(k)$. The system is now stochastic. It is assumed that no correlation with respect to the 'counter' k exists. In this situation, one may also have a preference for one of these three routings or another one. In this context four routings will be studied: the ones characterized by the matrices $A_i, i = 1, 2, 3$, and the routing in which the trains move in opposite directions compared with the routing characterized by A_3. The matrix corresponding to the latter routing, though not explicitly given, will be indicated by A_4. In fact, the results corresponding to this fourth routing will be obtained by using A_3 in which s_{13} is now the uncertain factor rather than s_{31}.

In Tables 8.3 and 8.4 the normalized states corresponding to the stationary situations of the four routings are given (one must check again in each of these four cases whether the set of normalized states in the stationary situation is unique, which turns out to be true). These states have been normalized in such a way that the least component equals zero. The transition matrices of these Markov chains follow directly from these tables. As an example, if $p = 0.2$, the transition matrix for the

Table 8.3: Markov chains of the first three routing schemes

Initial state	$s_{31} = 1$		$s_{31} = 2$	
First routing scheme				
$n_1 = \begin{pmatrix} 0 & 0 & 0 & 0 & 0 & 0 & 0 & 0 & 0 \end{pmatrix}'$	n_1	1	n_2	1
$n_2 = \begin{pmatrix} 1 & 0 & 0 & 0 & 0 & 0 & 1 & 0 & 0 \end{pmatrix}'$	n_3	1	n_3	1
$n_3 = \begin{pmatrix} 1 & 1 & 0 & 1 & 0 & 0 & 1 & 1 & 0 \end{pmatrix}'$	n_1	2	n_1	2
Second routing scheme				
$n_1 = \begin{pmatrix} 0 & 0 & 0 & 0 & 0 & 0 & 0 & 0 & 0 \end{pmatrix}'$	n_1	1	n_2	1
$n_2 = \begin{pmatrix} 1 & 0 & 0 & 1 & 0 & 0 & 1 & 0 & 0 \end{pmatrix}'$	n_1	2	n_1	2
Third routing scheme				
$n_1 = \begin{pmatrix} 0 & 0 & 0 & 0 & 0 & 0 & 0 & 0 & 0 \end{pmatrix}'$	n_1	1	n_2	1
$n_2 = \begin{pmatrix} 0 & 0 & 0 & 1 & 0 & 0 & 1 & 0 & 0 \end{pmatrix}'$	n_3	1	n_4	1
$n_3 = \begin{pmatrix} 1 & 0 & 0 & 1 & 0 & 0 & 0 & 0 & 1 \end{pmatrix}'$	n_5	1	n_6	1
$n_4 = \begin{pmatrix} 1 & 0 & 0 & 1 & 0 & 0 & 1 & 0 & 1 \end{pmatrix}'$	n_7	1	n_8	1
$n_5 = \begin{pmatrix} 0 & 1 & 1 & 0 & 0 & 1 & 0 & 1 & 1 \end{pmatrix}'$	n_9	1	n_2	2
$n_6 = \begin{pmatrix} 0 & 1 & 1 & 1 & 0 & 1 & 1 & 1 & 1 \end{pmatrix}'$	n_1	2	n_2	2
$n_7 = \begin{pmatrix} 1 & 1 & 1 & 1 & 0 & 1 & 0 & 1 & 1 \end{pmatrix}'$	n_9	1	n_2	2
$n_8 = \begin{pmatrix} 1 & 1 & 1 & 1 & 0 & 1 & 1 & 1 & 1 \end{pmatrix}'$	n_1	2	n_2	2
$n_9 = \begin{pmatrix} 1 & 1 & 1 & 0 & 1 & 1 & 1 & 1 & 1 \end{pmatrix}'$	n_1	2	n_2	2

Table 8.4: Markov chain of the fourth routing scheme

Initial state	$s_{31} = 1$		$s_{31} = 2$	
$n_1 = (0\ 0\ 0\ 0\ 0\ 0\ 0\ 0\ 0)'$	n_1	1	n_2	1
$n_2 = (0\ 0\ 0\ 0\ 0\ 0\ 0\ 0\ 1)'$	n_3	1	n_4	1
$n_3 = (0\ 0\ 1\ 0\ 0\ 1\ 0\ 0\ 0)'$	n_5	1	n_6	1
$n_4 = (0\ 0\ 1\ 0\ 0\ 1\ 0\ 0\ 1)'$	n_7	1	n_8	1
$n_5 = (1\ 0\ 0\ 0\ 0\ 0\ 1\ 1\ 0)'$	n_9	1	n_{10}	1
$n_6 = (1\ 0\ 0\ 0\ 0\ 0\ 1\ 1\ 1)'$	n_{11}	1	n_{12}	1
$n_7 = (1\ 0\ 1\ 0\ 0\ 1\ 1\ 1\ 0)'$	n_{13}	1	n_{14}	1
$n_8 = (1\ 0\ 1\ 0\ 0\ 1\ 1\ 1\ 1)'$	n_{15}	1	n_1	2
$n_9 = (1\ 1\ 0\ 1\ 1\ 0\ 0\ 1\ 0)'$	n_{16}	1	n_{17}	1
$n_{10} = (1\ 1\ 0\ 1\ 1\ 0\ 0\ 1\ 1)'$	n_{18}	1	n_{19}	1
$n_{11} = (1\ 1\ 1\ 1\ 1\ 1\ 0\ 1\ 0)'$	n_{20}	1	n_{21}	1
$n_{12} = (1\ 1\ 1\ 1\ 1\ 1\ 0\ 1\ 1)'$	n_{22}	1	n_{23}	1
$n_{13} = (1\ 1\ 0\ 1\ 1\ 0\ 1\ 1\ 0)'$	n_{24}	1	n_{25}	1
$n_{14} = (1\ 1\ 0\ 1\ 1\ 0\ 1\ 1\ 1)'$	n_1	2	n_2	2
$n_{15} = (1\ 1\ 1\ 1\ 1\ 1\ 1\ 1\ 0)'$	n_{24}	1	n_{25}	1
$n_{16} = (0\ 1\ 1\ 0\ 1\ 0\ 1\ 1\ 1)'$	n_{26}	1	n_{26}	1
$n_{17} = (0\ 1\ 1\ 0\ 1\ 0\ 1\ 1\ 2)'$	n_{27}	1	n_{27}	1
$n_{18} = (0\ 1\ 1\ 0\ 1\ 1\ 1\ 1\ 1)'$	n_1	2	n_1	2
$n_{19} = (0\ 1\ 1\ 0\ 1\ 1\ 1\ 1\ 2)'$	n_3	2	n_3	2
$n_{20} = (1\ 1\ 1\ 0\ 1\ 0\ 1\ 1\ 1)'$	n_1	2	n_1	2
$n_{21} = (1\ 1\ 1\ 0\ 1\ 0\ 1\ 1\ 2)'$	n_3	2	n_3	2
$n_{22} = (1\ 1\ 1\ 0\ 1\ 1\ 1\ 1\ 1)'$	n_1	2	n_1	2
$n_{23} = (1\ 1\ 1\ 0\ 1\ 1\ 1\ 1\ 2)'$	n_3	2	n_3	2
$n_{24} = (1\ 1\ 1\ 1\ 1\ 0\ 1\ 1\ 1)'$	n_1	2	n_2	2
$n_{25} = (1\ 1\ 1\ 1\ 1\ 0\ 1\ 1\ 2)'$	n_3	2	n_4	2
$n_{26} = (1\ 1\ 1\ 1\ 1\ 1\ 1\ 0\ 1)'$	n_{28}	1	n_{29}	1
$n_{27} = (1\ 1\ 2\ 1\ 1\ 2\ 1\ 0\ 1)'$	n_{30}	1	n_{31}	1
$n_{28} = (1\ 1\ 1\ 1\ 0\ 1\ 1\ 1\ 1)'$	n_1	2	n_2	2
$n_{29} = (1\ 1\ 1\ 1\ 0\ 1\ 1\ 1\ 2)'$	n_3	2	n_4	2
$n_{30} = (2\ 1\ 1\ 1\ 0\ 1\ 2\ 2\ 1)'$	n_9	2	n_{10}	2
$n_{31} = (2\ 1\ 1\ 1\ 0\ 1\ 2\ 2\ 2)'$	n_{11}	2	n_{12}	2

Markov chain of the first routing scheme becomes

$$\begin{pmatrix} -0.2 & 0 & 1 \\ 0.2 & -1 & 0 \\ 0 & 1 & -1 \end{pmatrix},$$

from which the stationary distribution can be calculated: it is $\begin{pmatrix} 5/7 & 1/7 & 1/7 \end{pmatrix}'$. The cycle time then becomes

$$(0.8 \times 1 + 0.2 \times 1) \times 5/7 + (0.8 \times 1 + 0.2 \times 1)/7 + (0.8 \times 2 + 0.2 \times 2)/7 = 8/7 \ .$$

The results for varying p are given in Figure 8.4 for all four routing schemes. Note

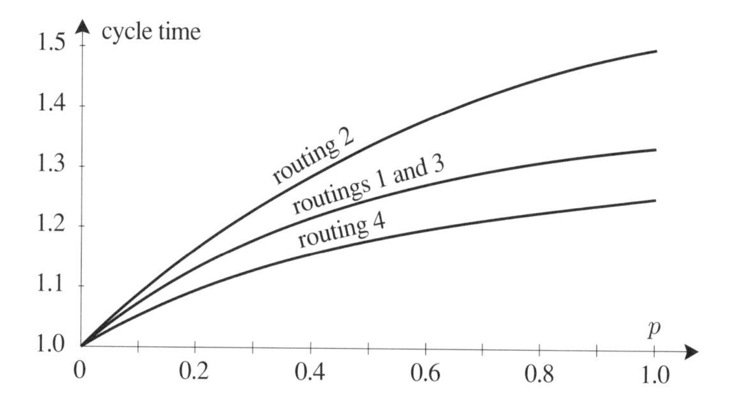

Figure 8.4: Cycle times of the four routings

that the cycle times of routings one and three completely coincide. If one could choose among the routings, then routing 4 would be preferred since it has the least cycle time for any p-value. ■

Example 8.54 (Example of Parallel Computation Revisited) The starting point for this subsection is Equation (1.33). This is a nonlinear equation which describes the evolution of the speed of a simple matrix multiplication on a wavefront array processor. Suppose that $\alpha_i, i = 1, 2$, are either 1 (ordinary multiplication) or 0 (multiplication by a 0 or a 1). Let $z(k)$ be defined as in Theorem 8.44. Using the same type of arguments as in the proof of this theorem, one shows that $z(k)$ is a Markov chain, provided that the random variables $\alpha_i(k)$ are independent.

It is possible to aggregate the state space of this Markov chain into twelve macro-states, as defined in Table 8.5, while preserving the Markov property (the transition probabilities satisfy the conditions of the 'lumping theorem' 6.3.2 in [74]).

The other steps of the analysis (computation of the transition matrix of the aggregated Markov chain and of the invariant measure) can then be carried out in the standard way. ■

Table 8.5: States of Markov chain; example of parallel computation; $l = 0, 1, 2, \ldots$

n_1	n_2	n_3	n_4	n_5	n_6	n_7	n_8	n_9	n_{10}	n_{11}	n_{12}
l	l	l	l	l	l	l	l	l	l	l	l
l	l	l	l	$l+1$	l	l	l	$l+1$	$l+1$	l	l
l	$l+1$	l	l	$l+1$	$l+1$	l	l	$l+1$	l	l	l
l	$l+1$	$l+1$	l	$l+1$	$l+1$	$l+1$	l	$l+1$	$l+1$	l	$l-1$
0	0	0	1	0	1	1	1	0	0	0	1
0	1	0	0	1	1	0	1	0	0	1	0

8.4.3 Continuous Distribution Functions

The starting point is Equation (8.32). It is assumed that the sequence of matrices $\{A(k)\}_{k\geq 0}$ is i.i.d., that for each k the entries $A_{ij}(k), i, j = 1, \ldots, n$, are mutually independent, and that the random variables $A_{ij}(0)$ all have the same distribution function on $\mathbb{R}^+$, which will be denoted F. We will assume that F admits a density. Under these assumptions, there is one m.s.c.s. Whenever the support of F is infinite, the increment process $\{\delta(k)\}$ of Chapter 7, and hence the process $\{z(k)\}$ couple in finite time with stationary processes which do not depend on the initial condition.

From Theorem 8.44, the variables $z(k)$ form a Markov chain on $\mathbb{R}^{n-1}$ (the first coordinate is zero), the transition matrix of which is characterized by the relation

$$z_i(k+1) = \frac{A_{i1}(k) \oplus \bigoplus_{j=2}^{n}(A_{ij}(k) \otimes z_j(k))}{A_{11}(k) \oplus \bigoplus_{j=2}^{n}(A_{1j}(k) \otimes z_j(k))} \ , \qquad i = 2, \ldots, n \ . \tag{8.34}$$

The transition kernel of the Markov chain, or equivalently the distribution of $\begin{pmatrix} z_2(k+1) & \ldots & z_n(k+1) \end{pmatrix}$ given $\begin{pmatrix} z_2(k) & \ldots & z_n(k) \end{pmatrix}$ is obtained from (8.34):

$$\begin{aligned} & K(x_2, \ldots, x_n; y_2, \ldots, y_n) \\ \stackrel{\text{def}}{=} \ & \mathbb{P}\left[z_2(k+1) \leq y_2, \ldots, z_n(k+1) \leq y_n \mid z_2(k) = x_2, \ldots, z_n(k) = x_n\right] \\ = \ & \mathbb{P}\left[X_2 \not{} X_1 \leq y_2, \ldots, X_n \not{} X_1 \leq y_n\right] \ , \end{aligned}$$

where the random variables $X_i, i = 2, \ldots, n$, defined by

$$X_i =_{\text{st}} A_{i1}(k) \oplus \bigoplus_{j=2}^{n} A_{ij}(k) x_j \ ,$$

are independent. The notation $=_{\text{st}}$ refers here to equality in distribution. From the last two equations and the fact that all $A_{ij}(k)$ possess the same distribution F, it follows that

$$K(x_2,\ldots,x_n;y_2,\ldots,y_n) = \int_{-\infty}^{\infty}\left[\prod_{j=2}^{n}\mathbb{P}[X_j \le y_j + t]\right]\mathrm{d}\mathbb{P}\left[X_1 \le t\right] =$$

$$\int_{-\infty}^{\infty}\left[\prod_{j=2}^{n}H(t+y_j, t+y_j-x_2,\ldots,t+y_j-x_n)\right]\left[\frac{\mathrm{d}}{\mathrm{d}t}H(t,t-x_2,\ldots,t-x_n)\right]\mathrm{d}t$$

where

$$H(u_1,u_2,\ldots,u_n) \stackrel{\text{def}}{=} \prod_{i=1}^{n}F(u_i) \ .$$

The distribution function

$$\zeta_k(x_2,\ldots,x_n) \stackrel{\text{def}}{=} \mathbb{P}\left[z_2(k)\le x_2,\ldots,z_n(k)\le x_n\right]$$

satisfies the functional equation

$$\zeta_{k+1}(y_2,\ldots,y_n) = \int_{-\infty}^{\infty}\cdots\int_{-\infty}^{\infty}K(x_2,\ldots,x_n;y_2,\ldots,y_n)\zeta_k(\mathrm{d}x_2,\ldots,\mathrm{d}x_n) \ .$$

Whenever the infinite support condition is satisfied, we know that the limit $\lim_{k\to\infty}\zeta_k = \zeta$ exists in the (weak convergence) distributional sense. This limit is a solution of the functional equation

$$\zeta(y_2,\ldots,y_n) = \int_{-\infty}^{\infty}\cdots\int_{-\infty}^{\infty}K(x_2,\ldots,x_n;y_2,\ldots,y_n)\,\zeta(\mathrm{d}x_2,\ldots,\mathrm{d}x_n) \ . \quad (8.35)$$

Let ζ be the unique solution of this equation. It is immediate to see that the distribution function $D(t)$ of the stationary ratio δ_{11} is given by the relation

$$\begin{aligned}
D(t) &= \lim_{k\to\infty}\mathbb{P}\left[x_1(k+1)-x_1(k)\le t\right] \\
&= \lim_{k\to\infty}\mathbb{P}\left[A_{11}(k)\le t, A_{12}(k)+z_2(k)\le t,\ldots,A_{1n}(k)+z_n(k)\le t\right] \\
&= \lim_{k\to\infty}\mathbb{P}\left[A_{11}(k)\le t, A_{12}(k)\le t-z_2(k),\ldots,A_{1n}(k)\le t-z_n(k)\right] \\
&= F(t)\lim_{k\to\infty}\int_{-\infty}^{\infty}\cdots\int_{-\infty}^{\infty}\prod_{i=2}^{n}F(t-y_i)\zeta_k(\mathrm{d}y_2,\ldots,\mathrm{d}y_n) \\
&= F(t)\int_{-\infty}^{\infty}\cdots\int_{-\infty}^{\infty}\prod_{i=2}^{n}F(t-y_i)\zeta(\mathrm{d}y_2,\ldots,\mathrm{d}y_n) \ . \qquad (8.36)
\end{aligned}$$

The limit and the integral in (8.36) can be interchanged by the definition of weak convergence, since F is continuous, see [21].

Example 8.55 Consider (8.32) with $n = 2$ and with the assumptions made in the previous subsection. The transition kernel $K(x; y)$ of the Markov chain $\{z_2(k)\}$ is given by

$$\begin{aligned} K(x;y) &\stackrel{\text{def}}{=} \mathbb{P}\left[z_2(k+1) \le y \mid z_2(k) = x\right] \\ &= \int_{-\infty}^{\infty} H(t+y, t+y-x) \left(\frac{\mathrm{d}}{\mathrm{d}t} H(t, t-x)\right) \mathrm{d}t \\ &= \int_{-\infty}^{\infty} F(t+y)F(t+y-x) \left(F(t-x)\frac{\mathrm{d}}{\mathrm{d}t}F(t) + F(t)\frac{\mathrm{d}}{\mathrm{d}t}F(t-x)\right) \mathrm{d}t \ . \end{aligned}$$

Explicit calculations will be made for

$$F(x) = (1 - \exp(-x))\mathbf{1}_{[0,\infty)}(x) \ . \tag{8.37}$$

It follows from (8.35) that the density d of the stationary distribution ζ satisfies the equation

$$d(y) = \int_{-\infty}^{\infty} \left(\frac{\mathrm{d}}{\mathrm{d}y} K(x;y)\right) d(x)\, \mathrm{d}x \ ,$$

or, equivalently, after some calculations,

$$\begin{aligned} d(y) = \ & \frac{1}{2}\int_{-\infty}^{\infty} \left[-\frac{1}{3}\exp\left(-|y| - 2|x|\right) + \frac{2}{3}\exp\left(-2|y| - 2|x|\right)\right. \\ & \left. + \frac{2}{3}\exp\left(-|y| - |x|\right) - \frac{4}{3}\exp\left(-2|y| - |x|\right) + \exp\left(-|y|\right)\right] d(x)\, \mathrm{d}x \ . \end{aligned} \tag{8.38}$$

This is an integral equation, the kernel of which is degenerate, see [90]. The solution $d(y)$ must be of the form

$$d(y) = c_1 \exp\left(-|y|\right) + c_2 \exp\left(-2|y|\right)$$

(see [90, Chapter 1, §4]), where the coefficients c_i still must be determined. Substitution of this form into the integral equation leads to $8c_1 + 23c_2 = 0$. This, together with the normalization condition $\int_{-\infty}^{\infty} d(y)\, \mathrm{d}y = 1$ results in $2c_1 + c_2 = 1$, which uniquely determines these coefficients. The stationary density is hence given by

$$d(y) = \frac{23}{38}\exp\left(-|y|\right) - \frac{4}{19}\exp\left(-2|y|\right) \ , \quad y \in (-\infty, \infty) \ . \tag{8.39}$$

It is easy to show that $d(y) \ge 0, \forall y \in (-\infty, \infty)$, and hence d is indeed a probability density function. With the aid of (8.39), one now obtains, after some straightforward analysis,

$$\lim_{k\to\infty} \mathbb{E}\left[x_1(k+1) \not/ x_1(k)\right] = \frac{407}{228} = 1.79 \ .$$

This expression also equals $\lim_{k\to\infty} \mathbb{E}\left[(x_i(k))^{1/k}\right]$, provided that the random variables $x_1(0)$ and $x_2(0)$ are integrable. ■

Remark 8.56 The fact that d_k, defined as the density of ζ_k, indeed approaches the limit d as k goes to infinity, can easily be illustrated for this example. If one starts with an arbitrary density d_0, then d_1 is already the sum of the two exponentials $\exp(-y)$ and $\exp(-2y)$, as follows from

$$d_{k+1}(y) = \int_{-\infty}^{\infty} K(x; y) d_k(x) \, \mathrm{d}x \ ,$$

where the kernel K is the same as in (8.38). In general,

$$d_k(y) = c_1(k) \exp\left(-|y|\right) + c_2(k) \exp\left(-2|y|\right) \ , \quad k \geq 1 \ ,$$

and the coefficients satisfy

$$\begin{pmatrix} c_1(k+1) \\ c_2(k+1) \end{pmatrix} = \begin{pmatrix} 11/9 & 23/36 \\ -4/9 & -5/18 \end{pmatrix} \begin{pmatrix} c_1(k) \\ c_2(k) \end{pmatrix} .$$

If one starts with with a probability density function ζ_0, then $2c_1(k) + c_2(k) = 1, \ k = 1, 2, \ldots$, and

$$\lim_{k\to\infty} c_1(k) = 23/38 \ , \qquad \lim_{k\to\infty} c_2(k) = -4/19 \ .$$

Hence the transient behavior converges to the limit (stationary) behavior. ■

Remark 8.57 The outcome of the above example will be compared with the outcomes of two other examples. The models of all three examples will be the same, i.e. (8.32) with $n = 2$. The difference is the stochastic behavior of A_{ij}. In the above example, it was characterized by (8.37). In the next two examples we have

Example 2: $\mu(A_{ij} = 0) = \mu(A_{ij} = 2) = 1/2$;

Example 3: A_{ij} is uniformly distributed on the interval $[0, 2)$.

In all these three examples, $\mathbb{E}[A_{ij}] = 1$. In spite of this, it will turn out that the throughput for all three examples is different. In the second example, the elements A_{ij} have a discrete distribution. The method of §8.4.2 can be applied, which results in

$$\lim_{k\to\infty} \mathbb{E}\left[x_1(k+1) \not/ x_1(k)\right] = \frac{12}{7} = 1.71 \ .$$

For the third example, the method described at the beginning of this subsection can be used. The same type of analysis leads to

$$\lim_{k\to\infty} \mathbb{E}\left[x_1(k+1) \not/ x_1(k)\right] = 1.44 \ .$$

The third example leads to the best throughput. This is not surprising: for instance, the comparison between the exponential case and the case of Example 3 follows from Theorem 8.3; indeed, the Karlin-Novikoff cut criterion [123, Proposition 1.5.1, p. 12] immediately implies that an exponential random variable of mean 1 is $\leq_{cx}$-bounded from below by a uniform random variable on $[0, 1]$. ■

8.5 APPENDIX

8.5.1 Stochastic Comparison

This subsection gathers a few basic properties of the three stochastic orders introduced in §8.2.1, and related definitions. For proofs and details, see [123] or [6].

From the very definitions, it should be clear that

$$x \leq_{\mathrm{st}} x^{\dagger} \Rightarrow f(x) \leq_{\mathrm{st}} f(x^{\dagger}) , \tag{8.40}$$

for all coordinatewise nondecreasing functions $f : \mathbb{R}^n \to \mathbb{R}^m$. In the same vein,

$$x \leq_{\mathrm{icx}} x^{\dagger} \Rightarrow f(x) \leq_{\mathrm{icx}} f(x^{\dagger}) , \tag{8.41}$$

for all nondecreasing and convex functions $f : \mathbb{R}^n \to \mathbb{R}^m$.

Remark 8.58 From Jensen's inequality, it is immediately checked that for all integrable random variables $x \in \mathbb{R}^n$, the following relation holds:

$$\mathbb{E}[x] \leq_{\mathrm{cx}} x . \tag{8.42}$$

■

Lemma 8.59 *If x and $x^{\dagger}$ are nonnegative, real-valued random variables, each of the properties $x \leq_{\mathrm{st}} x^{\dagger}$, $x \leq_{\mathrm{cx}}$ and $x \leq_{\mathrm{icx}} x^{\dagger}$ implies the moment relation $\mathbb{E}[x^n] \leq \mathbb{E}\left[(x^{\dagger})^n\right]$, for all $n \geq 0$.*

Consider an $\mathbb{R}^{\mathrm{n}}$-valued sequence x generated by the recursion

$$x(k+1) = a(x(k), u(k)) , \quad k \geq 0 , \tag{8.43}$$

for some Borel mapping $a : \mathbb{R}^{\mathrm{n}} \times \mathbb{R}^{\mathrm{p}} \to \mathbb{R}^{\mathrm{n}}$, for some given sequence of $\mathbb{R}^{\mathrm{p}}$-valued random variables $u = \{u(0), \ldots, u(k), \ldots\}$ and some initial condition $x(0)$, all defined on the probability space $(\Omega, \mathbb{F}, \mathbb{P})$. Let $x^{\dagger}$ be the sequence defined as above, but for the initial condition and the sequence, which are respectively replaced by $x^{\dagger}(0)$ and $u^{\dagger}$.

In what follows, ξ denotes the sequence

$$\xi = \{x(0), u(0), u(1), \ldots\} , \tag{8.44}$$

with a similar definition for $\xi^{\dagger}$. The proofs of the following results can be found in [6, Chapter 4].

Theorem 8.60 *Assume that the mapping $(X, U) \mapsto a(X, U)$ is nondecreasing; then $\xi \leq_{\mathrm{st}} \xi^{\dagger}$ implies that $x \leq_{\mathrm{st}} x$.*

Theorem 8.61 *Assume that the random variables $x(0)$ and $u(0), u(1), \ldots$ are integrable, that the mapping $(X, U) \mapsto a(X, U)$ is convex, and that the mapping $X \mapsto a(X, U)$ is nondecreasing for all U; then $\xi \leq_{\mathrm{cx}} \xi^{\dagger}$ implies that $x \leq_{\mathrm{icx}} x$.*

Theorem 8.62 *Assume that the random variables $x(0)$ and $u(0), u(1), \ldots$ are integrable, and that the mapping $(X, U) \mapsto a(X, U)$ is convex and nondecreasing; then $\xi \leq_{\text{icx}} \xi^{\dagger}$, implies that $x \leq_{\text{icx}} x$.*

Definition 8.63 (Stochastic convexity) *A collection of $\mathbb{R}^n$-valued random variables $\{Z(\rho)\}_{\rho \in \mathcal{R}}$ with a convex parameter set $\mathcal{R} \subset \mathbb{R}^m$ is said to be stochastically (increasing and) convex in ρ if $\mathbb{E}\left[\phi(Z(\rho))\right]$ is (nondecreasing and) convex in $\rho \in \mathcal{R}$, for all nondecreasing functions $\phi : \mathbb{R}^n \to \mathbb{R}$.*

The stochastic concavity with respect to a parameter is defined in a similar way.

Definition 8.64 (Association) *The (set of) $\mathbb{R}$-valued random variables $x_1, \ldots, x_n$, all defined on the same probability space, are (is) said to be associated if*

$$\mathbb{E}\left[f(x_1, \ldots, x_n) g(x_1, \ldots, x_n)\right] \geq \mathbb{E}\left[f(x_1, \ldots, x_n)\right] \mathbb{E}\left[g(x_1, \ldots, x_n)\right] \ ,$$

for all pairs of increasing functions $f, g : \mathbb{R}^n \to \mathbb{R}$ such that the integrals are well defined.

This definition is extended to sets of $\mathbb{R}^n$-valued random variables by requiring that the set of all coordinates be associated. It is also extended to sequences and to random processes in the usual way: the sequence is said to be associated if all finite subsequences are associated.

Remark 8.65 The association property can often be established without computing the joint distribution of the variables explicitly: for instance, the union of independent sets of associated random variables forms a set of associated random variables; as can easily be checked, for any nondecreasing function $\phi : \mathbb{R}^n \to \mathbb{R}$, and any set of associated random variables $\{x_1, \ldots, x_n\}$, the variables $\{\phi(x), x_1, \ldots, x_n\}$ are associated, where $\phi(x) \stackrel{\text{def}}{=} \phi(x_1, \ldots, x_n)$. ■

Lemma 8.66 *If the random variables $\{x_1, \ldots, x_n\}$ are associated, then*

$$\bigoplus_{i=1}^{n} x_i \leq_{\text{st}} \bigoplus_{i=1}^{n} \overline{x}_i \ ,$$

and

$$\bigwedge_{i=1}^{n} x_i \geq_{\text{st}} \bigwedge_{i=1}^{n} \overline{x}_i \ ,$$

where $\overline{x}$ is the product form version of x, namely the random vector such that

- *$\overline{x}_i =_{\text{st}} x_i$ for all $i = 1, \ldots, n$;*
- *the marginals of $\overline{x}$ are mutually independent.*

Note that the product form version $\overline{x}$ of x is only characterized through its probability law. Concerning recursions of the type (8.43), we also have the following theorem.

Theorem 8.67 *Assume that the function $(X, U) \to a(X, U)$ is nondecreasing. If the set $\{\xi(0), \xi(1), \ldots\}$ and the initial condition $x(0)$ form a set of associated random variables, then the random sequence x, ξ is also associated.*

8.5.2 Markov Chains

Sequences generated like in (8.43) satisfy the following property:

Theorem 8.68 *If the sequence* $u(k)$ *is i.i.d., then* $\{x(k)\}$ *forms a homogeneous Markov chain.*

For this and related results on Markov chain theory, see [121].

8.6 NOTES

A good survey on the methods for deriving stochastic monotonicity results for classical queuing systems can be found in the book by D. Stoyan [123]. The interest of these techniques to analyze synchronization constraints was first stressed by A.M. Makowski, Z. Liu and one of the coauthors, in [12] for queuing systems, and in [10] for stochastic event graphs. The uniformization method for proving the concavity of throughput generalizes an idea of L.E. Meester and J.G. Shanthikumar (see [89] and [10]).

The use of large deviation techniques for deriving growth rates for age-dependent branching processes was initiated by J.F. Kingman and D. Biggins [76] and [19]. The relation between event graphs and branching processes which is presented in §8.3, is that considered in [8]. This approach has interesting connections with the work of B. Derrida on directed polymers in a random medium (see [56], [48]), from which the idea of Remark 8.42 originates.

Most of the results mentioned in §8.4 come from [97], [101], [104] and [117]. In the latter reference, one can also find further results on the asymptotic normality of daters. The analysis of the finiteness of the state space in §8.4.2 is mainly drawn from [54]. The type of functional equation which is established in §8.4.3 is also considered in certain nonautonomous cases (see [9]).

PART V

Postface

CHAPTER 9

Related Topics and Open Ends

9.1 INTRODUCTION

In this chapter various items will be discussed which either did not find a natural place in one of the preceding chapters, which are only related to discrete events, or which are not yet fully grown in a scientific way. The personal bias of the authors will be clearly reflected in this chapter. There are no direct relations between the sections. Section 9.2 is concerned with various continuations of the linear theory developed throughout the book. Section 9.3 is devoted to the control of discrete event systems, whereas §9.4 gives a picture of the analogies between the theory of optimization and that of Markov chains. The last three sections are devoted to some (limited) incursions into the realm of general Petri nets and nonlinear systems.

9.2 ABOUT REALIZATION THEORY

9.2.1 The Exponential as a Tool; Another View on Cayley-Hamilton

If a and b are reals (or $-\infty$) then the following identities are easily verified:

$$a \oplus b = \max(a,b) = \lim_{s\to\infty} s^{-1}(\ln(\exp(as)+\exp(bs))) \ ,$$

$$a \otimes b = a + b = s^{-1}\ln(\exp(as)\exp(bs)) \ .$$

Rather than working in the max-plus algebra setting with variables $a, b, \ldots$, one can now envisage working with the variables $\exp(as), \exp(bs), \ldots$, where s is a positive real, in conventional algebra. After having obtained results in conventional algebra, we must translate these results back into corresponding results in the max-plus algebra by using careful limit arguments when $s \to \infty$. This procedure will be elucidated in this subsection and in §9.2.3. Instead of working with $\exp(as), \exp(bs), \ldots$, we will work with $z^a, z^b, \ldots$, z real, and study the behavior for $z \to \infty$.

In conventional calculus, the Cayley-Hamilton theorem states that every square matrix satisfies its own characteristic equation. To be more explicit, let A be an $n \times n$ matrix with entries in $\mathbb{R}$. If

$$\det(\lambda I - A) = \lambda^n + c_1\lambda^{n-1} + \cdots + c_{n-1}\lambda + c_n \ , \tag{9.1}$$

then

$$A^n + c_1A^{n-1} + \cdots + c_{n-1}A + c_nI = 0 \ .$$

In these equations I is the conventional identity matrix and 0 is the zero matrix. The coefficients $c_i, i = 1, \ldots, n$, in (9.1) satisfy

$$c_k = (-1)^k \sum_{i_1 < i_2 < \cdots < i_k} \begin{pmatrix} A_{i_1 i_1} & \cdots & A_{i_1 i_k} \\ \vdots & & \vdots \\ A_{i_k i_1} & \ldots & A_{i_k i_k} \end{pmatrix} . \tag{9.2}$$

Now consider the matrix $z^A \stackrel{\text{def}}{=} (z^{A_{ij}})$, i.e. the ij-th entry of z^A equals $z^{A_{ij}}$. The Cayley-Hamilton theorem applied to matrix z^A yields

$$(z^A)^n + \zeta_1 (z^A)^{n-1} + \cdots + \zeta_{n-1} z^A + \zeta_n I = 0 . \tag{9.3}$$

If the principal $k \times k$ submatrix occurring on the right-hand side of (9.2) is denoted $A(i_1, i_2, \ldots, i_k)$, then the coefficients ζ_k are given by

$$\zeta_k = (-1)^k \sum_{i_1 < i_2 < \cdots < i_k} \det z^{A(i_1, i_2, \ldots, i_k)} .$$

If we take the limit when $z \to \infty$, then we obtain

$$\zeta_k \approx (-1)^k \overline{\zeta}_k z^{\max_{i_1 < i_2 < \cdots < i_k} \operatorname{dom} A(i_1, i_2, \ldots, i_k)} , \tag{9.4}$$

where dom (for *dominant*) is a concept similar to per (for *permanent*); for the latter see [91]. For an arbitrary square matrix B, $\operatorname{dom}(B)$ is defined as

$$\operatorname{dom}(B) = \begin{cases} \text{greatest exponent in } \det(z^B) & \text{if } \det(z^B) \neq 0 , \\ \varepsilon & \text{otherwise.} \end{cases} \tag{9.5}$$

The coefficient $\overline{\zeta_k}$ in (9.4) equals the number of even permutations minus the number of odd permutations contributing to the highest-degree term in the exponents of z:

$$\max_{i_1 < i_2 < \cdots < i_k} \operatorname{dom} \left(A(i_1, i_2, \ldots, i_k) \right) .$$

Now let us consider the asymptotic behavior of $(z^A)^k$ as $z \to \infty$. One may easily understand that

$$(z^A)^k \approx z^{A^k} , \tag{9.6}$$

where A^k on the right-hand side denotes the k-th power of A for the matrix product in $\mathbb{R}_{\max}$. Define

$$\begin{aligned} \zeta_k^* &= (-1)^k \overline{\zeta}_k , \\ \mathrm{I} &= \{k \mid 1 \leq k \leq n, \zeta_k^* > 0\} , \\ c_k^* &= \bigoplus_{i_1 < i_2 < \cdots < i_k} \operatorname{dom}(A(i_1, i_2, \ldots, i_k)) . \end{aligned}$$

Substitution of (9.4) and (9.6) into (9.3) yields the following:

$$z^{A^n} + \sum_{k \in \mathrm{I}} \zeta_k^* z^{c_k^*} z^{A^{n-k}} \approx \sum_{k \notin \mathrm{I}} \zeta_k^* z^{c_k^*} z^{A^{n-k}} .$$

Since all terms now have positive coefficients, the comparison of the highest degree terms in both members of this approximation leads to the following identity in $\mathbb{R}_{\max}$:

$$A^n \oplus \bigoplus_{k \in \mathrm{I}} c_k^* A^{n-k} = \bigoplus_{k \notin \mathrm{I}} c_k^* A^{n-k} \ . \tag{9.7}$$

It is this identity that we consider as a version of the Cayley-Hamilton theorem in the max-plus algebra sense.

Remark 9.1 The dominant, as it appears implicitly in (9.7) through the coefficients c_k^*, can be directly obtained from A

$$\mathrm{dom}(A(i_1 i_2, \ldots, i_k)) = \bigoplus A_{i_1 j_1} \cdots A_{i_k j_k} \ ,$$

where $j_1, \ldots, j_k$ is a permutation of $i_1, \ldots, i_k$, and where the $\bigoplus$-symbol is with respect to all such permutations. ■

Remark 9.2 It is important to realize that this version of the Cayley-Hamilton theorem differs slightly from the one given in §2.5. The reason is that in the derivation of the current version terms have been canceled in the calculation of $\overline{\zeta}_k$ as it appears in (9.4). If terms of equal magnitude but of opposite signature (of the permutations) had been kept, then one would have obtained the 'original' Cayley-Hamilton theorem in the max-plus algebra. ■

Example 9.3 Consider

$$A = \begin{pmatrix} 1 & 2 & 3 \\ 4 & 1 & \varepsilon \\ e & 5 & 3 \end{pmatrix} ,$$

which was also considered in §2.5. First the coefficients c_k^* will be calculated:

$$\begin{aligned} c_1^* &= \textstyle\bigoplus_{i_1} \mathrm{dom} A(i_1) = \bigoplus_{i_1} \mathrm{dom}(A_{i_1 i_1}) = 1 \oplus 1 \oplus 3 = 3 \ , \\ c_2^* &= \textstyle\bigoplus_{i_1 < i_2} \mathrm{dom} A(i_1, i_2) = \mathrm{dom} A(1,2) \oplus \mathrm{dom} A(1,3) \oplus \mathrm{dom} A(2,3) \\ &= 6 \oplus 4 \oplus 4 = 6 \ , \\ c_3^* &= \mathrm{dom} A(1,2,3) = 12 \ . \end{aligned}$$

The quantity $\overline{\zeta}_1$ equals the number of even permutations minus the number of odd permutations needed to obtain c_1^*. The permutations of the diagonal elements are even end hence $\overline{\zeta}_1 = +1$. The permutation which realized $c_2^* = 6$, where the number 6 was obtained by $A_{12}A_{21}$, is odd and therefore $\overline{\zeta}_2 = -1$. Similarly, the permutation which realized $c_3^* = 12$, where the number 12 was obtained by $A_{13}A_{21}A_{32}$, is even and therefore $\overline{\zeta}_3 = +1$. Thus one obtains $\zeta_k^* = -1, k = 1, 2, 3$, and (9.7) becomes

$$A^3 = 3A^2 \oplus 6A \oplus 12e \ .$$

Note that this equation was also given in §2.5, with the actual A substituted. However, in that section the characteristic equation was first simplified before A was substituted into this characteristic equation. ■

From the above example it is clear that for any square matrix A, $\overline{\zeta}_1 = +1$ and hence A^n and A^{n-1} always appear on different sides of the equality symbol in the Cayley-Hamilton theorem. The lower order exponentials of A can appear at either side (but not on both sides simultaneously) of the equality symbol in the current version of the Cayley-Hamilton theorem.

9.2.2 Rational Transfer Functions and ARMA Models

In conventional discrete time system theory a rational transfer function can be expressed as the ratio of two polynomials $p(z) = \sum_{i=0}^{m} p_i z^i$ and $q(z) = \sum_{j=0}^{n} q_j z^j$ (z is the delay operator). Let $U(z)$ and $Y(z)$ denote the z-transforms of the input and of the output trajectories $u(\cdot)$ and $y(\cdot)$ respectively. We have

$$Y(z) = \frac{p(z)}{q(z)} U(z) \Leftrightarrow q(z)Y(z) = p(z)U(z) \Leftrightarrow \sum_{j=0}^{n} q_j y(t+j) = \sum_{i=0}^{m} p_i u(t+i) \ .$$

In Statistics the last equation is known as an 'ARMA' model: the 'autoregressive' (AR) part of the model corresponds to the left-hand side of the equation, whereas the 'moving average' (MA) part is the right-hand side.

In §5.7 rational transfer functions $H(\gamma, \delta) \in \mathcal{M}_{\text{in}}^{\text{ax}}[\![\gamma, \delta]\!]$ were identified with functions which can be written as CA^*B, where C (respectively B) is a row (respectively a column) vector and A is a square matrix. The entries of C and B may be restrained to be Boolean and those of A are elements of $\mathcal{M}_{\text{in}}^{\text{ax}}[\![\gamma, \delta]\!]$ which can be represented by polynomials of degree 1 in γ and δ (see Theorem 5.39). Our main objective here is to show that rational transfer functions are amenable to ARMA models as previously.

However, since there is no possibility of having 'negative' coefficients of polynomials, the AR and the MA part should both appear in both sides of the equation, which yields an implicit equation. This implicit equation in Y (U is given) may have several solutions in general, and among them, there is the 'true' solution of $Y = CA^*BU$. No results are available yet to select this true solution among the possibly many other solutions. Our purpose is just to show a utilization of the Cayley-Hamilton theorem to pass from the (C, A, B)-form to the ARMA form.

Lemma 9.4 *If Y is the output of a rational transfer function when U is the input, then there exist four polynomials $p_1, p_2, q_1, q_2 \in \mathcal{M}_{\text{in}}^{\text{ax}}[\![\gamma, \delta]\!]$, with* $\deg(p_1) < \deg(q_2)$ *and* $\deg(p_2) < \deg(q_1)$*, such that, for all $U \in \mathcal{M}_{\text{in}}^{\text{ax}}[\![\gamma, \delta]\!]$, Y satisfies*

$$q_1 Y \oplus p_1 U = q_2 Y \oplus p_2 U \ . \tag{9.8}$$

Proof Note that Y can be written as CX with $X = AX \oplus BU$ (conditions on (C, A, B) have been recalled earlier). In Theorem 2.22 it was shown that, in a commutative dioid such as $\mathcal{M}_{\rm in}^{\rm ax}[\![\gamma,\delta]\!]$, there exist two polynomials $p^+(z)$ and $p^-(z)$ of an abstract variable z with coefficients belonging to the dioid (here $\mathcal{M}_{\rm in}^{\rm ax}[\![\gamma,\delta]\!]$), such that

$$p^+(A) = p^-(A) \ .$$

The explicit form of these polynomials, given in Definition 2.21, shows that their coefficients are themselves polynomials in (γ, δ) since A is a polynomial matrix. Now, for any $k \in \mathbb{N}$, we have

$$X = A^k X \oplus \left(e \oplus A \oplus \cdots \oplus A^{k-1}\right) BU \ . \tag{9.9}$$

Let $p^+ = \bigoplus_{k=0}^{n_1} p_k^+ z^k$ and consider a similar expression for p^- (with degree n_2). One can multiply both sides of Equation (9.9) by p_k^+ and sum up all these equations for $k = 0, \ldots, n_1$. This yields

$$\underbrace{\left(\bigoplus_{k=0}^{n_1} p_k^+\right) X}_{a_1} = \underbrace{\left(p^+(A)\right) X}_{a_2} \oplus \underbrace{r^+(A)BU}_{a_3}$$

for some polynomial $r^+(z)$ of degree less than n_1, the form of which is not given in detail here. In a similar way, one can obtain

$$\underbrace{\left(\bigoplus_{l=0}^{n_2} p_l^-\right) X}_{a_4} = \underbrace{\left(p^-(A)\right) X}_{a_5} \oplus \underbrace{r^-(A)BU}_{a_6}$$

for some polynomial $r^-(z)$ of degree less than n_2. Note that $a_2 = a_5$ by the Cayley-Hamilton theorem. Then, we have

$$a_1 \oplus a_6 = a_2 \oplus a_3 \oplus a_6 = a_3 \oplus a_5 \oplus a_6 = a_3 \oplus a_4 \ .$$

To complete the proof, it suffices to multiply both sides of this equation by C (which commutes with 'scalars') to let Y appear ($Y = CX$). ■

9.2.3 Realization Theory

For this subsection the reader is assumed to be familiar with conventional realization theory, see e.g. [72]. In §1.3 the following question was posed: How do we obtain a time-domain representation, or equivalently, how do we find A, B and C, if the $p \times m$ transfer matrix

$$H(\gamma) = \gamma CB \oplus \gamma^2 CAB \oplus \gamma^3 CA^2 B \oplus \cdots$$

is given? For the sake of simplicity we will confine ourselves to SISO systems, i.e. matrices B and C are vectors ($m = p = 1$). One may be tempted to study the related semi-infinite Hankel matrix G defined by

$$G = \begin{pmatrix} g_1 & g_2 & g_3 & \cdots \\ g_2 & g_3 & g_4 & \cdots \\ g_3 & g_4 & g_5 & \cdots \\ g_4 & \vdots & \vdots & \\ \vdots & \vdots & & \end{pmatrix},$$

where $g_i = CA^{i-1}B$; these quantities are sometimes called *Markov parameters*. Only partial results, some of which will be shown now, have been obtained along this line. The matrix $G_{(\leq i)(\leq j)}$ is, as in Chapter 2, defined as the submatrix of G, consisting of the intersection of the first i columns and the first j rows of G. As an example, consider the Markov parameters

$$g_1 = 1, \; g_2 = 3, \; g_3 = 0, \; g_4 = 1, \; g_5 = -2, \; g_6 = -1, \; g_7 = -4, \ldots . \tag{9.10}$$

It is easily verified that for this series, $\mathrm{dom}G_{(\leq 1)(\leq 1)} = 1$, $\mathrm{dom}G_{(\leq 2)(\leq 2)} = 6$, $\mathrm{dom}G_{(\leq 3)(\leq 3)} = 0$, and $\mathrm{dom}G_{(\leq i)(\leq i)} = \varepsilon$ for $i \geq 4$, where dom was defined in (9.5). This, with the conventional theory in mind, might lead to the conclusion that the minimal realization would have order 3. This is false since the Markov parameters above were derived from the system with

$$A = \begin{pmatrix} \varepsilon & e \\ -2 & -3 \end{pmatrix}, \quad B = \begin{pmatrix} \varepsilon \\ e \end{pmatrix}, \quad C = \begin{pmatrix} 3 & 1 \end{pmatrix}, \tag{9.11}$$

and hence the minimal realization will maximally have order 2. Studying $\mathrm{dom}G_{(\leq i)(\leq i)}$ turns out not to be very fruitful. A better approach is to consider linear dependences among the rows of G. For the current example, for instance, we have

$$G_{\cdot i} = (-3)G_{\cdot i-1} \oplus (-2)G_{\cdot i-2} \,, \quad i = 3, 4, \ldots .$$

Now we can use the following theorem.

Theorem 9.5 *Given the series $\{g_i\}_{i=1}^{\infty}$ such that for the corresponding Hankel matrix,*

$$G_{\cdot i} = c_1 G_{\cdot i-1} \oplus \cdots \oplus c_n G_{\cdot i-n} \,, \quad i = n+1, n+2, \ldots ,$$

holds true for certain coefficients $c_1, \ldots, c_n$, and where n is the smallest integer for which this, or another linear dependence (see below), is possible, then the discrete-event system characterized by

$$A = \begin{pmatrix} \varepsilon & e & \varepsilon & \ldots & \varepsilon \\ \varepsilon & \varepsilon & e & \ldots & \varepsilon \\ \vdots & & & & \\ \varepsilon & \ldots & \ldots & \varepsilon & e \\ c_n & \ldots & \ldots & c_2 & c_1 \end{pmatrix}, \qquad B = \begin{pmatrix} g_1 \\ g_2 \\ \vdots \\ g_{n-1} \\ g_n \end{pmatrix}, \qquad C' = \begin{pmatrix} e \\ \varepsilon \\ \vdots \\ \vdots \\ \varepsilon \end{pmatrix},$$

is a minimal realization.

The proof can be found in [105]. The essence of the proof consists in converting the statement of the theorem into the conventional algebra setting by means of the exponential transformation as introduced in §9.2.1, giving the proof there and then returning to the max-plus algebra setting. In the statement of the theorem above, the notion of linear dependence of columns is used.

Definition 9.6 *Column vectors $v_1, \ldots, v_n$ are said to be linearly dependent if scalars $c_1, \ldots, c_n$, not all ε, and a subset $\mathrm{I} \in \{1, \ldots, n\}$ exist such that*

$$\bigoplus_{k \in \mathrm{I}} c_k v_k = \bigoplus_{k \notin \mathrm{I}} c_k v_k \ .$$

If this theorem is applied to the series in (9.10), then the result is

$$A = \begin{pmatrix} \varepsilon & e \\ -2 & -3 \end{pmatrix}, \quad B = \begin{pmatrix} 1 \\ 3 \end{pmatrix}, \quad C = \begin{pmatrix} e & \varepsilon \end{pmatrix},$$

which is different from (9.11), although both 3-tuples (A, B, C) characterize the same series of Markov parameters.

Unfortunately, Theorem 9.5 is of limited use. The reason is that it cannot deal with general linear dependences of column vectors. Take as an example

$$g_1 = 5 \ , \quad g_2 = 8 \ , \quad g_3 = 11.5 \ , \quad g_4 = 15.5 \ , \quad g_5 = 19.5 \ , \ldots . \qquad (9.12)$$

For the corresponding Hankel matrix the following dependence is true:

$$G_{\cdot i} \oplus 7G_{\cdot i-2} = 4G_{\cdot i-1} \ , \quad i = 3, 4, \ldots ,$$

but Theorem 9.5 does not cover this kind of linear dependence. The system characterized by

$$A = \begin{pmatrix} 3 & 7 \\ -2 & 4 \end{pmatrix}, \quad B = \begin{pmatrix} 5 \\ e \end{pmatrix}, \quad C = \begin{pmatrix} e & 3.5 \end{pmatrix},$$

however, is a minimal realization of the Markov parameters given in (9.12).

The conclusion of this subsection is that, given an arbitrary series of Markov parameters, it is not known how to obtain a minimal state space realization (if it exists). In the next subsection, however, the reader will find a recent development.

9.2.4 More on Minimal Realizations

R.A. Cuninghame-Green [50] has recently come up with a promising method to obtain a state space realization from a series of Markov parameters. The following two theorems are used, the proofs of which can be found in [49]. In these theorems, if K is a matrix, then $\overline{K}$ is the matrix obtained from K by transposition and a change of sign.

Theorem 9.7 *For a general matrix K, $(K \odot \overline{K}) \otimes K = K$.*

The symbol $\odot$ refers to the multiplication of two matrices (or of a matrix and a vector) in which the min-operation rather than the max-operation is used; it will be discussed more extensively in §9.6. The theorem just formulated states that all columns of K are eigenvectors of $K \odot \overline{K}$.

Theorem 9.8 *For a given matrix K, consider D, where $D = K \otimes K_d$ and K_d is derived from $\overline{K} \odot K$ by replacing all diagonal elements by e. Then a column of K is linearly dependent on the other columns, i.e. one column can be written as a linear combination of the other columns in the max-plus algebra sense, if and only if it is identical to the corresponding column of D. The corresponding column of K_d then yields the coefficients expressing the linear dependence.*

This theorem gives a routine method of finding linear dependences among the columns of a given matrix. This linear dependence is to be understood as one column being written as a linear combination of the others. Note that this definition of linear dependence is more restrictive than the definition used in §9.2.

For the realization one forms a Hankel matrix $G_{(\leq n+1)(\leq n+1)}$ for some n sufficiently large. From Theorem 9.7 we know that the columns of $G_{(\leq n+1)(\leq n+1)}$ are preserved by the action of $G_{(\leq n+1)(\leq n+1)} \odot (G_c)_{(\leq n+1)(\leq n+1)}$. It follows that, if A is the matrix obtained by dropping the first row and last column of $G_{(\leq n+1)(\leq n+1)} \odot (G_c)_{(\leq n+1)(\leq n+1)}$, then

$$A \otimes \begin{pmatrix} g_1 \\ g_2 \\ \vdots \\ g_n \end{pmatrix} = \begin{pmatrix} g_2 \\ g_3 \\ \vdots \\ g_{n+1} \end{pmatrix},$$

i.e. A moves the Markov parameters 'one position up'. A state space realization is now obtained by A, by B as the first column of $G_{(\leq m)(\leq m)}$ and by $C = \begin{pmatrix} e & \varepsilon & \dots & \varepsilon \end{pmatrix}$.

In general, the realization found will not have minimal dimension. In order to reduce the dimension, Theorem 9.8 is used. One searches for column linear dependences, as well as for row linear dependences of A. By simultaneously deleting dependent rows and columns of the same index from A, the state space dimension is reduced.

As an example, consider the Markov parameters $g_1 = 0, g_2 = 3, g_3 = 6, g_4 = 10, g_5 = 14, g_6 = 18, \dots$, and take $n = 3$. It is easily verified, by following the procedure described above, that

$$A = \begin{pmatrix} 3 & e & -4 \\ 6 & 3 & e \\ 10 & 7 & 4 \end{pmatrix}, \quad B = \begin{pmatrix} e \\ 3 \\ 6 \end{pmatrix}, \quad C = \begin{pmatrix} e & \varepsilon & \varepsilon \end{pmatrix}.$$

Since the second column of A depends linearly on the first one, and the second row depends linearly on the other rows (it is linearly dependent on the last row), the second row and second column can be deleted so as to obtain a state space realization of lower dimension:

$$A = \begin{pmatrix} 3 & -4 \\ 10 & 4 \end{pmatrix}, \quad B = \begin{pmatrix} e \\ 6 \end{pmatrix}, \quad C = \begin{pmatrix} e & \varepsilon \end{pmatrix}.$$

This latter realization turns out to have minimal dimension. It is left as an exercise to show that if started with $n = 2$ rather than with $n = 3$, one would have obtained the wrong result. This in spite of the fact that the Hankel matrix G has 'rank' 2; for $i \geq 1$ we have that $(-4)G_{\cdot i+2} \oplus 3G_{\cdot i} = G_{\cdot i+1}$.

9.3 CONTROL OF DISCRETE EVENT SYSTEMS

In this section special instances of nonlinear systems will be described, for which the max-plus setting is still appropriate. The system equations to be considered have the form

$$x(k+1) = A(u(k), u(k-1))x(k) \ . \tag{9.13}$$

Matrix A can be controlled by the decision variable u to be defined. In Chapter 1 a decision variable u was encountered also. There it had the function of an input to the system; nodes of the underlying network had to wait for external inputs. In the current setting as expressed by (9.13), u influences the entries of the system matrix A. For a motivation of the system described by (9.13), think of a production planning where the holding times at the nodes are zero and where the standard traveling time from node j to node i is indicated by A_{ij}. This traveling time can be reduced maximally by an amount c if an extra piece of equipment is used. Rather than A_{ij}, it then becomes $\overline{A}_{ij} \stackrel{\text{def}}{=} \max(A_{ij} - c, 0)$. It is assumed that only one such a piece of equipment is available and that it can be used only once (i.e. at one arc, connecting two nodes) during each k-step. One can envisage situations in which this piece of equipment could be used a number of times during the same k-step, at different arcs of the network. Although such a generalization can in principle be handled within the context of this section, the analysis becomes rather laborious and such a generalization will therefore not be considered.

Suppose we are given a network with two nodes. If no extra piece of equipment were available, the evolution of the state vector $x(k) \in \mathbb{R}$ is according to $x(k+1) = Ax(k)$ in $\mathbb{R}_{\max}$, where

$$A = \begin{pmatrix} 3 & 1 \\ 4 & 2 \end{pmatrix}.$$

Boolean variables $u_{ij}(k), i, j = 1, 2$, are now introduced to describe the control actions; they are defined subject to $\sum u_{ij}(k) = 0$ or $= 1$. Hence maximally one of the $u_{ij}(k)$ can be 1, which indicates that the piece of extra equipment has been set in for arc a_{ij}, from node j to node i during the k-th cycle. Thus there are five possibilities; the piece of equipment is not applied or it is applied to one of the arcs corresponding to $A_{ij}, i, j = 1, 2$. If $c = 2$, these possibilities result in the following matrices A:

$$\begin{pmatrix} 3 & 1 \\ 4 & 2 \end{pmatrix}, \quad \begin{pmatrix} 1 & 1 \\ 4 & 2 \end{pmatrix}, \quad \begin{pmatrix} 3 & 0 \\ 4 & 2 \end{pmatrix}, \quad \begin{pmatrix} 3 & 1 \\ 2 & 2 \end{pmatrix}, \quad \begin{pmatrix} 3 & 1 \\ 4 & 0 \end{pmatrix}.$$

Formally, we can write

$$x(k+1) = \begin{pmatrix} A_{11} \not{u}_{11}^2 & A_{12} \not{u}_{12} \\ A_{21} \not{u}_{21}^2 & A_{22} \not{u}_{22}^2 \end{pmatrix} x(k) \ .$$

This equation does not take into account the fact that the piece of equipment might not be available at the appropriate node at the right time. For this reason, an extra state variable x_3 will be introduced; $x_3(k)$ denotes the epoch of release of the equipment during the $(k-1)$-st cycle. The correct evolution equations then become

$$\begin{aligned} \begin{pmatrix} x_1(k+1) \\ x_2(k+1) \end{pmatrix} &= \begin{pmatrix} A_{11} \not{u}_{11}^2 & A_{12} \not{u}_{12} \\ A_{21} \not{u}_{21}^2 & A_{22} \not{u}_{22}^2 \end{pmatrix} \begin{pmatrix} x_1(k) \\ x_2(k) \end{pmatrix} \\ &\oplus \begin{pmatrix} \ln(u_{11}(k)) \oplus \ln(u_{21}(k)) \\ \ln(u_{12}(k)) \oplus \ln(u_{22}(k)) \end{pmatrix} x_3(k) \ , \qquad (9.14) \\ x_3(k) &= \bigoplus_{i=1}^{2} \bigoplus_{j=1}^{2} \overline{A}_{ji} \ln(u_{ji}(k-1)) x_i(k) \ , \qquad (9.15) \end{aligned}$$

where we made the convention that $\ln(1) = e$ and $\ln(0) = \varepsilon$. If (9.15) were substituted into (9.14), then an equation of the form (9.13) would result.

If we start at state $\begin{pmatrix} 0 & 0 & 0 \end{pmatrix}'$ for $k = 0$, then the five possible next states are respectively[1]

$$\begin{pmatrix} 3 \\ 4 \\ 3 \end{pmatrix}, \begin{pmatrix} 1 \\ 4 \\ 1 \end{pmatrix}, \begin{pmatrix} 3 \\ 4 \\ 3 \end{pmatrix}, \begin{pmatrix} 3 \\ 2 \\ 2 \end{pmatrix}, \begin{pmatrix} 3 \\ 4 \\ 3 \end{pmatrix}.$$

From these states, new states can be reached again. Thus a tree of states can be found. We will not count states as such if they are linearly dependent on an already existing state. The states will be normalized by adding a same scalar to all components of a state vector, such that the last component becomes zero. (other normalizations are possible, such as for instance setting the least component equal to zero). It turns out that five different normalized states exist. Some trial and error will show that whatever the initial condition is, the evolution will always end up in these five states in a finite number of steps. These states are indicated by $n_i, i = 1, \ldots, 5$, and are given, together with the possible follow-ups, in Table 9.1. In the same table the normalization factors are given. If according to this table, n_i is mapped to n_j with a normalization factor a, then the actual mapping is $n_i \mapsto a \otimes n_j$. This table defines a Markov chain the 'transition matrix' of which is given below. The ji-th entry equals the normalization factor corresponding to the mapping from n_i to n_j by means of an appropriate control, if this control exists. If it does not

[1]The first one is the result of applying $u_{ij} = 0$ for all i, j; for the second one, we used $u_{11} = 1$ and the other $u_{ij} = 0$; for the third, fourth and fifth ones, $u_{12} = 1$, $u_{12} = 1$ and $u_{22} = 1$, respectively, where the nonmentioned u-entries remain zero.

exist, then the entry is indicated by ε:

$$V = \begin{pmatrix} \varepsilon & 3 & 4 & \varepsilon & \varepsilon \\ 3 & 3 & 4 & 4 & 3 \\ 1 & \varepsilon & \varepsilon & 2 & \varepsilon \\ 2 & \varepsilon & \varepsilon & 3 & \varepsilon \\ \varepsilon & 2 & 3 & \varepsilon & \varepsilon \end{pmatrix} .$$

Suppose that in Table 9.1 an initial state n_i would have been mapped to another

Table 9.1: Possible transitions

Initial state	New states according to the five different controls									
$n_1 = \begin{pmatrix} 0 & 0 & 0 \end{pmatrix}'$	n_2	3	n_3	1	n_2	3	n_4	2	n_2	3
$n_2 = \begin{pmatrix} 0 & 1 & 0 \end{pmatrix}'$	n_2	3	n_5	2	n_2	3	n_1	3	n_2	3
$n_3 = \begin{pmatrix} 0 & 3 & 0 \end{pmatrix}'$	n_2	4	n_2	4	n_5	3	n_2	4	n_1	4
$n_4 = \begin{pmatrix} 1 & 0 & 0 \end{pmatrix}'$	n_2	4	n_3	2	n_2	4	n_4	3	n_2	4
$n_5 = \begin{pmatrix} 0 & 2 & 0 \end{pmatrix}'$	n_2	3	n_2	3	n_2	3	n_2	3	n_2	3

state n_j twice, with normalization factors a and b respectively, with $a < b$ (which does not occur in this example though). Matrix V should then have contained the smaller of the two factors a and b. Since the controls should be chosen in such a way that the network operates as fast as possible, this matrix V will be considered in the min-plus algebra; hence $\varepsilon = +\infty$. The eigenvalue of this matrix is found by applying Karp's algorithm; it turns out to be equal to 2.5. There are two critical circuits, namely

$$n_1 \rightarrow n_3 \rightarrow n_1 , \qquad n_2 \rightarrow n_5 \rightarrow n_2 .$$

There are two different periodic solutions to our problem; they are characterized by the two critical circuits. From Table 9.1, it will be clear how to control the network, i.e. where to use this extra piece of equipment, such that the evolution of the state equals one of these periodic solutions.

9.4 BROWNIAN AND DIFFUSION DECISION PROCESSES

We show the analogy between probability calculus and dynamic programming. In the former area, iterated convolutions of probability laws play a central role; in the latter area, this role is played by the inf-convolution of cost functions. The main analysis tool is the Fourier transform for the former situation, and it is the Fenchel transform for the latter. Quadratic forms, which form a stable set by inf-convolution, correspond to Gaussian laws, which are stable by convolution. Asymptotic theorems for the value function of dynamic programming correspond to the law of large numbers and the central limit theorem. Straight line optimal trajectories correspond to Brownian motion trajectories. The operator $v \mapsto \partial v/\partial t - (\partial v/\partial x)^2$, which will be appear as a min-plus linear operator, corresponds to the operator $v \mapsto \partial v/\partial t + \partial^2 v/\partial x^2$. The min-plus function $x^2/2t$ corresponds to the Green function $(1/\sqrt{2\pi t})\exp(-x^2/2t)$. A diffusion decision process with generator $v \mapsto \partial v/\partial t - b(x)\partial v/\partial x - a(x)(\partial v/\partial x)^2$ corresponds to the diffusion process with generator $\partial/\partial t + b(x)\partial/\partial x + a(x)\partial^2 v/\partial x^2$.

9.4.1 Inf-Convolutions of Quadratic Forms

For $m \in \mathbb{R}$ and $\sigma \in \mathbb{R}^+$, let $Q_{m,\sigma}(x)$ denote the quadratic form in x defined by

$$Q_{m,\sigma}(x) = \frac{1}{2}\left(\frac{x-m}{\sigma}\right)^2 \quad \text{for } \sigma \neq 0 \ ,$$

$$Q_{m,0}(x) = \delta_m(x) = \begin{cases} 0 & \text{for } x = m \ ; \\ +\infty & \text{otherwise.} \end{cases}$$

These quadratic forms take a zero value at m.

Given two mappings f and g from $\mathbb{R}$ into $\overline{\mathbb{R}}$, we define the inf-convolution of f and g [119] as the mapping from $\overline{\mathbb{R}}$ into $\overline{\mathbb{R}}$ (with the convention $\infty - \infty = \infty$) defined by

$$z \mapsto \inf_{x+y=z} (f(x) + g(y)) \ .$$

It is denoted $f \otimes g$.

Theorem 9.9 *We have*

$$Q_{m_1,\sigma_1} \otimes Q_{m_2,\sigma_2} = Q_{m_1+m_2,\sqrt{\sigma_1^2+\sigma_2^2}} \ .$$

This result is the analogue of the (conventional) convolution of Gaussian laws (denoted $*$):

$$\mathcal{N}(m_1,\sigma_1) * \mathcal{N}(m_2,\sigma_2) = \mathcal{N}(m_1+m_2, \sqrt{\sigma_1^2+\sigma_2^2}) \ ,$$

where $\mathcal{N}(m,\sigma)$ denotes the Gaussian law with mean m and standard deviation σ. Therefore, there exists a morphism between the set of quadratic forms endowed with the inf-convolution operator and the set of exponentials of quadratic forms endowed with the convolution operator.

Clearly this result can be generalized to the vector case.

9.4.2 Dynamic Programming

Given the simplest decision process:

$$x(n+1) = x(n) - u(n) \ , \quad x_0 \text{ given,}$$

for $x(n) \in \mathbb{R}, u(n) \in \mathbb{R}, n \in \mathbb{N}$, and the particular additive cost function

$$\min_{u(0),u(1),\ldots,u(N-1)} \left(\sum_{i=0}^{N-1} c(u(i)) + \phi(x(N)) \right) ,$$

where c and ϕ are mappings from $\mathbb{R}$ into $\overline{\mathbb{R}}$ which are supposed to be convex, lower-semicontinuous in the conventional sense, equal to zero at their minimum and thus nonnegative. Let m denote the abscissa where c achieves its minimum, then

$$\min c(\cdot) = c(m) = 0 \ .$$

The assumptions retained here are not minimal but they will simplify our discussion.

The value function defined by

$$v(n,x) = \min_{u(n),\ldots,u(N-1)} \left(\sum_{p=n}^{N-1} c(u(p)) + \phi(x(N)) \;\middle|\; x(n) = x \right)$$

satisfies the dynamic programming equation

$$v(n,x) = \min_u \left(c(u) + v(n+1, x-u) \right) \ , \quad v(N,x) = \phi(x) \ .$$

It can be written using the inf-convolution:

$$v(n,\cdot) = c \otimes v(n+1,\cdot) \ , \quad v(N,\cdot) = \phi \ ,$$

that is (with the change of time index $p = N - n$, and the choice $\phi = \delta_0$),

$$v(p,\cdot) = c^p(\cdot) \otimes \delta_0 = c^p(\cdot) \ .$$

This, in words, means that the solution of the dynamic programming equation in this particular case of an 'independent increment decision process' is obtained by iterated inf-convolutions of the instantaneous cost function.

In a more general case, the instantaneous cost c depends on the initial and the final states of a decision period, namely $x(n)$ and $x(n+1)$ (and not only on the state variation $u(n) = x(n+1) - x(n)$). Moreover the dynamics is a general Markovian process, namely, $x(n+1) \in \Gamma(x(n))$ (where Γ denotes a set-valued function from $\mathbb{R}$ into $2^{\mathbb{R}}$). Then the dynamic programming equation becomes

$$v(n,x) = \min_{y \in \Gamma(x)} \left(c(x,y) + v(n+1,y) \right) \ , \quad v(N,x) = \delta_0(x) \ ,$$

the solution of which can be written, with the same change of time, as

$$v(n,\cdot) = \left[c^n \otimes \delta_0 \right] (\cdot) \ ,$$

where the product of two kernels is now defined as

$$\left[c_1 \otimes c_2 \right] (x,z) = \min_{y \in \Gamma(x)} \left(c_1(x,y) + c_2(y,z) \right) \ .$$

This more general case is the analogue of the general Markov chain case.

In addition to the analogues of the law of large numbers and of the central limit theorem, we will show the analogue of the Brownian motion and of diffusion processes.

Before addressing this issue, let us recall, once more, that the role of the Fourier transform in probability theory is played by the Fenchel transform in dynamic programming as it was noticed for the first time in [17].

9.4.3 Fenchel and Cramer Transforms

Let f be a mapping from $\mathbb{R} \to \overline{\mathbb{R}}$, supposed to be convex, l.s.c. and proper (i.e. never equal to $-\infty$) and let $\widehat{f} : \overline{\mathbb{R}} \to \overline{\mathbb{R}}$ be its Fenchel transform (see Remark 3.36). Then it can be shown that $\widehat{f}$ is convex, l.s.c. and proper.

Example 9.10 The function defined by $\left[\mathcal{F}_{\mathrm{e}}\left(Q_{m,\sigma}\right)\right](p) = (1/2)p^2\sigma^2 + pm$ is the analogue of the characteristic function of a Gaussian law. ■

The transform $\mathcal{F}_{\mathrm{e}}$ behaves as an involution, that is, $\mathcal{F}_{\mathrm{e}}(\mathcal{F}_{\mathrm{e}}(f)) = f$ for all convex, proper, l.s.c. functions f.

As already noticed, the main interest of the Fenchel transform is its ability to convert inf-convolutions into sums, that is,

$$\mathcal{F}_{\mathrm{e}}(f \otimes g) = \mathcal{F}_{\mathrm{e}}(f) + \mathcal{F}_{\mathrm{e}}(g) \ .$$

Applying the Fenchel transform to the dynamic programming equation in the case when c depends only on x, we obtain $v(N, \cdot) = \mathcal{F}_{\mathrm{e}}\left(\widehat{\phi} + N\widehat{c}\right)$. Using the fast Fenchel algorithm [31], this formula gives a fast algorithm to solve this particular instance of the dynamic programming equation.

Moreover, let us recall that the Fenchel transform is continuous for the epigraph topology, that is, the epigraphs of the transformed functions converge if the epigraphs of the source functions converge for a well chosen topology. We can use, for example, the Hausdorff topology for the epigraphs which are closed convex sets of $\mathbb{R}^2$, but this may be too strong (see [71] and [2] for discussions of these topological aspects). Here we will be more concerned with the analogies between probability and deterministic control.

Example 9.11 Let $\ell_\nu : x \mapsto \nu x$. One has $\left[\mathcal{F}_{\mathrm{e}}(\ell_\nu)\right](p) = \delta_\nu(p)$. When $\nu \to 0$, then $\delta_\nu \to \delta_0$ in the epigraph sense, but it does not converge pointwise even if $\ell_\nu \to 0$ pointwise. ■

Moreover, the pointwise convergence of numerical convex, l.s.c. functions towards a function in the same class implies the convergence of their epigraphs.

The Cramer transform is defined by $\mathcal{F}_{\mathrm{e}} \circ \log \circ \mathcal{L}$, where $\mathcal{L}$ denotes the Laplace transform. Therefore, it transforms the convolutions into inf-convolutions. Thus it is exactly the morphism which we are interested in. Unfortunately, it is only a morphism for a set of functions endowed with one operation, the convolution. It is not a morphism for the sum (the pointwise sum of two functions is not transformed by the Cramer transform into the pointwise min of the transformed functions). Moreover, the Cramer transform convexifies the functions but the inf-convolution is defined on a more general set of functions. Nevertheless the mapping $\lim_{\nu\to 0}\log_\nu$ defines a morphism of algebra between the asymptotics (around zero) of positive real functions of a real variable and the real numbers endowed with the two operations min and plus. Indeed,

$$\lim_{\nu\to 0}\log_\nu(\nu^a+\nu^b)=\min(a,b)\ ,\qquad \log_\nu(\nu^a\nu^b)=a+b\ .$$

This transformation has been already utilized in §9.2 under a slightly different form.

We can now study the analogues of the limit theorems of probability calculus.

9.4.4 Law of Large Numbers in Dynamic Programming

Suppose we are given two numerical mappings c and ϕ which are nonnegative, convex, l.s.c. and which are equal to zero at their unique minimum. The first and second order derivatives of a function c are denoted $\dot{c}$ and $\ddot{c}$ respectively.

To simplify the discussion, let us suppose that $c\in C^2$ and $|1/\ddot{c}(u)|_\infty<\infty$ in a neighborhood of the minimum. Let m denote the abscissa where c achieves its minimum (zero) value. Let $w_N(x)$ be the mapping $x\mapsto v(N,Nx)$. For the value function, this scaling operation corresponds to the conventional averaging of the sample.

Theorem 9.12 (Weak law of large numbers for dynamic programming) *Under the foregoing assumptions, we have*

$$\lim_{N\to\infty} v(N,Nx)=\delta_m(x)\ ,$$

the limit being in the sense of the epigraph convergence.

Proof We have

$$\widehat{v}(N,p)=\widehat{\phi}(p/N)+N\widehat{c}(p/N)\ ,\qquad \lim_{N\to\infty}\widehat{\phi}(p/N)=\widehat{\phi}(0)=0\ ,$$

since ϕ admits a zero minimum by assumption. Moreover, $\widehat{c}(0)=0$ for the same reason. Then $\widehat{c}(p)$ admits a Taylor expansion around 0 of the form $pm+\mathrm{O}(p^2)$. Indeed,

$$\dot{\widehat{c}}(p)=x_{\mathrm{o}}(p)+\dot{x}_{\mathrm{o}}(p)(p-\dot{c}(x_{\mathrm{o}}(p)))=x_{\mathrm{o}}(p)=m+\mathrm{O}(p)\ ,$$

where $x_{\mathrm{o}}(p)$ denotes the point at which the maximum is achieved in the definition of the Fenchel transform of c. Therefore, $\widehat{v}(N,p)=pm+\mathrm{O}(1/N)$. Then, using the continuity of the Fenchel transform, we obtain

$$\lim_{N\to\infty}\mathcal{F}_{\mathrm{e}}\left(\widehat{v}(N,\cdot)\right)=\mathcal{F}_{\mathrm{e}}(pm)=\delta_m\ .$$

■

9.4.5 Central Limit Theorem in Dynamic Programming

We have the analogue of the central limit theorem of probability calculus. The value function, centered and normalized with the scaling $\sqrt{N}$, is asymptotically quadratic.

Theorem 9.13 (Central Limit Theorem) *Under the foregoing assumptions, we have*

$$\lim_{N\to\infty} v\left(N, \sqrt{N}(y+Nm)\right) = \frac{1}{2}\ddot{c}(m)y^2 \ .$$

The limit is in the sense of epigraph convergence.

Proof We make the expansion up to the second order of $p \mapsto \widehat{r}_N(p)$ where r_N is the mapping $y \mapsto v\left(N, \sqrt{N}(y+Nm)\right)$.

But

$$\widehat{r}_N(p) = \widehat{\phi}\left(\frac{p}{\sqrt{N}}\right) + N\widehat{c}_m\left(\frac{p}{\sqrt{N}}\right) \ ,$$

where $c_m(y) = c(y+m)$. Then we have $\widehat{\phi}(0) = 0$ and $\widehat{c}_m(0) = 0$ because the minima of ϕ and c_m are zero.

Let us expand $\widehat{c}_m$ up to the second order. We have seen that $\dot{\widehat{c}}_m(p) = x_{\text{o}}(p)$, and therefore $\ddot{\widehat{c}}_m(p) = \dot{x}_{\text{o}}(p)$. Moreover, we know that $x_{\text{o}}(p)$ is defined by $p - \dot{c}_m(x_{\text{o}}(p)) = 0$, and therefore $1 - \ddot{c}_m(x_{\text{o}}(p))\dot{x}_{\text{o}}(p) = 0$, that is, $\dot{x}_{\text{o}}(p) = 1/\ddot{c}_m(x_{\text{o}}(p))$. Finally,

$$\widehat{r}_N(p) = \frac{1}{2}\frac{p^2}{\ddot{c}_m(0)} + \text{o}(1) \ .$$

We obtain the result by passing to the limit using the continuity of the epigraph of the Fenchel transform. ■

These results can be extended to the vector case, to the case when c depends on time, etc.

9.4.6 The Brownian Decision Process

Let us consider the discrete time decision process

$$\min_u \left(\sum_{i=0}^{(T/h)-1} \frac{(u(ih))^2}{2h} + \Phi(x(T)) \right) , \qquad x(t+h) = x(t) - u(t) \ .$$

It satisfies the dynamic programming equation

$$v(t,x) = \min_u \left(\frac{u^2}{2h} + v(t+h, x-u) \right) , \quad v(T,\cdot) = \Phi \ .$$

The cost function $Q_{0,\sqrt{h}}$ is therefore the analogue of the increment of Brownian motion on a time step of h. The analogue of the independence of the increments of the Brownian motion is the independence of the instantaneous cost function u^2/h from the state variable x.

Let us make the change of control $u = wh$ in the dynamic programming equation. We obtain

$$v(t, x) = \min_w \left(\frac{hw^2}{2} + v(t + h, x - wh) \right) .$$

Passing to the limit when $h \to 0$, we obtain the Hamilton-Jacobi-Bellman (HJB) equation

$$\frac{\partial v}{\partial t} + \min_w \left(-w \frac{\partial v}{\partial x} + \frac{w^2}{2} \right) = 0, \quad v(T, \cdot) = \Phi \ ,$$

that is,

$$\frac{\partial v}{\partial t} - \frac{1}{2} (\frac{\partial v}{\partial x})^2 = 0 \ , \quad v(T, \cdot) = \Phi \ ,$$

which is the analogue of the heat equation

$$\frac{\partial v}{\partial t} + \frac{1}{2} \frac{\partial^2 v}{\partial x^2} = 0 \ , \quad v(T, \cdot) = \Phi \ .$$

Therefore, we can see the Brownian decision process as the Sobolev space $H^1(0, T)$ endowed with the cost function $W(\omega) = \int_0^T (\dot{\omega})^2 \, tt$ for any function $\omega \in H^1(0, T)$. Then the decision problem can be written

$$\mathbb{M}_W \Phi(x(T)) \stackrel{\text{def}}{=} \min_{\omega \in H^1(0,T)} \big(W(\omega) + \Phi(x(T; \omega)) \big) \tag{9.16}$$

by analogy with probability theory. The function W is the analogue of the Brownian measure, and it can be interpreted as the cost of choosing ω. Then $\Phi(x(T; \omega))$ is the cost of a decision function $\Phi(x(T; \cdot))$ once we have chosen ω. But the solution of the Hamilton-Jacobi equation

$$\frac{\partial v}{\partial t} - \frac{1}{2} \left(\frac{\partial v}{\partial x} \right)^2 = 0 \ , \quad v(T, \cdot) = \delta_y \ ,$$

is unique [86], and is explicitly given by

$$v(t, x) = \frac{(y - x)^2}{2(T - t)} \ , \quad t \leq T \ .$$

It can be considered as the min-plus Green kernel of the dynamic programming equation and as the analogue of the Green kernel of the Kolmogorov equation for the Brownian equation, namely

$$\frac{1}{\sqrt{2\pi(T - t)}} \exp \left(-\frac{(y - x)^2}{2(T - t)} \right) .$$

Therefore, by min-plus linearity, we can derive the solution of

$$\min\left(\frac{\partial v}{\partial t} - \frac{1}{2}\left(\frac{\partial v}{\partial x}\right)^2, c - v\right) = 0 \ , \quad v(T, \cdot) = \Phi \ ,$$

which is the solution of the control problem

$$v(t, y) = \mathbb{M}_W\left[\min\left(\min_{t \le s \le T} c(x_s(\omega)), \Phi(x(T;\omega))\right) \,\middle|\, x(t) = y\right],$$

where s denotes a stopping time that we also want to optimize. This cost is clearly the min-plus analogue of

$$v(t, y) = \mathbb{E}_W\left[\int_t^T c(x(s;\omega))\, \mathrm{d}s + \Phi(x(T;\omega)) \,\middle|\, x(t) = y\right].$$

The solution of the decision problem is

$$v(t, x) = \min\left(\min_y\left(\Phi(y) + \frac{(y-x)^2}{2(T-t)}\right), \min_{t \le s \le T}\min_y\left(c(y) + \frac{(y-x)^2}{2(s-t)}\right)\right).$$

This formula is the analogue of

$$v(t, x) = \int \Phi(y) \exp\left(-\frac{(y-x)^2}{2(T-t)}\right) \mathrm{d}y + \int_t^T ds \int c(y) \exp\left(-\frac{(y-x)^2}{2(s-t)}\right) \mathrm{d}y \ .$$

Using the change of time $s = T - t$, we can summarize this part by the following theorem.

Theorem 9.14 *We have*

$$\lim_{h \to 0} (Q_{0,\sqrt{h}})^{[s/h]} = Q_{0,\sqrt{s}} \ ,$$

where $[x]$ *denotes the integer part of* x. *Moreover,* $Q_{0,\sqrt{s}}$ *is the unique solution of*

$$\frac{\partial Q}{\partial s} + \frac{1}{2}\left(\frac{\partial Q}{\partial x}\right)^2 = 0 \ , \quad s \ge 0 \ , \quad Q_{0,0} = \delta_0 \ .$$

9.4.7 Diffusion Decision Process

In the previous subsection, the system dynamics was trivial and the instantaneous cost depended on the control only. Let us generalize this situation with a more general instantaneous cost, which will induce more complex optimal trajectories and which is the complete analogue of the diffusion process.

We consider the discrete decision process

$$\min_u\left(\sum_{i=0}^{(T/h)-1} \frac{(u(ih) - b(ih)h)^2}{2h(\sigma(ih))^2} + \Phi(x(T))\right), \qquad x(t+h) = x(t) - u(t) \ .$$

It satisfies the dynamic programming equation

$$v(t,x) = \min_u \left(\frac{(u - b(x)h)^2}{2h\sigma^2} + v(t+h, x-u) \right), \quad v(T,\cdot) = \Phi \ .$$

By the change of control $u = wh$ in the dynamic programming equation and by passing to the limit when $h \to 0$, we obtain the HJB equation defined, for $t \le T$, by

$$\frac{\partial v}{\partial t} - b(x)\frac{\partial v}{\partial x} - \frac{\sigma(x)^2}{2}\left(\frac{\partial v}{\partial x}\right)^2 = 0 \ , \quad v(T,\cdot) = \Phi \ .$$

This is the HJB equation corresponding to the variational problem

$$v(t,x) = \min_{x \in H^1} \left(\int_t^T \frac{1}{2}\left(\frac{\dot{x} - b}{\sigma}\right)^2 \mathrm{d}t + \Phi(x(T)) \right) . \tag{9.17}$$

This HJB equation is the analogue of the Kolmogorov equation

$$\frac{\partial v}{\partial t} + b(x)\frac{\partial v}{\partial x} + \frac{\sigma(x)^2}{2}\frac{\partial^2 v}{\partial x^2} = 0 \ , \quad v(T,\cdot) = \Phi \ .$$

It is not necessary that the instantaneous cost be quadratic for the discrete decision process to converge to the diffusion decision process.

Theorem 9.15 *The discrete decision process*

$$\min_u \left(\sum_{i=0}^{(T/h)-1} c_h(u(ih), x(ih)) + \Phi(x(T)) \right), \qquad x(t+h) = x(t) - u(t) \ ,$$

admits the discrete dynamic programming equation

$$v(t,x) = \min_u \left(c_h(u,x) + v(t+h, x-u) \right) \ , \quad v(T,\cdot) = \Phi \ ,$$

which converges to the continuous dynamic programming equation

$$\frac{\partial v}{\partial t} - b(x)\frac{\partial v}{\partial x} - \frac{\sigma(x)^2}{2}\left(\frac{\partial v}{\partial x}\right)^2 = 0 \ , \quad v(T,\cdot) = \Phi \ ,$$

as long as

$$\widehat{c}_h(x,p) = \left(b(x)p + \frac{\sigma(x)^2}{2}p^2 \right) h + \mathrm{o}(h) \ ,$$

where $\widehat{c}_h$ *denotes the Fenchel transform of the mapping* $u \mapsto c_h(u,x)$.

Table 9.2: Analogy between probability and dynamic programming

Probability	Dynamic programming
$+$	min
$\times$	$+$
$\mathcal{N}(m,\sigma)$	$Q_{m,\sigma}$
$\int \mathrm{d}F(x) = 1$ $\mathbb{E}_F f = \int f(x)\,\mathrm{d}F(x)$ Convolution	$\min_x c(x) = 0$ $\mathbb{M}_c f = \inf_x \{f(x) + c(x)\}$ Inf-convolution
Fourier: $\widehat{F}(s) = \mathbb{E}_F\left(\exp(jsX)\right)$ $\frac{\mathrm{d}}{\mathrm{d}s}\log(\widehat{F})(0) = j\int x\,\mathrm{d}F(x) = jm$ $-\frac{\mathrm{d}^2}{\mathrm{d}s^2}\log(\widehat{F})(0) = \int (x-m)^2\,\mathrm{d}F(x)$	Fenchel: $\widehat{c}(p) = -\mathbb{M}_c(-pX)$ $\dot{\widehat{c}}(0) = m : c(m) = \min_x c(x)$ $\ddot{\widehat{c}}(0) = 1/\ddot{c}(m)$
Brownian motion $v \mapsto \frac{\partial^2 v}{\partial x^2}$ $v \mapsto \frac{\partial v}{\partial t} + \frac{1}{2}\frac{\partial^2 v}{\partial x^2}$ $\left(1/\sqrt{2\pi t}\right)\exp(-x^2/(2t))$	Brownian decision process $v \mapsto \left(\frac{\partial v}{\partial x}\right)^2$ $v \mapsto \frac{\partial v}{\partial t} - \frac{1}{2}\left(\frac{\partial v}{\partial x}\right)^2$ $x^2/(2t)$
Diffusion process $v \mapsto \frac{\partial v}{\partial t} + b(x)\frac{\partial v}{\partial x} + a(x)\frac{\partial^2 v}{\partial x^2}$ Invariance principle	Diffusion decision process $v \mapsto \frac{\partial v}{\partial t} - b(x)\frac{\partial v}{\partial x} - a(x)\left(\frac{\partial v}{\partial x}\right)^2$ Min-plus invariance principle

The variational problem (9.17) was encountered by researchers in large deviation when they studied differential equations perturbed by a small Brownian noise. For example, we have the following estimate:

$$\lim_{\substack{\Delta \to 0 \\ \nu \to 0}} \nu \log \left(\mathbb{P}_\nu \left[x(T) \in (z - \Delta, z + \Delta) \mid x(0) = y \right] \right)$$
$$= \min_{x \in H^1(0,T), x(0)=y, x(T)=z} \int_t^T \frac{1}{2} \left(\frac{\dot{x} - b}{\sigma} \right)^2 dt \ ,$$

where $\mathbb{P}_\nu$ denotes the probability law of a diffusion process with drift term b and diffusion term $\nu\sigma$.

We conclude this section by summarizing the analogy between probability and dynamic programming in Table 9.2.

9.5 EVOLUTION EQUATIONS OF GENERAL TIMED PETRI NETS

The aim of this section is to provide the basic equations that govern the evolution of general Petri nets, when structural consumption conflicts are resolved by a predefined 'switching' mechanism. These equations can be viewed as a nonlinear extension of the evolution equations for event graphs (see §2.5). The system of notation concerning timed Petri nets is that introduced in Chapter 2.

9.5.1 FIFO Timed Petri Nets

We will adopt the following definition concerning the numbering of tokens traversing a place and the numbering of firings of a transition, that generalizes that of §2.5.

- The initial tokens of place p are numbered $1, \ldots, M_p$, whereas the n-th token, $n > M_p$, of place p is the $(n - M_p)$-th to *enter* p after the beginning of the network evolution. Tokens entering p at the same time are numbered arbitrarily.

- The n-th firing, $n \geq 1$, of transition q is the n-th firing of q to be enabled from the beginning of the network evolution. Firings of q enabled at the same time are numbered arbitrarily (nothing prevents the same transition from being enabled twice at the same epoch).

Timing is involved in the evolution of the system through the following two rules.

- The n-th initial token of place p, $n \leq M_p$, is not considered immediately available for downstream transitions. It is put in place p at time $z_p(n)$ (where the function z_p is given), and it has then to stay in p for a minimal holding time $\alpha_p(n)$ before enabling the transitions that follow p. Similarly, the n-th token of place p, $n > M_p$ (or equivalently the $(n - M_p)$-th to enter p) can only be taken into account by the transitions that follow p, $\alpha_p(n)$ units of time after its arrival.

- Each transition starts firing as soon as it is enabled (we will discuss the problem that arises with conflicts later on). Once transition q is enabled for the n-th time, the tokens that it intends to consume become reserved tokens (they cannot contribute to enabling another transition before being consumed by the firing of transition q). Once it is enabled, the time for transition q to complete its n-th firing takes $\beta_q(n)$ units. Once the firing time is completed, the transition completes its firing. This firing completion consists in withdrawing one token from each of the places that precede q (the reserved tokens), and adding one new token into the places that follow q. These two actions are supposed to be simultaneous.

FIFO places and transitions have been defined at §2.5.2.2.

Example 9.16 To make the following results more tangible, we deal throughout the section with the FIFO Petri net of Figure 9.1. This Petri net can be considered as

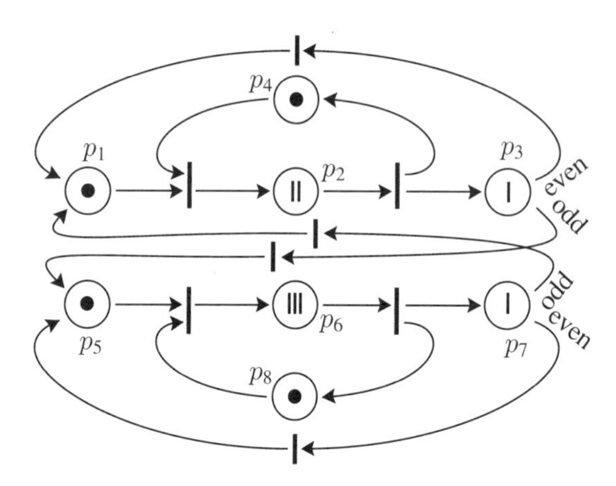

Figure 9.1: A FIFO Petri net

a closed queuing network with two servers and two infinite buffers. The customers served by server 1 (respectively 2) are routed to buffers 1 or 2 (respectively 2 or 1). In order to obtain simple results, we have chosen constant holding times on places and zero firing times (i.e. $\beta_q(n) = 0, \forall q \in \mathcal{Q}, n \in \mathbb{N}$ and $\alpha_p(n) = \alpha_p \in \mathbb{R}^+, \forall p \in \mathcal{P}, n \in \mathbb{N}$). ■

9.5.2 Evolution Equations

Let $U = (U_1, \ldots, U_n)$ be a vector of $\mathbb{R}^n$. The symbol $\mathcal{R}(U)$ denotes the vector

$$\mathcal{R}(U) = \left(U_{i(1)}, \ldots, U_{i(n)}\right) \in \mathbb{R}^n ,$$

where $i : \{1, \ldots, n\} \mapsto \{1, \ldots, n\}$ is a bijection such that

$$U_{i(1)} \leq U_{i(2)} \cdots \leq U_{i(n)} .$$

This notation is extended to vectors of $\mathbb{R}^{\mathbb{N}}$ whenever meaningful.

9.5.2.1 State Variables

Let

- $x_q(n), q \in \mathcal{Q}, n \geq 1$, denote the time when transition q starts firing for the n-th time, with the convention that for all $q \in \mathcal{Q}$, $x_q(n) = \infty$ if transition q never fires for the n-th time and $x_q(0) = -\infty$;
- $y_q(n), q \in \mathcal{Q}, n \geq 1$, denote the time when transition q completes its n-th firing, with the same convention as above;
- $v_p(n), p \in \mathcal{P}, n \geq 1$, denote the time when place p receives its n-th token, with the convention that for all $p \in \mathcal{P}$, $v_p(n) = \infty$ if the place never receives its n-th token and $v_p(0) = -\infty$;
- $w_p(n), p \in \mathcal{P}, n \geq 1$, denote the time when place p releases its n-th token, with the usual convention if the n-th token is never released and $w_p(0) = -\infty$.

Owing to our conventions, $\beta_q(n)$ denotes the firing time of q that starts at $x_q(n)$, $n \geq 1$, whereas $\alpha_p(n)$ denotes the holding time of the token that enters p at $v_p(n)$, $n \geq 1$.

If transition q is FIFO, we have the obvious relation

$$y_q(n) = x_q(n) + \beta_q(n) \ . \tag{9.18}$$

More generally,

$$(y_q(n))_{n\geq 1} = \mathcal{R}\left((x_q(n) + \alpha_q(n))_{n\geq 1}\right) \ .$$

If place p is FIFO, we can write

$$w_p(n) \geq v_p(n) + \alpha_p(n) \ ,$$

since the token that enters p at time $v_p(n)$ stays there for at least $\alpha_p(n)$ time units. More generally,

$$(w_p(n))_{n\geq 1} \geq \mathcal{R}\left((v_p(n) + \alpha_p(n))_{n\geq 1}\right) \ .$$

9.5.2.2 Initial Conditions

It is assumed that the origin of time and the initial marking have been fixed in such a way that the variables $v_p(n)$ and $w_p(n)$ satisfy the bounds

$$v_p(n) \begin{cases} = z_p(n) \leq 0 & \text{for } n = 1, \ldots, M_p, \text{ if } M_p \geq 1 \ ; \\ \geq 0, & \text{for } n > M_p \ , \end{cases}$$

and $w_p(n) \geq 0$ for $n \geq 1$. These conventions are natural: they mean that tokens that arrived in place p prior to the initial time and which left p before that initial time are not considered to belong to the initial marking. Similarly, tokens that arrived in p 'at or after' the initial time do not belong to the initial marking.

9.5.2.3 Upstream Equations Associated with Transitions

We first look at the relationships induced by a transition q due to the places preceding q. We first consider the case without *structural consumption conflicts*, namely for every place p preceding q, the set of transitions that follow p is reduced to q.

No Structural Consumption Conflicts For all $p \in \pi(q)$, one token leaves p at time $w_p(n)$. Since q is the only transition that can consume the tokens of p, this corresponds to the starting of the n-th firing of q. Hence,

$$x_q(n) = w_p(n) \ , \quad \forall p \in \pi(q) \ .$$

In the FIFO case, the n-th token of place p to become available for enabling q must be the n-th to enter p, so that

$$x_q(n) = \max_{p \in \pi(q)} (v_p(n) + \alpha_p(n)) \ .$$

More generally,

$$x_q(n) = \max_{p \in \pi(q)} U_p(n) \ ,$$

where

$$(U_p(n))_{n \geq 1} = \mathcal{R}\left((v_p(n) + \alpha_p(n))_{n \geq 1}\right) \ .$$

General Case Without further specifications on how the conflict is resolved, we can only state the following inequalities: in the FIFO case,

$$x_q(n) \geq \max_{p \in \pi(q)} (v_p(n) + \alpha_p(n)) \ , \tag{9.19}$$

and more generally

$$x_q(n) \geq \max_{p \in \pi(q)} U_p(n) \ .$$

These inequalities are not very satisfactory, and we will return to this point later on.

9.5.2.4 Downstream Equations Associated with Transitions

We now look at the relationships induced by a transition q due to the places following q. We first consider the case without structural supply conflicts, namely, for every place p following q, the set of transitions that precede p is reduced to q (that is, p is only fed by this q).

No Structural Supply Conflicts If no other transition than q can feed the places following q, the token entering place $p \in \sigma(q)$ with rank (M_p+n) has been produced by the n-th firing of transition q; therefore,

$$y_q(n) = v_p(M_p + n) \ , \quad \forall p \in \sigma(q) \ .$$

In the FIFO case, this leads to the relation

$$x_q(n) + \beta_q(n) = v_p(M_p + n) \ , \quad \forall p \in \sigma(q) \ ,$$

whereas in the general case

$$\mathcal{R}\left((x_q(k) + \beta_q(k))_{k\geq 1}\right)_n = v_p(M_p + n) \ , \quad \forall p \in \sigma(q) \ .$$

General Case Without further specifications, we can only state the following inequalities: in the FIFO case,

$$x_q(n) + \beta_q(n) \geq v_p(M_p + n) \ , \quad \forall p \in \sigma(q) \ ,$$

whereas, in the general case,

$$\mathcal{R}\left((x_q(k) + \beta_q(k))_{k\geq 1}\right)_n \geq v_p(M_p + n) \ , \quad \forall p \in \sigma(q) \ .$$

9.5.2.5 *Upstream Equations Associated with Places*

We now focus on the upstream relationships induced by place p. Consider the sequences $\{y_q(n)\}_{n\geq 1}$, for all $q \in \pi(p)$. With each of them, associate a point process on the real line, where the points are located at $y_q(n)$. We can look at the arrival process into p as the superimposition of these $|\pi(p)|$ point processes.

With all $q \in \pi(p)$, we associate an integer $i_q \in \mathbb{N}$ representing a number of complete firings of q. If transition q has completed exactly i_q firings for all $q \in \pi(p)$, then place p has received exactly $\sum_{q\in\pi(p)} i_q$ tokens. The set of vectors $(i) = (i_q)_{q\in\pi(p)}$ such that the n-th token has entered place p is hence

$$\mathcal{A}_n^p = \left\{ i \in \mathbb{N}^{|\pi(p)|} \ \middle| \ \sum_{q\in\pi(p)} i_q = n \right\} \ .$$

The last token produced by the transition firings specified by some $i \in \mathcal{A}_n^p$ enters p at time $\max_{q\in\pi(p)} y_q(i_q)$, where $y_q(0) = -\infty$ by convention. Since n tokens have reached p once all the firings specified by i have been completed, one obtains

$$v_p(n + M_p) \leq \inf_{i\in\mathcal{A}_n^p} \max_{q\in\pi(p)} y_q(i_q) \ . \tag{9.20}$$

But $v_p(n + M_p)$ should be equal to some $y_{t_0}(n_0)$ since at least one collection of events puts n tokens in place p (unless $\mathcal{A}_n^p$ is empty and $v_p(n + M_p) = \infty$). Hence equality must hold true in (9.20). We then obtain the following final relation:

$$v_p(n + M_p) = \inf_{\left\{i\in\mathbb{N}^{|\pi(p)|} \middle| \sum_{q\in\pi(p)} i_q = n\right\}} \max_{q\in\pi(p)} y_q(i_q) \ , \tag{9.21}$$

where $y_q(0) = -\infty$ by convention.

9.5.2.6 Downstream Equations Associated with Places

We now concentrate on the downstream relationships induced by a place p. It is in this type of equations that the structural consumption conflicts associated with general Petri nets become apparent.

Consider the sequences $\{x_q(n)\}_{n\geq 1}$, for all $q \in \sigma(p)$. With all $q \in \sigma(p)$, we associate an integer $i_q \in \mathbb{N}$ representing some number of firing initiations of q. If q has started exactly i_q firings for all $q \in \sigma(p)$, then exactly $\sum_{q\in\sigma(p)} i_q$ tokens have been withdrawn from p. The set of vectors $i = (i_q)_{q\in\sigma(p)}$ such that the n-th token has left place p is hence

$$\mathcal{B}_n^p = \left\{ i \in \mathbb{N}^{|\sigma(p)|} \;\middle|\; \sum_{q\in\sigma(p)} i_q = n \right\} .$$

For any i in this set, the last token to leave p leaves at time $\max_{q\in\sigma(p)} x_q(i_q)$. Hence

$$w_p(n) \leq \inf_{i\in\mathcal{B}_n^p} \max_{q\in\sigma(p)} x_q(i_q) .$$

Using a similar reasoning as previously, we obtain the final relation

$$w_p(n) = \inf_{\left\{i\in\mathbb{N}^{|\sigma(p)|} | \sum_{q\in\sigma(p)} i_q = n\right\}} \max_{q\in\sigma(p)} x_q(i_q) . \tag{9.22}$$

Relations (9.21) and (9.22) exhibit nothing but a superficial symmetry. Indeed, while (9.21) allows one to construct the sequence $\{v_p(n)\}$ from the knowledge of what happens upstream of p and earlier, this is not true at all for (9.22) which only provides some sort of *backward* property stating that the knowledge of what will happen following p in the future allows one to reconstruct what happens in p now. The reason for this is that the way the conflict is solved is not yet sufficiently precise. We show now one natural way of solving conflicts, which we will call *switching*. Several other ways are conceivable like *competition*, which we will also outline.

9.5.2.7 Switching

Within this setting, each place that has several transitions downstream receives a switching sequence $\{\rho_p(n)\}$ with values in $\sigma(p)^{\mathbb{N}}$. In the same way as the n-th token to enter place p receives a holding time $\alpha_p(n)$, it also receives a route to which it must be switched. This information is given by $\rho_p(n)$, which specifies which transition it must be routed to. In other words, only those tokens such that $\rho_p(n) = q$ should be taken into account by $q \in \sigma(p)$. By doing so, one completely specifies the behavior of the system. For instance, in the FIFO case, one obtains the inequality

$$x_q(n) \geq w_p(\eta_{p,q}(n)) , \quad \forall p \in \pi(q) ,$$

where the *switching function* $\eta_{p,q}$ is defined by

$$\eta_{p,q}(0) = 0, \;\; \eta_{p,q}(n) = \inf\left\{ m \geq 1 \;\middle|\; \sum_{k=1}^{m} 1\{\rho_p(k) = q\} \geq n \right\}, \quad n \geq 1 . \tag{9.23}$$

Whenever the behavior of the places upstream of q is specified, one can go further and obtain the desired *forward* equation, as we will see in the next section.

Example 9.17 In our example (see Figure 9.1), the switchings are deterministic. They are chosen as follows:

$$\rho_3(2n) = 1 \ , \quad \rho_3(2n+1) = 5 \ , \quad \rho_7(2n) = 5 \ , \quad \rho_7(2n+1) = 1 \ , \quad \forall n \in \mathbb{N} \ .$$

■

9.5.2.8 Competition

The places following p compete for the tokens of p on a First Come First Served (FCFS) basis: within this interpretation, the tokens that have been served in place p can be seen as building up some *queue of tokens*. Once a transition q following p is enabled except for the condition depending on p, it puts in a request for one token in some FCFS queue of requests. This request is served (and the corresponding transition enabled) as soon as it is at the head of the request line and there is one token in the token queue.

9.5.3 Evolution Equations for Switching

In this subsection it is assumed that all places receive some switching. For places with a single downstream transition, this sequence is trivial in the sense that it always routes tokens to this transition.

Theorem 9.18 *Under the foregoing assumptions, the state variables* $v_p(n), p \in \mathcal{P}, n \geq 1$, *of a FIFO Petri net satisfy the (nonlinear) recurrence equations*

$$v_p(n+M_p) = \bigwedge_{\left\{ i \in \mathbb{N}^{|\pi(p)|} \,\middle|\, \sum_{q \in \pi(p)} i_q = n \right\}} \ \bigoplus_{\{q \in \pi(p), t \in \pi(q)\}} v_t(\eta_{t,q}(i_q))\alpha_t(\eta_{t,q}(i_q))\beta_q(i_q) \ , \quad (9.24)$$

for $n \geq 1$, *with the initial condition* $v_p(n) = z_p(n)$ *for* $1 \leq n \leq M_p$, *if* $M_p \geq 1$.

Proof In addition of the variables $v_p(n)$, we will make use of the auxiliary variables $x_q(n), q \in \mathcal{Q}, n \geq 1$. Owing to the switching assumptions, Inequality (9.19) can be replaced by the relation

$$x_q(n) = \bigoplus_{p \in \pi(q)} \ \bigwedge_{\left\{ j_p \geq 1 \,\middle|\, \sum_{k=1}^{j_p} 1\{\rho_p(k)=q\}=n \right\}} v_p(j_p)\alpha_p(j_p) \ , \quad n \geq 1 \ ,$$

or, equivalently,

$$x_q(n) = \bigoplus_{p \in \pi(q)} v_p(\eta_{p,q}(n))\alpha_p(\eta_{p,q}(n)) \ , \quad n \geq 1 \ , \qquad (9.25)$$

where we used the switching function $\eta_{p,q}$ defined in (9.23), and the FIFO assumption, which implies that the mapping $i \mapsto v_p(i)\alpha_p(i)$ is nondecreasing.

Similarly, using (9.18) in (9.21) yields

$$v_p(n + M_p) = \bigwedge_{\left\{i\in\mathbb{N}^{|\pi(p)|} \middle| \sum_{q\in\pi(p)} i_q = n\right\}} \bigoplus_{q\in\pi(p)} x_q(i_q)\beta_q(i_q) \ , \quad n \geq 1 \ . \tag{9.26}$$

Equation (9.24) follows immediately from (9.25) and (9.26). ■

Remark 9.19 In the case when the Petri net is not FIFO, Equations (9.25) and (9.26) have to be replaced by

$$x_q(n) = \bigoplus_{p\in\pi(q)} \left(\mathcal{R}(v_p(m)\alpha_p(m))_{m\geq 1}\right)_{(\eta_{p,q}(n))} \ , \quad n \geq 1 \ ,$$

and

$$v_p(n + M_p) = \bigwedge_{\left\{i\in\mathbb{N}^{|\pi(p)|} \middle| \sum_{q\in\pi(p)} i_q = n\right\}} \bigoplus_{q\in\pi(p)} \left(\mathcal{R}(x_q(m)\beta_q(m))_{m\geq 1}\right)_{(i_q)} \ , \quad n \geq 1 \ ,$$

respectively. ■

Remark 9.20 In (9.24), we can get rid of the firing times $\beta_q(n)$ by changing the holding times $\alpha_p(\eta_{p,q}(n))$, $\forall p \in \pi(q)$, into $\alpha_p(\eta_{p,q}(n))\beta_q(n)$. Thus we obtain an equivalent net with $\beta_q(n) = 0$ and $\alpha_p(n) > 0, q \in \mathcal{Q}, p \in \mathcal{P}, n \geq 1$, where the equivalence means that the entrance times are the same in both systems. ■

Example 9.21 In our example, we obtain

$$\begin{aligned}
v_1(n+1) &= \bigwedge_{n_3+n_7=n} \left(1v_3(2n_3) \oplus 1v_7(2n_7+1)\right) \ , && v_1(1) = 0 \ , \\
v_2(n) &= v_1(n) \oplus v_4(n) \ , && \\
v_3(n) &= 2v_2(n) \ , && \\
v_4(n+1) &= 2v_2(n) \ , && v_4(1) = 0 \ , \\
v_5(n+1) &= \bigwedge_{n_3+n_7=n} \left(1v_3(2n_3+1) \oplus 1v_7(2n_7)\right) \ , && v_5(1) = 0 \ , \\
v_6(n) &= v_5(n) \oplus v_8(n) \ , && \\
v_7(n) &= 3v_6(n) \ , && \\
v_8(n+1) &= 3v_6(n) \ , && v_8(1) = 0 \ .
\end{aligned}$$

■

9.5.4 Integration of the Recursive Equations

We assume that the Petri net is FIFO. We use Remark 9.20 to assume (without loss of generality) that $\beta_q(n) = 0, q \in \mathcal{Q}, n \geq 1$. Finally, we assume that the switching is given as well as the holding times in the places and that in every circuit of the Petri net there is a place p with $0 < \alpha_p(n) < \infty, n \geq 1$.

In what follows we will use weighted trees where the weights are associated with the nodes. We call the *weight* of a directed path the sum of the weights of all its nodes but its source. A node N_1 is said to be *deeper* than a node N_2 if we can find a directed path from N_2 to N_1. Finally, the *depth* of a tree is the length of its longest directed path.

Definition 9.22 (Evolution tree) *Let $(p, n) \in \mathfrak{P} \times \mathbb{N}$. An evolution tree A associated with (p, n) is a tree with root (p, n) defined recursively as follows.*

- *If $n \leq M_p$, then A is reduced to a single node (p, n) with weight $\alpha_p(n)+z_p(n)$.*
- *If $n > M_p$, choose one $i \in \mathbb{N}^{|\pi(p)|}$ satisfying $\sum_{t \in \pi(p)} i_q = n - M_p$. Then A is the tree with root (p, n) and with $|\pi(\pi(p))|$ subtrees being evolution trees associated with the nodes $(q, \eta_{q,t}(i_q)), t \in \pi(p), q \in \pi(q)$. The root (p, n) is given a weight $\alpha_p(n)$.*

The set of all the evolution trees of the pair (p, n) will be denoted $\mathfrak{E}(p, n)$.

In Equation (9.24), we can replace the variables $v_t(\eta_{t,q}(n))$ by using Equation (9.24) once more. We obtain

$$v_p(n + M_p) = \bigwedge_{\left\{ i \in \mathbb{N}^{|\pi(p)|} \,\middle|\, \sum_{q \in \pi(p)} i_q = n \right\}} \bigoplus_{\{q \in \pi(p), t \in \pi(q)\}} \bigwedge_{\left\{ j \in \mathbb{N}^{|\pi(t)|} \,\middle|\, \sum_{s \in \pi(t)} j_s = \eta_{t,q}(i_t) - M_t \right\}} \bigoplus_{\{s \in \pi(t), r \in \pi(s)\}} v_r(\eta_{r,s}(j_s)) \alpha_r(\eta_{r,s}(j_s)) \alpha_t(\eta_{t,q}(i_t)) \ .$$

If we use the distributivity of $\oplus$ with respect to $\wedge$ (see (4.95)), this equality becomes:

$$v_p(n + M_p) = \bigwedge_{\substack{i \in I \\ j \in J_{tq}}} \bigoplus_{\substack{q \in \pi(p) \\ t \in \pi(q) \\ s \in \pi(q) \\ r \in \pi(s)}} v_r(\eta_{rs}(j_s^{tq})) \alpha_r(\eta_{rs}(j_s^{tq})) \alpha_t(\eta_{tq}(i_q)) \ ,$$

where

$$I \stackrel{\text{def}}{=} \left\{ i \in \mathbb{N}^{|\pi(p)|} \,\middle|\, \sum_{q \in \pi(p)} i_q = n \right\}, J^{tq} \stackrel{\text{def}}{=} \left\{ j^{tq} \in \mathbb{N}^{|\pi(t)|} \,\middle|\, \sum_{s \in \pi(t)} j_s^{tq} = \eta_{tq}(i_q) - M_t \right\} .$$

This equation represents the first step in the 'integration' of the recurrence equations. Indeed, we obtain a tree of depth 2 from the root $(p, n + M_p)$. If we continue to develop this equation, we obtain trees with increasing depths. We stop when each path ends with a leaf, namely, when it terminates with a node (q, m) with $m \leq M_q$. We eventually obtain the integration of Equation (9.24):

$$v_p(n) = \inf_{A \in \mathfrak{E}(p,n)} C(A) \ , \quad n \geq M_p \ , \quad \text{with} \quad C(A) = \sup_{T \in \mathfrak{T}(A)} (w(T)) \ .$$

The quantity $C(A)$ is the weight of tree A, $\mathcal{T}(A)$ is the set of all the directed paths from the root to any leaf of the tree A, and $w(T)$ is the weight of the directed path T (i.e. the sum of the weights of all its nodes except its root).

Remark 9.23 The set $\mathcal{E}(p,n)$ might contain infinite trees, thus $\mathcal{E}(p,n)$ is not constructible and this transformation of the recursive equations does not obviously give the 'constructiveness' character of these equations. However, it is useful for preliminary results. The reader is referred to [7] where this issue is further analyzed. ■

9.6 MIN-MAX SYSTEMS

In this section we will be concerned with systems of which the evolution is determined by three rather than two different operations, namely addition, maximization and minimization. Because these operations occur simultaneously, a different notation for max and min is necessary: $\oplus$ is reserved for max, and $\wedge$ will denote min. The most general system to be considered is of the form

$$\begin{aligned} x(k+1) &= A_1 \otimes x(k) \oplus B_1 \otimes y(k) \oplus C_1 \otimes v(k) \oplus D_1 \otimes w(k) , && (9.27)\\ y(k+1) &= A_2 \odot x(k) \wedge B_2 \odot y(k) \wedge C_2 \odot v(k) \wedge D_2 \odot w(k) , && (9.28)\\ v(k) &= A_3 \otimes x(k) \oplus B_3 \otimes y(k) \oplus C_3 \otimes v(k) \oplus D_3 \otimes w(k) , && (9.29)\\ w(k) &= A_4 \odot x(k) \wedge B_4 \odot y(k) \wedge C_4 \odot v(k) \wedge D_4 \odot w(k) . && (9.30) \end{aligned}$$

The notation $\odot$ here refers to the multiplication of two matrices (or a matrix and a vector) in which the $\wedge$-operation is used instead of $\otimes$ (see §6.6.1 and §9.2.4). The expressions $a \otimes b$ and $a \odot b$ are identical if at least either a or b is a scalar. The operation $\oplus$ has the neutral element ε whereas $\wedge$ has the neutral element $\top$. The following convention, in accordance with (5.7) and (5.8), is made:

$$\top \otimes \varepsilon = \varepsilon , \qquad \top \odot \varepsilon = \top .$$

In analogy with conventional system theory, system (9.27)–(9.30) is called a *descriptor system*. It is assumed that the vectors $x(k), y(k), v(k)$ and $w(k)$ are respectively n-,m-, p-, and q-dimensional. The matrices A_l, B_l, C_l and D_l, $l = 1, \ldots, 4$, have appropriate dimensions. The elements of the matrices with an odd index are either finite or ε and the elements of the matrices with an even index are either finite or $\top$.

Equations (9.29) and (9.30) are implicit equations in $v(k)$ and $w(k)$, respectively. It is assumed that the precedence graph $\mathcal{G}(E)$, where E is the matrix

$$E = \begin{pmatrix} C_3 & D_3 \\ C_4 & D_4 \end{pmatrix} ,$$

contains neither circuits nor loops. For later reference this condition is called Condition C1:

Condition C1 The graph $\mathcal{G}(E)$ contains neither circuits nor loops.

Because of this condition, a finite number of repeated substitutions of the whole right-hand side of (9.29) and (9.30) into these same equations leads to, respectively,

$$v(k) = C_3^* \otimes (A_3 \otimes x(k) \oplus B_3 \otimes y(k) \oplus D_3 \otimes w(k)) \ ,$$
$$w(k) = D_4^{*'} \odot (A_4 \odot x(k) \wedge B_4 \odot y(k) \wedge C_4 \odot v(k)) \ ,$$

where

$$C_3^* = e \oplus C_3 \oplus C_3^2 \oplus \cdots \ , \qquad D_4^{*'} = e \wedge D_4 \wedge D_4^2 \wedge \cdots \ .$$

In these equations, the matrix product (which is used in the power computation of matrices) must be understood as being $\otimes$, respectively $\odot$, when used in conjunction with $\oplus$, respectively $\wedge$. Similarly, the symbol e denotes the identity matrix in $\mathbb{R}_{\max}$, respectively $\mathbb{R}_{\min}$. Condition C1 is sufficient, but not necessary, for C_3^* and $D_4^{*'}$ to exist in the expressions above.

Now the equations in $v(k)$ and $w(k)$ can be solved in a suitable order, and the solutions can be expressed in terms of $x(k)$ and $y(k)$. These solutions are written symbolically as

$$v(k) = f_1(x(k), y(k)) \ , \qquad w(k) = f_2(x(k), y(k)) \ .$$

If these equations are substituted into (9.27) and (9.28), then the new expressions for $x(k+1)$ and $y(k+1)$ will show a finite nesting of max- and min-operations.

For later reference, Equations (9.27)–(9.30), defining a mapping from $\mathbb{R}^{n+m+p+q}$ to itself, will symbolically be denoted $\mathcal{M}$. Similarly, the mapping of the corresponding nested equations is denoted $\overline{\mathcal{M}}$ ($\overline{\mathcal{M}}$ maps $\mathbb{R}^{n+m}$ to itself). Hence,

$$\begin{aligned} \left(x(k+1)\ y(k+1)\ v(k)\ w(k) \right)' &= \mathcal{M}\left(\left(x(k)\ y(k)\ v(k)\ w(k) \right)' \right) , \\ \left(x(k+1)\ y(k+1) \right)' &= \overline{\mathcal{M}}\left(\left(x(k)\ y(k) \right)' \right) . \end{aligned} \tag{9.31}$$

9.6.1 General Timed Petri Nets and Descriptor Systems

Consider the network depicted in Figure 9.2. Each of the three nodes performs activities. The loops around these nodes, with time durations τ_1, τ_2 and τ_3, refer to processing or recycling times of one activity at the respective nodes. All other time durations are assumed to be zero. Node q_1 delivers products to nodes q_2 and q_3 simultaneously. Node q_3 starts processing on the first incoming product. To start an activity, each node must have delivered its product(s) of the previous activity to its destination node(s). If the destination node is q_2, its buffer, indicated by a rectangle in the figure, can store one incoming item (while q_2 works at the present activity). Hence, if this buffer is full, node q_1 cannot yet deliver a product and must wait until this buffer becomes empty. Similarly, there is a buffer just before node q_3 which can

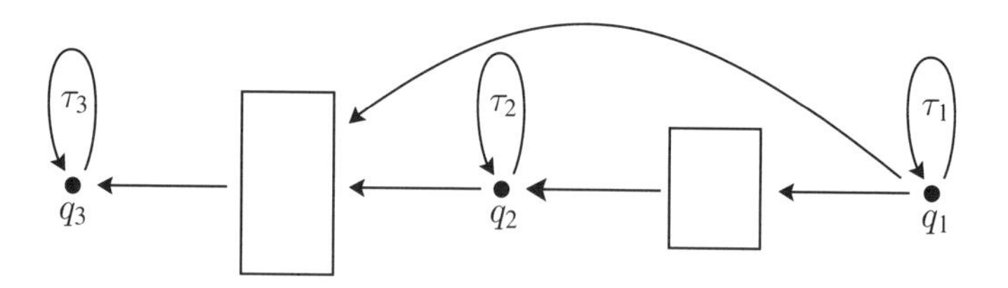

Figure 9.2: A network

contain two incoming items maximally. If each buffer contains one token initially, one may be tempted to model the succession of firing times as follows:

$$\left.\begin{aligned} x_1(k+1) &= \max(x_1(k)+\tau_1, x_2(k), x_3(k)) \ , \\ x_2(k+1) &= \max(x_1(k)+\tau_1, x_2(k)+\tau_2, x_3(k)) \ , \\ x_3(k+1) &= \max(\min(x_1(k)+\tau_1, x_2(k)+\tau_2), x_3(k)+\tau_3) \ , \end{aligned}\right\} \tag{9.32}$$

where the quantities $x_i(k), k = 1, 2, \ldots,$ are the successive firing times of node q_i. This model can be rewritten in the form (9.27)–(9.30) by adding $w(k) = \min(x_1(k), x_2(k))$ to (9.32) and by replacing the appropriate part in the last of the equations of (9.32) by $w(k)$. Indeed, node q_3 will process the first arriving k-th product, of either q_1 or q_2, first. The last arriving product at q_3, however, is not processed at all according to (9.32). It apparently leaves the system in some mysterious way. There is a discrepancy between the problem statement and its model (9.32). In order to model the processing of the last arriving product also, one can introduce a fictive node q_4, which is actually node q_3, and which takes care of the last arriving of the two products coming from q_1 and q_2. If this fictive node has firing times $x_4(k)$, then the new model becomes

$$\left.\begin{aligned} x_1(k+1) &= \max(x_1(k)+\tau_1, x_2(k), x_4(k)) \ , \\ x_2(k+1) &= \max(x_1(k)+\tau_1, x_2(k)+\tau_2, x_4(k)) \ , \\ x_3(k+1) &= \max(\min(x_1(k)+\tau_1, x_2(k)+\tau_2), x_4(k)+\tau_3) \ , \\ x_4(k+1) &= \max(\max(x_1(k)+\tau_1, x_2(k)+\tau_2), x_3(k+1)+\tau_3) \\ &= \max(x_1(k)+\tau_1, x_2(k)+\tau_2, x_4(k)+2\tau_3, \\ &\qquad \min(x_1(k)+\tau_1+\tau_3, x_2(k)+\tau_2+\tau_3)) \ . \end{aligned}\right\} \tag{9.33}$$

Model (9.33) assumes that the buffer just before q_3 must be emptied before a new cycle ($k \to k+1$) can be started. Note that x_3 does not appear on the right-hand side anymore and therefore the equation for $x_3(k+1)$ can be disregarded. It is obvious that this model can be rewritten in the form (9.27)–(9.30) also. Though model (9.33) does not throw away half-finished products, node q_3 still might not always take the product which arrives first. Node q_3 and its image node q_4 process the batch of the two arriving k-th products (from q_1 and q_2) according to first arrival. If the $(k+1)$-st product of q_1, say, arrive before the k-th product of q_2, it has to wait until this last k-th product has arrived.

Yet another remark with respect to (9.33) must be made. According to (9.33), nodes q_1 and q_2 can start another cycle only after q_4 has started its current activity. However, the performance of the network can be increased if $x_4(k)$ in either the first or the second equation of (9.33) is replaced by $x_3(k)$, depending on whether $x_1(k) + \tau_1 < x_2(k) + \tau_2$ or not. Such a conditional dependence can neither be expressed in terms of the operations min, max and $+$, nor can it be shown graphically by means of a Petri net as introduced in this book. This dependence can be expressed in Petri nets in which so-called inhibitor arcs are allowed. The reader is referred to [1] about such arcs.

One can enlarge the batch size from which node q_3 takes its products according to the FIFO priority rule. If, for instance, one introduces two fictive nodes, one for q_1 and one for q_2, and another pair of two fictive nodes, one node for q_3 and one for q_4, then one can construct a model which has a batch size of four. The original products numbered k and $k+1$ coming from q_1 and the original products numbered k and $k+1$ coming from q_2 are processed by q_3, or one of its images, according to FIFO. The next batch will then consist of the four original products numbered $k+2$ and $k+3$ coming from both q_1 or its image and q_2 or its image. The corresponding model will not be written down explicitly; its (eight) scalar equations become rather unwieldy expressions with nested forms of the operations max and min.

9.6.2 Existence of Periodic Behavior

In the following definition, the symbol $\mathcal{M}$ and $\overline{\mathcal{M}}$ are those of (9.31).

Definition 9.24 *A scalar λ, $\varepsilon \le \lambda \le \top$, is called an eigenvalue of the mapping $\mathcal{M}$, respectively $\overline{\mathcal{M}}$, if a vector $\begin{pmatrix} x' & y' & v' & w' \end{pmatrix}'$, respectively $\begin{pmatrix} x' & y' \end{pmatrix}'$, exists, where either x or y has at least one finite element, such that*

$$\begin{pmatrix} \lambda \otimes x' & \lambda \odot y' & v' & w' \end{pmatrix}' = \mathcal{M}\left(\begin{pmatrix} x' & y' & v' & w' \end{pmatrix}'\right),$$

respectively,

$$\begin{pmatrix} \lambda \otimes x' & \lambda \odot y' \end{pmatrix}' = \overline{\mathcal{M}}\left(\begin{pmatrix} x(k) & y(k) \end{pmatrix}'\right).$$

Such a vector is called an eigenvector of $\mathcal{M}$, respectively $\overline{\mathcal{M}}$.

It will be clear that, provided Condition C1 holds, see §9.3, an eigenvalue of $\mathcal{M}$ is also an eigenvalue of $\overline{\mathcal{M}}$ and vice versa. A motivation to study eigenvectors is that the system has a very regular behavior if the initial condition coincides with an eigenvector. In fact, the firing times of the $(k+1)$-st activities take place exactly λ time units later than the firing times of the k-th activities. Conditions will be given under which the eigenvalue and a corresponding eigenvector exist.

System (9.27)–(9.30) can be written as

$$\begin{pmatrix} x(k+1) \\ v(k) \end{pmatrix} = \begin{pmatrix} A_1 & C_1 \\ A_3 & C_3 \end{pmatrix} \otimes \begin{pmatrix} x(k) \\ v(k) \end{pmatrix} \oplus \begin{pmatrix} B_1 & D_1 \\ B_3 & D_3 \end{pmatrix} \otimes \begin{pmatrix} y(k) \\ w(k) \end{pmatrix},$$

$$\begin{pmatrix} y(k+1) \\ w(k) \end{pmatrix} = \begin{pmatrix} B_2 & D_2 \\ B_4 & D_4 \end{pmatrix} \odot \begin{pmatrix} x(k) \\ v(k) \end{pmatrix} \wedge \begin{pmatrix} A_2 & C_2 \\ A_4 & C_4 \end{pmatrix} \odot \begin{pmatrix} x(k) \\ v(k) \end{pmatrix},$$

the two 'autonomous' equations of which are

$$\begin{pmatrix} x(k+1) \\ v(k) \end{pmatrix} = \begin{pmatrix} A_1 & C_1 \\ A_3 & C_3 \end{pmatrix} \otimes \begin{pmatrix} x(k) \\ v(k) \end{pmatrix}, \tag{9.34}$$

$$\begin{pmatrix} y(k+1) \\ w(k) \end{pmatrix} = \begin{pmatrix} B_2 & D_2 \\ B_4 & D_4 \end{pmatrix} \odot \begin{pmatrix} x(k) \\ v(k) \end{pmatrix}. \tag{9.35}$$

These two sets of autonomous equations can be considered as two subsystems of (9.27)–(9.30), connected by means of the matrices

$$\begin{pmatrix} B_1 & D_1 \\ B_3 & D_3 \end{pmatrix}, \quad \begin{pmatrix} A_2 & C_2 \\ A_4 & C_4 \end{pmatrix}. \tag{9.36}$$

Condition C2 The first matrix in (9.36) is not identically $\top$ and the second one is not identically ε.

This amounts to saying that the two connections are actual.

If Condition C1 is satisfied, then $v(k)$ can be solved from (9.34) and subsequently be substituted into the right-hand side of (9.34):

$$x(k+1) = (A_1 \oplus C_1 \otimes C_3^* \otimes A_3) \otimes x(k) \ . \tag{9.37}$$

Similarly, we obtain

$$y(k+1) = (B_2 \wedge D_2 \odot D_4^{*'} \odot B_4) \odot y(k) \ . \tag{9.38}$$

Condition C3 The transition matrices of (9.37) and (9.38) are irreducible.

If Conditions C1 and C3 hold, then the matrices which govern the evolution of the systems in (9.37) and (9.38) have unique eigenvalues, denoted $\lambda_{\max}$ and $\lambda_{\min}$, respectively. The existence and uniqueness of these eigenvalues is a direct consequence of the theory of Chapter 3. Now the following theorem, proved in [102], holds.

Theorem 9.25 *Assume Conditions C1, C2 and C3 are fulfilled. The operator* $\overline{\mathcal{M}}$ *has an eigenvalue* λ *and a corresponding eigenvector* $\begin{pmatrix} x' & y' \end{pmatrix}'$ *all of which components are finite, i.e.*

$$\lambda \otimes \begin{pmatrix} x \\ y \end{pmatrix} = \overline{\mathcal{M}} \left(\begin{pmatrix} x \\ y \end{pmatrix} \right), \tag{9.39}$$

if and only if $\lambda_{\max} \leq \lambda_{\min}$. *Under these conditions,* λ *is unique and satisfies* $\lambda_{\max} \leq \lambda \leq \lambda_{\min}$.

The condition that the components of the eigenvector must all be finite is essential for the statement of this theorem to hold. As a counterexample, consider

$$x(k+1) = 2x(k) \oplus 3y(k) \ , \qquad y(k+1) = 4x(k) \wedge 5y(k) \ .$$

The unique eigenvalue which falls within the scope of the above theorem is $\lambda = 3.5$ with corresponding eigenvector $\begin{pmatrix} 0.5 & 1 \end{pmatrix}'$. However, $\lambda = \lambda_{\max} = 2$ is also an eigenvalue with eigenvector $\begin{pmatrix} e & \varepsilon \end{pmatrix}'$. Similarly, $\lambda = \lambda_{\min} = 5$ is an eigenvalue with eigenvector $\begin{pmatrix} \top & e \end{pmatrix}'$.

In Chapter 3 we saw that, within the max-plus algebra setting, the evolution of a linear system, such as (9.37), converges in a finite number of steps to a periodic behavior, the period being related to the length(s) of the critical circuit(s). Such a property has not been shown for systems within the min-max algebra setting, though simulations do point in this direction.

9.6.3 Numerical Procedures for the Eigenvalue

Three numerical procedures for the calculation of the eigenvalue and corresponding eigenvector of $\overline{\mathcal{M}}$ will be discussed briefly by means of examples. Of course, these procedures can also be applied to systems in $\mathbb{R}_{\max}$ only.

Procedure 1 Consider (9.27) and (9.28) with

$$A_1 = \begin{pmatrix} \varepsilon & 1 & \varepsilon \\ \varepsilon & e & 1 \\ 2 & 1 & \varepsilon \end{pmatrix}, \quad B_1 = \begin{pmatrix} 3 & 3 & \varepsilon \\ 3 & \varepsilon & \varepsilon \\ \varepsilon & \varepsilon & 1 \end{pmatrix}, \quad A_2 = \begin{pmatrix} \top & \top & 3 \\ \top & 3 & \top \\ \top & 3 & \top \end{pmatrix}, \quad B_2 = \begin{pmatrix} \top & 4 & 3 \\ 6 & \top & \top \\ \top & 9 & 6 \end{pmatrix},$$

and $C_3 = C_4 = (\varepsilon)$, $B_3 = B_4 = (\top)$. The evolution of this system will be studied by starting with an arbitrary initial vector. If $\begin{pmatrix} x(0)' & y(0)' \end{pmatrix}' = \begin{pmatrix} 1 & 2 & 3 & 4 & 5 & 6 \end{pmatrix}'$, then

$$\begin{aligned}
\begin{pmatrix} x'(0) & y'(0) \end{pmatrix} &= \begin{pmatrix} 1 & 2 & 3 & 4 & 5 & 6 \end{pmatrix} , \\
\begin{pmatrix} x'(1) & y'(1) \end{pmatrix} &= \begin{pmatrix} 8 & 7 & 7 & 6 & 5 & 6 \end{pmatrix} , \\
&\vdots \\
\begin{pmatrix} x'(12) & y'(12) \end{pmatrix} &= \begin{pmatrix} 38 & 37 & 37 & 37 & 37 & 37 \end{pmatrix} , \\
\begin{pmatrix} x'(13) & y'(13) \end{pmatrix} &= \begin{pmatrix} 40 & 40 & 40 & 40 & 40 & 40 \end{pmatrix} , \\
\begin{pmatrix} x'(14) & y'(14) \end{pmatrix} &= \begin{pmatrix} 43 & 43 & 42 & 43 & 43 & 43 \end{pmatrix} , \\
\begin{pmatrix} x'(15) & y'(15) \end{pmatrix} &= \begin{pmatrix} 46 & 46 & 45 & 45 & 46 & 45 \end{pmatrix} , \\
\begin{pmatrix} x'(16) & y'(16) \end{pmatrix} &= \begin{pmatrix} 49 & 48 & 48 & 48 & 49 & 48 \end{pmatrix} , \\
\begin{pmatrix} x'(17) & y'(17) \end{pmatrix} &= \begin{pmatrix} 52 & 51 & 51 & 51 & 51 & 51 \end{pmatrix} \ldots .
\end{aligned}$$

This evolution is continued until $x(k)$ becomes linearly dependent on one of the previous states $\left(x'(l) \ y'(l) \right)'$, $l = 1, \ldots, k-1$. For this example, this occurs for $k = 17$: $\left(x'(17) \ y'(17) \right)' = 14 \otimes \left(x'(12) \ y'(12) \right)'$. It is now claimed that $\lambda = 14/(17-12) = 14/5$ and that $\left(\sum_{j=12}^{16} \left(x'(j) \ y'(j) \right)' \right) /5$ is the eigenvector. Note that in this expression for the eigenvector, the conventional operations addition and division occur. These are nonlinear operations within the min-max algebra! For the example, the eigenvector thus becomes

$$\frac{1}{5} \left(216 \ 214 \ 212 \ 213 \ 215 \ 213 \right)' .$$

It can be verified by means of substitution that the quantities thus obtained are indeed the eigenvalue and eigenvector. No general proof exists of the fact that this method indeed yields the correct answers, however. If the same method is used for systems in the max-plus algebra only, it is known that it does not always give the correct results. In situations where it does not, a slightly more complicated algorithm exists which does give the correct results (see [28]).

Procedure 2 Consider (9.27)–(9.30) with sizes $n = 2, m = 0, p = 0, q = 1$. The matrices concerned are given by

$$A_1 = \begin{pmatrix} 2 & \varepsilon \\ 2 & 3 \end{pmatrix}, \quad D_1 = \begin{pmatrix} e \\ e \end{pmatrix}, \quad A_4 = \begin{pmatrix} 5 & 3 \end{pmatrix}, \quad D_4 = (\top) .$$

If the exponential approach of §9.2 is applied to the definition of the eigenvalue given in Definition 9.24, we obtain

$$z^{\lambda+x_1} = z^{2+x_1} + z^{w_1} , \tag{9.40}$$

$$z^{\lambda+x_2} = z^{2+x_1} + z^{3+x_2} + z^{w_1} , \tag{9.41}$$

$$z^{-w_1} = z^{-5-x_1} + z^{-3-x_2} . \tag{9.42}$$

The quantities z^{x_1} and z^{x_2} can be solved from (9.40) and (9.41), and expressed in z^{w_1}. These solutions can be substituted into (9.42), which yields

$$z^{-w_1} = z^{-w_1} z^{-5} (z^\lambda - z^2) + z^{-w_1} z^3 \frac{z^\lambda - z^3}{z^2 (z^\lambda - z^2)^{-1} + 1} .$$

Dividing this expression by z^{-w_1} and after some rearranging, we obtain

$$z^{2\lambda+3} + z^{2\lambda-5} + z^{-1} + z^8 = z^0 + z^2 + 2z^{\lambda-3} + z^{\lambda+6} + z^{\lambda+5} . \tag{9.43}$$

The essence of this arrangement is that all the exponential terms have been moved to that side of the equality symbol in such a way that only positive coefficients remain. Equation (9.43) must be valid as $z \to \infty$. Hence λ must satisfy

$$\max(2\lambda + 3, 2\lambda - 5, -1, 8) = \max(0, 2, \lambda - 3, \lambda + 6, \lambda + 5) .$$

This equation is most easily solved graphically. The result is $\lambda = 3$ and thus the eigenvalue has been found.

This method is only suitable when n, m, p and q are small. Essential is that an explicit equation in z^λ must be obtained.

Procedure 3 This procedure, which always works for systems which satisfy the conditions of Theorem 9.25, will be described by means of an algorithm. For an efficient way of explanation, (9.39) is rewritten as $\lambda \otimes a = \overline{\mathcal{M}}(a)$, where $a \in \mathbb{R}^{n+m}$. The vector function $\overline{\mathcal{M}}$ has components $\overline{\mathcal{M}}_i$, $i = 1, \ldots, n+m$. If the eigenvector is a, then $\overline{\mathcal{M}}_i(a) - a_i$ must be equal to λ for all i. We will say that the accuracy η is achieved if we find an a such that $\max_i \left(\overline{\mathcal{M}}_i(a) - a_i\right) - \min_i \left(\overline{\mathcal{M}}_i(a) - a_i\right) < \eta$. We then use the following algorithm.

1. Choose an arbitrary $a \in \mathbb{R}^{n+m}$ with all components finite.

2. Calculate $c_i = \overline{\mathcal{M}}_i(a) - a_i$, $i = 1, \ldots, n+m$. Define $\underline{c} = \min_i c_i$, $\overline{c} = \max_i c_i$.

3. If $\overline{c} - \underline{c} < \eta$, then stop.

4. Construct disjoint subsets Υ_i, $i = 1, 2, 3$, of $\Upsilon \stackrel{\text{def}}{=} \{1, \ldots, n+m\}$ such that
   - $\Upsilon = \Upsilon_1 \cup \Upsilon_2 \cup \Upsilon_3$,
   - $j \in \Upsilon_1 \Leftrightarrow c_j < \underline{c} + \eta/2$,
   - $j \in \Upsilon_2 \Leftrightarrow \eta/2 \le c_j - \underline{c} < \eta$,
   - $j \in \Upsilon_3 \Leftrightarrow c_j \ge \underline{c} + \eta$.

5. Change a_j into $a_j - \eta/2$ for all $j \in \Upsilon_1$. Do not change the other a-components.

6. Go to step 2.

This algorithm always ends in a finite number of steps. If k denotes the iteration index of the algorithm, then this k will, as an argument, specify the quantities related to the k-th iteration of the algorithm, and

$$\begin{cases} c_i(k+1) \ge c_i(k) & \text{for } i \in \Upsilon_1(k) \ , \\ c_i(k+1) \le c_i(k) & \text{for } i \in \Upsilon_2(k) \cup \Upsilon_3(k) \ . \end{cases}$$

Therefore $\overline{c}(k)$ is a nonincreasing function of k and similarly $\underline{c}(k)$ is nondecreasing. At each iteration of the algorithm some elements of Υ_1 may have moved to Υ_2 or vice versa. Some elements of Υ_3 may have moved to Υ_2, but *not* vice versa. As k increases, Υ_1 and Υ_2 will ultimately catch all c_i.

By means of the following example, it will be shown how the algorithm works:

$$\begin{aligned} a_1(k+1) &= \max(a_1(k)+1, a_2+2, a_3(k)) \ , \\ a_2(k+1) &= \max(a_1(k)+2, a_2, a_3(k)+1) \ , \\ a_3(k+1) &= \min(a_1(k)+2, a_2+4, a_3(k)+3) \ . \end{aligned}$$

We take $\eta = 0.2$ and start with $a(0) = \begin{pmatrix} 1 & 2 & 3 \end{pmatrix}'$. Application of the algorithm yields the following results:

$$a(0) = \begin{pmatrix} 1 \\ 2 \\ 3 \end{pmatrix}, \quad c(0) = \begin{pmatrix} 3 \\ 2 \\ 0 \end{pmatrix}, \quad a(1) = \begin{pmatrix} 1 \\ 2 \\ 2.9 \end{pmatrix}, \quad c(1) = \begin{pmatrix} 3 \\ 1.9 \\ 0.1 \end{pmatrix}, \ldots,$$

$$a(9) = \begin{pmatrix} 1 \\ 2 \\ 2.1 \end{pmatrix}, \quad c(9) = \begin{pmatrix} 3 \\ 1.1 \\ 0.9 \end{pmatrix}, \quad a(10) = \begin{pmatrix} 1 \\ 2 \\ 2 \end{pmatrix}, \quad c(10) = \begin{pmatrix} 3 \\ 1 \\ 1 \end{pmatrix},$$

$$a(11) = \begin{pmatrix} 1 \\ 1.9 \\ 1.9 \end{pmatrix}, \quad c(11) = \begin{pmatrix} 2.9 \\ 1.1 \\ 1.1 \end{pmatrix}, \ldots, a(20) = \begin{pmatrix} 1 \\ 1 \\ 1 \end{pmatrix}, \quad c(20) = \begin{pmatrix} 2 \\ 2 \\ 2 \end{pmatrix}.$$

For this example, even the exact results are obtained: the eigenvector is $a(20) = \begin{pmatrix} 1 & 1 & 1 \end{pmatrix}'$ and the eigenvalue is 2.

9.6.4 Stochastic Min-Max Systems

We are given the system described by (9.27)–(9.30). In contrast to the previous subsections, it is now assumed that the matrices in these formulæ are event-dependent; $A_l(k)$, $B_l(k)$, etc. The reason of this dependence is that (some of) the entries of these matrices will be assumed stochastic. For each k, the stochastic entries are assumed to be mutually independent and moreover, it is assumed that there is no correlation for different k-values. The underlying probability distributions are assumed to be finite, i.e. the entries can only assume a finite number of values. For the calculation of the average throughput, the same technique as used in Chapter 8 for max-plus algebra systems will be used. As an example, consider

$$\left.\begin{aligned} x_1(k+1) &= \max(x_1(k) + \tau_1(k), x_2(k), x_3(k)) \ , \\ x_2(k+1) &= \max(x_1(k) + \tau_1(k), x_2(k) + \tau_2(k), x_3(k)) \ , \\ x_3(k+1) &= \max(\min(x_1(k) + \tau_1(k), x_2(k) + \tau_2(k)), x_3(k) + 1) \ , \end{aligned}\right\} \tag{9.44}$$

which resembles (9.32). The stochastic quantities $\tau_i(k)$ are supposed to be independent of each other (i.e. for all i and all k). Assume that $\tau_i(k) = 0$ or $\tau_i(k) = 2$, both with probability 0.5. Starting from an 'arbitrary' x_0-vector, say $x_0 = \begin{pmatrix} 0 & 0 & 0 \end{pmatrix}'$, we will set up the reachability tree of all possible states of (9.44). From x_0, four new states can be reached in one step in principle, since there are four possibilities for the combination $(\tau_1(0), \tau_2(0))$. Actually, one of these states, $\begin{pmatrix} 2 & 2 & 2 \end{pmatrix}'$, is a translation of x_0 and hence is not considered to be a new state. For this example, it turns out that the reachability tree consists of ten states which will be denoted

$n_1, \ldots, n_{10}$. Here $n_1 = \begin{pmatrix} 0 & 0 & 0 \end{pmatrix}'$. If for instance $\tau_1(0) = \tau_2(0) = 0$, then we get to state $n_2 = \begin{pmatrix} 0 & 0 & 1 \end{pmatrix}'$. The ten states, together with their immediate successors, are given in Table 9.3. It is not difficult to show that, from any initial condition, the

Table 9.3: Reachable states

Initial state	$\tau_1 = 0$, $\tau_2 = 0$		$\tau_1 = 0$, $\tau_2 = 2$		$\tau_1 = 2$, $\tau_2 = 0$		$\tau_1 = 2$, $\tau_2 = 2$	
$n_1 = \begin{pmatrix} 0 & 0 & 0 \end{pmatrix}'$	n_2	e	n_3	e	n_4	1	n_1	2
$n_2 = \begin{pmatrix} 0 & 0 & 1 \end{pmatrix}'$	n_2	1	n_5	1	n_1	2	n_1	2
$n_3 = \begin{pmatrix} 0 & 2 & 1 \end{pmatrix}'$	n_1	2	n_6	2	n_1	2	n_6	2
$n_4 = \begin{pmatrix} 1 & 1 & 0 \end{pmatrix}'$	n_1	1	n_6	1	n_7	1	n_1	3
$n_5 = \begin{pmatrix} 0 & 1 & 1 \end{pmatrix}'$	n_2	1	n_3	1	n_1	2	n_8	2
$n_6 = \begin{pmatrix} 0 & 2 & 0 \end{pmatrix}'$	n_4	1	n_9	1	n_1	2	n_6	2
$n_7 = \begin{pmatrix} 2 & 2 & 0 \end{pmatrix}'$	n_1	2	n_6	2	n_7	2	n_1	4
$n_8 = \begin{pmatrix} 0 & 1 & 0 \end{pmatrix}'$	n_1	1	n_6	1	n_4	1	n_8	2
$n_9 = \begin{pmatrix} 1 & 3 & 0 \end{pmatrix}'$	n_7	1	n_{10}	1	n_1	3	n_6	3
$n_{10} = \begin{pmatrix} 2 & 4 & 0 \end{pmatrix}'$	n_7	2	n_{10}	2	n_1	4	n_6	4

state x_k will converge in a finite number of steps to the Markov chain consisting of the given 10 nodes. Since the probabilities with which the different $(\tau_1(k), \tau_2(k))$ occur are known, the transition probabilities of a Markov chain in which the ten states n_i are the nodes can be calculated. Subsequently, the stationary behavior of this Markov chain can be calculated. Once this stationary distribution is known, it is not difficult to calculate

$$\lim_{k\to\infty} \mathbb{E}(x_i(k+1) - x_i(k)) \ , \tag{9.45}$$

which turns out to be independent of the subscript i. This method, together with its properties, has been described more extensively in Chapter 8 for only max-plus systems. As long as the reachability-tree consists of a finite number of states, the method works equally well for min-max systems. For the example treated, it turns out that the expression in (9.45) equals 1.376 time units. This quantity can be considered as the average cycle time of system (9.44).

9.7 ABOUT CYCLE TIMES IN GENERAL PETRI NETS

In this section we are interested in finding how fast each transition can initiate firing in a periodically operated Petri net (not necessarily an event graph). Connections with Chapter 2 will be made. It will not be assumed here that each transition fires as soon as it is enabled. The order and timing of the initiation of firings of enabled transitions must be chosen in such a way (if at all possible) that a periodic behavior is possible. During each period, each transition must fire at least once. The thus smallest possible period τ is called the cycle time and it is defined as the time to complete a firing sequence leading back to the initial marking. Therefore we will confine ourselves to consistent Petri nets, i.e.

$$\exists x > 0 \ , \quad G'x = 0 \ ,$$

where G was defined in § 2.4.1. Later on in this section, we will narrow down the consistent Petri nets to event graphs.

It is assumed that the holding and firing times are constant in time. Suppose there is a firing time of at least β_i time units associated with transition $q_i, i = 1, \ldots, n$. This means that when q_i is enabled and it initiates a firing, it takes β_i time units before the firing is completed. Hence one token is reserved in each place $p_j \in \pi(q_i)$ for at least β_i time units before the transition completes its firing. The 'resource-time multiplication' ϱ_j is defined as (the number of tokens in place p_j) $\times$ (the length of time that these tokens remain in that place). A popular interpretation is: if a token represents a potential buyer who has to wait, then ϱ_j is proportional to the amount of coffee you will offer to all these buyers. In matrix form, $\varrho = (G^{\text{in}})'Dx$, where the m-dimensional vector ϱ has an entry per place and where D is the diagonal matrix with the elements β_i on the diagonal and 0's elsewhere (in conventional algebra). Here we only considered the reserved tokens (reserved for the duration of the firing time). Now suppose that there are on average $(\mu_a)_j$ tokens in place p_j during one period (this average is with respect to clock time). Then the corresponding ϱ is given by the vector $\mu_a\tau$ (popular again: the amount of coffee which you will need during one cycle for all waiting clients). Since this latter ϱ was calculated for both the reserved and not reserved tokens, the following inequality holds:

$$\mu_a\tau \geq (G^{\text{in}})'Dx \ . \tag{9.46}$$

Since μ_a is the average of a finite number of markings μ, and since $\mu'y$, with y satisfying $Gy = 0$, does not depend on the particular μ (see §2.4.1), we obtain

$$y'\mu_0\tau = y'\mu_a\tau \geq y'(G^{\text{in}})'Dx$$

(where μ_0 is the initial marking or any other one), provided $y \geq 0$. Hence

$$\tau \geq \max_y \frac{y'(G^{\text{in}})'Dx}{y'\mu_0} \ , \quad \text{subject to} \quad Gy = 0 \ , \quad y \geq 0 \ , \quad y \neq 0 \ . \tag{9.47}$$

It is tacitly assumed that the denominator of (9.47) is strictly greater than zero. In fact, it is not difficult to derive an upper bound for τ and hence the right-hand side of (9.47) must be finite.

An upper bound for τ is given by $2\beta_{\text{max}} \sum_i x_i$, where β_{max} is an upper bound for the holding and firing times. If a transition starts firing, then any other transition, if it is enabled at all, will be enabled within $2\beta_{\text{max}}$ units of time, the factor 2 coming from the firing time of the first transition and from the holding time between the two transitions. Hence such a transition can initiate firing within $2\beta_{\text{max}}$ time units of the initiation of firing of the first transition. The total number of firings in a circuit is $\sum x_i$. Thus the upper bound has been obtained.

For event graphs the analysis related to the lower bound on τ can be made more explicit. Take those indices of the y-vector which correspond to the indices of the transitions of an arbitrary elementary circuit of the event graph equal to 1 and the other indices equal to 0. It can be shown that the y-vector thus constructed satisfies $Gy = 0, y \geq 0, y \neq 0$. If this y is used in (9.47) one obtains

$$\tau \geq \max_k \left[\left(\sum_{i \in \zeta_k} \beta_i \right) / \mu_0(\zeta_k) \right], \tag{9.48}$$

where $\zeta_k, k = 1, \ldots,$ are the elementary circuits (the number of such circuits is finite), and where $\mu_0(\zeta_k)$ denotes the (initial) number of tokens in circuit ζ_k. According to Theorem 2.37, this number of tokens in a circuit of an event graph does not change after a firing. Equation (9.48) will be elucidated by means of the following example.

Consider Figure 9.3 in which an event graph is shown with five transitions and

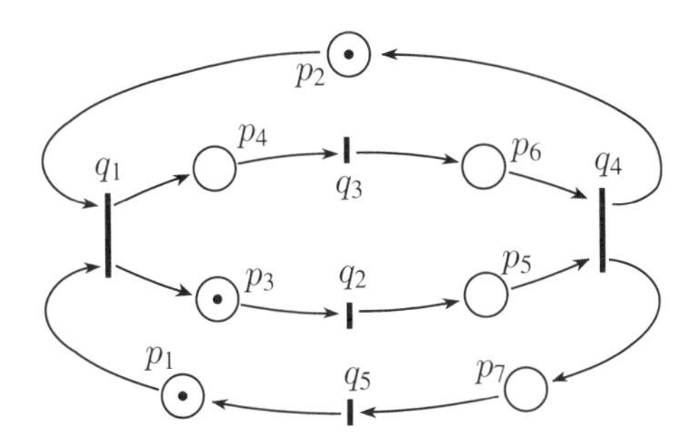

Figure 9.3: Event graph with four elementary circuits

seven places. The matrix G of this event graph is

$$G = \begin{pmatrix} -1 & -1 & 1 & 1 & \varepsilon & \varepsilon & \varepsilon \\ \varepsilon & \varepsilon & -1 & \varepsilon & 1 & \varepsilon & \varepsilon \\ \varepsilon & \varepsilon & \varepsilon & -1 & \varepsilon & 1 & \varepsilon \\ \varepsilon & 1 & \varepsilon & \varepsilon & -1 & -1 & 1 \\ 1 & \varepsilon & \varepsilon & \varepsilon & \varepsilon & \varepsilon & -1 \end{pmatrix}.$$

The columns and the rows of this matrix are numbered as the places and transitions in Figure 9.3 respectively. It follows that $G'x = 0$ for $x = \begin{pmatrix} k & k & k & k & k \end{pmatrix}'$, where k is a positive integer. A vector y, $y \geq 0$, $y \neq 0$, which satisfies $Gy = 0$ is found as follows. Take an arbitrary elementary circuit in the event graph, for instance the circuit formed by the places p_1, p_3, p_5 and p_7. Consider $y = \begin{pmatrix} 1 & 0 & 1 & 0 & 1 & 0 & 1 \end{pmatrix}'$, where the 1-elements correspond to the indices i of the places in the circuit. Indeed $Gy = 0$. A lower bound for the cycle time is $(\beta_1 + \beta_2 + \beta_4 + \beta_5)/2$.

If we were to deal exclusively with event graphs, it can be shown by continuing along the lines set out above, see [115], that the $\geq$-symbol in (9.48) becomes the $=$-symbol and that then all transitions will initiate firing as soon as they are enabled. This result has already been established in Chapter 3 of this book.

9.8 NOTES

Section 9.2 is based on [100, 103, 105, 50] (except for §9.2.2 with was unpublished yet). Section 9.3 on control of discrete event systems is believed to be original. One can imagine various ways to 'control' discrete event systems. One such a way has been given in §9.3. Another possibility is discussed in [68]. A relation between controlled event graphs and automata-based models is given in [78].

Section 9.4 is based on [113]. The Cramer transform is an important tool for people interested in large deviations and researchers in this field know the morphism between conventional algebra and min-plus algebra. But in general they are more interested in the probability side than in the optimization one. Moreover, they are not much concerned with the algebraic point of view [3], [60], [125]. Bellman and Karush [17] were aware of the interest of the Fenchel transform (which they call max transform). Maslov has also clearly understood the analogy between probability and dynamic programming and has developed a theory of idempotent integration [88]. The analogue of some jump processes can be found in [55].

The law of large numbers, the central limit theorem, the Brownian decision process, the diffusion decision process and the min-plus invariance principle do not seem to have been written down explicitly before. Some work on the min-plus analogue of stochastic integrals, and more generally on the analogy between probability and dynamic programming, has been done by Bellalouna under the supervision of M. Viot [16]. Some comments by P.L. Lions and R. Azencott on this morphism have been included in this section.

The first attempt to derive equations for general timed Petri nets can be traced back to [39]. However, the approach proposed in that paper did not lead to explicit equations, owing to the problem of consumption conflicts. The setting that is summarized in §9.5 is that of [7]. The evolution equations established in this section can be shown to be 'constructive' whenever the Petri net is live, and a computational scheme can be obtained that allows one to determine the firing times of the transitions iteratively (see [7]). Section 9.6 is based on [98] and [102].

Section 9.7 is based on [96]. As explained, the results of this section can be narrowed down to event graphs (see [115]), so as to get back the results which have been derived in a different way in Chapter 3.

Bibliography

[1] M. Ajmone Marsan, G. Balbo, and G. Conte. *Performance Models of Multiprocessor Systems.* The MIT Press, Cambridge, Mass., USA, 1986.

[2] H. Attouch and R.J.B. Wets. Isometries of the Legendre-Fenchel transform. *Transactions of the American Mathematical Society*, 296:33–60, 1986.

[3] R. Azencott, Y. Guivarc'h, and R.F. Gundy. In *Ecole d'été de Saint Flour 8*. Springer-Verlag, Berlin, 1978.

[4] F. Baccelli, N. Bambos, and J. Walrand. Flow analysis of stochastic marked graphs. In *Proceedings of the 28-th Conference on Decision and Control*. IEEE, 1989.

[5] F. Baccelli and P. Brémaud. *Palm Probabilities and Stationary Queues*. Number 41 in Lecture Notes in Statistics. Springer-Verlag, Berlin, 1987.

[6] F. Baccelli and P. Brémaud. *Elements of Queueing Theory*. Applications of Mathematics. Springer-Verlag, Berlin, 1992.

[7] F. Baccelli, G. Cohen, and B. Gaujal. Recursive equations and basic properties of timed Petri nets. *Journal of Discrete Event Dynamic Systems*, 1-4:415–439, 1992.

[8] F. Baccelli and T. Konstantopoulos. Estimates of cycle times in stochastic Petri nets. In I. Karatzas, editor, *Proceedings of Workshop on Stochatic Analysis*, Rutgers University, 1991. Springer-Verlag, Berlin.

[9] F. Baccelli and Z. Liu. On the executions of parallel programs on multiprocessor systems—a queueing theory approach. *Journal of the ACM*, 37(2):373–414, 1990.

[10] F. Baccelli and Z. Liu. Comparison properties of stochastic decision free Petri nets. Technical Report 1433, INRIA, Sophia-Antipolis, 06565 Valbonne, France, 1991. To appear in *IEEE Transactions on Automatic Control*.

[11] F. Baccelli and Z. Liu. On a class of stochastic evolution equations. *the Annals of Probability*, 20-1:350–374, 1992.

[12] F. Baccelli and A.M. Makowski. Queueing models for systems with synchronization constraints. *Proceedings of the IEEE*, 77:138–161, 1989. Special Issue on Dynamics of Discrete Event Systems.

[13] François Baccelli. Ergodic theory of stochastic Petri networks. *the Annals of Probability*, 20-1:375–396, 1992.

[14] Jean Le Bail, Hassane Alla, and René David. Hybrid Petri nets. In *Proceedings of the 1-st European Control Conference*, pages 1472–1477, Grenoble, France, 1991. Hermès, Paris.

[15] R. Barlow and F. Proschan. *Statistical Theory of Reliability and Life Testing*. Holt, Rinehart and Winston, 1975.

[16] F. Bellalouna. *Processus de décision min-markoviens*. PhD thesis, University of Paris-Dauphine, 1992. To appear.

[17] R. Bellman and W. Karush. Mathematical programming and the maximum transform. *SIAM Journal of Applied Mathematics*, 10, 1962.

[18] Dimitri P. Bertsekas. *Dynamic Programming*. Prentice-Hall, Englewood Cliffs, N.J., 1987.

[19] J.D. Biggins. The first and last birth problem for a multitype age-dependent branching process. *Advances in Applied Probability*, 8:446–459, 1976.

[20] P. Billingsley. *Ergodic Theory and Information*. John Wiley and Sons, New York, 1965.

[21] P. Billingsley. *Convergence of Probability Measures*. John Wiley and Sons, New York, 1968.

[22] G. Birkhoff. *Lattice Theory*. Amer. Math. Soc. Coll. Pub., Providence, 1967 (3rd Ed.).

[23] T.S. Blyth. Matrices over ordered algebraic structures. *Journal of London Mathematical Society*, 39:427–432, 1964.

[24] T.S. Blyth and M.F. Janowitz. *Residuation Theory*. Pergamon Press, Oxford, 1972.

[25] Dieter Bochmann and Christian Posthoff. *Binaere dynamische Systeme*. Akademie-Verlag, Berlin, 1981.

[26] A. Borovkov. *Asymptotic Methods in Queueing Theory*. John Wiley and Sons, New York, 1984.

[27] J.G. Braker. Max-algebra modelling and analysis of time-table dependent networks. In *Proceedings of the 1-st European Control Conference*, pages 1831–1836, Grenoble, France, 1991. Hermès, Paris.

[28] J.G. Braker and G.J. Olsder. The power algorithm in max-algebra. Technical Report 90-78, Dept. of Mathematics, Delft University of Technology, 1990. To appear in Linear Algebra and its Applications.

[29] W. Brauer, W. Reising, and G. Rozenberg, editors. *Petri Nets: Applications and Relationships to Other Models of Concurrency*, volume 255 of *Lecture notes in Computer Science*. Springer-Verlag, Berlin, 1987.

[30] W. Brauer, W. Reising, and G. Rozenberg, editors. *Petri Nets: Central Models and Their Properties*, volume 254 of *Lecture notes in Computer Science*. Springer-Verlag, Berlin, 1987.

[31] Y. Brenier. Un algorithme rapide pour le calcul des transformées de Legendre-Fenchel discrètes. *Comptes Rendus à l'Académie des Sciences*, 308:587–589, 1989.

[32] Roger W. Brockett. *Finite Dimensional Linear Systems*. John Wiley and Sons, New York, 1970.

[33] R. A. Brualdi and H. J. Ryser. *Combinatorial Matrix Theory*. Cambridge University Press, Cambridge, 1991.

[34] Z.-Q. Cao, K.H. Kim, and F.W. Roush. *Incline Algebra and Applications*. Ellis Horwood Limited, Chichester, 1984.

[35] P. Caspi and N. Halbwachs. A functional model for describing and reasoning about time behaviour of computing systems. *Acta Informatica*, 23:595–627, 1986.

[36] P. Chrétienne. *Les Réseaux de Pétri Temporisés*. Thèse d'état, University of Paris 6, Paris, 1983.

[37] G. Cohen, D. Dubois, J.P. Quadrat, and M. Viot. Analyse du comportement périodique de systèmes de production par la théorie des dioïdes. Technical Report 191, INRIA, Rocquencourt, 78153 Le Chesnay, France, February 1983.

[38] G. Cohen, D. Dubois, J.P. Quadrat, and M. Viot. A linear system-theoretic view of discrete event processes. In *Proceedings of the 22-nd Conference on Decision and Control*. IEEE, 1983.

[39] G. Cohen, D. Dubois, J.P. Quadrat, and M. Viot. A linear system-theoretic view of discrete event processes and its use for performance evaluation in manufacturing. *IEEE Transactions on Automatic Control*, AC-30:210–220, 1985.

[40] G. Cohen, S. Gaubert, R. Nikoukhah, and J.P. Quadrat. Convex analysis and spectral analysis of timed event graphs. In *Proceedings of the 28-th Conference on Decision and Control*. IEEE, 1989.

[41] G. Cohen, P. Moller, J.P. Quadrat, and M. Viot. Linear system theory for discrete event systems. In *Proceedings of the 23-rd Conference on Decision and Control*. IEEE, 1984.

[42] G. Cohen, P. Moller, J.P. Quadrat, and M. Viot. Une théorie linéaire des systèmes à événements discrets. Technical Report 362, INRIA, Rocquencourt, 78153 Le Chesnay, France, January 1985.

[43] G. Cohen, P. Moller, J.P. Quadrat, and M. Viot. Dating and counting events in discrete event systems. In *Proceedings of the 25-th Conference on Decision and Control*. IEEE, 1986.

[44] Guy Cohen, Pierre Moller, Jean-Pierre Quadrat, and Michel Viot. Algebraic tools for the performance evaluation of discrete event systems. *Proceedings of the IEEE*, 77:39–58, 1989. Special Issue on Dynamics of Discrete Event Systems.

[45] J. Cohen, H. Kesten, and C. Newman. Random matrices and their applications. In *Proceedings of a Summer Research Conference*. AMS, June 1984.

[46] J.W. Cohen. *The Single Server Queue*. North-Holland, Amsterdam, 1969.

[47] F. Commoner, A. W. Holt, S. Even, and A. Pnuelly. Marked directed graphs. *Computer and System Science*, 5:511–523, 1971.

[48] J. Cook and B. Derrida. Directed polymers in a random medium: $1/d$ expansion and the n-tree approximation. *Journal of Physics, Series A: Mathematical and General*, 23:1523–1553, 1990.

[49] R.A. Cuninghame-Green. *Minimax Algebra*. Number 166 in Lecture Notes in Economics and Mathematical Systems. Springer-Verlag, Berlin, 1979.

[50] R.A. Cuninghame-Green. Algebraic realization of discrete dynamical systems. In *Proceedings of the 1991 IFAC Workshop on Discrete Event System Theory and Applications in Manufacturing and Social Phenomena*, pages 11–15, Beijing, China, 1991. International Academic Publishers.

[51] R.A. Cuninghame-Green and P.F.J. Meijer. An algebra for piecewise-linear minimax problems. *Discrete Applied Mathematics*, 2:267–294, 1980.

[52] Y. Dallery, Z. Liu, and D. Towsley. Equivalence, reversibility and symmetry properties in assembly/disassembly networks. Technical Report 1267, INRIA, Rocquencourt, 78153 Le Chesnay, France, 1990.

[53] R. David and H. Alla. Continuous Petri nets. In *Proceedings of the 8th European Workshop on Applications and Theory of Petri Nets, Saragossa, Spain*, pages 275–294, 1987.

[54] R.E. de Vries. *On the asymptotic behavior of discrete event systems*. PhD thesis, Delft University of Technology, 1992.

[55] P. Del Moral, T. Thuillet, G. Rigal, and G. Salut. Optimal versus random processes: the non-linear case. Technical report, LAAS, Toulouse, France, 1990.

[56] B. Derrida. Mean field theory of directed polymers in a random medium and beyond. *Physica Scripta*, T38:6–12, 1990.

[57] P. Dubreil and M.L. Dubreil-Jacotin. *Leçons d'Algèbre Moderne.* Dunod, Paris, France, 1964.

[58] W. Fenchel. On the conjugate convex functions. *Canadian Journal of Mathematics*, 1:73–77, 1949.

[59] Forney, Jr. and G. David. The Viterbi algorithm. *Proceedings of the IEEE*, 61:268–278, 1973.

[60] M.I. Freidlin and A.D. Wentzell. *Random Perturbations of Dynamical Systems.* Springer-Verlag, Berlin, 1979.

[61] F.R. Gantmacher. *The Theory of Matrices.* Chelsea Publishing Company, England, 1959.

[62] Stéphane Gaubert. *Théorie Linéaire des Systèmes dans les Dioïdes.* Thèse, École des Mines de Paris, Paris, 1992.

[63] P.W. Glynn. A gsmp formalism for discrete event systems. *Proceedings of the IEEE*, 77, 1989. Special Issue on Dynamics of Discrete Event Systems.

[64] M. Gondran and M. Minoux. Valeurs et vecteurs propres dans les dioïdes et leur interpretation en théorie des graphes. *Bulletin de la Direction des Études et Recherches, Série C (Mathématiques, Informatique)*, (2):25–41, 1977. E.D.F., Clamart, France.

[65] M. Gondran and M. Minoux. L'indépendance linéaire dans les dioïdes. *Bulletin de la Direction des Études et Recherches, Série C (Mathématiques, Informatique)*, (1):67–90, 1978. E.D.F., Clamart, France.

[66] M. Gondran and M. Minoux. Linear algebra in dioids: a survey of recent results. *Annals of Discrete Mathematics.*, 19:147–164, 1984.

[67] M. Gondran and M. Minoux. *Graphs and Algorithms.* John Wiley and Sons, New York, 1986.

[68] Hans-Michael Hanisch. Dynamik von koordinierungssteuerungen in diskontinuierlichen verfahrentechnischen systemen. *Automatisierungstechnik at*, 38:399–405, 1990.

[69] S. Helbig. Optimization problems on extremal algebras: Necessary and sufficient conditions for optimal points. In B. Borowski and F. Deutsch, editors, *Parametric Optimization and Approximation*, Lecture Notes in Control and Information Sciences, pages 166–184. Birkhäuser-Verlag, Basel, 1985.

[70] Y.C. Ho and X.R. Cao. *Perturbation Analysis of Discrete Event Dynamic Systems.* Kluwer Academic Publishers, 1991.

[71] J.L. Joly. Une famille de topologies sur l'ensemble des fonctions convexes pour lesquelles la polarité est bicontinue. *Journal de Mathématiques Pures et Appliquées*, 52:421–441, 1973.

[72] Thomas Kailath. *Linear systems*. Prentice-Hall, Englewood Cliffs, N.J., 1980.

[73] Richard M. Karp. A characterization of the minimum cycle mean in a digraph. *Discrete Mathematics*, 23:309–311, 1978.

[74] J.L. Kemeny and J.L. Snell. *Finite Markov chains*. Springer-Verlag, Berlin, 1960, reprint 1976.

[75] J.F.C. Kingman. Subadditive ergodic theory. *Annals of Probability*, 1:883–909, 1973.

[76] J.F.C. Kingman. Subadditive processes. In P.-L. Hennequin, editor, *École d'été de probabilité de Saint-Flour*, volume 539 of *Lecture Notes in Mathematics*, pages 165–223. Springer-Verlag, Berlin, 1976.

[77] Leonard Kleinrock. *Queueing Systems*. John Wiley & Sons, New York, 1976.

[78] Bruce H. Krogh, Jan Magott, and Lawrence E. Holloway. On the complexity of forbidden state problems for controlled marked graphs. In *Proceedings of the 30-th Conference on Decision and Control*, pages 85–91. IEEE, 1991.

[79] S.Y. Kung. *VLSI Array Processors*. Prentice-Hall, Englewood Cliffs, N.J., 1988.

[80] J. Kuntzmann. *Théorie des Réseaux et Graphes*. Dunod, Paris, France, 1972.

[81] Huibert Kwakernaak and Raphael Sivan. *Linear Optimal Control Systems*. Wiley-Interscience, New York, 1972.

[82] Stephane Lafortune and Hyuck Yoo. Some results on Petri net languages. *IEEE Transactions on Automatic Control*, AC-35:482–485, 1990.

[83] S. Lavenberg. Maximum departure rate of certain open queueing networks having finite capacity constraints. *RAIRO série bleue*, 12(4), 1978. Dunod, Paris.

[84] Charles E. Leiserson and James B. Saxe. Optimizing synchronous systems. *Journal of VLSI and Computer Systems*, 1:41–67, 1983.

[85] Rudolf Lidl and Gunther Pilz. *Abstract Applied Algebra*. Springer-Verlag, Berlin, 1984.

[86] P.L. Lions. *Generalized Solutions of Hamilton-Jacobi Equation*. Pitman, 1982.

[87] R. M. Loynes. The stability of queues with nonindependent inter-arrival and service times. *Proceedings of Cambridge Philosophical Society*, 58:497–520, 1962.

[88] V. Maslov. *Méthodes opératorielles*. Éditions MIR, Moscow, 1987. French translation.

[89] L.E. Meester and J.G. Shanthikumar. Concavity of the throughput of tandem queueing systems with finite storage space. *Advances in Applied Probability*, 22:764–767, 1990.

[90] S.G. Mikhlin. *Integral Equations*. Pergamon Press, Oxford, 1957.

[91] Henryk Minc. *Nonnegative Matrices*. John Wiley and Sons, New York, 1988.

[92] Debasis Mitra and Isi Mitrani. Analysis for a novel discipline for cell coordination in production lines, 1. Technical report, AT&T Bell Laboratories, Murray Hill, New Jersey 07974, USA, June 1988.

[93] Pierre Moller. *Théorie Algèbrique des Systèmes à Événements Discrets*. Thèse, École des Mines de Paris, Paris, 1988.

[94] Michael K. Mollow. Performance analysis using stochastic Petri nets. *IEEE Transactions on Computers*, C-31:913–917, 1982.

[95] T. Murata. Circuit theoretic analysis and synthesis of marked graphs. *IEEE Transactions on Circuits and Systems*, CAS-24:400–405, 1977.

[96] Tadao Murata. Petri nets: Properties, analysis and applications. *Proceedings of the IEEE*, 77:541–580, 1989.

[97] Geert Jan Olsder. Applications of the theory of stochastic discrete-event systems to array processors and scheduling in public transportation. In *Proceedings of the 28-th Conference on Decision and Control*. IEEE, 1989.

[98] Geert Jan Olsder. Eigenvalues of dynamic max-min systems. *Journal on Discrete Event Dynamic Systems*, 1:177–207, 1991.

[99] Geert Jan Olsder. Synchronized continuous flow systems. Technical report, Faculty of Technical Mathematics and Informatics, Delft University of Technology, P.O. Box 5031, 2600GA Delft, the Netherlands, 1991.

[100] G.J. Olsder. On the characteristic equation and minimal realizations for discrete event systems. In A. Bensoussan and J.L. Lions, editors, *Analysis and Optimization of Systems*, pages 189–201. Springer-Verlag, Berlin, 1986.

[101] G.J. Olsder. Performance analysis of data-driven networks. In John McCanny, John McWhirter, and Earl Swartzlander Jr., editors, *Systolic Array Processors*, pages 33–41. Prentice-Hall, Englewood Cliffs, N.J., 1989.

[102] G.J. Olsder. Descriptor systems in min-max algebra. In *Proceedings of the 1-st European Control Conference*, pages 1825–1830, Grenoble, France, 1991. Hermès, Paris.

[103] G.J. Olsder and R.E. de Vries. On an analogy of minimal realizations in conventional and discrete event dynamic systems. In P.Varaiya and A.B. Kurzhanski, editors, *Discrete Event Systems: Models and Applications*, volume 103 of *Lecture Notes in Control and Information Sciences*, pages 149–161. Springer-Verlag, Berlin, 1988.

[104] G.J. Olsder, J.A.C. Resing, R.E. de Vries, M.S. Keane, and G. Hooghiemstra. Discrete event systems with stochastic processing times. *IEEE Transactions on Automatic Control*, AC-35:299–302, 1990.

[105] G.J. Olsder and C. Roos. Cramer and Cayley-Hamilton in the max-algebra. *Linear Algebra and its Applications*, 101:87–108, 1988.

[106] V.I. Oseledeç. A multiplicative ergodic theorem. Lyapounov characteristic numbers for dynamical systems. *Transactions of Moscow Mathematical Society*, 19:197–231, 1968.

[107] E.P. Patsidou and J.C. Kantor. Application of minimax algebra to the study of multipurpose batch plants. *Computers and Chemical Engineering*, 15-1:35–46, 1991.

[108] James L. Peterson. *Petri Net Theory and the Modeling of Systems*. Prentice-Hall, Englewood Cliffs, N.J., 1981.

[109] Max Plus. L'algèbre $(\max, +)$ et sa symétrisation ou l'algèbre des équilibres. *Comptes Rendus à l'Académie des Sciences*, 311:443–448, 1990.

[110] Max Plus. Linear systems in $(\max, +)$ algebra. In *Proceedings of the 29-th Conference on Decision and Control*. IEEE, 1990.

[111] Max Plus. A linear system theory for systems subject to synchronization and saturation constraints. In *Proceedings of the 1-st European Control Conference*, pages 1022–1033, Grenoble, France, 1991. Hermès, Paris.

[112] Max Plus. Second order theory of min-linear systems and its application to discrete event systems. In *Proceedings of the 30-th Conference on Decision and Control*. IEEE, 1991.

[113] J.P. Quadrat. Théorèmes asymptotiques en programmation dynamique. *Comptes Rendus à l'Académie des Sciences*, 311:745–748, 1990.

[114] P.J. Ramadge and W.M. Wonham. Supervisory control of a class of discrete event processes. *SIAM Journal of Control and Optimization*, 25(1):206–230, Jan 1987.

[115] C.V. Ramamoorthy and Gary S. Ho. Performance evaluation of asynchronous concurrent systems using Petri nets. *IEEE Transactions on Software Engineering*, 6:440–449, 1980.

[116] Kurt J. Reinschke. *Multivariable Control, a Graph-Theoretic Approach*. Akademie-Verlag, Berlin, 1988.

[117] J.A.C. Resing, R.E. de Vries, M.S. Keane, G. Hooghiemstra, and G.J. Olsder. Asymptotic behaviour of random discrete event systems. *Stochastic Processes and Their Applications*, 36:195–216, 1990.

[118] C. Reutenauer and H. Straubing. Inversion of matrices over commutative semiring. *Journal of Algebra*, 88:350–360, 1984.

[119] R.T. Rockafellar. *Convex Analysis*. Princeton University Press, Princeton, N.J., 1970.

[120] D.E. Rutherford. Inverses of Boolean matrices. *Proceedings of the Glasgow Mathematical Association*, 6:49–53, 1963.

[121] A.N. Shiryayev. *Probability*. Springer-Verlag, Berlin, 1984.

[122] E.D. Sontag. *Mathematical Control Theory*. Springer-Verlag, Berlin, 1990.

[123] D. Stoyan. *Comparison Methods for Queues and Other Stochastic Models*. John Wiley and Sons, New York, 1984. English translation (D.J. Daley editor).

[124] H. Straubing. A combinatorial proof of the Cayley-Hamilton theorem. *Discrete Mathematics*, 43:273–279, 1983.

[125] S.R.S. Varadhan. In *Large Deviations and Applications*, number 46 in CBMS-NSF Regional Conference Series in Applied Mathematics. SIAM, Philadelphia, Penn., USA, 1984.

[126] Edouard Wagneur. Moduloids and pseudomodules: 1. dimension theory. *Discrete Mathematics*, 98:57–73, 1991.

[127] L. Wang, C. Liu, and X. Xu. Some results on eigensystems of deterministic discrete event systems. Technical report, North-East University of Technology, Shenyang, China, 1990.

[128] J.H. Wedderburn. Boolean linear associative algebra. *Annals of Mathematics*, 35:185–194, 1934.

[129] Doron Zeilberger. A combinatorial approach to matrix algebra. *Discrete Mathematics*, 56:61–72, 1985.

[130] U. Zimmermann. *Linear Combinatorial Optimization in Ordered Algebraic Structures*. North-Holland, Amsterdam, 1981.

Notation

deg	degree
dom	dominant
epi	epigraph
gcd	greatest common divisor
hypo	hypograph
lcm	least common multiple
supp	support
val	valuation
2^S	collection of subsets of S (including $\emptyset$ and S itself)
$\mathbb{N}$	natural numbers
$\mathbb{Z}$	relative integer numbers
$\mathbb{R}$	real numbers
$\mathbb{B}$	the Boolean dioid
$\overline{\mathbb{R}}$	$\mathbb{R} \cup \{-\infty\} \cup \{+\infty\}$ (similar definitions for $\overline{\mathbb{N}}$, $\overline{\mathbb{Z}}$, etc.)
$\mathbb{R}_{\max}$	dioid $\{\mathbb{R}, \max, +\}$ (similar definitions for $\mathbb{R}_{\min}$, $\mathbb{Z}_{\max}$, etc.)
$\mathbb{S}$	symmetrized dioid of $\mathbb{R}_{\max}$
$\mathbb{S}^{\oplus}$	see Equation (3.15)
$\mathbb{S}^{\ominus}$	see Equation (3.15)
$\mathbb{S}^{\bullet}$	see Equation (3.15)
$\mathbb{S}^{\vee}$	$\mathbb{S}^{\vee} = \mathbb{S}^{\oplus} \cup \mathbb{S}^{\ominus}$
$\mathbb{S}^{\vee}_{\star}$	$\mathbb{S}^{\vee}_{\star} = \mathbb{S}^{\vee} \setminus \{\varepsilon\}$
$\langle \cdot , \cdot \rangle$	scalar or duality product
$\oplus$	addition in a dioid (pronounced 'oplus')
$\otimes$	multiplication in a dioid (pronounced 'otimes')
$\odot$	in the context of $\mathbb{R}_{\max}$, $(A \odot B)_{ij} = \inf_k \big(A_{ik} + B_{kj}\big)$
$\wedge$	(greatest) lower bound in a lattice or dioid (pronounced 'wedge')
$\vee$	(least) upper bound in a lattice or dioid (pronounced 'vee')
$\ominus$	minus sign in a symmetrized dioid (pronounced 'ominus')
$-\!\!\!\circ$	$x \mapsto x -\!\!\!\circ\, a$ is the dual residual of $y \mapsto y \oplus a$
$\not\circ$	$x \mapsto x \not\circ a$ (pronounced 'x (right) divided by a') is the residual of $y \mapsto y \otimes a$ (in $\mathbb{R}_{\max}$, this is $x - a$)

$\frac{\cdot}{\cdot}$	$\frac{x}{a}$ is a two-dimensional display notation for $x \phi a$
$\backslash\!\!\backslash$	$x \mapsto a \backslash\!\!\backslash x$ (pronounced 'a (left) dividing x') is the residual of $y \mapsto a \otimes y$
$\frac{\cdot}{\cdot}$	$\frac{x}{a}$ is a two-dimensional display notation for $a \backslash\!\!\backslash x$
∇	balance (see Definition 3.63)
$\emptyset$	empty set
ε	zero element in a dioid
$\acute{\varepsilon}$	zero element in the dater dioid (see (5.9))
$\top$	top (largest) element in a lattice or a (complete) dioid
e	identity element in a dioid
$\acute{e}$	identity element in the dater dioid (see (5.9)) (also, integrator)
$\acute{u}$	best approximation from above of a signal u ($\acute{u} = \acute{e}u$)
$x^\bullet$	$x^\bullet = x \ominus x$ (pronounced 'x bullet')
$\dot{c}$	first order derivative of function c
$\ddot{c}$	second order derivative of function c
γ	shift operator in the event domain (formally $\gamma x(k) = x(k-1)$)
δ	shift operator in the time domain (formally $\delta x(t) = x(t-1)$)
θ	shift operator on $(\Omega, \mathbb{F}, \mathbb{P})$
T_a	translation by a
L_a	left multiplication by a
R_a	right multiplication by a
Γ^g	shift (by g)
Δ^d	gain (by d)
Φ_a	flow limiter (by a)
Σ^w	local integrator (over a window of width w)
$\oint$	if f is a mapping from $\mathbb{R}$ into $\mathbb{R}_{\max}$, $\oint_a^b f(s) = \sup_{a \leq s \leq b} f(s)$
$\lvert \cdot \rvert$	if $\mathfrak{P}$ is a set, $\lvert\mathfrak{P}\rvert$ is the cardinality of $\mathfrak{P}$; if $x \in \mathbb{S}$, $\lvert x\rvert$ is the absolute value (see §3.4.1.1); if A is a matrix, $\lvert A\rvert$ is its determinant
$\lvert \cdot \rvert_\mathrm{l}$	$\lvert\rho\rvert_\mathrm{l}$ is the length of path ρ (total number of arcs)
$\lvert \cdot \rvert_\mathrm{w}$	$\lvert\rho\rvert_\mathrm{w}$ is the weight of path ρ (total holding time)
$\lvert \cdot \rvert_\mathrm{t}$	$\lvert\rho\rvert_\mathrm{t}$ is the total number of tokens along path ρ
$\lvert \cdot \rvert_\oplus$	see Equation (7.23)
$\lvert \cdot \rvert_\wedge$	see Equation (7.24)
$\leq_{\mathrm{st}}$	stochastic ordering
$\leq_{\mathrm{cx}}$	convex ordering
$\leq_{\mathrm{icx}}$	increasing convex ordering
$[\leftarrow, x]$	closed lower set generated by x
$[x, \rightarrow]$	closed upper set generated by x

$\mathcal{D}_\Pi^\natural$	$\mathcal{D}_\Pi^\natural = \{x \mid \Pi(x) = x\}$
$\mathcal{D}_\Pi^\flat$	$\mathcal{D}_\Pi^\flat = \{x \mid \Pi(x) \le x\}$
$\mathcal{D}_\Pi^\sharp$	$\mathcal{D}_\Pi^\sharp = \{x \mid \Pi(x) \ge x\}$
Π^*	if Π is a mapping, $\Pi^* = e \oplus \Pi \oplus \Pi^2 \oplus \cdots$ where e is the identity mapping and, e.g. $\Pi^2 = \Pi \circ \Pi$
Π_*	if Π is a mapping, $\Pi_* = e \wedge \Pi \wedge \Pi^2 \wedge \cdots$
$\Pi^\sharp$	(pronounced 'pi sharp') residual of mapping Π
$\Pi^\flat$	(pronounced 'pi flat') dual residual of mapping Π
$\mathcal{F}$	evaluation homomorphism
$\mathcal{F}_\mathrm{e}$	Fenchel transform
$\mathcal{G}$	(directed) graph
$\mathcal{E}$	set of arcs of a directed graph
$\mathcal{V}$	set of nodes of a directed graph
$\mathcal{G}(A)$	precedence graph associated with matrix A
N_A	number of m.s.c.s.'s of $\mathcal{G}(A)$
$\mathcal{G}^\mathrm{c}(A)$	critical graph associated with matrix A
$\mathcal{V}^\mathrm{c}$	set of nodes of $\mathcal{G}^\mathrm{c}(A)$
N_A^c	number of m.s.c.s.'s of $\mathcal{G}^\mathrm{c}(A)$
$\mathcal{S}(A, y)$	saturation graph associated with matrix A and vector y
$c(\mathcal{G})$	cyclicity of graph $\mathcal{G}$
π	$\pi(i)$ is the set of immediate predecessors of node i
σ	$\sigma(i)$ is the set of immediate successors of node i
π^+	$\pi^+(i)$ is the set of all predecessors of node i (not including i itself)
π^*	$\pi^*(i)$ is the set of all predecessors of node i (including i itself)
σ^+	$\sigma^+(i)$ is the set of all successors of node i (not including i itself)
σ^*	$\sigma^*(i)$ is the set of all successors of node i (including i itself)
$[i]$	m.s.c.s. containing node i; more generally, equivalence class of i for a given equivalence relation
$[< i]$	see Notation 2.3
$[\le i]$	see Notation 2.3
A_{ij}	entry in row i and column j of matrix A
$x_{(i)}$	see Notation 2.5
$A_{(i)(j)}$	see Notation 2.5
$A_{\cdot j}$	column j of matrix A
A'	transpose of matrix A
A^*	if A is a matrix, $A^* = e \oplus A \oplus A^2 \oplus \cdots$

A^+	if A is a matrix, $A^+ = AA^* = A \oplus A^2 \oplus \cdots$
$\mathrm{cof}_{ij}(A)$	cofactor of matrix A associated with entry (i, j)
$A^\natural$	transpose of the matrix of cofactors of A
$\mathcal{P}$	in a Petri net, set of places
$\mathcal{Q}$	in a Petri net, set of transitions
I	in a Petri net, set of input transitions
μ	vector such that μ_i equals the number of tokens in place i in the initial marking of a Petri net
α	vector such that α_i is the holding time of place i of a (timed) Petri net
β	vector such that β_i is the firing time of transition i of a (timed) Petri net
$w_i(j)$	lag times of the j-th initial token in place p_i
$v_j(k)$	see Equation (2.16)
$z_j(k)$	see Definition 2.61
$\mathcal{H}$	algebra of impulse responses
$\mathcal{S}$	algebra of shift-invariant impulse responses
$\acute{\mathcal{S}}$	algebra of nondecreasing shift-invariant impulse responses
$\mathcal{S}_{\mathrm{cx}}$	subset of convex elements of $\mathcal{S}$
$\mathcal{S}_{\mathrm{cv}}$	subset of concave elements of $\mathcal{S}$
$\mathcal{C}_{\mathrm{cx}}$	dioid of closed convex functions from $\mathbb{R}_{\max}$ into $\overline{\mathbb{R}}_{\max}$
$\mathcal{D}^{n \times n}$	dioid of $n \times n$ matrices with entries in $\mathcal{D}$
$\mathcal{D}^\diamond$	dioid closure of $\mathcal{D}$
$\mathcal{D}^\star$	rational closure of $\mathcal{D}$
$\mathcal{D}[\alpha]$	dioid of formal polynomials in α with coefficients in $\mathcal{D}$
$\mathcal{P}(\mathcal{D})$	dioid of polynomial functions over $\mathcal{D}$
$\mathcal{D}[\![\alpha]\!]$	dioid of formal power series in α with coefficients in $\mathcal{D}$
$\mathcal{D}(\alpha)$	rational closure of $\mathcal{D} \cup \alpha$
$\mathcal{D}\{\alpha\}$	algebraic closure of $\mathcal{D} \cup \alpha$
$\mathcal{D}\{\!\{\alpha\}\!\}$	topological closure of $\mathcal{D} \cup \alpha$
$\mathcal{M}_{\mathrm{in}}^{\mathrm{ax}}[\![\gamma, \delta]\!]$	(pronounced 'min max gamma delta') the quotient of $\mathbb{B}[\![\gamma, \delta]\!]$ by the equivalence relation $x \equiv y \Leftrightarrow \gamma^*(\delta^{-1})^* x = \gamma^*(\delta^{-1})^* y$
$(\Omega, \mathbb{F}, \mathbb{P})$	probability space
$\mathbb{F}$	σ-algebra in a probability space
$\mathbb{P}$	probability
$\mathbb{E}$	mathematical expectation
$\mathbb{M}_W$	see Equation (9.16)
$1_{\mathcal{A}}$	indicator function of subset (or event) $\mathcal{A}$
$Q_{m,\sigma}(x)$	quadratic form such that $\dot{Q}_{m,\sigma}(m) = 0$ and $\ddot{Q}_{m,\sigma}(x) = \sigma^{-1}$
$\mathcal{N}(m, \sigma)$	Gaussian law with mean m and standard deviation σ

Index